Introduction to General Relativity

A student-friendly style, over 100 illustrations, and numerous exercises are brought together in this textbook for advanced undergraduate and beginning graduate students in physics and mathematics.

Lewis Ryder develops the theory of General Relativity in detail. Covering the core topics of black holes, gravitational radiation and cosmology, he provides an overview of General Relativity and its modern ramifications. The book contains a chapter on the connections between General Relativity and the fundamental physics of the microworld, explains the geometry of curved spaces and contains key solutions of Einstein's equations – the Schwarzschild and Kerr solutions.

Mathematical calculations are worked out in detail, so students can develop an intuitive understanding of the subject, as well as learn how to perform calculations. Password-protected solutions for instructors are available at www.cambridge.org/Ryder.

Lewis Ryder is an Honorary Senior Lecturer in Physics at the University of Kent, UK. His research interests are in geometrical aspects of particle theory and its parallels with General Relativity.

Introduction to General Relativity

Lewis Ryder
University of Kent, UK

CAMBRIDGE
UNIVERSITY PRESS

6/4/13
WW
$85-

CAMBRIDGE UNIVERSITY PRESS
Cambridge, New York, Melbourne, Madrid, Cape Town, Singapore, São Paulo, Delhi

Cambridge University Press
The Edinburgh Building, Cambridge CB2 8RU, UK

Published in the United States of America by Cambridge University Press, New York

www.cambridge.org
Information on this title: www.cambridge.org/9780521845632

First published 2009

Printed in the United Kingdom at the University Press, Cambridge

A catalogue record for this publication is available from the British Library

Library of Congress Cataloguing in Publication data
Ryder, Lewis H., 1941–
Introduction to general relativity / Lewis Ryder.
p. cm.
Includes bibliographical references and index.
ISBN 978-0-521-84563-2
1. General relativity (Physics) I. Title.
QC173.6.R93 2009
530.11–dc22
2009005862

ISBN 978-0-521-84563-2 hardback

For Mildred Elizabeth Ryder

It is always a source of pleasure when a great and beautiful idea proves to be correct in actual fact.

Albert Einstein [letter to Sigmund Freud]

The answer to all these questions may not be simple. I know there are some scientists who go about preaching that Nature always takes on the simplest solutions. Yet the simplest by far would be nothing, that there would be nothing at all in the universe. Nature is far more interesting than that, so I refuse to go along thinking it always has to be simple.

Richard Feynman

Contents

Preface

This book is designed for final year undergraduates or beginning graduate students in physics or theoretical physics. It assumes an acquaintance with Special Relativity and electromagnetism, but beyond that my aim has been to provide a pedagogical introduction to General Relativity, a subject which is now – at last – part of mainstream physics. The coverage is fairly conventional; after outlining the need for a theory of gravity to replace Newton's, there are two chapters devoted to differential geometry, including its modern formulation in terms of differential forms and coordinate-free vectors, then the Einstein field equations, the Schwarzschild solution, the Lense–Thirring effect (recently confirmed observationally), black holes, the Kerr solution, gravitational radiation and cosmology. The book ends with a chapter on field theory, describing similarities between General Relativity and gauge theories of particle physics, the Dirac equation in Riemannian space-time, and Kaluza–Klein theory.

As a research student I was lucky enough to attend the Les Houches summer school in 1963 and there, in the magnificent surroundings of the French alps, began an acquaintance with many of the then new aspects of this subject, just as it was entering the domain of physics proper, eight years after Einstein's death. A notable feature was John Wheeler's course on gravitational collapse, before he had coined the phrase 'black hole'. In part I like to think of this book as passing on to the community of young physicists, after a gap of more than 40 years, some of the excitement generated at that school.

I am very grateful to the staff at Cambridge University Press, Tamsin van Essen, Lindsay Barnes and particularly Simon Capelin for their unfailing help and guidance, and generosity over my failure to meet deadlines. I also gratefully acknowledge helpful conversations and correspondence with Robin Tucker, Bahram Mashhoon, Alexander Shannon, the late Jeeva Anandan, Brian Steadman, Daniel Ryder and especially Andy Hone, who have all helped to improve my understanding. Finally I particularly want to thank my wife, who has supported me throughout this long project, with constant good humour and generous and selfless encouragement. To her the book is dedicated.

Notation, important formulae and physical constants

Latin indices i, j, k, and so on run over the three spatial coordinates 1, 2, 3 or x, y, z or r, θ, φ

Greek indices $\alpha, \beta, \gamma, \ldots \kappa, \lambda, \mu, \ldots$ and so on run over the four space-time coordinates 0, 1, 2, 3 or ct, x, y, z or ct, r, θ, ϕ

Minkowski space-time: metric tensor is $\eta_{\mu\nu} = \mathrm{diag}\,(-1, 1, 1, 1)$, $\mathrm{d}s^2 = -c^2\,\mathrm{d}t^2 + \mathrm{d}x^2 + \mathrm{d}y^2 + \mathrm{d}z^2$ in Cartesian coordinates

Riemannian space-time: $\mathrm{d}s^2 = g_{\mu\nu}\,\mathrm{d}x^\mu\,\mathrm{d}x^\nu = -\,c^2\,\mathrm{d}\tau^2$

The Levi-Cività totally antisymmetric symbol (in Minkowski space) is

$$\varepsilon^{0123} = -\varepsilon_{0123} = 1$$

Connection coefficients: $\Gamma^\nu_{\mu\kappa} = {}^1\!/_2\, g^{\nu\rho}(g_{\mu\rho,\kappa} + g_{\kappa\rho,\mu} - g_{\mu\kappa,\rho})$

Riemann tensor: $R^\kappa{}_{\lambda\mu\nu} = \Gamma^\kappa{}_{\lambda\nu,\mu} - \Gamma^\kappa{}_{\lambda\mu,\nu} + \Gamma^\kappa{}_{\rho\mu}\Gamma^\rho{}_{\lambda\nu} - \Gamma^\kappa{}_{\rho\nu}\Gamma^\rho{}_{\lambda\mu}$

Ricci tensor: $R_{\mu\nu} = R^\rho{}_{\mu\rho\nu}$

Curvature scalar: $R = g^{\mu\nu}R_{\mu\nu}$

Field equations: $G_{\mu\nu} = R_{\mu\nu} - {}^1\!/_2\, g_{\mu\nu}R = \dfrac{8\pi G}{c^2}T_{\mu\nu}$

Covariant derivatives:

$$\frac{\mathrm{D}V^\mu}{\mathrm{d}x^\nu} = \frac{\partial V^\mu}{\partial x^\nu} + \Gamma^\mu{}_{\lambda\nu}V^\lambda \quad \text{or} \quad V^\mu{}_{;\nu} = V^\mu{}_{,\nu} + \Gamma^\mu{}_{\lambda\nu}V^\lambda$$

$$\frac{\mathrm{D}W_\mu}{\mathrm{d}x^\nu} = \frac{\partial W_\mu}{\partial x^\nu} - \Gamma^\lambda{}_{\mu\nu}W_\lambda \quad \text{or} \quad W_{\mu;\nu} = W_{\mu,\nu} - \Gamma^\lambda{}_{\mu\nu}W_\lambda$$

Speed of light	$c = 3.00 \times 10^8\ \mathrm{m\,s^{-1}}$
Gravitational constant	$G = 6.67 \times 10^{-11}\ \mathrm{N\,m^2\,kg^{-1}}$
Planck's constant	$\hbar = 1.05 \times 10^{-34}\ \mathrm{J\,s}$
	$= 6.58 \times 10^{-22}\ \mathrm{MeV\,s}$
Electron mass	$m_\mathrm{e} = 9.11 \times 10^{-31}\ \mathrm{kg}$
	$m_\mathrm{e}c^2 = 0.51\ \mathrm{MeV}$
Proton mass	$m_\mathrm{p} = 1.672 \times 10^{-27}\ \mathrm{kg}$
	$m_\mathrm{p}c^2 = 938.3\ \mathrm{MeV}$
Neutron mass	$m_\mathrm{n} = 1.675 \times 10^{-27}\ \mathrm{kg}$
	$m_\mathrm{n}c^2 = 939.6\ \mathrm{MeV}$
Boltzmann constant	$k = 1.4 \times 10^{-23}\ \mathrm{J\,K^{-1}}$
	$= 8.6 \times 10^{-11}\ \mathrm{MeV\,K^{-1}}$
Solar mass	$M_\mathrm{S} = 1.99 \times 10^{30}\ \mathrm{kg}$

Solar radius	$R_S = 6.96 \times 10^8$ m
Earth mass	$M_E = 5.98 \times 10^{24}$ kg
Earth equatorial radius	$R_E = 6.38 \times 10^6$ m
Mean Earth–Sun distance	$R = 1.50 \times 10^{11}$ m = 1 AU
Schwarzschild radius of Sun	$2m = \frac{2M_S G}{c^2} = 2.96$ km
Stefan–Boltzmann constant	$\sigma = 5.67 \times 10^{-8}$ W m^{-2} K^{-4}

1 light year (ly) = 9.46×10^{15} m
1 pc = 3.09×10^{16} m = 3.26 ly
1 radian = 2.06×10^5 seconds of arc

1 Introduction

Einstein's General Theory of Relativity, proposed in 1916, is a theory of gravity. It is also, as its name suggests, a generalisation of Special Relativity, which had been proposed in 1905. This immediately suggests two questions. Firstly, why was a new theory of gravity needed? Newton's theory was, to put it mildly, perfectly good enough. Secondly, why is it that a generalisation of Special Relativity yields, of all things, a theory of *gravity*? Why doesn't it give a theory of electromagnetism, or the strong or weak nuclear forces? Or something even more exotic? What is so special about gravity, that generalising a theory of space and time (because that is what Special Relativity is) gives us an account of it? We begin this chapter by answering the first question first. By the end of the chapter we shall also have made a little bit of headway in the direction of answering the second one.

1.1 The need for a theory of gravity

Newton's theory of gravitation is a spectacularly successful theory. For centuries it has been used by astronomers to calculate the motions of the planets, with a staggering success rate. It has, however, the fatal flaw that it is inconsistent with Special Relativity. We begin by showing this.

As every reader of this book knows, Newton's law of gravitation states that the force exerted on a mass m by a mass M is

$$\mathbf{F} = -\frac{MmG}{r^3}\mathbf{r}. \tag{1.1}$$

Here M and m are not necessarily point masses; r is the distance between their centres of mass. The vector $\mathbf{r}$ has a direction from M to m. Now suppose that the mass M depends on time. The above formula will then become

$$\mathbf{F}(t) = -\frac{M(t)mG}{r^3}\mathbf{r}. \tag{1.2}$$

This means that the force felt by the mass m at a time t depends on the value of the mass M *at the same time t*. There is no allowance for time delay, as Special Relativity would require. From our experience of advanced and retarded potentials in electrodynamics, we can say that Special Relativity would be satisfied if, in the above equation, $M(t)$ were modified to $M(t - r/c)$. This would reflect the fact that the force felt by the small mass at time t depended on the value of the large mass at an *earlier* time $t - r/c$; assuming, that is, that the relevant gravitational 'information' travelled at the speed of light. But this would then not be Newton's law. Newton's law is Equation (1.2) which allows for no time delay, and therefore implicitly suggests that the *information* that the mass M is changing travels with infinite velocity, since the effect of a changing M is felt at the same instant by the mass m. Since Special Relativity implies that nothing can travel faster than light, Equations (1.1) and (1.2) are incompatible with it. If two theories are incompatible, at least one of them must be wrong. The only possible attitude to adopt is that Special Relativity must be kept intact, so Newton's law has to be changed.

Faced with such a dramatic situation – not to say crisis – the instinctive, and perfectly sensible, reaction of most physicists would be to try to 'tinker' with Newton's law; to change it slightly, in order to make it compatible with Special Relativity. And indeed many such attempts were made, but none were successful.[1] Einstein eventually concluded that nothing less than a complete 'new look' at the problem of gravitation had to be taken. We shall return to this in the next section, but before leaving this one it will be useful to rewrite the above equations in a slightly different form; it should be clear that, although Newton's equations are 'wrong', they are an extremely good approximation to whatever 'correct' theory is eventually found, so this theory should then give, as a first approximation, Newton's law. We have by no means finished with Newton!

Let us define $\mathbf{g} = \mathbf{F}/m$, the gravitational *field intensity*. This is a parallel equation to $\mathbf{E} = \mathbf{F}/q$ in electrostatics; the electric field is the force per unit charge and the gravitational field the force per unit mass. Mass is the 'source' of the gravitational field in the same way that electric charge is the source of an electric field. Then Equation (1.1) can be written

$$\mathbf{g} = -\frac{GM}{r^2}\hat{r}, \tag{1.3}$$

which gives an expression for the gravitational field intensity at a distance r from a mass M. This expression, however, is of a rather special form, since the right hand side is a *gradient*. We can write

$$\mathbf{g} = -\nabla\phi, \quad \phi(r) = -\frac{GM}{r}. \tag{1.4}$$

The function $\phi(r)$ is the gravitational *potential*, a scalar field. Newton's theory is then described simply by *one function*. (In contrast, as we shall see in due course, the gravitational field in General Relativity is described by *ten* functions, the ten components of the metric tensor. The non-relativistic limit of *one* of these components is, in essence, the Newtonian potential.) A mass, or a distribution of masses, gives rise to a scalar gravitational potential that completely determines the gravitational field. The potential ϕ in turn satisfies *field equations*. These are Laplace's and Poisson's equations, relevant, respectively, to the cases where there is a vacuum, or a matter density ρ:

$$\text{(Laplace)} \qquad \nabla^2\phi = 0 \qquad \text{(vacuum)}, \tag{1.5}$$

$$\text{(Poisson)} \qquad \nabla^2\phi = 4\pi G\rho \qquad \text{(matter)}. \tag{1.6}$$

[1] For references to these see 'Further reading' at the end of the chapter.

In the case of a point mass, of course, we have $\rho(r) = M\delta^3(\mathbf{r})$, and by virtue of the identity

$$\nabla^2(1/r) = -4\pi\delta^3(\mathbf{r}) \tag{1.7}$$

Equations (1.4) and (1.6) are in accord.

This completes our account of Newtonian gravitational theory. The field $\mathbf{g}$ depends on r but not on t. Such a field is incompatible with Special Relativity. It is not a Lorentz covariant field; such a field would be a *four*-vector rather than a three-vector and would depend on t as well as on r, so that the equations of gravity looked the same in all frames of reference related by Lorentz transformations. This is not the case here. Since Newton's theory is inconsistent with Special Relativity it must be abandoned. This is both a horrifying prospect and a slightly encouraging one; horrifying because we are having to abandon one of the best theories in physics, and encouraging because Newton's theory is so precise and so successful that any new theory of gravity will immediately have to fulfil the very stringent requirement that in the non-relativistic limit it should yield Newton's theory. This will provide an immediate test for a new theory.

1.2 Gravitation and inertia: the Equivalence Principle in mechanics

Einstein's new approach to gravity sprang from the work of Galileo (1564–1642; he was born in the same year as Shakespeare and died the year Newton was born). Galileo conducted a series of experiments rolling spheres down ramps. He varied the angle of inclination of the ramp and timed the spheres with a water clock. Physicists commonly portray Galileo as dropping masses from the Leaning Tower of Pisa and timing their descent to the ground. Historians cast doubt on whether this happened, but for our purposes it hardly matters whether it did or didn't; what matters is the conclusion Galileo drew. By extrapolating to the limit in which the ramps down which the spheres rolled became vertical, and therefore that the spheres fell freely, he concluded that *all bodies fall at the same rate in a gravitational field*. This, for Einstein, was a crucially important finding. To investigate it further consider the following 'thought-experiment', which I refer to as 'Einstein's box'. A box is placed in a gravitational field, say on the Earth's surface (Fig. 1.1(a)). An experimenter in the box releases two objects, made of different materials, from the same height, and measures the times of their fall in the gravitational field $\mathbf{g}$. He finds, as Galileo found, that they reach the floor of the box at the same time. Now consider the box in free space, completely out of the reach of any gravitational influences of planets or stars, but subject to an *acceleration* $\mathbf{a}$ (Fig. 1.1(b)). Suppose an experimenter in this box also releases two objects at the same time and measures the time which elapses before they reach the floor. He will find, of course, that they take the same time to reach the floor; he *must* find this, because when the two objects are released, they are then subject to no force, because no acceleration, and it is the floor of the box that accelerates up to meet them. It clearly reaches them at the same time. We conclude that this experimenter, by releasing objects and timing their fall, *will not be able to tell whether he is in a gravitational field or being accelerated through*

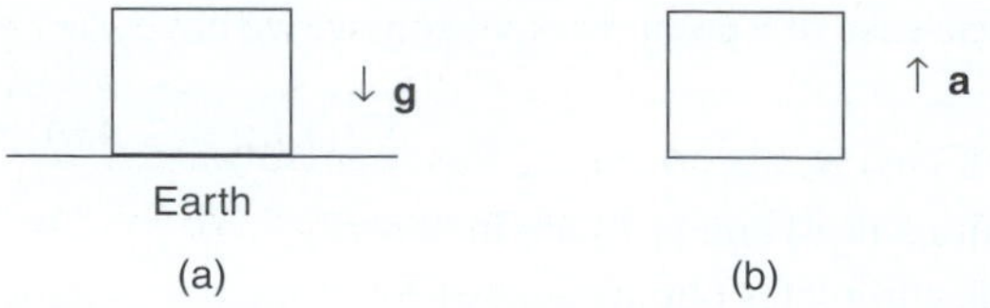

Fig. 1.1 The Einstein box: a comparison between a gravitational field and an accelerating frame of reference.

empty space. The experiments will give identical results. A gravitational field is therefore *equivalent* to an accelerating frame of reference – at least, as measured in this experiment. This, according to Einstein, is the significance of Galileo's experiments, and it is known as the *Equivalence Principle*. Stated in a more general way, the Equivalence Principle says that *no experiment in mechanics can distinguish between a gravitational field and an accelerating frame of reference*. This formulation, the reader will note, already goes beyond Galileo's experiments; the claim is made that *all* experiments in mechanics will yield the same results in an accelerating frame and in a gravitational field. Let us now analyse the consequences of this.

We begin by considering a particle subject to an acceleration **a**. According to Newton's second law of motion, in order to make a particle accelerate it is necessary to apply a force to it. We write

$$\mathbf{F} = m_{\rm i}\mathbf{a}. \tag{1.8}$$

Here $m_{\rm i}$ is the *inertial mass* of the particle. The above law states that the reason a particle needs a force to accelerate it is that the particles possesses *inertia*. A very closely related idea is that acceleration is *absolute*; (constant) velocity, on the other hand, is *relative*. Now consider a particle falling in a gravitational field **g**. It will experience a force (see (1.2) and (1.3) above) given by

$$\mathbf{F} = m_{\rm g}\mathbf{g}. \tag{1.9}$$

Here $m_{\rm g}$ is the *gravitational mass* of the particle. It measures the response of a particle to a *gravitational field*. It is very important to appreciate that gravitational mass and inertial mass are conceptually *entirely distinct*. Acceleration in free space is an entirely different thing from a gravitational field, and we make this distinction clear by distinguishing gravitational and inertial mass, as in the two equations above. Now, however, consider a particle falling freely in a gravitational field, as in the Einstein box experiments. Both equations above apply. Because the particle is in a gravitational field it will experience a force, given by (1.9); and because a force is acting on the particle it will accelerate, the acceleration being given by (1.8). These two equations then give

$$\mathbf{a} = \frac{\mathbf{F}}{m_{\rm i}} = \frac{m_{\rm g}}{m_{\rm i}}\mathbf{g}; \tag{1.10}$$

the acceleration of a particle in a gravitational field **g** is the ratio of its gravitational and inertial masses times **g**. Galileo's experiments therefore imply that $m_{\rm g}/m_{\rm i}$ *is the same for all* *materials*. Without loss of generality we may put $m_g = m_i$ for all materials; this is because the formula for **g** contains G (see (1.3)), so by scaling G, m_g/m_i can be made equal to unity. (In fact, of course, historically G was found by *assuming* that $m_g = m_i$; no distinction was made between gravitational and inertial masses. We are now 'undoing' history.) We conclude that the Equivalence Principle states that

$$m_g = m_i. \tag{1.11}$$

Gravitational mass is the *same as* inertial mass for *all materials*. This is an interesting and non-trivial result. Some very sensitive experiments have been performed, and continue to be performed, to test this equality to higher and higher standards of accuracy. After Galileo, the most interesting experiment was done by Eötvös and will be described below. Before that, however, it is worth devoting a few minutes' thought to the significance of the equality (1.11) above.

The inertial mass of a piece of matter has contributions from two sources; the mass of the 'constituents' and the binding energy, expressed in mass units ($m = E/c^2$). This is the case no matter what the type of binding. So for example the mass of an atom is the sum of the masses of its constituent protons and neutrons *minus the nuclear binding energy* (divided by c^2). In the case of nuclei, the binding energy makes a contribution of the order of 10^{-3} to the total mass. Atoms are bound together by electromagnetic forces and stars and planets are bound by gravitational forces. In all of these cases, the binding energy, as well as the inertial mass of the constituents, contributes to the overall inertial mass of the sample. The statement (1.11) above then implies that the binding energy of a body will *also* contribute to its *gravitational* mass, so binding energy (in fact, energy in general) has a *gravitational effect* since its mass equivalent will in turn give rise to a gravitational field. The gravitational force itself, by virtue of the binding it gives rise to, also gives rise to further gravitational effects. In this sense gravity is *non-linear*. Electromagnetism, on the other hand, is linear; electromagnetic forces give rise to (binding) energy, which acts as a source of gravity, but *not* as a source of further electromagnetic fields, since electromagnetic energy possesses *no charge*. Gravitational energy, however, possesses an effective mass and therefore gives rise to further gravitational fields.

Now let us turn to experiments to test the Equivalence Principle. The simplest one to imagine is simply the measurement of the displacement from the vertical with which a large mass hangs, in the gravitational field of the (rotating) Earth. From Problem 1.1 we see that this displacement is (in Budapest) of the order of 6 minutes of arc multiplied by m_g/m_i. To see whether m_g/m_i is the same for all substances, then, involves looking for tiny variations in this angle, for masses made of differing materials. This is a very difficult measurement to make, not least because it is *static*.

A better test for the constancy of m_g/m_i relies on the gravitational attraction of the *Sun*, whose position relative to the Earth varies with a 24 hour period. We are therefore looking for a periodic signal, which stands more chance of being observed above the noise than does a static one. The simplest version of this is the *Eötvös* or *torsion balance*; the original torsion balance was invented by Coulomb and by Mitchell, and was used by Cavendish to verify the inverse square law of gravity. For the purposes of this experiment the torsion balance takes the form shown in Fig. 1.2.

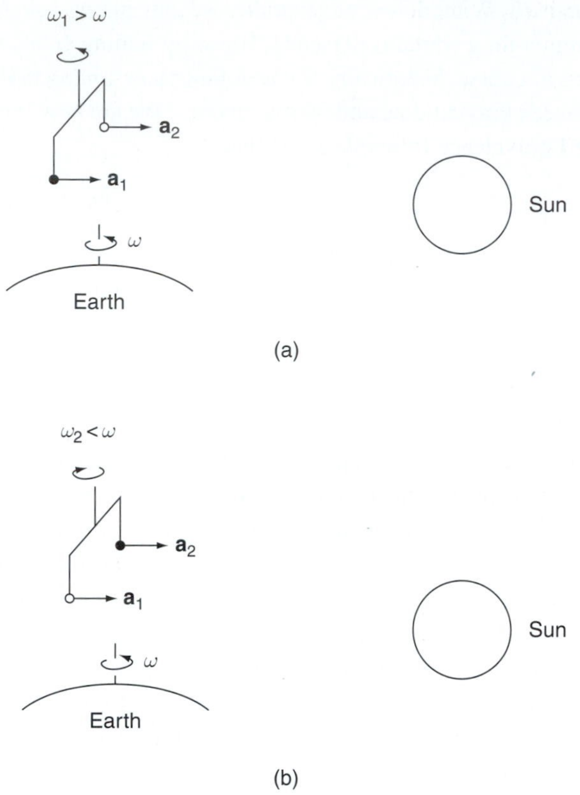

Fig. 1.2 A torsion balance at the North Pole. (a) and (b) represent two situations with a 12 hour time separation. The Earth is rotating with angular velocity ω and $\mathbf{a}_1$ and $\mathbf{a}_2$ are the accelerations of the gold and aluminium masses towards the Sun. Assuming that $\mathbf{a}_1 > \mathbf{a}_2$ the resulting torques are of opposite sign.

Two masses, one of gold (shaded) one of aluminium (not shaded), hang from opposite ends of an arm suspended by a thread in the gravitational field of the Earth. Consider such a balance at the North Pole, with the Sun in some assigned position to the right of the diagram. Then at 6 a.m., say, the situation is as shown in (a), the Earth rotating with angular velocity ω. The force exerted by the Sun on the gold mass is (M is the mass of the Sun and r the Earth–Sun distance)

$$F_{\mathrm{Au}} = \frac{GM(m_{\mathrm{g}})_{\mathrm{Au}}}{r^2} \tag{1.12}$$

and hence its acceleration towards the Sun is

$$a_{\mathrm{Au}} = \frac{GM}{r^2}\left(\frac{m_{\mathrm{g}}}{m_{\mathrm{i}}}\right)_{\mathrm{Au}}. \tag{1.13}$$

A similar formula holds for the aluminium mass. Putting

$$\frac{m_{\mathrm{g}}}{m_{\mathrm{i}}} = 1 + \delta, \tag{1.14}$$

then if $\delta_{\rm Au} \neq \delta_{\rm Al}$ a *torque* is exerted on the balance, of magnitude ($2l$ is the length of the arm)

$$T = \frac{GMl}{r^2}[(m_{\rm g})_{\rm Au} - (m_{\rm g})_{\rm Al}]. \tag{1.15}$$

This results in an angular acceleration α given by $T = I\alpha$, with I, the moment of inertia, given by $I = m_{\rm i}l^2$, so we have, at 6 a.m.,

$$(\alpha)_{\rm 6am} = \frac{GM}{lr^2}(\delta_{\rm Au} - \delta_{\rm Al}) \equiv \frac{GM}{lr^2}\Delta, \tag{1.16}$$

where $\Delta = \delta_{\rm Au} - \delta_{\rm Al}$. In diagram (a) we suppose that $\Delta > 0$, i.e. the acceleration of the gold mass is greater than that of the aluminium mass. This in effect causes the torsion balance to rotate with angular velocity $\omega_1 > \omega$. At 6 p.m., however, the situation is reversed (Fig. 1.2(b)) so the direction of the torque will be reversed, and

$$(\alpha)_{\rm 6pm} = -\frac{GM}{lr^2}\Delta. \tag{1.17}$$

Thus there would be a periodic variation in the torque, with a period of 24 hours. No such variation has been observed,[2] allowing the conclusion that

$$\delta < 10^{-11}; \tag{1.18}$$

gravitational mass and inertial mass are equal to one part in 10^{11} – at least as measured using gold and aluminium.

1.1.1 A remark on inertial mass

The Equivalence Principle states the equality of gravitational and inertial mass, as we have just seen above. It is worthwhile, however, making the following remark. The inertial mass of a particle refers to its mass (deduced, for example, from its behaviour analysed according to Newton's laws) when it undergoes non-uniform, or non-inertial, motion. There are, however, two different types of such motion; it may for instance be acceleration in a straight line, or circular motion with constant speed. In the first case the magnitude of the velocity vector changes but its direction remains constant, while in the second case the magnitude is constant but the direction changes. In each of these cases the motion is non-inertial, but there is a conceptual distinction to be made. To be precise we should observe this distinction and denote the two types of mass $m_{\rm i,acc}$ and $m_{\rm i,rot}$. We believe, without, as far as I know, proper evidence, that they are equal

$$m_{\rm i,acc} = m_{\rm i,rot}. \tag{1.19}$$

The interesting thing is that Einstein's formulation of the Equivalence Principle referred to inertial mass measured in an accelerating frame, $m_{\rm i,acc}$, whereas the Eötvös experiment, described above, establishes the equality (to within the stated bounds) of $m_{\rm i,rot}$ and the

[2] Roll *et al.* (1964), Braginsky & Panov (1972).

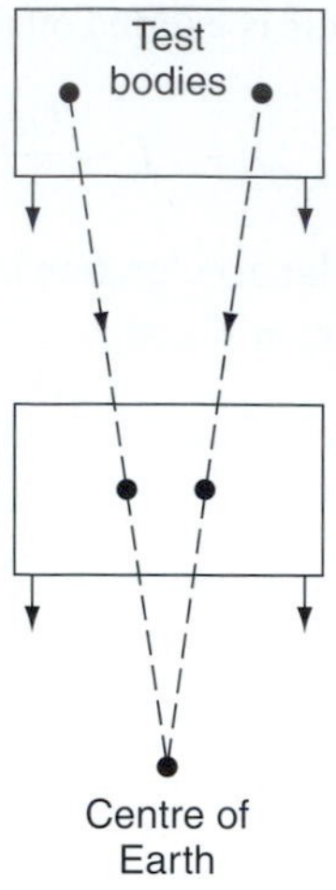

Fig. 1.3 **Test bodies falling to the centre of the Earth.**

gravitational mass. The question is: can an experiment be devised to test the equality of $m_{\mathrm{i,acc}}$ and m_{g}? Or even to test (1.19)?

1.1.2 Tidal forces

The Principle of Equivalence is a *local* principle. To see this, consider the Einstein box in the gravitational field of the Earth, as in Fig. 1.3. If the box descends over a large distance towards the centre of the Earth, it is clear that two test bodies in the box will approach one another, so over this *extended* journey it is clear that they are in a genuine gravitational field, and *not* in an accelerating frame (in which they would stay the same distance apart). In other words, the Equivalence Principle has broken down. We conclude that this principle is only valid as a *local* principle. Over small distances a gravitational field is equivalent to an acceleration, but over larger distances this equivalence breaks down. The effect is known as a *tidal effect*, and ultimately is due to the *curvature* produced by a real gravitational field.

Another way of stating the situation is to note that an object in free fall is in an inertial frame. The effect of the gravitational field has been *cancelled* by the acceleration of the elevator (the 'acceleration due to gravity'). The accelerations required to annul the gravitational fields of the two test bodies, however, are slightly different, because they are directed along the radius vectors. So the inertial frames of the two bodies differ slightly. The frames are 'locally inertial'. The Equivalence Principle treats a gravitational field *at a single point* as equivalent to an acceleration, but it is clear that no gravitational fields encountered in nature give rise to a *uniform* acceleration. Most real gravitational fields are produced by more or less spherical objects like the Earth, so the equivalence in question is only a local one.

We may find an expression for the tidal forces which result from this non-locality. Figure 1.4 shows the forces exerted on the two test bodies – call them A and B – in the gravitational field of a body at O. They both experience a force towards O of magnitude

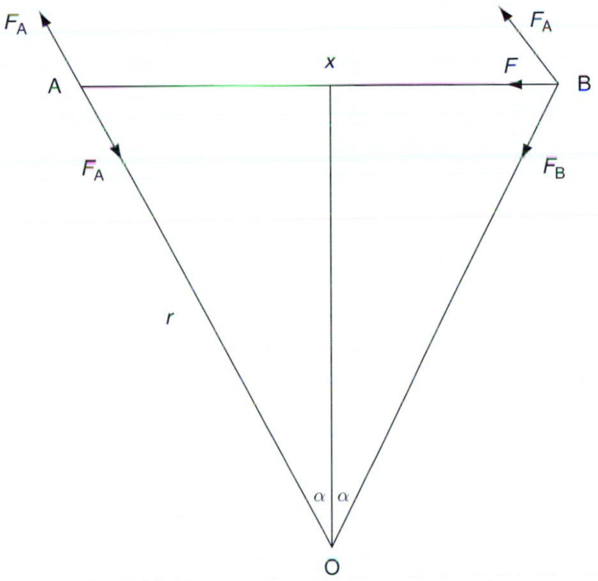

Fig. 1.4 Tidal effect: forces on test bodies A and B.

$$F_A = F_B = \frac{mMG}{r^2}$$

where m is the mass of A and B, M is the mass of the Earth and r the distance of A and B from its centre. In addition, let the distance between A and B be x. Consider the frame in which A is at rest. This frame is realised by applying a force equal and opposite to F_A, to both A and B, as shown in Fig. 1.4. In this frame, B experiences a force F, directed towards A, which is the vector sum of F_B and $-F_A$:

$$F = 2F_A \sin\alpha = 2F_A \cdot \frac{x}{2r} = \frac{mMG}{r^3}x.$$

A then observes B to be accelerating towards him with an acceleration given by $F = -m\,d^2x/dt^2$, i.e.

$$\frac{d^2x}{dt^2} = -\frac{MG}{r^3}x. \tag{1.20}$$

The $1/r^3$ behaviour is characteristic of tidal forces.

1.3 The Equivalence Principle and optics

The Equivalence Principle is a principle of *indistinguishability*; it is impossible, using any experiment in mechanics, to *distinguish* between a gravitational field and an accelerating frame of reference. To this extent it is a *symmetry principle*. If a symmetry of nature is *exact*,

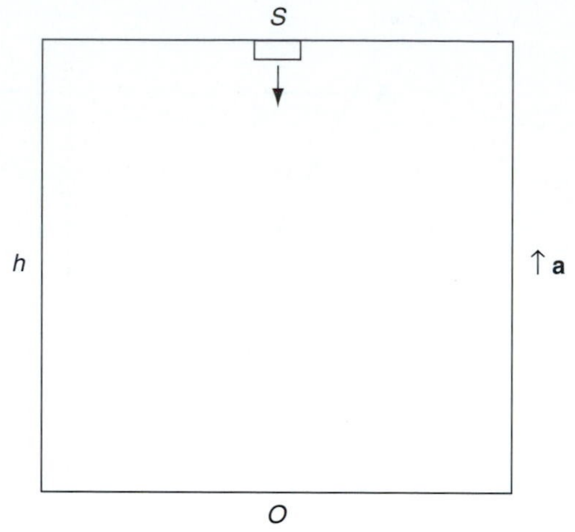

Fig. 1.5 **Light propagating downwards in a box accelerating upwards.**

this means that various situations are experimentally indistinguishable. If, for example, parity were an exact symmetry of the world (which it is not, because of beta decay), it would be impossible to distinguish left from right. The fact that it *is* possible to distinguish them is a direct indication of the breaking of the symmetry.

No experiment in mechanics, then, can distinguish a gravitational field from an accelerating frame. What about other areas of physics? Let us generalise the Equivalence Principle to *optics*, and consider the idea that no experiment in optics could distinguish a gravitational field from an accelerating frame.[3] To make this concrete, return to the Einstein box and consider the following simple two experiments. The first one is to release monochromatic light (of frequency ν) from the ceiling of the accelerating box, and receive it on the floor (Fig. 1.5). The light is released from the source S at $t=0$ towards the observer O. At the same instant $t=0$ the box begins to accelerate upwards with acceleration a. The box is of height h. Light from S reaches O after a time interval $t=h/c$, at which time O is moving upwards with speed $u=at=ah/c$.

Now consider the emission of two successive crests of light from S. Let the time interval between the emission of these crests be $\mathrm{d}t$ in the frame of S. Then

$$\mathrm{d}t = \frac{1}{\nu} \quad \text{in frame } S, \tag{1.21}$$

where ν is the frequency of the light in frame S. Arguing *non-relativistically*, the time interval between the *reception* of these crests at O is

$$\mathrm{d}t' = \mathrm{d}t - \Delta t = \mathrm{d}t - u\frac{\mathrm{d}t}{c} = \mathrm{d}t\left(1 - \frac{u}{c}\right) = \frac{1}{\nu'},$$

[3] This generalisation is sometimes characterised as a progression from a Weak Equivalence Principle (which is the statement $m_\mathrm{i} = m_\mathrm{g}$) to a Strong Equivalence Principle, according to which all the laws of nature (not just those of freely falling bodies) are affected in the same way by a gravitational field and a constant acceleration.

hence

$$\frac{\nu'}{\nu} = \frac{1}{1 - u/c} = 1 + \frac{u}{c} + \mathrm{O}\left(\frac{u}{c}\right)^2 > 1; \tag{1.22}$$

the light is Doppler (blue) shifted. With $\nu' = \nu + \Delta\nu$ we have

$$\frac{\Delta\nu}{\nu} = \frac{u}{c} + \mathrm{O}\left(\frac{u}{c}\right)^2 = \frac{ah}{c^2} + \mathrm{O}\left(\frac{ah}{c^2}\right)^2. \tag{1.23}$$

Arguing *relativistically*, the above result is unchanged to order $(ah/c^2)^2$; the equation above becomes (with $\gamma = (1 - u^2/c^2)^{-1/2}$)

$$\mathrm{d}t' = \gamma(\mathrm{d}t - \Delta t) = \gamma\,\mathrm{d}t(1 - u/c) = \frac{1}{\nu'},$$

hence

$$\frac{\nu'}{\nu} = \frac{1}{\gamma\left(1 - \frac{u}{c}\right)} = \sqrt{\frac{1 + \frac{u}{c}}{1 - \frac{u}{c}}} = 1 + \frac{u}{c} + \mathrm{O}\left(\frac{u}{c}\right)^2,$$

which is the same as (1.22), to the given order. The Equivalence Principle then implies that this is the relativistic frequency shift of light *in a gravitational field*. That is to say, if light is emitted at a point S in a gravitational field and observed at a point O closer to the source of the field, the measured frequency of the light at O is greater than that at S; light 'falling into' a gravitational field is *blue-shifted*. By the same token, if light moves 'out of' a gravitational field its frequency is decreased – it is red-shifted. To get an order of magnitude estimate for this effect, it follows directly from (1.23) that for light travelling 10 metres vertically downwards in the Earth's gravitational field, $h = 10$ m, $a = 10\,\mathrm{m\,s^{-2}}$, we have

$$\frac{\Delta\nu}{\nu} \approx 10^{-15}. \tag{1.24}$$

Of course in the gravitational case the frequency shift described above is *not* a Doppler shift. It is a purely gravitational effect, in which the source and the detector are *not* in relative motion. The formula was, however, derived from the hypothesis that the physical consequences of observing light frequency in a gravitational field are the same as those of observing it in an accelerating frame; and this *is* a Doppler shift, because in this case the source and detection point are in relative motion. This concludes the first thought-experiment on the Equivalence Principle and optics.

The second such thought-experiment is also concerned with light propagation; this time the light travels from left to right across the Einstein box. Consider the situations drawn schematically in Fig. 1.6. In (a) a beam of laser light travels *in an inertial frame* (that is, in neither a gravitational field nor an accelerating frame) across the box. It leaves the laser on the left hand wall and is detected on the right hand wall, after travelling in a straight line. In (b) the box is accelerating upwards with an acceleration a; this acceleration commences at the same time that the light leaves the laser. After a time Δt the light has travelled in the x direction a distance $\Delta x = c\,\Delta t$, while the box has moved upwards a distance $\Delta y = \frac{1}{2}a(\Delta t)^2$, from which

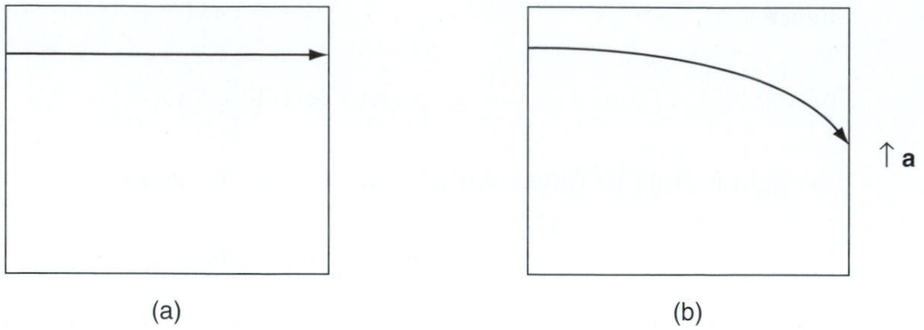

Fig. 1.6 Light travelling across a box (a) in an inertial frame, (b) in an accelerating frame, or equivalently a gravitational field.

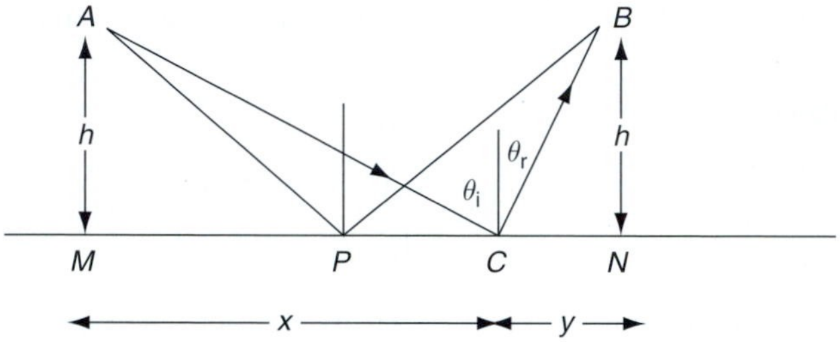

Fig. 1.7 Fermat's Principle: light reaches *A* from *B* after reflection at a mirror surface.

$$\Delta y = \frac{a}{2c^2}(\Delta x)^2. \tag{1.25}$$

Since Δy and Δx are the coordinates of the light as measured in the box, it follows that the light describes a *parabolic path* if the box is accelerated. It will therefore be detected at a detector nearer the floor of the box than the laser is. The Equivalence Principle then implies that *light follows a curved path in a gravitational field*, since it does so in an accelerating frame.

This conclusion is extremely far-reaching; even more so than the prediction of a gravitational frequency shift. Fermat (1601–1665) postulated that light takes a *minimum time* to travel from one point to another. For example, consider (Fig. 1.7) the passage of light from A to B, after reflection in a mirror. Let an arbitrary path be ACB, where C is the point where the light beam strikes the mirror, and let the angles of incidence and reflection be θ_i and θ_r, as marked. For simplicity, let A and B each be a perpendicular distance h from the mirror ($AM = BN = h$) and let x and y be the horizontal distances MC and CN respectively. Then, since A and B are fixed points, $x + y = d$ (fixed). The distance s travelled by the light is

$$s = AC + CB = \sqrt{x^2 + h^2} + \sqrt{y^2 + h^2} = \sqrt{x^2 + h^2} + \sqrt{(d - x)^2 + h^2}.$$

The *time* taken to travel a distance s is then s/c, with c the speed of light; more generally, the time taken to travel over a given path is $\int ds/c$. The requirement that the time taken be a minimum is, since c is constant, clearly the same as the requirement that the distance travelled be a minimum. In the example of reflection of light at the mirror above, if s is a minimum then $\mathrm{d}s/\mathrm{d}x = 0$, which is easily seen to give

$$\mathrm{d}s/\mathrm{d}x = 0 \Rightarrow x = y = \frac{d}{2};$$

in other words, the light beam strikes the mirror at P, at which $\theta_\mathrm{i} = \theta_\mathrm{r}$. Fermat's requirement of *least time* yields Snell's law, that the angles of incidence and reflection are equal.

Fermat's principle is, however, much more far-reaching than this. To begin with, in a general sense, the demand that light propagation takes a minimum time is the requirement that $\int \mathrm{d}s$ be a minimum; or that, in the sense of the *variational principle*,[4]

$$\delta \int \mathrm{d}s = 0. \tag{1.26}$$

The *bending* of light in a gravitational field then implies, if we take Fermat's principle seriously, that the shortest path between two points in a gravitational field is not a straight line. In any flat (Euclidean) space, however, the shortest path between two points *is* a straight line. We therefore conclude that the effect of the gravitational field is to make space *curved*. This is Einstein's conclusion: to study gravity we have to study *curved spaces*. The motion of particles in a gravitational field is to be formulated as the motion of particles in curved spaces. And more generally we can then learn to formulate *any* laws of physics in a gravitational field; for example, the study of electrodynamical effects in a gravitational field is 'simply' arrived at by writing Maxwell's equations in a curved space. The study of curved spaces is, however, not easy, and this is precisely why General Relativity is so difficult. On the other hand, in a *qualitative* sense some results become immediately 'comprehensible' in this new language. For example, the reason that planetary orbits are curves (ellipses, in general) is that planets travel in *free-fall* motion, so they trace out the shortest path they can. In a flat space this would be a straight line, but the effect of the Sun's gravity is to make the space surrounding it curved, making the planetary orbits curved paths (there are no straight lines in a curved space.). Newton's account of gravity, involving a *force*, becomes replaced by an entirely different account, involving a curved space. This is an absolutely totally different vision! It is, however, worth remarking again that the effects of a curved space are not going to be easy to detect on Earth; the deflection of a light beam on Earth, travelling over a distance of 100 km (an order of magnitude larger than, for example, SLAC, the particle accelerator at Stanford), is, from (1.25), with $a = g = 10\,\mathrm{m\,s^{-2}}$, about 10^{-3} mm.

The reader who has followed the logic so far will agree that the plan of action is now, in principle, clear. We have to learn about curved spaces; and this includes learning how to describe vectors in curved spaces, and how to *differentiate* them, which we must do if we are to carry over ideas such as Gauss's theorem and Stokes' theorem into curved spaces. The task is large, not to say daunting, but thanks to the efforts of differential geometers and theoretical relativists over a long period of time (from before Einstein's birth to after his

[4] The variational principle continues to play a crucial role in the formulation of fundamental theories in physics, from classical mechanics and quantum mechanics to General Relativity and gauge field theory. For an introduction to the central role of the variational principle see Yourgrau & Mandelstam (1968).

death) it is not impossible; and, like the task of climbing a mountain, great efforts are rewarded with excellent views. The chapters ahead chart, I hope, a sensible way through this rather complex material, but we close this chapter by making some simple observations and calculations about curved surfaces.

1.4 Curved surfaces

A surface is a 2-dimensional space. It has the distinct advantage that we can imagine it easily, because we see it (as the mathematicians say) 'embedded' in a 3-dimensional space – which also happens to be flat (I mean, of course, Euclidean 3-space). What I want to demonstrate, however, is that there are measurements *intrinsic to a surface* that may be performed to see whether it is flat or not. It is not necessary to embed a surface in a 3-dimensional space in order to see whether or not the surface is curved; we can tell just by performing measurements on the surface itself. The reader will appreciate that this is a necessary exercise; for if we are to make the statement that *3-dimensional space* is curved, this statement must have an *intrinsic* meaning. There is no fourth dimension into which our 3-dimensional space may be embedded (time does not count here).

To begin, consider the three surfaces illustrated in Fig. 1.8. They are a plane, a sphere and a saddle. On each surface draw a circle of radius a and measure its circumference C and area A. On the plane, of course, $C = 2\pi a$ and $A = \pi a^2$, but our claim is that these relations do not hold on the curved surfaces. In fact we have

$$\begin{array}{llll} \text{Plane:} & C = 2\pi a & A = \pi a^2 & \text{flat (zero curvature),} \\ \text{Sphere:} & C < 2\pi a & A < \pi a^2 & \text{curved (positive curvature),} \\ \text{Saddle:} & C > 2\pi a & A > \pi a^2 & \text{curved (negative curvature).} \end{array} \tag{1.27}$$

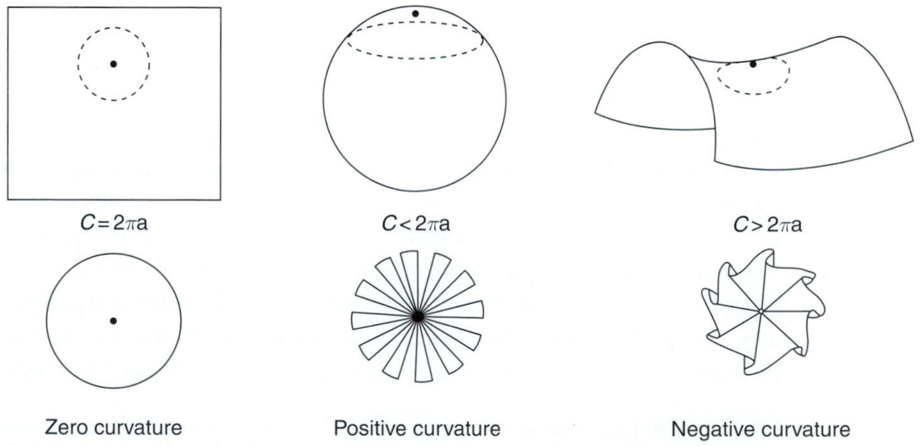

Fig. 1.8 Circles inscribed on a plane, on a sphere and on a saddle.

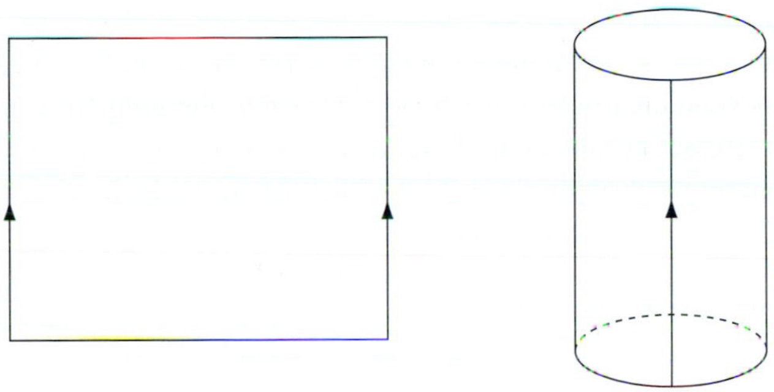

Fig. 1.9 A cylinder is made by joining the edges of a plane (those marked with arrows). No cutting or stretching is involved.

In the case of the sphere, for example, to see that $C < 2\pi a$ imagine cutting out the circular shape which acts as a 'cap' to the sphere. This shape cannot be pressed flat. In order to make it flat some radial incisions must be inserted, but this then has the consequence that the *total* circumference C of the dotted circle, which is equal to the sum of the arcs of all the incisions in the diagram below, is less than $2\pi a$. In the case of the saddle the opposite thing happens; in order to get the circular area to lie flat, we have to *fold* parts of it back on itself, so that the true circumference C is greater than $2\pi a$.

It is important to bear in mind in the above reasoning that a, the radius of the circle, is the actual distance from the centre to the perimeter, *as measured in the space*. For simplicity imagine drawing a circle of radius 1000 km on the surface of the Earth, with the centre of the circle at the North Pole. This could be done by having a piece of wire 1000 km long, fixing one end at the North Pole and walking round in a circle, with the wire kept taught. The distance travelled before returning to one's starting point is the circumference C of the circle. The radius of this circle, 1000 km, is the length of the wire, which is laid out along the curved surface. One might feel tempted to point out that one could define the radius of the circle as the 'straight' distance between a point on the circumference and the North Pole, measured by tunnelling through the Earth. But this would be cheating, because it would involve *leaving the space*. We are to imagine the surface as being a world in itself, which we do not leave; we are insisting, in other words, on making measurements *intrinsic* to the space. It should now be clear that the statements (1.27) above constitute a way of telling whether a space – in this case a 2-dimensional one – is flat or curved (and if curved, whether open or closed), and this by means of measurements made entirely within the space. A corollary of this is that, on this definition, a cylinder is *flat*; for a cylinder can be made by joining together the edges of a flat piece of paper, without stretching or tearing (see Fig. 1.9, where the edges with arrows are joined together). Since $C = 2\pi a$ before joining the edges, the same relation holds after joining them, so a cylinder is not *intrinsically* curved. It is said that a cylinder has zero intrinsic curvature but non-zero *extrinsic* curvature.

It is interesting to make one final observation about the exercise of drawing circles on spheres. As the circle S^1 is lowered over the sphere, becoming further and further south, its circumference increases as its radius increases (with $C < 2\pi a$ always holding).[5] This happens until the circle becomes the Equator. This is the circle with a *maximum* circumference in S^2; beyond the Equator, when the circle enters the southern hemisphere, its radius continues to increase, but its circumference actually *decreases*. This continues to be the case until the circle itself approaches the South Pole, at which its circumference tends to zero. This is the limit of a circle with maximum radius (which is the maximum distance attainable in the space) but with a circumference approaching zero. These observations are, in a sense, obvious, but they become interesting and physically relevant in a particular cosmological model, in which the geometry of 3-dimensional space is S^3, the 3-sphere. The above exercise can then be rehashed, increasing the dimension of everything by 1; that is, to discuss surfaces S^2 in S^3, rather than lines S^1 in S^2. This model describes a 'closed' universe, and is described further in Section 10.2 below.

Further reading

Accounts of the various attempts to construct relativistic theories of gravity (other than General Relativity) are outlined in Pauli (1958) pp. 142–5, Mehra (1973), Pais (1982) Chapter 13, Torretti (1996) Chapter 5 and Cao (1997) Chapter 3.

For details of the Eötvös experiment on the torsion balance, see Dicke (1964) and Nieto *et al.* (1989). A modern assessment of the experimental evidence for the Equivalence Principle is contained in Will (2001). The reference to Einstein's seminal paper on the Equivalence Principle and optics is Einstein (1911).

Good introductory accounts of General Relativity and curved spaces are to be found in Hoffmann (1983) Chapter 6, and in Harrison (2000) Chapters 10 and 12. See also, for a slightly more advanced treatment, Ellis & Williams (1988).

Problems

1.1 Find an expression for the angle of displacement from the vertical with which a mass hangs in the gravitational field of the Earth as a function of latitude λ, and calculate its value at Budapest (latitude 47.5° N).

1.2 Suppose that mass, like electric charge, can take on both positive and negative values, but with Newton's laws continuing to hold. Consider two masses, m_1 and m_2, a distance r apart. Describe their motion in the cases (i) $m_1 = m_2 = m$ ($m > 0$), (ii) $m_1 = m_2 = -m$, (iii) $m_1 = m$, $m_2 = -m$. Is momentum conserved in all these cases?

[5] S^n is an n-dimensional subspace of the $(n+1)$-dimensional Euclidean space, given by the formula ${x_1}^2 + {x_2}^2 + \cdots + {x_{n+1}}^2 = \text{const}$. So S^1 is a circle, S^2 (the surface of) a sphere, etc.

1.3 The usual formula for the period T of a simple pendulum of length l is

$$T = 2\pi\sqrt{\frac{l}{g}},$$

where g is the acceleration due to gravity. Denoting the inertial mass of the pendulum bob by m_{i} and its gravitational mass by m_{g}, derive an alternative expression for T in terms of these masses, the radius R of the Earth and its mass M_{g}.

1.4 By employing spherical polar coordinates show that the circumference C of a circle of radius R inscribed on a sphere S^2 (as in Fig. 1.8) obeys the inequality $C < 2\pi R$.

2 Special Relativity, non-inertial effects and electromagnetism

General Relativity is a generalisation of Special Relativity, and this chapter begins with a brief summary of the special theory, with which the reader is assumed already to have some familiarity. After an account of the famous 'Einstein train' thought experiment, the more formal matters of Minkowski space-time and Lorentz transformations are discussed. We then consider some non-inertial effects in the shape of the twin paradox and the Sagnac effect. Mach's Principle, which concerns itself with the origin of inertia, is considered, and this is followed by a section on Thomas precession; an effect derivable from Special Relativity alone, but associated with forces, and therefore with non-inertial frames. The chapter finishes with a brief treatment of electrodynamics – which was Einstein's starting point for Special Relativity.

2.1 Special Relativity: Einstein's train

We are concerned with the laws of transformation of coordinates between frames of reference in (uniform) relative motion. Two frames, S and S', both inertial, move relative to one another with (constant) speed v, which we may take to be along their common x axis. The space-time coordinates in each frame are then

$$S : (x, y, z, t); \quad S' : (x', y', z', t').$$

What is the relation between these? In the physics of Galileo and Newton it is

$$x' = x - vt, \quad y' = y, \quad z' = z, \quad t' = t \tag{2.1}$$

whose inverse is

$$x = x' + vt', \quad y = y', \quad z = z', \quad t = t'; \tag{2.2}$$

S and S' have a common origin at $t = 0$. There is an infinite number of inertial frames and the laws of Newtonian mechanics *are the same in all of them*. There is no such thing as absolute velocity; we can only meaningfully talk about the *relative velocity* of one inertial frame relative to another one. This is the Newtonian–Galilean *Principle of Relativity*. Under the above transformations the laws of Newtonian mechanics are *covariant* (of the same form). These transformations form a group – the Galileo group – which is the symmetry group of Newtonian mechanics. Its actions take one from one frame of reference S to another one S', in which the laws of mechanics are the same. If there is a frame S'', moving relative to S' with speed u along their common x axis, then the speed of S'' relative to S is

$$w = u + v. \tag{2.3}$$

This is the law of addition of velocities in the Newtonian-Galilean Principle of Relativity.

Can this Principle of Relativity be generalised from mechanics to all of physics? This is surely a worthy aim, but a strong hint of trouble came when Maxwell, in his theory of electromagnetism, showed that the speed of light (electromagnetic waves) was given by the formula

$$c = (\varepsilon_0 \mu_0)^{-1/2}, \tag{2.4}$$

where ε_0 is the electric permitivity and μ_0 the magnetic permeability of free space. When the values are inserted this gives $c \approx 3 \times 10^8\,\mathrm{m\,s^{-1}}$ – the observed speed of light. So in Maxwell's electrodynamics the speed of light (in a vacuum) depends only on electric and magnetic properties of the vacuum, and is therefore *absolute*; this clearly contradicts the Principle of Relativity above. It must be the same in all frames of reference and Equation (2.3) must therefore break down (at least when applied to light).

The most famous demonstration of this is the Michelson–Morley experiment, which showed that the speed of light is indeed the same in different frames of reference. It is therefore clear that Equations (2.1) and (2.2) must be revised. It was in fact already known that the transformations which left Maxwell's equations invariant were the Lorentz transformations, which for relative motion along the x axis take the form

$$x' = \gamma(x - vt), \quad y' = y, \quad z' = z, \quad t' = \gamma(t - vx/c^2) \tag{2.5}$$

with inverse

$$x = \gamma(x' + vt'), \quad y = y', \quad z = z', \quad t = \gamma(t' + vx'/c^2), \tag{2.6}$$

where

$$\gamma = (1 - v^2/c^2)^{-1/2}. \tag{2.7}$$

Einstein interpreted these equations not just as a mathematical curiosity, but as a demonstration that time, like space, is relative: $x' \neq x$, $t' \neq t$. Let us illustrate this by considering the 'Einstein train'.

Trains A and B, with the same length L, pass one another with relative speed v in the x direction. How long does this take? Let us consider two events:

Event 1 : front of train B passes front of train A

Event 2 : rear of train B passes front of train A

These are illustrated in Fig. 2.1.

Let us adopt the coordinates

$$\begin{aligned} &(x', t') : \ \text{coordinates in moving frame} \\ &(x, t) : \ \text{coordinates in stationary frame} \end{aligned} \tag{2.8}$$

What is the time interval between these events, as measured in the two coordinate systems? To be definite, let us consider train A as stationary and train B moving. We take the origins ($x = 0$, $x' = 0$) at the right hand ends of the trains and synchronise the clocks so that event 1 happens at $t = 0$, $t' = 0$. Then, for event 1

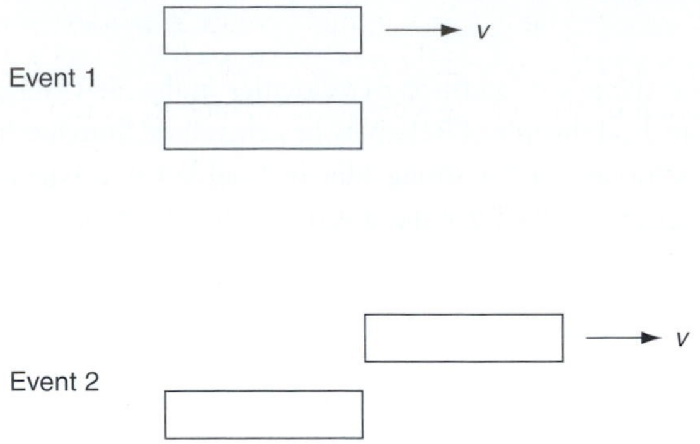

Fig. 2.1 **Event 1: front of train B passes front of train A; Event 2: rear of train B passes front of train A.**

$$(x_1', t_1') = (0, 0), \quad (x_1,\ t_1) = (0, 0). \tag{2.9}$$

If event 2 happens after a time interval T in the stationary frame (train A) and after an interval T' in the moving frame (train B) then we have

$$(x_2', t_2') = (-L, T'), \quad (x_2, t) = (0, T). \tag{2.10}$$

The Lorentz transformation (2.5) applied to event 2 gives $-L = \gamma(-v\,T)$, $T' = \gamma\,T$, or

$$L = \gamma v T, \quad T' = \gamma T. \tag{2.11}$$

Since $\gamma > 1$, then $T' > T$; the time interval between events 1 and 2 in the moving frame is greater than in the stationary frame – 'time goes slower in moving frames'. So when Andrei (in train A) looks at Bianca's clock (in train B), he sees it goes slower than his own. It is also true that when Bianca looks at Andrei's clock, she sees it goes slower than her own (since of course the whole sequence of events can be considered in the frame in which B is at rest). One is tempted to ask the question, whose clock is *really* going slower? But this is a bit like asking, when walking along a road, is the house on the left or the right hand side of the road? It all depends in which direction you are walking; and in our case it all depends who is looking at the two clocks: if Andrei is looking, Bianca's clock is going slower, and if Bianca is looking, Andrei's clock is lagging. This is, after all, a theory of *relativity* – only relative motion has physical significance. It has no meaning to say that A is at rest and B is moving, any more than it has to say that B is at rest and A is moving. Since only relative motion has significance, anything observed by A must also be observed by B; the situation is symmetrical. Einstein's train gives a neat demonstration of the relativity of time – to be precise, of time intervals.

There is, as the reader will know, a similar result for space intervals: what is the length of train B as viewed from train A? Call it L'. It is of course Tv:

$$L' = T\,v = L/\gamma = L\,(1 - v^2/c^2)^{1/2} < L. \tag{2.12}$$

A measures B's train as being shorter than his own. Similarly, B measures A's train as being shorter than her own: moving objects appear contracted. This is the Fitzgerald–Lorentz contraction.

We see that time intervals and lengths are not invariant under Lorentz transformations. Infinitesimally, $dx^2+dy^2+dz^2$ is not invariant, and neither is dt^2. The quantity which *is* invariant between the events (x, y, z, t) and $(x+dx, y+dy, z+dz, t+dt)$ is

$$ds^2 = -c^2\,dt^2 + dx^2 + dy^2 + dz^2. \tag{2.13}$$

In the present case of the train, $dy=dz=0$, so $ds^2=-c^2\,dt^2+dx^2$. This should be the same in all frames of reference, where dt and dx refer to the time and space separation of the two events above. We then have, in the rest frame S with coordinates (x, t)

$$ds^2 = -c^2\,dt^2 + dx^2 = -c^2(t_2-t_1)^2 + (x_2-x_1)^2 = -c^2T^2, \tag{2.14}$$

while in the moving frame S', with coordinates (x', t')

$$\begin{aligned} ds^2 &= -c^2\,dt'^2 + dx'^2 = c^2(t_2'-t_1')^2 + (x_2'-x_1')^2 \\ &= -c^2T'^2 + L^2 \\ &= -c^2\gamma^2T^2 + \gamma^2v^2T^2 = -c^2\gamma^2T^2(1-v^2/c^2) \\ &= -c^2T^2, \end{aligned} \tag{2.15}$$

where (2.11) and (2.12) have been used. We see that ds^2 is the same in the two frames. We also see the force of Minkowski's remark,[1] 'Henceforth space by itself, and time by itself, are doomed to fade away into mere shadows, and only a kind of union of the two will preserve an independent reality.'

2.1.1 Minkowski space-time

We now formalise Special Relativity as follows. Space and time become a 4-dimensional manifold, Minkowski space-time. Points in this space-time ('events') have coordinates x^μ ($\mu=0, 1, 2, 3$), with, in Cartesian coordinates $(x^0, x^1, x^2, x^3)=(ct, x, y, z)$, and in spherical polars $(x^0, x^1, x^2, x^3)=(ct, r, \theta, \phi)$. We also adopt the notation that while Greek suffices take on the values (0, 1, 2, 3), Latin suffices take on the values (1, 2, 3) for space variables only; $x^\mu=(x^0, x^i)$. The invariant distance, or 'separation' between two events (in Cartesian coordinates), $ds^2=-c^2\,dt^2+dx^2+dy^2+dz^2$, is written in the form

$$ds^2 = \eta_{\mu\nu}\,dx^\mu\,dx^\nu, \tag{2.16}$$

where the *summation convention* has been used: repeated indices are summed over the values 0, 1, 2, 3. Thus (2.16) is short-hand for

$$ds^2 = \eta_{00}(dx^0)^2 + \eta_{01}\,dx^0\,dx^1 + \eta_{02}\,dx^0\,dx^2 + \cdots \text{ (16 terms)},$$

and $\eta_{\mu\nu}$ has the following values, in *Cartesian coordinates*:

$$ds^2 = -c^2\,dt^2 + dx^2 + dy^2 + dz^2, \tag{2.17}$$

[1] In Lorentz *et al.*, 1952, p. 75.

hence

$$\eta_{00} = -1, \quad \eta_{11} = \eta_{22} = \eta_{33} = 1, \quad \eta_{\mu\nu} = 0, \quad \mu \neq \nu;$$

or in matrix form

$$\eta_{\mu\nu} = \begin{pmatrix} -1 & 0 & 0 & 0 \\ 0 & 1 & 0 & 0 \\ 0 & 0 & 1 & 0 \\ 0 & 0 & 0 & 1 \end{pmatrix}; \tag{2.18}$$

and in *spherical polar coordinates*:

$$ds^2 = -c^2\, dt^2 + dr^2 + r^2\, d\theta^2 + r^2 \sin^2\theta\, d\phi^2 \tag{2.19}$$

hence

$$\eta_{00} = -1, \quad \eta_{11} = 1, \quad \eta_{22} = r^2, \quad \eta_{33} = r^2 \sin^2\theta, \quad \eta_{\mu\nu} = 0, \quad \mu \neq \nu;$$

$$\eta_{\mu\nu} = \begin{pmatrix} -1 & 0 & 0 & 0 \\ 0 & 1 & 0 & 0 \\ 0 & 0 & r^2 & 0 \\ 0 & 0 & 0 & r^2 \sin^2\theta \end{pmatrix}. \tag{2.20}$$

The object $\eta_{\mu\nu}$ is the *metric tensor* of Minkowski space; it makes the space a *metric space*, one in which distance is defined.

A useful concept in Special Relativity is that of *proper time* τ. It is defined by

$$ds^2 = -c^2\, d\tau^2. \tag{2.21}$$

In a particle's own rest frame – in which, of course, $dx = dy = dz = 0$, τ coincides with t, so proper time is simply time as measured in the rest frame, or time recorded on one's own clock.

2.1.2 Lorentz transformations

Lorentz transformations are transformations between coordinates labelling space-time events recorded by two inertial observers in uniform relative motion. They take a system from one inertial frame to another one, and consist of rotations and Lorentz 'boost' transformations.[2] Under a general Lorentz transformation

[2] The *maximal* set of transformations leaving ds^2 invariant includes, in addition to rotations and boosts, also translations in space and time, $x^i \to x^i + a^i$, $t \to t + t_0$ (or simply $x^\mu \to x^\mu + a^\mu$). These are *inhomogeneous* transformations and, corresponding to the philosophy outlined above, their inclusion represents the fact that the laws of physics are invariant under space and time translations; there is no absolute origin in space, nor in time (the Big Bang is not relevant here; firstly, we are not considering cosmology, and secondly, we are concerned with the laws of physics themselves, not with whatever state the Universe happens to be in). Enlarging the group of Lorentz transformations to include these translations produces the *inhomogeneous Lorentz group*, or *Poincaré group*. The importance of the Poincaré group as the maximal invariance group in Minkowski space was emphasised particularly by Wigner, whose analysis remains of fundamental importance in particle physics. For more details, see Wigner (1939, 1964), Wightman (1960), Sexl & Urbantke (1976), Tung (1985), Doughty (1990), Ryder (1996), Cao (1997).

$$x^\mu \to x'^\mu = \Lambda^\mu{}_\nu x^\nu \tag{2.22}$$

so $dx'^\mu = \Lambda^\mu{}_\nu\, dx^\nu$ and the invariance of ds^2 gives

$$\eta_{\mu\nu}\, dx'^\mu\, dx'^\nu = \eta_{\mu\nu}\, dx^\mu dx^\nu,$$

hence

$$\eta_{\mu\nu}\Lambda^\mu{}_\rho\Lambda^\nu{}_\sigma\, dx^\rho\, dx^\sigma = \eta_{\rho\sigma}\, dx^\rho\, dx^\sigma$$

or

$$\eta_{\mu\nu}\Lambda^\mu{}_\rho\Lambda^\nu{}_\sigma = \eta_{\rho\sigma}. \tag{2.23}$$

Let us now check that this holds for some specific Lorentz transformations. First consider a rotation about the z axis through an angle θ:

$$x' = x\cos\theta + y\sin\theta, \quad y' = -x\sin\theta + y\cos\theta.$$

The corresponding matrix is

$$\Lambda^\mu{}_\nu = \begin{pmatrix} 1 & 0 & 0 & 0 \\ 0 & \cos\theta & \sin\theta & 0 \\ 0 & -\sin\theta & \cos\theta & 0 \\ 0 & 0 & 0 & 1 \end{pmatrix}. \tag{2.24}$$

Equation (2.23) with $\rho = \sigma = 1$ then gives $\eta_{\mu\nu}\,\Lambda^\mu{}_1\Lambda^\nu{}_1 = 1$, i.e. (summation convention!)

$$-\Lambda^0{}_1\Lambda^0{}_1 + \Lambda^1{}_1\Lambda^1{}_1 + \Lambda^2{}_1\Lambda^2{}_1 + \Lambda^3{}_1\Lambda^3{}_1 = 1,$$

or $\cos^2\theta + \sin^2\theta = 1$, which is correct. Taking different values for ρ and σ also gives consistency with (2.23), as may easily be checked.

Now consider a Lorentz boost along the x direction. With $x^0 = ct$ (and replacing $-v$ by $+v$), Equation (2.5) corresponds to the Lorentz matrix

$$\Lambda^\mu{}_\nu = \begin{pmatrix} \gamma & \gamma v/c & 0 & 0 \\ \gamma v/c & \gamma & 0 & 0 \\ 0 & 0 & 1 & 0 \\ 0 & 0 & 0 & 1 \end{pmatrix}. \tag{2.25}$$

Now put, for example, $\rho = \sigma = 0$. With $\Lambda^0{}_0 = \gamma$, $\Lambda^1{}_0 = \gamma v/c$, $\Lambda^2{}_0 = \Lambda^3{}_0 = 0$, we have

$$\eta_{00}(\Lambda^0{}_0)^2 + \eta_{11}(\Lambda^1{}_0)^2 = -1,$$

or $\gamma^2(1 - v^2/c^2) = 1$, which is correct.

For future reference it is convenient to give the most general form of a Lorentz (boost) transformation, from frame S to frame S' moving with relative velocity $\mathbf{v}$:

$$\mathbf{x}' = \mathbf{x} + (\gamma - 1)\frac{\mathbf{x}\cdot\mathbf{v}}{v^2}\mathbf{v} - \gamma\mathbf{v}t; \quad t' = \gamma\left(t - \frac{\mathbf{x}\cdot\mathbf{v}}{c^2}\right), \tag{2.26}$$

with inverse

$$\mathbf{x} = \mathbf{x}' + (\gamma - 1)\frac{\mathbf{x}' \cdot \mathbf{v}}{v^2}\mathbf{v} + \gamma\mathbf{v}t'; \quad t = \gamma\left(t' - \frac{\mathbf{x}' \cdot \mathbf{v}}{c^2}\right), \tag{2.27}$$

and, as usual, $\gamma = (1 - v^2/c^2)^{-1/2}$.

The matrix (2.25) may be written in a 'trigonometric' form, similar to (2.24). Defining the hyperbolic angle ϕ by ($\beta = v/c$)

$$\gamma = \cosh\phi, \quad \gamma\beta = \sinh\phi, \tag{2.28}$$

the Lorentz transformation given by (2.25) may be written

$$\begin{pmatrix} x'^0 \\ x'^1 \\ x'^2 \\ x'^3 \end{pmatrix} = \begin{pmatrix} \cosh\phi & \sinh\phi & 0 & 0 \\ \sinh\phi & \cosh\phi & 0 & 0 \\ 0 & 0 & 1 & 0 \\ 0 & 0 & 0 & 1 \end{pmatrix} \begin{pmatrix} x^0 \\ x^1 \\ x^2 \\ x^3 \end{pmatrix}. \tag{2.29}$$

We now define the *generator* of Lorentz boosts along the x axis by

$$K_x = \frac{1}{\mathrm{i}}\frac{\partial\Lambda}{\partial\phi}\bigg|_{\phi=0} = -\mathrm{i}\begin{pmatrix} 0 & 1 & 0 & 0 \\ 1 & 0 & 0 & 0 \\ 0 & 0 & 0 & 0 \\ 0 & 0 & 0 & 0 \end{pmatrix}, \tag{2.30}$$

where Λ is the matrix in (2.29). It may then easily be checked that

$$\exp(\mathrm{i}K_x\phi) = \begin{pmatrix} \cosh\phi & \sinh\phi & 0 & 0 \\ \sinh\phi & \cosh\phi & 0 & 0 \\ 0 & 0 & 1 & 0 \\ 0 & 0 & 0 & 1 \end{pmatrix}. \tag{2.31}$$

The generators of boosts along the y and z axes are defined analogously and turn out as

$$K_y = -\mathrm{i}\begin{pmatrix} 0 & 0 & 1 & 0 \\ 0 & 0 & 0 & 0 \\ 1 & 0 & 0 & 0 \\ 0 & 0 & 0 & 0 \end{pmatrix}, \quad K_z = -\mathrm{i}\begin{pmatrix} 0 & 0 & 0 & 1 \\ 0 & 0 & 0 & 0 \\ 0 & 0 & 0 & 0 \\ 1 & 0 & 0 & 0 \end{pmatrix}. \tag{2.32}$$

Generators of rotations may be defined similarly. The matrix (2.24) represents a rotation about the z axis, whose generator is defined as $J_z = \frac{1}{\mathrm{i}}\frac{\partial\Lambda}{\partial\theta}\Big|_{\theta=0}$. This and analogous definitions for J_x and J_y yield

$$J_x = -\mathrm{i}\begin{pmatrix} 0 & 0 & 0 & 0 \\ 0 & 0 & 0 & 0 \\ 0 & 0 & 0 & 1 \\ 0 & 0 & -1 & 0 \end{pmatrix}, \quad J_y = -\mathrm{i}\begin{pmatrix} 0 & 0 & 0 & 0 \\ 0 & 0 & 0 & -1 \\ 0 & 0 & 1 & 0 \\ 0 & 0 & 0 & 0 \end{pmatrix}, \quad J_z = -\mathrm{i}\begin{pmatrix} 0 & 0 & 0 & 0 \\ 0 & 0 & 1 & 0 \\ 0 & -1 & 0 & 0 \\ 0 & 0 & 0 & 0 \end{pmatrix}. \tag{2.33}$$

These six generators obey the commutation relations ($[\mathrm{A,B}] \equiv \mathrm{AB} - \mathrm{BA}$)

$$
\begin{aligned}
[J_x, J_y] &= \mathrm{i} J_z \text{ and cyclic perms} \\
[K_x, K_y] &= -\mathrm{i} J_z \text{ and cyclic perms} \\
[J_x, K_y] &= \mathrm{i} K_z \text{ and cyclic perms} \\
[J_x, K_x] &= 0 \text{ etc.}
\end{aligned}
\tag{2.34}
$$

Equivalently, relabelling the subscripts x, y, z as 1, 2, 3,

$$[J_i, J_k] = \mathrm{i}\, \varepsilon_{ikm} J_m, \tag{2.35a}$$

$$[J_i, K_k] = \mathrm{i}\, \varepsilon_{ikm} K_m, \tag{2.35b}$$

$$[K_i, K_k] = \mathrm{i}\, \varepsilon_{ikm} J_m, \tag{2.35c}$$

where ε_{ikm} is the totally antisymmetric symbol

$$
\varepsilon_{ikm} = \begin{cases} 1\ (ikm) \text{ even permutation of } (123), \\ -1\ (ikm) \text{ odd permutation of } (123), \\ 0 \text{ otherwise.} \end{cases}
\tag{2.36}
$$

In terms of these six generators a general Lorentz boost transformation is

$$\Lambda(\boldsymbol{\phi}) = \exp(\mathrm{i}\mathbf{K} \cdot \boldsymbol{\phi}); \tag{2.37}$$

a general rotation is represented by

$$\Lambda(\boldsymbol{\theta}) = \exp(\mathrm{i}\mathbf{J} \cdot \boldsymbol{\theta}); \tag{2.38}$$

while a general Lorentz transformation, comprising both a boost and a rotation is given by

$$\Lambda(\boldsymbol{\phi}, \boldsymbol{\theta}) = \exp(\mathrm{i}\mathbf{K} \cdot \boldsymbol{\phi} + \mathrm{i}\mathbf{J} \cdot \boldsymbol{\theta}). \tag{2.39}$$

The relations (2.35) define the *Lie algebra* of the Lorentz group, involving three generators K_i of Lorentz 'boosts' (or 'pure' Lorentz transformations) and three generators J_i of rotations in space. The algebra is closed, corresponding to the fact that Lorentz transformations form a group. Rotations in space form a subgroup of the Lorentz group, as may be seen from the fact that the generators J_i form by themselves a closed algebra. The boost generators K_i however do not generate a closed system, as is seen from (2.35c); pure Lorentz transformations do not form a group. As a simple consequence of this, the product of two Lorentz boosts in different directions is *not* a single Lorentz boost, but also involves a rotation. It is this fact which is responsible for Thomas precession (see Section 2.5 below) – and which, as far as I can tell, seems to have been unknown to Einstein.

We finish this section with an additional remark about notation. In Equation (2.16),

$$\mathrm{d}s^2 = \eta_{\mu\nu}\, \mathrm{d}x^\mu\, \mathrm{d}x^\nu,$$

it was pointed out that the summation convention is understood. To be more precise, indices to be summed over appear twice, *once in a lower and once in an upper position*. We may write (2.16), however, in an alternative way. Defining

$$x_\mu = \eta_{\mu\nu} x^\nu,$$

we may put

$$ds^2 = dx^\mu \, dx_\mu, \tag{2.40}$$

where still the summation convention holds, and the repeated index – only one index now – appears once in a lower and once in an upper position. Note that the components x_μ and x^μ are quite different: in Cartesian coordinates we have

$$(x^0, x^1, x^2, x^3) = (ct, x, y, z); \quad (x_0, x_1, x_2, x_3) = (-ct, x, y, z);$$

and in spherical polar coordinates

$$(x^0, x^1, x^2, x^3) = (ct, r, \theta, \phi); \quad (x_0, x_1, x_2, x_3) = (-ct, r, r^2, \theta, \{r^2 \sin^2\theta\}\phi).$$

In General Relativity the position of indices on vectors is important. Vectors with an *upper* index, V^μ, are called *contravariant vectors*, and those with a lower index, V_μ, *covariant vectors*. In the modern mathematical formulation, these vectors actually arise in conceptually different ways, as will be explained in the next chapter.

2.2 Twin paradox: accelerations

The so-called twin paradox is not a paradox. It is the following statement: if A and B are twins and A remains on Earth while B goes on a long trip, say to a distant star and back again, then on return B is younger than A. Suppose the star is a distance l away and B travels with speed v there and back. Then, as measured in A's frame, B is away for a time $2l/v$, and that is how much A has aged when B returns. When A looks at B's clock, however, there is a time dilation factor of $\gamma = (1 - v^2/c^2)^{-1/2}$, so B's clock – including her biological clock – has only registered a passage of time $2l/\gamma v = (2l/v)(1 - v^2/c^2)^{1/2}$; on return, therefore, she is younger than A. This is the true situation. It *appears* paradoxical because one is tempted to think that 'time is relative', so that while A reckons B to be younger on return, as argued above, B should also reckon A to be younger; so in actual fact, one might think, they are the same age after the trip, just as before it. This, however, is wrong, and the reason is that while A remains in an inertial frame (or at least the *approximately* inertial frame of the Earth), B does not, since B has to reverse her velocity for the return trip, and that means she undergoes an *acceleration*. There is no reason why the twins should be the same age after B's space trip, and they are not.

It may be useful to consider some numbers. Suppose the star is 15 light years away and B (Bianca) travels at speed $v = (3/5)c$. Then, measured by A (Andrei), Bianca reaches the star in $\dfrac{15}{3/5} = 25$ years, so Andrei is 50 years older when Bianca returns (see Fig. 2.2). The time dilation factor is $1/\gamma = (1 - v^2/c^2)^{1/2} = 4/5$, so, as seen by Andrei, Bianca takes a time $25 \times (4/5) = 20$ years to reach the star, and will therefore be 40 years older when she returns. She will therefore be 10 years younger than Andrei after the trip. Of course, this is an approximation, since we have ignored the time taken for Bianca to change her velocity from

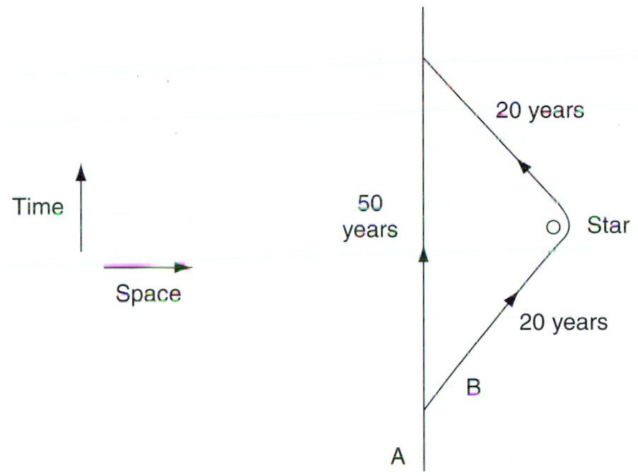

Fig. 2.2 A stays on Earth while B travels to a star and back.

$+v$ to $-v$; this is indicated by B's 'smoothed out' world-line near the star in Fig. 2.2. We now show, however, that this time may be as short as desired, if B is subjected to a large enough acceleration. We must therefore consider the treatment of accelerations in Minkowski space-time.

First define the 4-velocity u^{μ}:

$$u^{\mu} = \frac{\mathrm{d}x^{\mu}}{\mathrm{d}\tau} = \left(c\frac{\mathrm{d}t}{\mathrm{d}\tau}, \frac{\mathrm{d}x}{\mathrm{d}\tau}, \frac{\mathrm{d}y}{\mathrm{d}\tau}, \frac{\mathrm{d}z}{\mathrm{d}\tau}\right). \tag{2.41}$$

In view of (2.16) and (2.17) we have

$$\eta_{\mu\nu}u^{\mu}u^{\nu} = u^{\mu}u_{\mu} = -c^2; \tag{2.42}$$

the 4-velocity has constant length. Differentiating this gives $\left(\text{with } \dot{u}^{\mu} = \frac{\mathrm{d}u^{\mu}}{\mathrm{d}\tau}\right)$

$$\frac{\mathrm{d}}{\mathrm{d}\tau}(u^{\mu}u_{\mu}) = 0 = 2\dot{u}^{\mu}u_{\mu},$$

or, defining the *acceleration four-vector* $a^{\mu} = \dot{u}^{\mu}$,

$$\eta_{\mu\nu}a^{\mu}u^{\nu} = a^{\mu}u_{\mu} = 0. \tag{2.43}$$

Now consider a particle moving in the x^1 direction with constant acceleration g. The velocity and acceleration 4-vectors are

$$c\frac{\mathrm{d}t}{\mathrm{d}\tau} = u^0, \quad \frac{\mathrm{d}x^1}{\mathrm{d}\tau} = u^1; \quad \frac{\mathrm{d}u^0}{\mathrm{d}\tau} = a^0, \quad \frac{\mathrm{d}u^1}{\mathrm{d}\tau} = a^1;$$

(both vectors have vanishing 2- and 3-components). Equations (2.42) and (2.43) give

$$-(u^0)^2 + (u^1)^2 = -c^2; \quad -u^0a^0 + u^1a^1 = 0. \tag{2.44}$$

In addition

$$a^{\mu}a_{\mu} = -(a^0)^2 + (a^1)^2 = g^2; \tag{2.45}$$

this last equation *defines* the constant acceleration g. These two equations have the solutions

$$a^0 = \frac{g}{c}u^1, \quad a^1 = \frac{g}{c}u^0, \tag{2.46}$$

from which $\frac{\mathrm{d}a^0}{\mathrm{d}\tau} = \frac{g}{c}\frac{\mathrm{d}u^1}{\mathrm{d}\tau} = \frac{g}{c}a^1 = \frac{g^2}{c^2}u^0$ and hence

$$\frac{\mathrm{d}^2 u^0}{\mathrm{d}\tau^2} = \frac{g^2}{c^2}u^0. \tag{2.47}$$

Similarly,

$$\frac{\mathrm{d}^2 u^1}{\mathrm{d}\tau^2} = \frac{g^2}{c^2}u^1. \tag{2.48}$$

The solution to (2.48) is

$$u^1 = \mathrm{A}e^{g\tau/c} + \mathrm{B}e^{-g\tau/c},$$

hence

$$\frac{\mathrm{d}u^1}{\mathrm{d}\tau} = \frac{g}{c}(\mathrm{A}e^{g\tau/c} - \mathrm{B}e^{-g\tau/c}).$$

With the boundary conditions $t=0$, $\tau=0$; $u^1=0$, $\frac{\mathrm{d}u^1}{\mathrm{d}\tau} = a^1 = g$ we find $\mathrm{A} = -\mathrm{B} = c/2$ and hence $u^1 = \frac{\mathrm{d}x}{\mathrm{d}\tau} = c\sinh(g\tau/c)$. Equation (2.46) then gives

$$a^0 = \frac{\mathrm{d}u^0}{\mathrm{d}\tau} = g\sinh(g\tau/c),$$

hence $u^0 = c\frac{\mathrm{d}t}{\mathrm{d}\tau} = c\cosh(g\tau/c)$, and finally

$$x = \frac{c^2}{g}\cosh(g\tau/c), \quad ct = \frac{c^2}{g}\sinh(g\tau/c). \tag{2.49}$$

The space and time coordinates then fall on the hyperbola

$$x^2 - c^2t^2 = \frac{c^4}{g^2} \tag{2.50}$$

sketched in Fig. 2.3. The non-relativistic limits (i.e. $g\tau/c \ll 1$) of x and t above are

$$t = \tau, \quad x = c^2/g + \tfrac{1}{2}gt^2.$$

We may now return to the twin paradox. We saw that the amount of proper time elapsing for B, travelling at a constant speed $v = 3c/5$, was 20 years for each of the journeys to and from the star. The remaining question was, how much proper time elapses while B reverses her velocity from $+v$ to $-v$? If this is achieved with a constant acceleration a, then we have from (2.49) $\frac{\mathrm{d}x}{\mathrm{d}\tau} = c\sinh(a\tau/c) = \frac{3}{5}c$, hence

$$\tau = \frac{c}{a}\sinh^{-1}0.6 \approx \frac{0.55c}{a}.$$

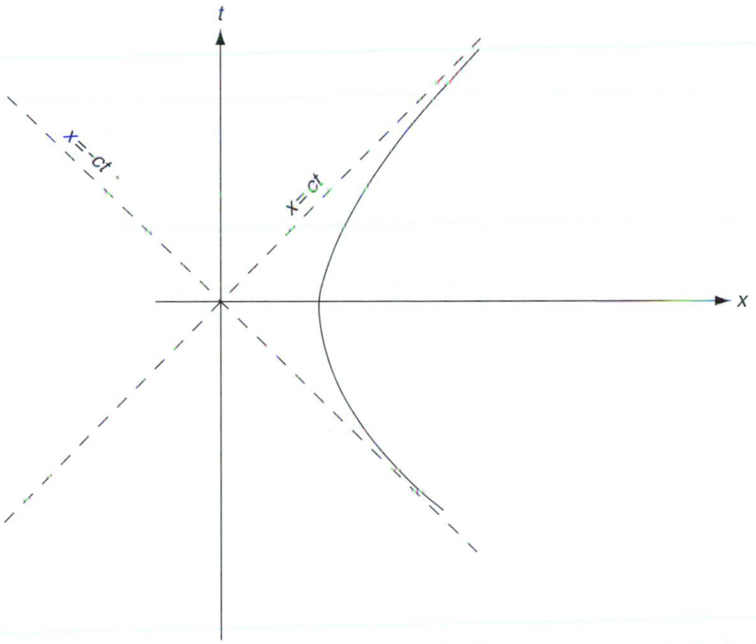

Fig. 2.3

The hyperbola $x^2 - c^2t^2 = c^4/g^2$.

Hence even with only the acceleration of the Earth's gravity, $a \approx g \approx 10\,\mathrm{m\,s^{-2}}$, $\tau \approx 1.6 \times 10^7$ s, which is only about 6 months. The proper time that elapses during the period of deceleration and acceleration is negligible, and we conclude that the twin paradox is not a paradox at all; after her space journey Bianca is younger than Andrei – almost ten years younger!

2.3 Rotating frames: the Sagnac effect

The twin paradox arises when one of two observers (twins), but not the other one, undergoes an *acceleration* in Minkowski space-time; that is, motion in which $\frac{d\mathbf{v}}{dt} \neq 0$, where $\mathbf{v}$ is the relative velocity of the twins. Putting $\mathbf{v}=\mathbf{n}v$, however, we may distinguish in general two cases in which $\frac{d\mathbf{v}}{dt} \neq 0$, (i) $\frac{d\mathbf{n}}{dt} = 0$, $\frac{dv}{dt} \neq 0$; this is the case of acceleration in a straight line, (ii) $\frac{d\mathbf{n}}{dt} \neq 0$, $\frac{dv}{dt} = 0$; this is the case of motion with changing direction but constant speed, for example, in a rotating frame. Let us now consider this case; we expect to find, and do find, some interesting new effects.

Let us suppose that S' rotates relative to S around their common z axis. It is convenient to use cylindrical coordinates, so that in S

$$\mathrm{d}s^2 = -c^2\,\mathrm{d}t^2 + \mathrm{d}r^2 + r^2\,\mathrm{d}\phi^2 + \mathrm{d}z^2. \tag{2.51}$$

The coordinates in S' are related to those in S by

$$t' = t, \quad r' = r, \quad \phi' = \phi - \omega t, \quad z' = z \tag{2.52}$$

and hence

$$\begin{aligned}\mathrm{d}s'^2 &= -c^2\,\mathrm{d}t'^2 + \mathrm{d}r'^2 + r'^2(\mathrm{d}\phi' + \omega\,\mathrm{d}t')^2 + \mathrm{d}z'^2\\ &= -(c^2 - \omega^2 r'^2)\mathrm{d}t'^2 + 2\omega r'^2\,\mathrm{d}\phi'\,\mathrm{d}t' + \mathrm{d}r'^2 + r'^2\,\mathrm{d}\phi'^2 + \mathrm{d}z'^2.\end{aligned}$$

Dropping the primes we then have for the invariant space-time interval in a rotating frame

$$\mathrm{d}s^2 = -(c^2 - \omega^2 r^2)\,\mathrm{d}t^2 + 2\omega r^2\,\mathrm{d}\phi\,\mathrm{d}t + \mathrm{d}r^2 + r^2\,\mathrm{d}\phi^2 + \mathrm{d}z^2. \tag{2.53}$$

We write this in the form

$$\mathrm{d}s^2 = g_{\mu\nu}\,\mathrm{d}x^\mu\,\mathrm{d}x^\nu = g_{00}(\mathrm{d}x^0)^2 + 2\,g_{0i}\,\mathrm{d}x^0\,\mathrm{d}x^i + g_{ik}\,\mathrm{d}x^i\,\mathrm{d}x^k, \tag{2.54}$$

where, as usual, i and k are summed over spatial indices 1–3 only. With $x^\mu = (x^0, x^1, x^2, x^3) = (ct, r, \phi, z)$, $g_{\mu\nu}$ takes on the form

$$g_{\mu\nu} = \begin{pmatrix} -\left(1 - \dfrac{\omega^2 r^2}{c^2}\right) & 0 & \dfrac{\omega r^2}{c} & 0 \\ 0 & 1 & 0 & 0 \\ \dfrac{\omega r^2}{c} & 0 & r^2 & 0 \\ 0 & 0 & 0 & 1 \end{pmatrix}. \tag{2.55}$$

The fact that the frame is rotating shows up in the g_{02} and g_{20} terms; and also g_{00} is affected. In the next chapter we shall use $g_{\mu\nu}$ to indicate the metric tensor in a general Riemannian (curved) space. Here $g_{\mu\nu}$ refers to Minkowski space-time only, but in a rotating frame. Nevertheless, some of the observations made in this section will resurface in our future considerations of static and stationary space-times.

Let us first consider the definitions of time intervals and distances. There is a *time interval* between the events (x^0, x^i) and $(x^0 + \mathrm{d}x^0, x^i)$. The invariant interval is

$$\mathrm{d}s^2 = g_{00}c^2\,\mathrm{d}t^2 = -(c^2 - \omega^2 r^2)\mathrm{d}t^2. \tag{2.56}$$

The parameter t is *world time*. In contrast *proper time* τ is defined by (see (2.21))

$$\mathrm{d}s^2 = -c^2\,d\tau^2, \tag{2.57}$$

so the relation between world time and proper time is in the general case

$$\mathrm{d}\tau = \sqrt{-g_{00}}\,\mathrm{d}t \tag{2.58}$$

and in our particular case

$$\mathrm{d}\tau = \sqrt{1 - \frac{\omega^2 r^2}{c^2}}\mathrm{d}t. \tag{2.59}$$

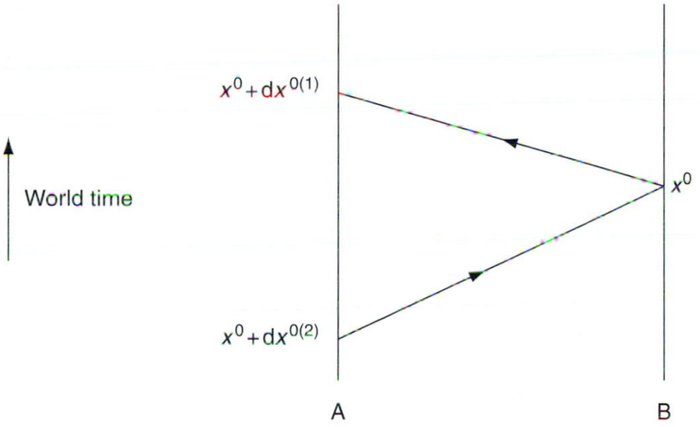

Fig. 2.4 A light signal sent from A to B and back to A.

It is seen that time 'goes slower' in a rotating frame, $d\tau < dt$. A clock on the rim of a rotating disc will measure the elapse of a smaller amount of proper time than a clock at the centre, given precisely by the formula above.

Now consider measuring the *spatial* separation between two points; call them A and B. Familiarity with the methods of Special Relativity suggests that we use light – we send a light signal from A to B, where it is reflected and returns to A. The distance between A and B is then defined in terms of the total proper time elapsing at A between emission and reception of the light signals. This is illustrated in Fig. 2.4 which shows the world-lines of A and B with the light signals drawn between them.[3] The light leaves A at world time $x^0 + dx^{0(2)}$, reaches B at time x^0 and returns to A at $x^0 + dx^{0(1)}$; of course we expect $dx^{0(2)}$ to be negative and $dx^{0(1)}$ positive. Using the fact that light obeys $ds^2 = 0$, in the notation of (2.54) we have

$$ds^2 = 0 = g_{00}\,(dx^0)^2 + 2g_{0i}\,dx^0\,dx^i + g_{ik}\,dx^i\,dx^k, \tag{2.60}$$

with the solutions

$$dx^0 = \frac{1}{g_{00}}\left[-g_{0i}\,dx^i \pm \sqrt{(g_{0i}\,g_{0k} - g_{00}\,g_{ik})dx^i\,dx^k}\right]. \tag{2.61}$$

These solutions correspond to $dx^{0(1)}$ and $dx^{0(2)}$ (note that $g_{00} < 0$). The world time interval between the emission and reception of the signal is then

$$\Delta x^0 = dx^{0(1)} - dx^{0(2)} = -\frac{2}{g_{00}}\sqrt{(g_{0i}\,g_{0k} - g_{00}\,g_{ik})dx^i\,dx^k}, \tag{2.62}$$

a positive quantity. The corresponding proper time interval between these events is $c\Delta\tau = \sqrt{-g_{00}}\,\Delta x^0$, and the distance between A and B is then given by $dl = \frac{c}{2}\Delta\tau$, or

$$dl^2 = \left(g_{ik} - \frac{g_{0i}\,g_{0k}}{g_{00}}\right)dx^i\,dx^k. \tag{2.63}$$

[3] This treatment follows that of Landau & Lifshitz (1971), Section 84.

As an illustration of this, let us consider the points (r, ϕ, z) and $(r, \phi+\mathrm{d}\phi, z)$ on a rotating disc. If the disc is *not* rotating the separation between these is $\mathrm{d}l^2=r^2\,\mathrm{d}\phi^2$, or $\mathrm{d}l=r\,\mathrm{d}\phi$. The *circumference* C of the circle joining all such points is

$$C=\int \mathrm{d}l=\int_0^{2\pi} r\,\mathrm{d}\phi=2\pi r, \tag{2.64}$$

entirely as expected. On the *rotating* disc we now apply Equation (2.63) with the values of g_{00}, $g_{02}=g_{20}$ and g_{22} read off from (2.55) to give

$$\mathrm{d}l^2=\left(g_{22}-\frac{(g_{02})^2}{g_{00}}\right)\mathrm{d}\phi^2=\left(r^2+\frac{\dfrac{\omega^2 r^4}{c^2}}{1-\dfrac{\omega^2 r^2}{c^2}}\right)\mathrm{d}\phi^2\approx r^2\left(1+\frac{\omega^2 r^2}{c^2}\right)\mathrm{d}\phi^2,$$

hence $\mathrm{d}l\approx\left(1+\frac{\omega^2 r^2}{2c^2}\right)r\,\mathrm{d}\phi$. The circle joining all such points has circumference

$$C=\int \mathrm{d}l\approx 2\pi r\left(1+\frac{\omega^2 r^2}{2c^2}\right)$$

or

$$\frac{C}{r}=2\pi\left(1+\frac{\omega^2 r^2}{2c^2}\right)>2\pi; \tag{2.65}$$

exactly the criterion indicating a *curved space*, as discussed in Chapter 1. We conclude that the *space* of a rotating disc is *curved*. This is *not* true, however, of the *space-time*, which is Minkowski space-time. This latter claim must for the present be taken on trust, but the point is that a rotating frame is given simply by a coordinate transformation – Equation (2.52) above – and it will be shown in due course that a flat space remains flat under any coordinate transformation. So even in a rotating frame the background space-time (Minkowski) remains flat, but the *spatial section of it* becomes curved. Not only is the spatial section radically changed; so also is the measure of *time*, as we now investigate.

2.3.1 Clock synchronisation

Consider events at two nearby points A and B. How do we decide if they are simultaneous? How do we *define* simultaneity? As usual, we use light; we send a light signal from A to B and back to A. Referring to Fig. 2.4, the time on A's world-line which is simultaneous with the event at world time coordinate x^0 on B's world-line is defined to be half-way between the emission and reception of the light signals, i.e. at (NB: Δx^0 below is different from that in (2.62))

$$x^0+\frac{\mathrm{d}x^{0(1)}+\mathrm{d}x^{0(2)}}{2}=x^0-\frac{g_{0i}\,\mathrm{d}x^i}{g_{00}}\equiv x^0+\Delta x^0. \tag{2.66}$$

In a rotating frame $g_{0i}\neq 0$, so $\Delta x^0\neq 0$. Using the above formula simultaneity may be defined – and clocks therefore synchronised – at points along any open line. An attempt to synchronise clocks at all points along a *closed* line will in general *fail*, however, since on return to the starting point the difference in world time recorded will be

$$\Delta x^0 = -\oint \frac{g_{0i}}{g_{00}} \, dx^i. \tag{2.67}$$

In a rotating frame of reference this integral will not vanish, so clock synchronisation is not possible: time is *not a single-valued parameter* in such a situation.

The experimental consequences of this were first revealed in the *Sagnac effect*. Sagnac found that the interference pattern changed when an interferometer was set in uniform rotation.[4] Beams of light traverse a closed path in opposite directions, then meet again at the starting point and interfere (we may refer to Fig. 2.4 again). Now arrange for the whole apparatus to rotate. We may derive an expression for the fringe shift using the formulae above. If the axis of rotation is the z axis then the time discrepancy integrated over one circuit is (see (2.55))

$$\begin{aligned} \Delta t = \frac{1}{c}\Delta x^0 &= \oint \frac{g_{02}\, dx^2}{g_{00}} = \frac{1}{c^2}\oint \frac{\omega r^2 \, d\phi}{1-(\omega^2 r^2)/c^2} \\ &\approx \frac{\omega}{c^2}\int r^2 \, d\phi = \frac{2\pi\omega r^2}{c^2} = \frac{2A\omega}{c^2}, \end{aligned} \tag{2.68}$$

where $A = \pi r^2$ is the area enclosed by the path. The associated discrepancy in *proper time* is

$$\Delta\tau = \sqrt{-g_{00}}\, \Delta t \approx \Delta t \tag{2.69}$$

to leading order. This results in an *optical path length*

$$l + \Delta l = 2\pi r + c\,\Delta t = L + \frac{2A\omega}{c}, \tag{2.70}$$

where L is the 'undisturbed' path length. For the light beam travelling in the opposite direction the optical path length is

$$l - \Delta l = L - \frac{2A\omega}{c}.$$

When these beams interfere the difference in optical path length is $\frac{4A\omega}{c}$, giving a fringe shift

$$\Delta N = \frac{4A\omega}{c\lambda}. \tag{2.71}$$

In Sagnac's experiment $\omega = 14$ rad s^{-1}, $A = 0.0863$ m^2 and $\lambda = 0.436 \times 10^{-6}$ m, giving $\Delta N = 0.036$, in agreement with his findings; or, as he put it, 'bien visible sur les photographies que je joins à cette Note' – despite the fact that the relevant photographs were not attached to the published version!

[4] Sagnac (1913a, b). Sagnac's finding was expressed in terms of the ether. A more modern account would be simply to state that his experiment demonstrated the absolute nature of rotation.

2.4 Inertia: Newton versus Mach

> One recalls in memory the feeling of anxious, heart-constricting solitude and emptiness that I daresay has crept over everyone on first comprehending the description given by Kirchhoff and Mach of the task of physics (or science in general): a description of facts that is as far as possible complete and as far as possible economical of thought. (E. Schrödinger[5])

Newton's first law of motion states that a body subject to no forces remains at rest or continues to move in a straight line with constant speed. (Note that by virtue of Newton's Principle of Relativity (or of course Special Relativity), these situations are equivalent.) Let us rephrase this by saying 'remains at rest or continues to move in a straight line with constant speed *with respect to X*'. Then what is X? Newton replied 'Absolute space' and demonstrated the existence of absolute space with his famous bucket experiment. A bucket of water is held at rest, but hanging by a highly coiled (twisted) rope. The water surface is, of course, flat. The bucket is then released so that the rope begins to uncoil, and the bucket starts to turn. After some time the water also starts to rotate, as it begins to partake of the motion of the bucket. This makes the surface of the water *concave*, because of the 'centrifugal' force on the water. Eventually the rope becomes untwisted and the bucket stops turning; the water, however, is still rotating and has a curved surface. At the beginning of this experiment there is no relative motion between the bucket and the water, and the water surface is flat. Later on, however, when both the water and the bucket are turning, there is also no relative motion, but the surface of the water is curved. The centrifugal force felt by the water is *not* due to its motion relative to the bucket; it must be caused by its motion relative to *absolute space*. Inertia results from acceleration or rotation relative to absolute space.

The Austrian physicist, mathematician and philosopher Ernst Mach (1838–1916), however, took a different view, one we should now describe as 'positivist'; space is not 'real', only matter is real. Space is simply an abstraction taken from the set of distance relations between material *objects* on this view. X cannot be 'absolute space', it must be *matter* – what is more, matter on the cosmological scale. 'When ... we say that a body preserves unchanged its direction and velocity in *space*, our assertion is nothing more or less than an abbreviated reference to *the entire universe*',[6] by which he meant, in effect, heavenly bodies at large distances, commonly referred to as the 'fixed stars'. These are thought of as defining a rigid system, while the motion of nearby stars averages out to zero. Our knowledge of the Universe is, of course, more detailed and more sophisticated than that obtaining in Mach's day. In particular we know that the distribution of matter in the Universe is, to a very good approximation, homogeneous, so that we are not 'at the centre' and the 'fixed stars' are not 'near the edge' of the Universe; moreover, the whole distribution of matter is expanding. Nevertheless, despite our more sophisticated perspective, we may still entertain Mach's original, and highly interesting, suggestion by identifying X with an average distribution of masses in the Universe.

[5] Quoted in Moore (1989).
[6] Mach (1919), Chapter 2, Section 6.7.

This whole question is, of course, one about the origin of *inertia*, since the inertia of a body (its inertial mass) is that property of the body which *resists* any motion except uniform motion relative to X. If we want to accelerate the body, to give it *non-uniform* motion, we must exert a force on it; the minute we stop exerting the force, the body will revert to uniform motion. Then the question is, 'Why do bodies possess inertia?'. If Mach is right, it must be due to some sort of *interaction* between individual masses and the 'rest of the Universe'. What sort of interaction? Gravitational? Unfortunately this is not usually specified in discussions of Mach's ideas, making the whole subject much more difficult to discuss; it becomes a qualitative rather than quantitative matter. It might also be objected that the question is more philosophical than physical, since we cannot 'do experiments on' the Universe; we cannot remove all the other galaxies and stars, to see if inertia disappears. Mach would predict, of course, that in an *empty universe*, particles would have *no inertia*, and so could move in any sort of way, including accelerated motion.

Einstein was, at least initially, much influenced by Mach's ideas: 'In a consistent theory of relativity there can be no inertia *relatively to "space"*, but only an inertia of masses *relatively to one another*. If, therefore, I have a mass at a sufficient distance from all other masses in the universe, its inertia must fall to zero.'[7] He referred to the whole complex of Mach's ideas as Mach's Principle and tried to incorporate this into General Relativity. It is now generally considered that he did not succeed, but one interesting consequence of General Relativity, the so-called Lense–Thirring effect (discussed in Section 6.2 below), predicts that inertial reference frames outside a rotating body are actually affected by the rotation – causing, for example, a gyroscope to precess. Newton's hypothesis of absolute space – duly upgraded to absolute space-time, in the spirit of Special Relativity – would certainly not allow a phenomenon like this.

To conclude this section we consider a very interesting simple experiment that has a direct bearing on Mach's ideas. If the inertia (inertial mass) of a particle (for example a proton) is affected *in any way at all* by mass distributions on the cosmological scale, then we would expect an *anisotropy* in the inertia of any proton on Earth, as a consequence of the anisotropy of the mass distribution of *our galaxy*. Then if a proton were subjected to an acceleration towards the galactic centre its mass would be different from what it would be if the acceleration were in a different direction. Atomic physics experiments involving the Mössbauer effect and nuclear magnetic resonance have been performed since 1960 to find $\delta m/m$, where m is the proton mass and δm its variation, which will result in changed frequencies of spectral lines in hyperfine transitions, observable as the apparatus, over a 24-hour period of the Earth's rotation, traces out different directions in the Galaxy.[8] A recent experiment gives $\delta m < 2 \times 10^{-21}$ eV, or $\delta m/m < 10^{-30}$; which would seem to offer little support for Mach's Principle.[9]

There is, however, another interpretation of this seemingly null result.[10] The Equivalence Principle implies that in any gravitational field there is a frame of reference in which its

[7] Einstein (1917).

[8] Hughes *et al.* (1960), Drever (1960); Weinberg (1972) pp. 86–88, gives a good account of these experiments. For more recent references see Ciufolini & Wheeler (1995), Section 3.2.4.

[9] Lamoreaux *et al.* (1986).

[10] What follows is based on an observation made by Weinberg (1972), p. 87.

effects are nullified. This is the state of 'free fall'. For example, in the case of the elevator, described in Section 1.2, when the supporting cable is severed, the elevator plunges down to the centre of the Earth in free fall. It finds itself in an *inertial* frame of reference – the gravitational field of the Earth and the acceleration of the box relative to the Earth cancel each other out. Now consider an experiment performed on Earth, such as the one mentioned above, in which an anisotropy in inertial mass is looked for. It could be argued that the Equivalence Principle makes it *impossible* to detect any such effect since the experimental apparatus will find for itself a locally inertial frame, in which, by definition, any gravitational field (and therefore, presumably, any effect) of the 'rest of the Universe' is nullified. No wonder a null result is found! From this perspective, Mach's Principle is *not necessarily* disproved by the experiment.

Let me conclude this section by taking these considerations slightly further, and approaching Mach's Principle from the perspective of particle physics. If Mach is right, electrons at different locations in the Universe might be expected to have different masses, but this runs completely counter to the general understanding of mass in particle physics. Mass is, in effect, one of the Casimir operators of the Poincaré group (Wigner (1939)) and the Poincaré group is the isometry group of Minkowski space-time, so in a locally inertial frame the mass of an electron is fixed. It should also be noted that the idea of inertial mass 'falling to zero' in an empty universe does not stand up well to investigation; for by virtue of *Special Relativity* particles may possess *zero (rest)mass* (like photons do), but it does not follow that they possess zero inertia, as Einstein himself pointed out.[11] The only entity that might be said to possess zero inertia is the vacuum, and from quantum field theory we learn that the vacuum is filled with particle–antiparticle pairs, to say nothing of a non-zero expectation value of a Higgs field! While Mach's Principle might have proved very valuable as a stimulant to thought, it appears to be very tricky to pin down an actual demonstration that it is correct. Perhaps the moral to be drawn is to note that the Equivalence Principle is a *local* one, whereas Mach's ideas are distinctly *global*. If Mach's Principle could be implemented by way of a *gravitational (or some other) interaction*, that would presumably be equivalent to a *local* formulation – which does not work.[12] Perhaps what is needed is to find a mathematical way to describe a global principle.

2.5 Thomas precession

The phenomenon known as Thomas precession was first studied in the context of atomic physics.[13] Since it intimately involves the theory of relativity in a rather surprising way, it is appropriate to discuss it here; and although the original setting in atomic physics might seem to be far removed from the preoccupations of a general relativist, it is worth at least giving this a very brief description, since the orbiting of electrons round the atomic nucleus has,

[11] Einstein (1905b).

[12] An attempt was made by Sciama (1953), who proposed a Maxwell-type of theory to implement Mach's Principle, but this theory turns out to be incompatible with Special Relativity.

[13] Thomas (1926, 1927).

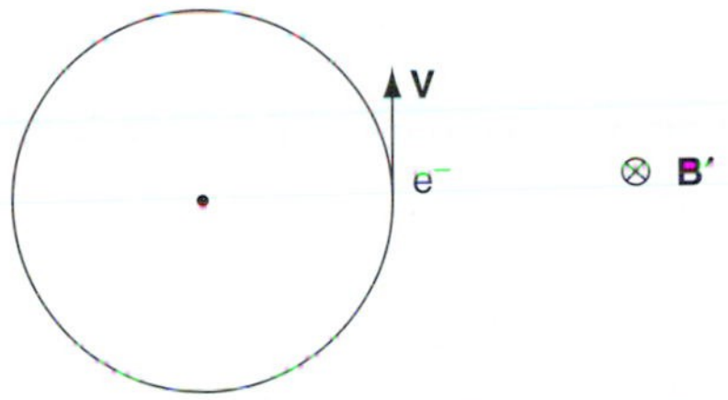

Fig. 2.5 An electron orbiting a proton. $\mathbf{B}'$ is the magnetic field produced by the proton in the frame in which the electron is at rest.

despite the vast difference in scale, some similarities with the orbiting of planets round the Sun and of satellites around planets. Indeed the whole phenomenon of spin precession of an orbiting body is at the heart of one of the most recent and delicate tests of General Relativity, the Lense–Thirring effect (see Section 6.2 below).

Consider the simplest case of the hydrogen atom, in which an electron orbits a proton, to which it is bound by electromagnetic forces. The Coulomb field of the proton is

$$\mathbf{E} = \frac{e}{4\pi\varepsilon_0 r^3}\mathbf{r} \tag{2.72}$$

and in this field the electron moves with velocity $\mathbf{v} = \frac{\mathrm{d}\mathbf{r}}{\mathrm{d}t}$ – see Fig. 2.5. Let us define the frames in which the electron and proton are at rest:

$$\begin{aligned} &S: \text{p at rest}, \\ &S': \text{e}^- \text{ at rest}. \end{aligned} \tag{2.73}$$

In S' the proton is moving so produces a magnetic field $\mathbf{B}'$ in which the electron spin (= intrinsic angular momentum) precesses, according to the well-known classical formula

$$\frac{\mathrm{d}\mathbf{S}}{\mathrm{d}t} = \boldsymbol{\mu} \times \mathbf{B}', \tag{2.74}$$

where $\boldsymbol{\mu}$ is the magnetic moment of the electron. What is $\mathbf{B}'$? From the relativistic transformation laws (see Section 2.6 below) the relation between the electric and magnetic fields in frames S and S' is

$$\mathbf{B}' = \gamma\left(\mathbf{B} - \frac{\mathbf{v}\times\mathbf{E}}{c^2}\right) = -\frac{\gamma}{c^2}\mathbf{v}\times\mathbf{E},$$

where the second equality follows since $\mathbf{B} = 0$. In the non-relativistic approximation $\gamma \approx 1$, so $\mathbf{B}' \approx -\frac{1}{c^2}\mathbf{v}\times\mathbf{E}$ and, using Equation (2.72) and putting $\mathbf{L} = \mathbf{r}\times\mathbf{p}$ with $\mathbf{p} = m\mathbf{v}$,

$$\mathbf{B}' = \left(\frac{e}{4\pi\varepsilon_0 mc^2 r^3}\right)\mathbf{L}. \tag{2.75}$$

The 'interaction energy' of a magnetic moment $\boldsymbol{\mu}$ in a magnetic field – call it $\mathbf{B}'$ – is

$$U = -\boldsymbol{\mu}\cdot\mathbf{B}'.$$

What is $\boldsymbol{\mu}$? For a charge q in a circular orbit, $\boldsymbol{\mu} = \dfrac{q}{2m}\mathbf{L}$ where $\mathbf{L}$ is the angular momentum of the orbiting body. According to the Uhlenbeck–Goudsmit hypothesis for *intrinsic spin*, however, the proportionality factor is different and we have

$$\boldsymbol{\mu} = \frac{q}{m}\mathbf{S}; \tag{2.76}$$

so putting together the above equations we have for the interaction energy

$$U = -\frac{e^2}{4\pi\varepsilon_0 m^2 c^2 r^3}(\mathbf{L}\cdot\mathbf{S}). \tag{2.77}$$

This is a form of binding energy of the electron in orbit, which results from the interaction between the magnetic field of the proton and the electron's magnetic moment, proportional to its spin. It is known as a 'spin-orbit' interaction and is responsible for the so-called 'fine structure' in atomic spectral lines. The details need not concern us, except to point out the crucial fact that the above expression does *not* account satisfactorily for the fine structure phenomena. The correct expression is one-half of this:

$$U = -\frac{e^2}{8\pi\varepsilon_0 m^2 c^2 r^3}(\mathbf{L}\cdot\mathbf{S}). \tag{2.78}$$

It was Thomas who satisfactorily explained the occurrence of this extra factor of 2, and we now outline his reasoning.

In Fig. 2.6 the electron is shown with velocity $\mathbf{v}$ at A, and a small time δt later with velocity $\mathbf{v}+\delta\mathbf{v}$ at B. In addition to the frames S and S' let us introduce a frame S'':

$$\begin{aligned} &S\text{: rest frame of p,}\\ &S'\text{: rest frame of e}^-\text{ at A,}\\ &S''\text{: rest frame of e}^-\text{ at B.}\end{aligned} \tag{2.79}$$

The corresponding Lorentz transformations are (see (2.37))

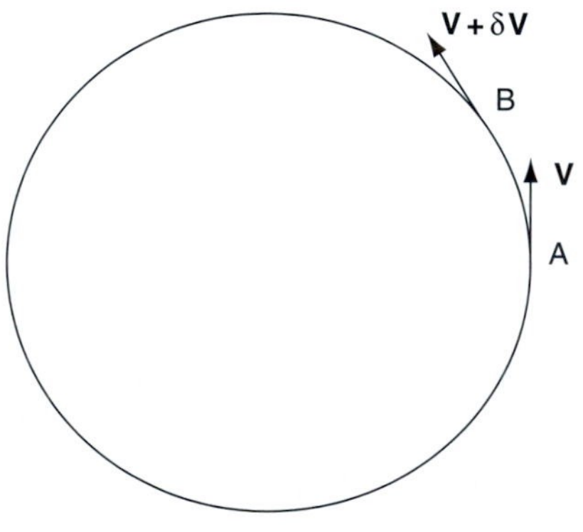

Fig. 2.6 **The electron at A with velocity v and at B with velocity v + δv.**

$$S' \to S: \exp(-i\mathbf{K}.\boldsymbol{\phi})$$
$$S \to S'': \exp[i\mathbf{K}.(\boldsymbol{\phi} + \delta\boldsymbol{\phi})],$$

hence the transformation linking the frames S' and S'' is

$$S' \to S'': \exp(-i\mathbf{K}.\boldsymbol{\phi})\exp[i\mathbf{K}.(\boldsymbol{\phi} + \delta\boldsymbol{\phi})] \equiv T(\boldsymbol{\phi}, \delta\boldsymbol{\phi}).$$

T is a transformation matrix between the coordinates of the electron at A and those at B: $x''^{\mu} = T^{\mu}{}_{\nu} x'^{\nu}$; and it is our contention that T consists of a *rotation* as well as a pure boost. This is straightforward to prove, using the Baker–Campbell–Hausdorf formula (in which A and B are operators; in the present context, matrices)

$$e^A e^B = e^{A+B+1/2[A,B]+\cdots} \tag{2.80}$$

to give

$$T = \exp\left\{i\mathbf{K}.\delta\boldsymbol{\phi} + \frac{1}{2}[-i\mathbf{K}.\boldsymbol{\phi}, i\mathbf{K}.\delta\boldsymbol{\phi}] + \cdots\right\}. \tag{2.81}$$

We ignore the higher order terms. The commutator term above is

$$\tfrac{1}{2}[K_i, K_k]\phi_i\delta\phi_k = -\tfrac{1}{2}\, i\varepsilon_{ikm}J_m\phi_i\,\delta\phi_k = i\mathbf{J}\cdot(\tfrac{1}{2}\,\delta\boldsymbol{\phi}\times\boldsymbol{\phi}),$$

where the commutation relation (2.35) has been used. This is a pure rotation, through an angle $\delta\boldsymbol{\alpha} = \tfrac{1}{2}\delta\boldsymbol{\phi}\times\boldsymbol{\phi}$. Since the angle is infinitesimal we may find an expression for it by noting that for infinitesimal boost parameters (i.e. small velocities – see (2.28)) $\phi \approx \sinh\phi = \gamma v/c \approx v/c$, so

$$\delta\boldsymbol{\alpha} = \frac{1}{2c^2}\mathbf{v}\times\delta\mathbf{v}.$$

It is then clear that, in addition to the expected Lorentz boost in T (the first term in (2.81)), there is a rotation. As the electron orbits the proton its rest frame *rotates* relative to the laboratory frame and the resulting precession rate $\boldsymbol{\Omega} = \frac{\delta\alpha}{\delta t}$ is

$$\boldsymbol{\Omega}_{\text{Thomas}} \approx \frac{1}{2c^2}\mathbf{v}\times\mathbf{a} \tag{2.82}$$

where $\mathbf{a} = \frac{\delta\mathbf{v}}{\delta t}$ is the *acceleration* of the electron in orbit. The Thomas precession is purely *kinematic* in origin; the acceleration above may have any cause. In this case, of course, it is the electromagnetic force exerted by the proton (or nucleus), but any force will have the same result.

To complete our task we must make a connection between this precession, or rotation of the inertial frame, and the interaction energy U introduced above, in Equations (2.76) and (2.77). This connection comes about because of the relation between magnetic fields and rotating frames. As every student of classical mechanics knows, the rate of change of a quantity (let us call it spin $\mathbf{S}$) in a rotating frame is related to its rate of change in a non-rotating one by the formula[14]

[14] See for example Goldstein (1950), Kibble & Berkshire (1996) or Morin (2007).

$$\left(\frac{\mathrm{d}S}{\mathrm{d}t}\right)_{\mathrm{rot}} = \frac{\mathbf{DS}}{\mathrm{d}t} = \frac{\mathrm{d}\mathbf{S}}{\mathrm{d}t} - \boldsymbol{\Omega} \times \mathbf{S}. \tag{2.83}$$

Substituting from (2.74) and (2.76) gives

$$\frac{\mathbf{DS}}{\mathrm{d}t} = \frac{e}{m}\mathbf{S} \times \left(\mathbf{B}' + \frac{m}{e}\,\boldsymbol{\Omega}\right) \equiv \frac{e}{m}\mathbf{S} \times \mathbf{B}_{\mathrm{eff}}, \tag{2.84}$$

which defines the *effective magnetic field* $\mathbf{B}_{\mathrm{eff}}$;

$$\mathbf{B}_{\mathrm{eff}} = \left(\frac{e}{4\pi\varepsilon_0 mc^2 r^3}\right)\mathbf{L} + \left(\frac{m}{e}\frac{1}{2c^2}\right)\mathbf{v} \times \mathbf{a},$$

where Equations (2.75) and (2.82) have been used. The acceleration $\mathbf{a}$, however, caused by an electrostatic force, is given by

$$\mathbf{a} = \frac{\mathbf{F}}{m} = \frac{e\mathbf{E}}{m} = \left(\frac{e^2}{4\pi\varepsilon_0 m r^3}\right)\mathbf{r},$$

giving

$$\mathbf{B}_{\mathrm{eff}} = \left(\frac{e}{4\pi\varepsilon_0 mc^2 r^3}\right)\mathbf{L} - \left(\frac{e}{8\pi\varepsilon_0 mc^2 r^3}\right)\mathbf{L} = \left(\frac{e}{8\pi\varepsilon_0 mc^2 r^3}\right)\mathbf{L}$$

and a spin-orbit interaction of

$$U = -\boldsymbol{\mu} \cdot \mathbf{B}_{\mathrm{eff}} = -\left(\frac{e^2}{8\pi\varepsilon_0 m^2 c^2 r^3}\right)\mathbf{L} \cdot \mathbf{S},$$

as desired, and as in (2.78), where (2.76) has been used. The Lie algebra relating Lorentz boosts with rotations gives the factor of 2 necessary for agreement with experiment.

2.6 Electromagnetism

The approach to relativity in this chapter has been anti-historical. It is well known that the paper which sprung Special Relativity on the world was entitled 'On the Electrodynamics of Moving Bodies'.[15] Einstein exploited the fact that Maxwell's equations kept the same form under *Lorentz*, rather than *Galilean*, transformations, to propose a new theory of space and time; in a word, that they defined what we now call a 4-vector, the components of which get mixed up in a Lorentz transformation. It is now time to return to electrodynamics and review how it is formulated in Special Relativity. Since Maxwell's electromagnetism is already consistent with Special Relativity, this is only a matter of getting a neat and consistent notation. The important observation is that just as space and time become the components of a 4-vector 'space-time' in Special Relativity, so all 'geometric objects' (for example fields) also become either 4-vectors or quantities clearly related to them; the language of 3-vectors has in effect disappeared.

[15] Einstein (1905a).

We define the 4-vector potential A^μ

$$A^\mu = (A^0, A^1, A^2, A^3) = (\phi, A_x, A_y, A_z) = (\phi, \boldsymbol{A}), \tag{2.85}$$

where ϕ and $\boldsymbol{A}$ are the usual scalar and vector potentials of electromagnetism. Note the positions of the indices in (2.85); we adopt the convention that indices 1, 2 and 3 may be raised or lowered, whereas x, y and z always appear in a lower position. Lowering the index with the Minkowski metric tensor (2.18) gives

$$A_\mu = (A_0, A_1, A_2, A_3) = (-\phi, A_x, A_y, A_z) = (-\phi, \boldsymbol{A}). \tag{2.86}$$

Under a Lorentz transformation Λ we have (cf. (2.22))

$$A^\mu \to A'^\mu = \Lambda^\mu{}_\nu A^\nu. \tag{2.87}$$

The electric and magnetic fields are defined by

$$\mathbf{E} = -\frac{1}{c}\frac{\partial \mathbf{A}}{\partial t} - \nabla\phi, \quad \mathbf{B} = \nabla \times \mathbf{A}. \tag{2.88}$$

Hence

$$E_x = -\frac{1}{c}\frac{\partial Ax}{\partial x} - \frac{\partial \phi}{\partial x} = -\partial_0 A^1 - \partial_1 A^0 = \partial^0 A^1 - \partial^1 A^0, \tag{2.89}$$

$$B_x = -\frac{\partial Az}{\partial y} - \frac{\partial Ay}{\partial z} = \partial_2 A^3 - \partial_3 A^2 = \partial^2 A^3 - \partial^3 A^2, \tag{2.90}$$

and similarly for the other components. Then defining the rank 2 tensor (i.e. an object with two indices)

$$F^{\mu\nu} = \partial^\mu A^\nu - \partial^\nu A^\mu, \tag{2.91}$$

we have, in matrix form, labelling the rows and columns of the matrix by 0, 1, 2 and 3,

$$F^{\mu\nu} = \begin{pmatrix} 0 & E_x & E_y & E_z \\ -E_x & 0 & B_z & -B_y \\ -E_y & -B_z & 0 & B_x \\ -E_z & B_y & -B_x & 0 \end{pmatrix}, \tag{2.92}$$

so for example $F^{01} = E_x$, $F^{21} = -B_z$ and so on. It is clear from the definition (2.91) that $F^{\mu\nu}$ is an antisymmetric tensor, i.e. (^{T} = transpose)

$$(F^{\mu\nu})^{\mathrm{T}} = F^{\nu\mu} = -F^{\mu\nu}. \tag{2.93}$$

Lowering the indices yields $F_{\mu\nu} = \eta_{\mu\rho}\,\eta_{\nu\sigma}\,F^{\rho\sigma}$, hence

$$F_{\mu\nu} = \begin{pmatrix} 0 & -E_x & -E_y & -E_z \\ E_x & 0 & B_z & -B_y \\ E_y & -B_z & 0 & B_x \\ E_z & B_y & -B_x & 0 \end{pmatrix}. \tag{2.94}$$

Under a Lorentz transformation

$$F^{\mu\nu} \to F'^{\mu\nu} = \Lambda^{\mu}{}_{\rho}\Lambda^{\nu}{}_{\sigma}F^{\rho\sigma}. \tag{2.95}$$

Consider, for example, a boost along the x axis (see (2.25) above). Let us find out how E_x and E_y transform. We have

$$\begin{aligned} E'_x = F'^{01} &= \Lambda^0{}_\rho\,\Lambda^0{}_\sigma\,F^{\rho\sigma} \\ &= \Lambda^0{}_0\Lambda^1{}_0F^{00} + \Lambda^0{}_0\Lambda^1{}_1F^{01} + \Lambda^0{}_1\Lambda^1{}_0F^{10} + \Lambda^0{}_1\Lambda^1{}_1F^{11} \\ &= \gamma^2F^{01} + \left(\frac{\gamma v}{c}\right)^2 F^{10} = \gamma^2E_x\left(1-\frac{v^2}{c^2}\right) = E_x; \\ E'_x &= E_x. \end{aligned} \tag{2.96}$$

In the summation above use was made of the fact that only $\rho=0,1$ and $\sigma=0,1$ give non-zero contributions. In a similar fashion

$$\begin{aligned} E'_y = F'^{02} &= \Lambda^0{}_\rho\Lambda^2{}_\sigma F^{\rho\sigma} = \Lambda^0{}_0\Lambda^2{}_2F^{02} + \Lambda^0{}_1\Lambda^2{}_2F^{12} \\ &= \gamma F^{02} + \frac{\gamma v}{c}F^{12} = \gamma E_y + \frac{\gamma v}{c}B_z; \end{aligned}$$

or

$$E'_y = \frac{E_y + \dfrac{v}{c}B_z}{\sqrt{1-\dfrac{v^2}{c^2}}}. \tag{2.97}$$

This is an example of how electric and magnetic fields transform in a moving frame. They do not transform as 3-vectors (as remarked above, 3-vectors are non-kosher objects in relativity), but as the six components of an antisymmetric second rank tensor $F^{\mu\nu}$.

It is useful to define the *dual* field strength tensor $\tilde{F}_{\mu\nu}$:

$$\tilde{F}^{\mu\nu} = \tfrac{1}{2}\varepsilon^{\mu\nu\rho\sigma}F_{\rho\sigma} \tag{2.98}$$

where $\varepsilon^{\mu\nu\rho\sigma}$, known as the *Levi–Cività symbol*, is totally antisymmetric in its indices and has the values

$$\varepsilon^{\mu\nu\rho\sigma} = \begin{cases} +1 \text{ if } (\mu\nu\rho\sigma) \text{ is an even permutation of } (0123), \\ -1 \text{ if } (\mu\nu\rho\sigma) \text{ is an odd permutation of } (0123), \\ 0 \text{ otherwise (i.e. if two or more indices are equal).} \end{cases} \tag{2.99}$$

For example,

$$\tilde{F}^{01} = \tfrac{1}{2}\varepsilon^{01ik}F_{ik} = \tfrac{1}{2}(\varepsilon^{0123}F_{23} + \varepsilon^{0132}F_{32}) = \tfrac{1}{2}(F_{23} - F_{32}) = F_{23} = B_x.$$

The other components follow in the same way, giving

$$\tilde{F}^{\mu\nu} = \begin{pmatrix} 0 & B_x & B_y & B_z \\ -B_x & 0 & -E_z & E_y \\ -B_y & E_z & 0 & -E_x \\ -B_z & -E_y & E_x & 0 \end{pmatrix}. \tag{2.100}$$

It is clear that $\tilde{F}^{\mu\nu}$ is also antisymmetric:

$$(\tilde{F}^{\mu\nu})^T = -\tilde{F}^{\mu\nu}. \tag{2.101}$$

In addition, the substitution $F \to \tilde{F}$ is equivalent to

$$F \to \tilde{F}: \mathbf{E} \to \mathbf{B}, \ \ \mathbf{B} \to -\mathbf{E}. \tag{2.102}$$

2.6.1 Maxwell's equations

We close by expressing Maxwell's equations in terms of the field tensor $F^{\mu\nu}$ and its dual. The equations are in two sets, homogeneous and inhomogeneous:

$$\begin{aligned} &\text{homogeneous: } \nabla \cdot \mathbf{B} = 0, \ \nabla \times \mathbf{E} + \frac{\partial \mathbf{B}}{\partial t} = 0, \\ &\text{inhomogeneous: } \nabla \cdot \mathbf{E} = \rho, \ \nabla \times \mathbf{B} + \frac{\partial \mathbf{E}}{\partial t} = \mathbf{j}. \end{aligned} \tag{2.103}$$

As the reader will easily verify, the inhomogeneous equations are

$$\partial_\nu F^{\mu\nu} = j^\mu \text{ where } j^\nu = (\rho, \mathbf{j}) \tag{2.104}$$

and the homogeneous ones are

$$\partial_\mu \tilde{F}^{\mu\nu} = 0. \tag{2.105}$$

It will be noted that each of these is actually four equations; μ is summed and therefore a dummy index, while ν takes on the four values 0, 1, 2, 3. Most people will agree that the Equations (2.104) and (2.105) are slightly neater in form than (2.103). A further degree of elegance is achieved when Maxwell's equations are written in terms of differential forms, which they are in the next chapter.

2.7 Principle of General Covariance

The Principle of General Covariance is a mathematical statement of the Equivalence Principle. Consider 'at random' a typical equation in physics

$$\nabla \times \mathbf{E} = -\frac{\partial \mathbf{B}}{\partial t}; \tag{2.106}$$

one of Maxwell's equations. The right and left hand sides are both vectors – more precisely, 3-vectors; and we may take the definition of a 3-vector to be a quantity with a well-defined transformation under *rotations* in 3-dimensional space. Under a general rotation the coordinates x^i ($i = 1, 2, 3$) change to x'^i given by

$$x'^i = R^i{}_k x^k, \tag{2.107}$$

where $R^i{}_k$ is a rotation matrix (and summation over k is understood). Under rotations, distance from the origin is preserved: $x^2+y^2+z^2=x'^2+y'^2+z'^2$, or

$$x'^i x'_i = x^i x_i,$$

which with (2.107) gives

$$R^i{}_k R_i{}^m = \delta^m{}_k;$$

the matrix R is *orthogonal*. (It may be pointed out that an example of R is the 3×3 part of the Lorentz matrix (2.24), whose orthogonality is easy to check.) A *vector* is by definition a quantity whose components transform in the same way as x^i under a coordinate transformation:

$$V^i \to V'^i = R^i{}_k V^k. \tag{2.108}$$

We are at present thinking in terms of a Minkowski world, a flat space-time in which it is conceivable to have the rotation matrix R a *constant* – the same rotation is to be performed at each point in space-time. Equivalently the transformation is linear, and in that case Equation (2.107) may be differentiated to give

$$\frac{\partial x'^i}{\partial x^j} = R^i{}_j$$

and (2.108) becomes

$$V'^i = \frac{\partial x'^i}{\partial x^j} V^j; \tag{2.109}$$

the law of transformation for contravariant vectors, as we shall see below.

Maxwell's equation (2.106) is a vector equation: so if an experimentalist finds this equation to be valid in a frame of reference S, another experimentalist working in the relatively rotated frame S' will find the *same* equation

$$\nabla' \times \mathbf{E}' = -\frac{\partial \mathbf{B}'}{\partial t};$$

the time parameter t is unaffected by the rotation. This is simply an example of the invariance of the laws of physics (in this case one of Maxwell's equations) under the rotation of reference frames (to be distinguished, of course, from *rotating* reference frames, which are non-inertial). And this requirement in turn follows from the isotropy of space – there is no distinguished direction in space.[16]

As spelled out in the previous section, this invariance under rotations may be extended to include Lorentz transformations, and then the set of all Maxwell's equations takes on the form (2.104) and (2.105). The point being made here is that all these equations may be expressed in the form

$$T = 0,$$

[16] The connection between symmetries, conservation laws and invariance principles is beautifully and carefully examined in several essays by Wigner; see for example the collection Wigner (1967).

where T is a quantity with a well-defined transformation property under rotations and/or Lorentz transformations; in formal terms T is a *tensor* of rank n (where $n=0$ corresponds to a scalar, $n=1$ to a vector – either a 3-vector, for rotations, or a 4-vector to include Lorentz transformations, $n=2$ to a rank 2 tensor, such as $F^{\mu\nu}$; and so on). To return, for the sake of definiteness, to our example, if T is the vector quantity

$$\mathbf{A} = \nabla \times \mathbf{E} + \frac{\partial \mathbf{B}}{\partial t}$$

then Equation (2.106) is the statement that $\mathbf{A}=\mathbf{0}$; and if this holds in frame S, then in S', obtained from S by a rotation, we also have $\mathbf{A}'=\mathbf{0}$. This is nothing but relativity.

We may now consider how to generalise this observation. We have seen that a uniform acceleration produces, locally, the same effect as a gravitational field. The coordinate transformation for an acceleration along the x axis,

$$x' = x + at^2,$$

is, however, non-linear, so our generalisation might be that the laws of nature in the presence of a gravitational field are tensor equations under *any* coordinate transformations, linear or non-linear. This is the so-called Principle of General Covariance, and the logic just described would justify the term 'general' in 'General Relativity'; generalising linear to non-linear transformations in order to include gravity. The term 'covariance' means, as usual, that the laws of physics take the same form for observers in relative motion. The Principle of General Covariance does not, however, fully account for General Relativity. General Relativity describes gravitational fields in terms of a curved space-time, as we shall see below, and as was indicated by the thought-experiment described in Section 1.3.

Further reading

Recent reviews of the Sagnac effect may be found in Post (1967), Chow *et al.* (1985), Anderson, Bilger & Stedman (1994) and Stedman (1997). A remarkable experiment involving interference of *neutrons*, rather than light, to detect the rotation of the Earth was performed by Colella *et al.* (1975). For an up-to-date review of these matters see Rauch & Werner (2000), Chapter 7.

Readable accounts of Mach's Principle can be found in Bondi (1960), Sciama (1969), Weinberg (1972), Misner *et al.* (1973), Rindler (1977, 2001), Ciufolini & Wheeler (1995), Cao (1997), Harrison (2000) and Jammer (2000). A good review is Raine (1981) and a useful more recent reference is Barbour and Pfister (1995). An interesting, if brief, account of Mach's ideas on the philosophy of physics may be found in Moore (1989).

Thomas precession is discussed in Robertson & Noonan (1968), Møller (1972), Jackson (1975) and Bacry (1977), and a rather interesting account is given in Rindler (2001). For its role in the context of atomic physics, see for example Cohen-Tannoudji *et al.* (1977) or Shankar (1980).

Problems

2.1 Consider the following experiment in the Einstein train. A light, situated at the centre of train A, is switched on and light beams travel up the train and down the train to reach the front and rear doors of the train at the same time, as viewed by an observer on the train. Show that, as viewed from a moving train, the arrival of the light at these two points is not simultaneous, but reaches the rear of the train at a time $\gamma vL/c^2$ *before* reaching the front. This is an example of the *relativity of simultaneity*.

2.2 Show that $F^{\mu\nu}F_{\mu\nu} = 2(-\mathbf{E}\cdot\mathbf{E} + \mathbf{B}\cdot\mathbf{B})$ and $\tilde{F}^{\mu\nu}F_{\mu\nu} = -4\mathbf{E}\cdot\mathbf{B}$.

3 Differential geometry I: vectors, differential forms and absolute differentiation

It is almost impossible for me to read contemporary mathematicians who, instead of saying, 'Petya washed his hands', write 'There is a $t_1 < 0$ such that the image of t_1 under the natural mapping $t_1 \to \text{Petya}(t_1)$ belongs to the set of dirty hands, and a t_2, $t_1 < t_2 \leq 0$, such that the image of t_2 under the above-mentioned mappings belongs to the complement of the set defined in the preceding sentence ...

V. I. Arnol'd

In this chapter we introduce the mathematical language which is used to express the theory of General Relativity. A student coming to this subject for the first time has to become acquainted with this language, which is initially something of a challenge. Einstein himself had to learn it (from his friend Marcel Grossmann). The subject is widely known as *tensor calculus*; it is concerned with tensors and how to define and differentiate them in curved spaces. In more recent times tensor calculus has been recast using a more sophisticated formalism, based on coordinate-free notation and differential forms. At first physicists were disinclined to learn this higher grade language, since it involved more work, without, perhaps, any reward in terms of mathematical or physical insight. Eventually, however, sceptical minds became convinced that there were indeed pay-offs in learning this new formalism, and knowledge of it is now almost essential to read research papers in the field of General Relativity. This chapter covers the notions of vectors, differential forms, tensors and covariant (or absolute) differentiation, and the next one goes on to consider curvature, which, of course, as a property of space-time, plays a starring role in Einstein's theory. We begin by considering the general notion of a differentiable manifold.

3.1 Space-time as a differentiable manifold

A simple mathematical model of space-time is a Cartesian product of three spatial variables and one time variable, and a point in this 4-dimensional space may be considered to represent an 'event' (t, x, y, z) in space-time. This, however, is slightly too simple: the Newtonian–Galilean Principle of Relativity tells us that an observer (in an inertial frame) regarding himself as stationary will mark the 'same' point in space by for instance tapping at it every second. A second observer (also inertial) moving relative to him will however see these points as *different* points in space. The idea of a 'point in space' therefore has to be replaced by an *equivalence class* of points related by a group of transformations $x' = x - vt$ etc. – the Galileo group. Time is taken as absolute in the Newtonian–Galilean scheme.

In Special Relativity this group is enlarged to the Lorentz group, in which time is no longer absolute, and one consequence of the relativity of time is the twin paradox, considered in Section 2.2. In this 'paradox' we noted (see Fig. 2.2) that a closed path in space (consisting of A's journey followed by the reverse of B's journey) results in a change of time, showing that space and time are not independent of each other. In more technical language, space-time is an example of a *fibre bundle*, in which a closed path in the *base space* ('space' in this example) results in movement along the *fibre* ('time'). Locally, however, a fibre bundle is simply a direct product of two manifolds, even if this is not true globally.

In General Relativity we have the additional feature that space-time becomes curved. In such places as the Solar System the curvature will only be slight but in some circumstances, for example near the horizon of a black hole, it may become highly curved, and even topologically non-trivial. We shall be wanting, of course, to denote points in space-time by a system of *coordinatisation*: how does this work in a curved space? Consider for example the sphere S^2: it is a 2-dimensional curved surface, in general coordinatised by θ, ϕ. Locally, in the neighbourhood of any point P on S^2 we may introduce *Cartesian* coordinates (x, y) but these cannot be made to cover the whole sphere without ambiguity. Neither, in fact, may θ and ϕ be used to cover the sphere without ambiguity: at the north pole $(\theta=0)$ ϕ is undefined, as is also the unit vector $\hat{\phi}$ – at the north pole, every direction is south! These problems may be overcome by having *two* coordinate systems; so on the Earth, for example, we would have the usual one, with $\theta=0, \pi$ at the N and S poles. A second coordinate system would have $\theta=0$ at for example Concordia, Argentina, on the Uruguay border, at 32° S, 58° W, and $\theta=\pi$ Nanjing, 32° N, 118° E. Then the one coordinate system may be employed unambiguously everywhere except at the N and S poles, and the other one everywhere except at Concordia and Nanjing. At all other points on the Earth's surface there is a smooth *transformation* from one coordinate system to the other one. We generalise this construction to define a *manifold* as an (n-dimensional) space of arbitrary curvature which may be coordinatised by a series of *charts*, as shown in Fig. 3.1. The charts are for simplicity taken to be (open subsets of) R^n and in the overlapping regions of M there is a transformation between them.

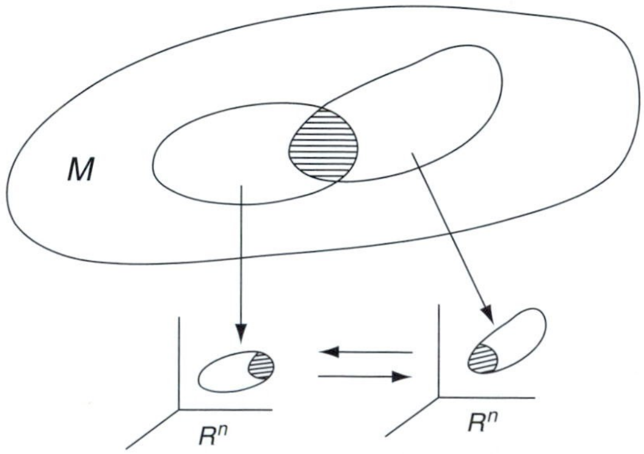

Fig. 3.1 A region of an n-dimensional manifold M is considered to consist of two overlapping regions, each coordinated by a chart in R^n. In the area of overlap there is a transformation from one coordinatisation to the other.

We saw in the last chapter that there is an equivalence between accelerating frames and gravitational fields, as illustrated in the Einstein box experiment. But as was pointed out there only homogeneous gravitational fields with constant magnitude and direction can be described in this way. Real gravitational fields – for example at different points on the Earth or in the Solar System – cannot be eliminated by any single coordinate transformation. To preserve the usefulness of the equivalence principle we need different transformations at different points, like the chart above.

3.2 Vectors and vector fields

Let us recall how to describe vectors in (flat) 3-dimensional Euclidean space R^3. We may write them in different forms, for example

$$\mathbf{V} = V_x\mathbf{i} + V_y\mathbf{j} + V_z\mathbf{k} \quad \text{or} \quad \mathbf{V} = V_r\hat{\mathbf{r}} + V_\theta\hat{\boldsymbol{\theta}} + V_\phi\hat{\boldsymbol{\phi}}, \tag{3.1}$$

where **i**, **j**, **k** are unit vectors in the x, y, z directions, and $\hat{\mathbf{r}}, \hat{\boldsymbol{\theta}}, \hat{\boldsymbol{\phi}}$ unit vectors in the r, θ, ϕ directions. The coefficients of these unit vectors are the *components* of **V** with respect to the different *basis vectors*. We may write the Equations (3.1) in the generic form

$$\mathbf{V} = V^i\mathbf{e}_i, \tag{3.2}$$

where the summation convention is used: i is summed from 1 to 3. The vector **V** has components V^i in the basis with basis vectors $\mathbf{e}_i$.

We want to generalise our formalism to describe vector *fields* in a space-time with arbitrary curvature. How do we do it? We want to retain the feature above, of describing vectors themselves, as well as their components in a particular coordinate system. First, the generalisation from 3-dimensional space to space-time is straightforward: (3.2) is replaced by

$$\mathbf{V} = V^\mu\mathbf{e}_\mu \tag{3.3}$$

where μ is summed over the values 0, 1, 2, 3.[1] We next introduce the key concept of a *curve* in a manifold. A *parametrised curve* in a manifold M is a mapping of an interval in R^1 into M – see Fig. 3.2 – while a local chart on M is mapped into R^n, as explained above. The interval in R^1, and therefore the curve into which it is mapped, is parametrised by λ, taking on values from 0 to 1. A point P on the curve may then be assigned local coordinates $x^\mu(\lambda)$. Suppose $f(x^\mu)$ is a function on M (i.e. a mapping from M to R^1). Then we may put

$$f(x^\mu(\lambda)) = g(\lambda).$$

The differential of the function along the curve is

[1] We adopt the convention that Greek indices are used in the general case, as well as in 4-dimensional space-time. Latin indices take on the values 1, 2, 3 and are therefore used in the specifically *space* section of space-time, as well as in examples of 2-and 3-dimensional spaces.

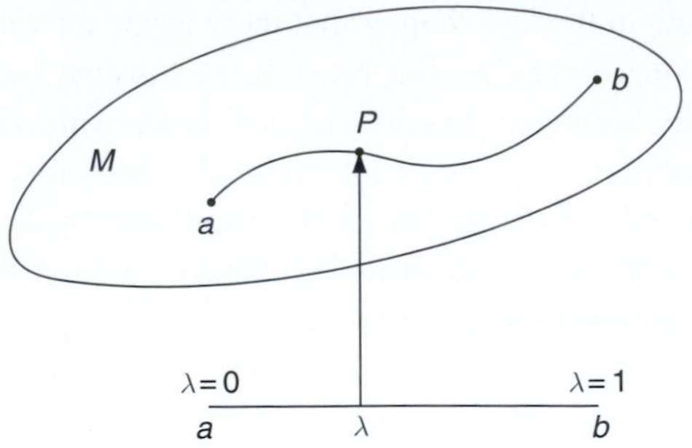

Fig. 3.2 A parametrised curve in M.

$$\frac{dg}{d\lambda} = \sum_{\mu} \frac{dx^{\mu}}{d\lambda}\frac{\partial f}{\partial x^{\mu}},$$

which may be expressed as an operator identity

$$\frac{d}{d\lambda} = \sum_{\mu} \frac{dx^{\mu}}{d\lambda}\frac{\partial}{\partial x^{\mu}},$$

which in turn, adopting the summation convention, may be written

$$\frac{d}{d\lambda} = \frac{dx^{\mu}}{d\lambda}\frac{\partial}{\partial x^{\mu}}.$$

Comparing this with Equation (3.3) above we may identify $\frac{dx^{\mu}}{d\lambda}$ with the components of a vector tangent to the curve, $\frac{\partial}{\partial x^{\mu}}$ as the basis vectors, and $\frac{d}{d\lambda}$ as the tangent vector to the curve at P. But really these are *vector fields* since they vary with P; so $\frac{d}{d\lambda}$ is a tangent vector field which changes with λ. We are here creating a *correspondence* between two languages:

$$\text{vector} \leftrightarrow \text{tangent to curve;}$$

or, to emphasise the different areas of mathematics involved,

$$\begin{array}{cc} \text{vector} \leftrightarrow & \text{derivative.} \\ \text{(geometry)} & \text{(analysis)} \end{array}$$

Vectors defined at P lie in the *tangent space* to M at P, denoted T_P. This may be visualised as the whole plane of vectors tangent to a curved surface, drawn in Fig. 3.3 as a sphere. We should note some features of the above construction:

- vectors lie in the *tangent space* to M at P;
- the picture of the tangent plane just given relies on an 'embedding' – the 2-sphere S^2 is embedded in R^3. This helps visualisation but is not actually necessary. The tangent space

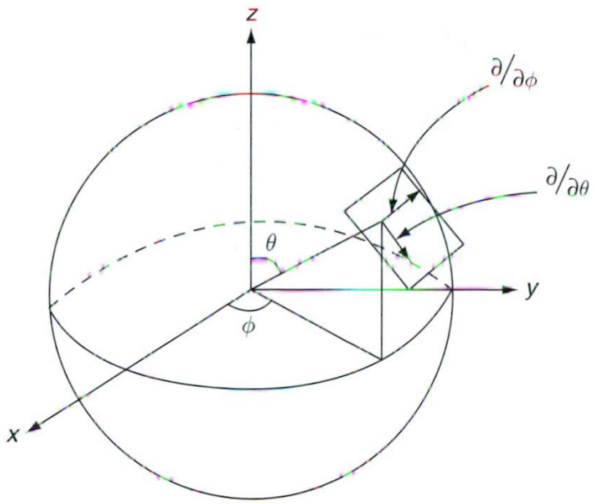

Fig. 3.3 Tangent plane to the sphere S^2.

T_P is obtained by taking *all possible* curves through P, and evaluating $\left.\frac{\mathrm{d}}{\mathrm{d}\lambda}\right|_P$ with $\frac{\mathrm{d}}{\mathrm{d}\lambda} = \frac{\mathrm{d}x^i}{\mathrm{d}\lambda}\frac{\partial}{\partial x^i}$;

- vectors defined at two different points have no relation to each other. We can only add or subtract vectors at the same point P.

Let us illustrate the above ideas by considering vectors in 3-dimensional Euclidean space R^3, in spherical polar coordinates. In the usual notation we have

$$\mathbf{r} = r\sin\theta\cos\phi\,\mathbf{i} + r\sin\theta\sin\phi\,\mathbf{j} + r\cos\theta\,\mathbf{k}$$

to denote the position of some point P. Then we define

$$\begin{aligned} \mathbf{e}_r &= \frac{\partial\mathbf{r}}{\partial r} = \sin\theta\cos\phi\,\mathbf{i} + \sin\theta\sin\phi\,\mathbf{j} + \cos\theta\,\mathbf{k}, \\ \mathbf{e}_\theta &= \frac{\partial\mathbf{r}}{\partial\theta} = r\cos\theta\cos\phi\,\mathbf{i} + r\cos\theta\sin\phi\,\mathbf{j} - r\sin\theta\,\mathbf{k}, \\ \mathbf{e}_\phi &= \frac{\partial\mathbf{r}}{\partial\phi} = -r\sin\theta\sin\phi\,\mathbf{i} + r\sin\theta\cos\phi\,\mathbf{j}. \end{aligned} \tag{3.4}$$

Appealing to Fig. 3.3 we have the directions $\frac{\partial}{\partial r}, \frac{\partial}{\partial\theta}, \frac{\partial}{\partial\phi}$, which are respectively perpendicular to the surface (and not shown in the diagram), due south and due east (viewing the sphere as a globe). With the (standard basis) normalisation

$$\mathbf{i}\cdot\mathbf{i} = \mathbf{j}\cdot\mathbf{j} = \mathbf{k}\cdot\mathbf{k} = 1; \quad \mathbf{i}\cdot\mathbf{j} = \mathbf{j}\cdot\mathbf{k} = \mathbf{k}\cdot\mathbf{i} = 0$$

we have

$$\mathbf{e}_r\cdot\mathbf{e}_r = 1, \quad \mathbf{e}_\theta\cdot\mathbf{e}_\theta = r^2, \quad \mathbf{e}_\phi\cdot\mathbf{e}_\phi = r^2\sin^2\theta, \quad \mathbf{e}_r\cdot\mathbf{e}_\theta = \mathbf{e}_r\cdot\mathbf{e}_\phi = \mathbf{e}_\theta\cdot\mathbf{e}_\phi = 0. \tag{3.5}$$

Making the identifications $\mathbf{i} = \partial/\partial x$, $\mathbf{j} = \partial/\partial y$, $\mathbf{k} = \partial/\partial z$, we may rewrite Equations (3.4) as

$$\mathbf{e}_r = \frac{\partial}{\partial r} = \sin\theta\cos\phi\frac{\partial}{\partial x} + \sin\theta\sin\phi\frac{\partial}{\partial y} + \cos\theta\frac{\partial}{\partial z},$$
$$\mathbf{e}_\theta = \frac{\partial}{\partial\theta} = r\cos\theta\cos\phi\frac{\partial}{\partial x} + r\cos\theta\sin\phi\frac{\partial}{\partial y} - r\sin\theta\frac{\partial}{\partial z}, \tag{3.6}$$
$$\mathbf{e}_\phi = \frac{\partial}{\partial\phi} = -r\sin\theta\sin\phi\frac{\partial}{\partial x} + r\sin\theta\cos\phi\frac{\partial}{\partial y},$$

with the orthogonality conditions (3.5). The system is clearly not orthonormal, but it is straightforward to define an *orthonormal* set of vectors, denoted with a hat

$$\hat{\mathbf{e}}_r = \mathbf{e}_r, \quad \hat{\mathbf{e}}_\theta = \frac{1}{r}\mathbf{e}_\theta, \quad \hat{\mathbf{e}}_\phi = \frac{1}{r\sin\theta}\mathbf{e}_\phi, \tag{3.7}$$

with

$$\hat{\mathbf{e}}_r\cdot\hat{\mathbf{e}}_r = \hat{\mathbf{e}}_\theta\cdot\hat{\mathbf{e}}_\theta = \hat{\mathbf{e}}_\phi\cdot\hat{\mathbf{e}}_\phi = 1; \quad \hat{\mathbf{e}}_r\cdot\hat{\mathbf{e}}_\theta = \hat{\mathbf{e}}_r\cdot\hat{\mathbf{e}}_\phi = \hat{\mathbf{e}}_\theta\cdot\hat{\mathbf{e}}_\phi = 0. \tag{3.8}$$

This has given us a new language in which to describe vectors. We now turn to the notion of *vector fields*; we must generalise the idea of a vector at P to a vector defined at each point of the manifold M – to take a simple example, the electric field vector $\mathbf{E}$ due to some space charge density $\rho(\mathbf{r})$. Since there is a tangent space at every point of M, a vector field will select one vector in the tangent space at each point of M. The question is, can all these tangent vectors be written as tangents to some *curve*? The converse is certainly true – given some curve, we may define a vector field, since the curve has a tangent at every point. And indeed the answer to our question is 'yes': given a vector field, a curve does (locally) exist such that the tangent to it is the vector at each point (see for example Choquet-Bruhat (1968), p. 21). In mathematical language the question is, do the equations

$$\frac{\mathrm{d}x^i}{\mathrm{d}\lambda} = v^i(x^m)$$

have a solution? These are first order differential equations, for which in general a local solution exists. Given v^i, we can find $x^i(\lambda)$, which is called the *integral curve* of the vector field. This holds at every point P – at all x^i – so that at every point in M we may draw the the integral curve of a vector field through that point, and these curves 'fill' M. Such a manifold-filling set of curves is called a *congruence*.

Figure 3.4 displays examples of some vector fields, or rather of the integral curves of vector fields. The respective fields are

(i) $\dfrac{\partial}{\partial\theta} = -y\dfrac{\partial}{\partial x} + x\dfrac{\partial}{\partial y}$,

(ii) $r\dfrac{\partial}{\partial\theta} = x\dfrac{\partial}{\partial x} + y\dfrac{\partial}{\partial y}$,

(iii) $y\dfrac{\partial}{\partial y}$,

(iv) $\dfrac{\partial}{\partial x}$.

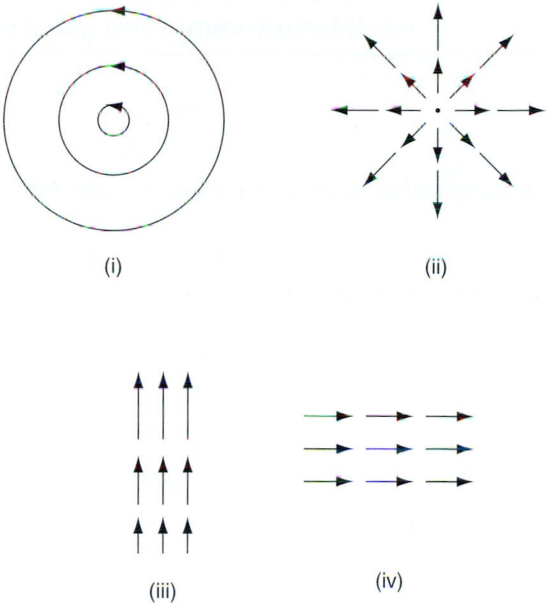

Fig. 3.4 Examples of vector fields.

Two-dimensional space R^2 may be filled with congruences of vector fields of any of these types. Let us now observe that the *commutator* of two vector fields is itself a vector field. Putting

$$\mathbf{V} = V^i \partial_i, \quad \mathbf{U} = U^m \partial_m, \tag{3.9}$$

we have, operating on any function f,

$$[\mathbf{V}, \mathbf{U}]f = V^i \partial_i (U^m \partial_m f) - U^m \partial_m (V^i \partial_i f) = \left(V^i \frac{\partial U^m}{\partial x^i} - U^i \frac{\partial V^m}{\partial x^i} \right) \partial_m f,$$

where in the second equality we have relabelled the indices – the second order terms cancel. Hence we have the operator identity

$$[\mathbf{V}, \mathbf{U}] = \mathbf{W}, \tag{3.10}$$

with

$$\mathbf{W} = W^n \partial_n, \quad W^n = V^i \frac{\partial U^n}{\partial x^i} - U^i \frac{\partial V^n}{\partial x^i}. \tag{3.11}$$

The commutator [**V**, **U**] is called the *Lie bracket* or *commutator* of the respective vector fields.

Lie brackets may be used to make a distinction between types of basis. Put $\mathbf{V} = V^i \mathbf{e}_i$, where $\mathbf{e}_i$ denotes the set of basis vectors. (In the following we consider the plane R^2, but the generalisation to other spaces is obvious.)

3.2.1 Holonomic and anholonomic bases

If

$$[\mathbf{e}_i, \mathbf{e}_k] = 0, \tag{3.12}$$

the basis is called a *coordinate* or *holonomic* one. For example, in R^2, if

$$\mathbf{e}_1 = \frac{\partial}{\partial x}, \quad \mathbf{e}_2 = \frac{\partial}{\partial y},$$

the basis is holonomic. Similarly if

$$\mathbf{e}_1 = \frac{\partial}{\partial r}, \quad \mathbf{e}_2 = \frac{\partial}{\partial \theta},$$

it is likewise holonomic. It is clear that if $\mathbf{e}_i$ is simply a derivative with respect to a coordinate the basis is holonomic; hence the alternative name *coordinate* basis.

If, on the other hand, for some $i \neq k$

$$[\mathbf{e}_i, \mathbf{e}_k] \neq 0, \tag{3.13}$$

the basis is called a *non-coordinate* or *anholonomic* one. For example (again confining ourselves to R^2) if, in the polar coordinate system, we put

$$\mathbf{e}_1 = \frac{\partial}{\partial r}, \quad \mathbf{e}_2 = \frac{1}{r}\frac{\partial}{\partial \theta}, \tag{3.14}$$

then

$$[\mathbf{e}_1, \mathbf{e}_2] = -\frac{1}{r^2}\frac{\partial}{\partial \theta} = -\frac{1}{r}\mathbf{e}_2. \tag{3.15}$$

It is usual to write the commutators as linear combinations of the basis elements, thus:

$$[\mathbf{e}_i, \mathbf{e}_k] = C^m{}_{ik}\, \mathbf{e}_m, \tag{3.16}$$

where the coefficients $C^m{}_{ik}$ are called the *structure constants* of the Lie algebra. In an anholonomic basis, therefore, not all the structure constants are zero. In the above example $C^2{}_{12} = -C^2{}_{21} = -\frac{1}{r}$ and the others are zero. Clearly in a holonomic basis all the structure constants are zero.

We now have a language in which to describe vectors, exemplified by Equations (3.2) and (3.3). It remains for us to demonstrate that such expressions are *independent of coordinate systems*. This is completely straightforward, since under a transformation $x^\mu \to x'^\mu$ we have

$$V^\mu \to V'^\mu = \frac{\partial x'^\mu}{\partial x^\lambda} V^\lambda, \tag{3.17}$$

and, if $\{\mathbf{e}_\mu\}$ and $\{\mathbf{e}'_\mu\}$ are coordinate bases, i.e. $\mathbf{e}_\mu = \frac{\partial}{\partial x^\mu}$, $\mathbf{e}'_\mu = \frac{\partial}{\partial x'^\mu}$

$$\mathbf{e}_\mu \to \mathbf{e}'_\mu = \frac{\partial x^\nu}{\partial x'^\mu} \mathbf{e}_\nu, \tag{3.18}$$

and hence

$$\mathbf{V} \to V'^{\mu}\mathbf{e}'_{\mu} = \frac{\partial x'^{\mu}}{\partial x^{\lambda}}\frac{\partial x^{\nu}}{\partial x'^{\mu}} V^{\lambda}\,\mathbf{e}_{\nu} = V^{\lambda}\,\mathbf{e}_{\lambda} = \mathbf{V}, \tag{3.19}$$

where we have used the identity (which follows from the fact that $x^{\nu} = x^{\nu}(x'^{\mu})$)

$$\frac{\partial x^{\nu}}{\partial x^{\lambda}} = \delta^{\nu}{}_{\lambda} = \frac{\partial x'^{\mu}}{\partial x^{\lambda}}\frac{\partial x^{\nu}}{\partial x'^{\mu}}. \tag{3.20}$$

3.3 One-forms

The definition of vectors given above involved the specifying of basis vectors, which are in essence directions along coordinate lines – $\partial/\partial x$, $\partial/\partial r$ etc. This gives a sense to the idea that a vector is a quantity 'with magnitude and direction'. But there is another type of 'vector' quantity defined in elementary calculus, which is the *gradient* of a (scalar) function. For example the function $f(x, y, z)$ has a gradient

$$\nabla f = \frac{\partial f}{\partial x}\mathbf{i} + \frac{\partial f}{\partial y}\mathbf{j} + \frac{\partial f}{\partial z}\mathbf{k}.$$

As a simple illustration consider the surface S^2(sphere) $r = \text{const}$ in R^3. Taking the normal to the surface we have

$$\nabla r = \frac{\partial r}{\partial x}\mathbf{i} + \frac{\partial r}{\partial y}\mathbf{j} + \frac{\partial r}{\partial z}\mathbf{k} = \frac{x}{r}\mathbf{i} + \frac{y}{r}\mathbf{j} + \frac{z}{r}\mathbf{k} = \frac{1}{r}\mathbf{r} = \mathbf{e}_r;$$

$$\nabla \theta = \frac{\partial \theta}{\partial x}\mathbf{i} + \frac{\partial \theta}{\partial y}\mathbf{j} + \frac{\partial \theta}{\partial z}\mathbf{k} = \frac{\cos\theta\cos\phi}{r}\mathbf{i} + \frac{\cos\theta\sin\phi}{r}\mathbf{j} - \frac{\sin\theta}{r}\mathbf{k} = \frac{1}{r^2}\mathbf{e}_\theta;$$

$$\nabla \phi = \frac{\partial \phi}{\partial x}\mathbf{i} + \frac{\partial \phi}{\partial y}\mathbf{j} + \frac{\partial \phi}{\partial z}\mathbf{k} = -\frac{\sin\phi}{r\sin\theta}\mathbf{i} + \frac{\cos\phi}{r\sin\theta}\mathbf{j} = \frac{1}{r^2\sin^2\theta}\,\mathbf{e}_\phi.$$

In this case the gradients are proportional to the vectors $\mathbf{e}_r$, $\mathbf{e}_\theta$, $\mathbf{e}_\phi$, with $\mathbf{e}_\theta$, $\mathbf{e}_\phi$ spanning the tangent space to S^2. This, however, is not typical, since a spherical surface in a flat 3-dimensional space has a very high degree of symmetry. *Conceptually*, gradients and basis vectors are distinct; basis vectors are associated with coordinate lines, but gradients with 'lines of steepest descent' from one surface to another, as sketched in Fig. 3.5. This difference may be shown up by considering a non-orthogonal coordinate system in R^3; define u, v and w by

$$x = u + v, \quad y = u - v, \quad z = 2uv + w$$

with inverse

$$u = \tfrac{1}{2}(x + y), \quad v = \tfrac{1}{2}(x - y), \quad w = z - \tfrac{1}{2}(x^2 - y^2).$$

It is clear that $u = \text{const}$ and $v = \text{const}$ are plane surfaces, whereas $w = \text{const}$ is a hyperboloid. These surfaces are sketched in Fig. 3.6. With $\mathbf{r} = (u + v)\mathbf{i} + (u - v)\mathbf{j} + (2uv + w)\mathbf{k}$ we find

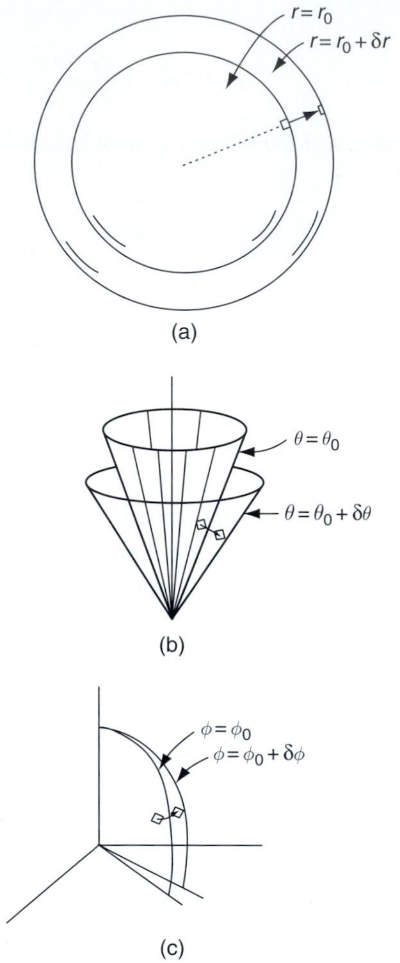

Fig. 3.5 Lines of steepest descent from one surface to another in R^3: (a) surfaces of constant r – spheres, (b) surfaces of constant θ – cones, (c) surfaces of constant ϕ – planes.

$$\mathbf{e}_u = \frac{\partial \mathbf{r}}{\partial u} = \mathbf{i} + \mathbf{j} + 2v\mathbf{k}; \quad \mathbf{e}_v = \mathbf{i} - \mathbf{j} + 2u\mathbf{k}; \quad \mathbf{e}_w = \mathbf{k}. \tag{3.21}$$

It is clear that $\mathbf{e}_u$ and $\mathbf{e}_v$ are not unit vectors, whereas $\mathbf{e}_w$ is. In addition,

$$\mathbf{e}_u \cdot \mathbf{e}_v = 4uv, \quad \mathbf{e}_u \cdot \mathbf{e}_w = 2v, \quad \mathbf{e}_v \cdot \mathbf{e}_w = 2u; \tag{3.22}$$

the vectors are not orthogonal. The direction of $\mathbf{e}_u$, for example, is given by a line of intersection of the planes $v = \text{const}$ and $w = \text{const}$ – a line along which only u changes. Thus it is pictorially clear that $\mathbf{e}_w$ is proportional to $\mathbf{k}$, as in (3.21) above; the intersection of the planes $u = \text{const}$ and $v = \text{const}$ is parallel to the z axis. On the other hand this is *not* perpendicular to the surface $w = \text{const}$. It is straightforward to calculate the gradient vectors, which we now denote with an upper suffix:

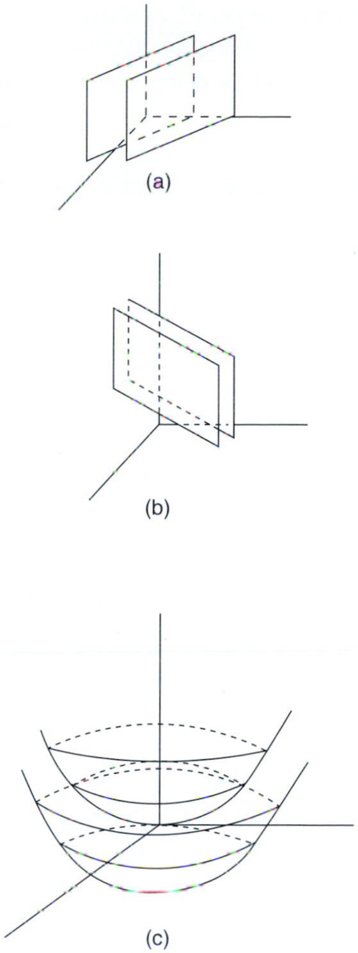

Fig. 3.6 Surfaces (a) u = constant (plane), (b) v = constant (plane), (c) w = constant (hyperboloid).

$$\begin{aligned} \nabla u = \frac{\partial u}{\partial x}\mathbf{i} + \frac{\partial u}{\partial y}\mathbf{j} + \frac{\partial u}{\partial z}\mathbf{k} = {}^1\!/_2\,\mathbf{i} + {}^1\!/_2\,\mathbf{j} = \mathbf{e}^u; \\ \nabla v = {}^1\!/_2\,\mathbf{i} - {}^1\!/_2\,\mathbf{j} = \mathbf{e}^v; \\ \nabla w = -(u+v)\,\mathbf{i} + (u-v)\,\mathbf{j} + \mathbf{k} = \mathbf{e}^w. \end{aligned} \tag{3.23}$$

As is easily checked from (3.21)–(3.23) these gradient vectors obey

$$\mathbf{e}^u \cdot \mathbf{e}_u = \mathbf{e}^v \cdot \mathbf{e}_v = \mathbf{e}^w \cdot \mathbf{e}_w = 1; \quad \mathbf{e}^u \cdot \mathbf{e}_v = \mathbf{e}^v \cdot \mathbf{e}_w = \mathbf{e}^w \cdot \mathbf{e}_u = 0. \tag{3.24}$$

Gradients are thus *dual* to basis vectors. They are conceptually distinct entities, and in fact are somewhat analogous to reciprocal lattice vectors in solid state physics; indeed Pauli[2] refers to them in this way. They are now however referred to as 'one-forms' or 1-forms.

[2] Pauli (1958), Section 10.

Instead of denoting them as $\mathbf{e}^\mu$, as above, it is common to use Greek letters for 1-forms. We shall adopt this convention, and therefore have the following situation. In a manifold we may define vector fields $\mathbf{V}$ in terms of a basis set $\mathbf{e}_\mu$

$$\mathbf{V} = V^\mu \mathbf{e}_\mu \tag{3.25}$$

and dual to these we define 1-forms $\boldsymbol{\omega}$, also in terms of a basis set $\boldsymbol{\theta}^\mu$

$$\boldsymbol{\omega} = \omega_\mu \boldsymbol{\theta}^\mu, \tag{3.26}$$

and the duality of the basis sets is commonly expressed using angular brackets:

$$\langle \boldsymbol{\theta}^\mu, \mathbf{e}_\nu \rangle = \delta^\mu{}_\nu. \tag{3.27}$$

Equation (3.27) is simply the general form of (3.24) in the particular example considered above, expressing the duality of vectors and 1-forms. Following from this we have

$$\langle \boldsymbol{\omega}, \mathbf{V} \rangle = \omega_\nu V^\mu \langle \boldsymbol{\theta}^\nu, \mathbf{e}_\mu \rangle = \omega_\mu V^\mu. \tag{3.28}$$

As may be seen from Fig. 3.5, a 1-form may be viewed as a series of parallel surfaces, which may be regarded as being 'pierced' by vectors. Misner *et al.* (1973) (see particularly Chapter 8) use the memorable phrase 'bongs of the bell' to describe a relation such as (3.28); as the vector $\mathbf{V}$ pierces the surfaces given by $\boldsymbol{\omega}$ the number $\omega_\mu V^\mu$ is the number of bongs of the bell, each 'bong' corresponding to one surface being pierced by a unit vector.

Just as the vectors $\mathbf{e}_\mu$ are basis vectors in the tangent space T_P at P, where the vectors are defined, so $\boldsymbol{\theta}^\mu$, the basis 1-forms, are the basis forms in the *cotangent* space $T_P{}^*$ at P. These are defined, however, not just at the single point P but in the tangent and cotangent spaces at all points in the manifold, which is to say the tangent and cotangent *bundles*:

$$\begin{aligned} &\mathbf{e}_\mu \text{ are basis vectors in tangent bundle,} \\ &\boldsymbol{\theta}^\mu \text{ are basis 1-forms in contangent bundle.} \end{aligned} \tag{3.29}$$

It is perhaps useful to remark here that this subject, like many others in physics and mathematics, is beset by different notation conventions. The duality (3.28) is also commonly written as

$$\langle \boldsymbol{\omega}, \mathbf{V} \rangle = \boldsymbol{\omega}(\mathbf{V}) = \omega_\mu V^\mu. \tag{3.30}$$

The components of a vector $\mathbf{V}$ and a 1-form $\boldsymbol{\omega}$ may then variously be written as

$$\begin{aligned} V^\mu &= \langle \boldsymbol{\theta}^\mu, \mathbf{V} \rangle = \boldsymbol{\theta}^\mu(\mathbf{V}); \\ \omega_\nu &= \langle \boldsymbol{\omega}, \mathbf{e}_\nu \rangle = \boldsymbol{\omega}(\mathbf{e}_\nu); \end{aligned} \tag{3.31}$$

so a 1-form basis $\boldsymbol{\theta}^\mu$ may be used to map a vector $\mathbf{V}$ into one of its components V^μ (in other words onto R^1); and a vector basis $\mathbf{e}_\mu$ maps a 1-form $\boldsymbol{\omega}$ into one of its components. This is the sense in which the basis of 1-forms is dual to the basis $\mathbf{e}_\mu$ of vectors, and the n-dimensional space of 1-forms is the *dual space* $T_P{}^*$ to the tangent space T_P.

This language is not so strange as it may seem. There are several analogous concepts with which the reader must already be familiar:

(i) Corresponding to a (column) vector in R^3, $\begin{pmatrix} a \\ b \\ c \end{pmatrix}$ is the dual (row) vector $(p\ q\ r)$ such that $(p\,q\,r)\begin{pmatrix} a \\ b \\ c \end{pmatrix} = pa + qb + rc$, a real number (belonging to R^1).

(ii) In the complex 2-dimensional space C^2, corresponding to the vector $z = x + iy$ is the complex conjugate $z^* = x - iy$, such that $z^*z = x^2 + y^2$, a real number.

(iii) In quantum mechanics a vector in Hilbert space is denoted by the Dirac ket vector $|\psi\rangle$. Its dual is the bra vector $\langle\psi|$ so that $\langle\psi|\psi\rangle$ is a real number (a c-number).

There is one further, and rather crucial, piece of notation to introduce. In the previous section we introduced the notion that vectors, in particular basis vectors, may be represented as operators, so that $\mathbf{e}_1 = \mathbf{e}_x = \partial/\partial x$ etc. It is then a natural question, what corresponding notational development is made for the dual one-form $\mathbf{e}^1 = \boldsymbol{\theta}^1$? The answer is that it is denoted $\mathbf{d}x$. The duality relation (3.25) then reads

$$\langle \mathbf{d}x, \partial/\partial x\rangle = 1 \tag{3.32}$$

or, in general

$$\left\langle \mathbf{d}x^\mu, \frac{\partial}{\partial x^\nu}\right\rangle = \delta^\mu{}_\nu. \tag{3.33}$$

The 1-form $\mathbf{d}x^\mu$ is *not* the same as the infinitesimal $\mathrm{d}x^\mu$; it is not a 'number', but a member of the cotangent space $T_P{}^*$. This mathematical language is more sophisticated than the language of 'classical' calculus. Though it might seem in danger of becoming unnecessarily difficult, it actually gets rid of some embarrassments encountered there, concerning the idea of 'infinitesimally small' changes $\mathrm{d}x^i$ in the coordinates x^i: to obtain the gradient of a function $f(x^i)$, for example, as we were taught at school, we consider an infinitesimally small change in x^i, $\mathrm{d}x^i$, find the corresponding 'infinitesimally small' change in f, which we denote $\mathrm{d}f$, and *divide them*, to get $\partial f/\partial x^i$! This is hardly a very satisfactory procedure, but, as remarked by Spivak:[3]

> No-one wanted to admit this was nonsense because true results were obtained when these infinitely small quantities were divided into each other (provided one did it in the right way). Eventually it was realised that the closest one can come to describing an infinitely small change is to describe a *direction* in which this change is supposed to occur, i.e. a tangent vector. Since $\mathrm{d}f$ is supposed to be the infinitesimal change in f under an infinitesimal change of the point, $\mathrm{d}f$ must be a function of this change, which means that $\mathrm{d}f$ *must be a function on tangent vectors*. The $\mathrm{d}x^i$ themselves then metamorphosed into functions, and it became clear that they must be distinguished from the tangent vectors $\partial/\partial x^i$. Once this realisation came, it was only a matter of making new definitions, which preserved the *old* notation, and waiting for everybody to catch up. In short, all classical notions involving infinitely small quantities became functions on tangent vectors, like $\mathrm{d}f$, except for quotients of infinitely small quantities, which became tangent vectors, like $\mathrm{d}f/\mathrm{d}t$.

[3] Spivak (1970), pp. 4–6. The italics are mine.

(In this last quantity c denotes a curve, with t a parameter along it; we would simply denote this d/dt.) The remark that df is a function on tangent vectors is then to be understood as follows:

$$\mathbf{d}f = \frac{\partial f}{\partial x^\mu}\mathbf{d}x^\mu; \quad \text{then} \left\langle \mathbf{d}f, \frac{\partial}{\partial x^\mu}\right\rangle = \frac{\partial f}{\partial x^\mu}; \tag{3.34}$$

the contraction of $\mathbf{d}f$ with a tangent vector yields the quantity $\partial f/\partial x^\mu$.

3.3.1 Transformation rules

One-forms, like vectors, have the same form in different coordinate systems. Under a transformation $x^\mu \to x^{\mu'}$ the (coordinate) basis $\boldsymbol{\theta}^\mu$ transforms as

$$\boldsymbol{\theta}^\mu \to \boldsymbol{\theta}'^\mu = \frac{\partial x'^\mu}{\partial x^\nu}\boldsymbol{\theta}^\nu \tag{3.35}$$

and the components ω_ν as

$$\omega_\nu \to \omega'_\nu = \frac{\partial x^\lambda}{\partial x'^\nu}\omega_\lambda \tag{3.36}$$

so that $\boldsymbol{\omega} = \omega_\mu \boldsymbol{\theta}^\mu$ is invariant:

$$\boldsymbol{\omega} \to \omega'_\mu \boldsymbol{\theta}'^\mu = \frac{\partial x^\lambda}{\partial x'^\mu}\frac{\partial x'^\mu}{\partial x^\kappa}\omega_\lambda \boldsymbol{\theta}^\kappa = \omega_\lambda \boldsymbol{\theta}^\lambda = \boldsymbol{\omega}, \tag{3.37}$$

in analogy with (3.17–3.19). In more old-fashioned language the transformation rules (3.17) and (3.36) are respectively the transformation rules for *covariant* and *contravariant vector fields*. What we now call vectors have components in a given coordinate system which transform 'contravariantly', and what we now call 1-forms have components in a given coordinate system which transform 'covariantly'. For convenience we state the formulae again:

$$\begin{aligned} &\text{covariant vector:} \quad V^\mu(x) \to V'^\mu(x) = \frac{\partial x'^\mu}{\partial x^\lambda}(x)\, V^\lambda(x), \\ &\text{contravariant vector:} \quad V_\nu(x) \to V'_\nu(x) = \frac{\partial x^\lambda}{\partial x'^\nu}(x)\, V_\lambda(x). \end{aligned} \tag{3.38}$$

These formulae emphasise, by including the arguments (x), that these vector fields are defined at a specific point x, and the transformation coefficients are evaluated at the same point. The reader will appreciate that for non-linear transformations the coefficients $\frac{\partial x'^\mu}{\partial x^\lambda}$ and $\frac{\partial x^\lambda}{\partial x'^\nu}$ will themselves depend on x. The above transformation laws are specific to one particular point in the manifold.

3.3.2 A note on orthogonal coordinate systems

It may be instructive to illustrate the above ideas with the simple case of the polar coordinate system (r; θ) in the plane R^2. From the equations $x = r\cos\theta$, $y = r\sin\theta$ and their inverses

$r=(x^2+y^2)^{-1/2}$, $\theta=\tan^{-1}(y/x)$, and recalling that $\frac{\partial x}{\partial r}$ means $\left(\frac{\partial x}{\partial r}\right)_\theta$, while $\frac{\partial r}{\partial x}$ means $\left(\frac{\partial r}{\partial x}\right)_y$, it is straightforward to find that

$$\frac{\partial x}{\partial r}=\frac{\partial r}{\partial x}=\cos\theta,\qquad \frac{\partial y}{\partial r}=\frac{\partial r}{\partial y}=\sin\theta,\ \text{etc.}$$

If a vector then has the *contravariant* components (V^x, V^y), (V^r, V^θ), the following relations will then hold

$$\begin{aligned}V^r &= \frac{\partial r}{\partial x}V^x+\frac{\partial r}{\partial y}V^y = V^x\cos\theta+V^y\sin\theta,\\ V^\theta &= \cdots = \frac{1}{r}(-V^x\sin\theta+V^y\cos\theta).\end{aligned}\tag{3.39}$$

The *covariant* components on the other hand obey

$$\begin{aligned}V_r &= \frac{\partial x}{\partial r}V_x+\frac{\partial y}{\partial r}V_y = V_x\cos\theta+V_y\sin\theta,\\ V_\theta &= r(-V_x\sin\theta+V_y\cos\theta).\end{aligned}\tag{3.40}$$

Neither (3.39) nor (3.40), however, describes the 'normal' relations between the Cartesian and polar components of a vector. The 'ordinary' components $\bar{V}_r$, $\bar{V}_\theta$ are given by[4]

$$\bar{V}_r = h_r V^r,\quad \bar{V}_\theta = h_\theta V^\theta$$

where

$$ds^2 = dx^2+dy^2 = h_r{}^2\,dr^2 + h_\theta{}^2\,d\theta^2 = dr^2+r^2\,d\theta^2,$$

and hence $h_r=1$, $h_\theta=r$. Then with $\bar{V}_x=V^x, \bar{V}_y=V^y$ we have

$$\bar{V}_r = V^r = \bar{V}_x\cos\theta+\bar{V}_y\sin\theta,\quad \bar{V}_\theta = rV^\theta = -\bar{V}_x\sin\theta+\bar{V}_y\cos\theta.\tag{3.41}$$

In addition, the relation between the ordinary and the covariant components is

$$\begin{aligned}\bar{V}_r &= \frac{1}{h_r}V_r = V_r = \bar{V}_x\cos\theta+\bar{V}_y\sin\theta,\\ \bar{V}_\theta &= \frac{1}{h_\theta}V_\theta = \frac{1}{r}V_\theta = -\bar{V}_x\sin\theta+\bar{V}_y\cos\theta.\end{aligned}\tag{3.42}$$

Equations (3.39–3.42) summarise the relations between the contravariant, covariant and 'ordinary' components of a vector. It should be remarked that the quantities h_i, introduced above, depend on the *metric*, i.e. on ds^2, a quantity which remains to be introduced – see Section 3.8 below.

3.4 Tensors

A vector is

$$\mathbf{V} = V^\mu \mathbf{e}_\mu,\tag{3.43}$$

[4] See, e.g. Arfken (1970), Chapter 2, Panofsky & Phillips (1962), Appendix III, or Weinberg (1972), p. 108.

where V^μ are the components in the given basis $\mathbf{e}_\mu$. A 1-form is

$$\boldsymbol{\omega} = \omega_\nu \boldsymbol{\theta}^\nu, \tag{3.44}$$

where ω_ν are the components in the basis $\boldsymbol{\theta}^\nu$, which is dual to the basis $\mathbf{e}_\mu$. We can then define a *tensor* – more precisely, an $\begin{pmatrix} r \\ s \end{pmatrix}$ tensor – as the geometric object that has components in the space which is the Cartesian product of r basis vectors and s basis 1-forms:

$$\mathbf{T} = T^{\alpha\ldots\beta}{}_{\lambda\ldots\mu}\, \mathbf{e}_\alpha \otimes \cdots \otimes \mathbf{e}_\beta \otimes \boldsymbol{\theta}^\lambda \otimes \cdots \otimes \boldsymbol{\theta}^\mu; \tag{3.45}$$

$T^{\alpha\ldots\beta}{}_{\lambda\ldots\mu}$ are the components in the given basis. A vector, then, is a $\begin{pmatrix} 1 \\ 0 \end{pmatrix}$ tensor and a 1-form a $\begin{pmatrix} 0 \\ 1 \end{pmatrix}$ tensor. It is common to omit the Cartesian product signs $\otimes$ in the above formula and simply write

$$\mathbf{T} = T^{\alpha\ldots\beta}{}_{\lambda\ldots\mu}\, \mathbf{e}_\alpha \ldots \mathbf{e}_\beta \boldsymbol{\theta}^\lambda \ldots \boldsymbol{\theta}^\mu. \tag{3.46}$$

(We should also note that instead of $\boldsymbol{\theta}^\mu$, the 1-form basis is sometimes written $\mathbf{e}^\mu$.) In a different basis $\mathbf{e}'_\alpha$, $\boldsymbol{\theta}'^\lambda$ etc., the same tensor is written

$$\mathbf{T} = T'^{\alpha\ldots\beta}{}_{\lambda\ldots\mu}\, \mathbf{e}'_\alpha \otimes \cdots \otimes \mathbf{e}'_\beta \otimes \boldsymbol{\theta}'^\lambda \otimes \cdots \otimes \boldsymbol{\theta}'^\mu.$$

The components of a tensor in a particular basis may, using the notation of Equation (3.31), be written as

$$T^{\alpha\ldots\beta}{}_{\lambda\ldots\mu} = \mathbf{T}(\boldsymbol{\theta}^\alpha, \ldots, \boldsymbol{\theta}^\beta,\ \mathbf{e}_\lambda, \ldots, \mathbf{e}_\mu); \tag{3.47}$$

i.e. on inserting the bases $\mathbf{e}_\lambda$ and the dual bases $\boldsymbol{\theta}^\alpha$ into the tensor we may obtain its components in this basis. This is analogous to (3.31).

Analogous to the formulae (3.38), the relation between the components of a tensor, with both contravariant and covariant indices, in two different bases is given by

$$T'^{\alpha\ldots\beta}{}_{\lambda\ldots\mu}(x) = \frac{\partial x'^\alpha}{\partial x^\rho}(x)\ldots\frac{\partial x'^\beta}{\partial x^\sigma}(x)\frac{\partial x^\kappa}{\partial x'^\lambda}(x)\ldots\frac{\partial x^\nu}{\partial x'^\mu}(x)\ T^{\rho\ldots\sigma}{}_{\kappa\ldots\nu}(x), \tag{3.48}$$

where again the arguments (x) are included, as in (3.38) above. It is worth observing that this is a *homogeneous* transformation law. If all the components of a tensor are zero in one frame, they are also zero in any other frame – we may say that the tensor 'is zero'. An important application of this remark will be the characterisation of *curvature*. It is important to say of a space whether it is curved or flat – whether the curvature is non-zero or zero; this must be a statement about the space itself, and not about a particular coordinate system. We shall therefore define a curvature tensor, with the important property that if it is zero (non-zero) in one coordinate system it is zero (non-zero) in all coordinate systems. It is also worth remarking that whereas tensors, like vectors and 1-forms, may be given a neat geometric definition (Equation (3.45) above), for many practical purposes it is best to consider the *components* of the tensor in a particular coordinate system; and then, on transforming to another coordinate system, the components reorganise themselves as in (3.48). Let us now consider various operations on tensors. We begin with contraction.

3.4.1 Contraction

Consider $T^{\alpha\ldots\beta}{}_{\kappa\ldots\lambda}$, an $\begin{pmatrix} r \\ s \end{pmatrix}$ tensor, with r upper and s lower indices. It is said to be of *rank* $r+s$. If one lower index is put equal to one upper one and a *summation* is performed over all the relevant components, what results is an $\begin{pmatrix} r-1 \\ s-1 \end{pmatrix}$ tensor; its rank has been reduced by 2. So for example $T^{\alpha\ldots\beta}{}_{\kappa\ldots\beta}$ is an $\begin{pmatrix} r-1 \\ s-1 \end{pmatrix}$ tensor (since the summation convention means that the summation is carried out without explicit indication). The proof of this is straightforward, but let us consider for simplicity a particular case. $T^{\alpha\beta}{}_{\mu}$ is a $\begin{pmatrix} 2 \\ 1 \end{pmatrix}$ tensor. Our claim is that $T^{\alpha\beta}{}_{\alpha}$ is a $\begin{pmatrix} 1 \\ 0 \end{pmatrix}$ tensor – in other words a vector. The proof depends on the transformation formula (3.48). We have, from (3.48) and (3.20)

$$T'^{\alpha\beta}{}_{\alpha} = \frac{\partial x'^{\alpha}}{\partial x^{\kappa}}\frac{\partial x'^{\beta}}{\partial x^{\lambda}}\frac{\partial x^{\mu}}{\partial x'^{\alpha}}T^{\kappa\lambda}{}_{\mu} = \frac{\partial x'^{\beta}}{\partial x^{\lambda}}\,T^{\mu\lambda}{}_{\mu}$$

which is precisely the transformation law for a vector, Equation (3.7). This proves our claim.

3.4.2 Symmetry and antisymmetry

The symmetric or antisymmetric part of a tensor with respect to *either* its upper *or* its lower indices may be defined. The convention is to use round brackets for symmetrisation, so that for example, if **T** is an $\begin{pmatrix} r \\ s \end{pmatrix}$ tensor then

$$\begin{aligned} &T^{(\kappa\ldots\lambda)}{}_{\mu\ldots\nu} \\ &\quad = \frac{1}{r!}\{\text{sum over all permutations of the } r \text{ indices } \kappa\ldots\lambda \text{ of } T^{\kappa\ldots\lambda}{}_{\mu\ldots\nu}\} \end{aligned} \tag{3.49}$$

is symmetric on interchange of any of its upper indices. Square brackets are used for antisymmetrisation, so that for example

$$\begin{aligned} &T^{\kappa\ldots\lambda}{}_{[\mu\ldots\nu]} \\ &\quad = \frac{1}{s!}\{\text{alternating sum over all permutations of the } s \text{ indices } \mu\ldots\nu \text{ of } T^{\kappa\ldots\lambda}{}_{\mu\ldots\nu}\}. \end{aligned} \tag{3.50}$$

To take some simple examples,

$$\begin{aligned} T^{(\kappa\lambda)}{}_{\mu} &= 1/2(T^{\kappa\lambda}{}_{\mu} + T^{\lambda\kappa}{}_{\mu}), \\ T^{[\kappa\lambda]}{}_{\mu} &= 1/2(T^{\kappa\lambda}{}_{\mu} - T^{\lambda\kappa}{}_{\mu}), \\ T^{[\kappa\lambda\mu]}{}_{\rho\sigma} &= 1/6(T^{\kappa\lambda\mu}{}_{\rho\sigma} + T^{\lambda\mu\kappa}{}_{\rho\sigma} + T^{\mu\kappa\lambda}{}_{\rho\sigma} - T^{\kappa\mu\lambda}{}_{\rho\sigma} - T^{\mu\lambda\kappa}{}_{\rho\sigma} - T^{\lambda\kappa\mu}{}_{\rho\sigma}). \end{aligned} \tag{3.51}$$

A tensor is called *symmetric* in a given set of (upper or lower – contravariant or covariant) indices if it is equal to its symmetric part on these indices; and it is called *antisymmetric* if it is equal to its antisymmetric part. It is important that the properties of symmetry or antisymmetry are intrinsic to the tensor – they are the same in all bases. For example if $S^{\kappa\lambda} = -S^{\lambda\kappa}$ in some basis (**S** is an antisymmetric $\begin{pmatrix} 2 \\ 0 \end{pmatrix}$ tensor), then

$$S'^{\mu\nu} = \frac{\partial x'^{\mu}}{\partial x^{\kappa}}\frac{\partial x'^{\nu}}{\partial x^{\lambda}}S^{\kappa\lambda} = -\frac{\partial x'^{\mu}}{\partial x^{\kappa}}\frac{\partial x'^{\nu}}{\partial x^{\lambda}}S^{\lambda\kappa} = -S'^{\nu\mu};$$

it is antisymmetric in all other bases. In terms of the tensor itself $\mathbf{S} = S^{\mu\nu}\mathbf{e}_{\mu}\mathbf{e}_{\nu}$, we have (see for example (3.47)) $\mathbf{S}(\boldsymbol{\theta}^{\mu}, \boldsymbol{\theta}^{\nu}) = S^{\mu\nu}$, and the statement of antisymmetry is the statement

$$S^{\mu\nu} = -S^{\nu\mu} \Rightarrow \mathbf{S}(\boldsymbol{\theta}^{\mu}, \boldsymbol{\theta}^{\nu}) = -\mathbf{S}(\boldsymbol{\theta}^{\nu}, \boldsymbol{\theta}^{\mu})$$

and since this is true for a basis it is true for all $\boldsymbol{\omega}$, $\boldsymbol{\sigma}$, i.e.

$$\mathbf{S}(\boldsymbol{\omega}, \boldsymbol{\sigma}) = -\mathbf{S}(\boldsymbol{\sigma}, \boldsymbol{\omega}). \tag{3.52}$$

3.4.3 Quotient theorem

It is clear from the transformation rules above that the *sum* and the *difference* of two tensor fields of the same type – say rank $\begin{pmatrix} p \\ q \end{pmatrix}$ tensors – and defined *at the same point* – is also a tensor field of that type. It is also clear that the *outer product* (product involving no contractions) of two tensors, one of type $\begin{pmatrix} p \\ q \end{pmatrix}$ and the other of type $\begin{pmatrix} r \\ s \end{pmatrix}$, is itself a tensor, of rank $\begin{pmatrix} p+r \\ q+s \end{pmatrix}$. To complete these findings there is a further result, which goes by the name of the *quotient theorem*, according to which if we have an (outer) product of two objects and it is known that the product is a tensor, say of type $\begin{pmatrix} p+r \\ q+s \end{pmatrix}$, and *one* of the constituent objects is also a tensor, say of type $\begin{pmatrix} r \\ s \end{pmatrix}$, then the other object in the product is itself a tensor, of rank $\begin{pmatrix} p \\ q \end{pmatrix}$. This theorem is clearly a sort of inverse of the previous result concerning outer products, but the proof is not quite so straightforward. We here consider a special case, from which the reader will have no difficulty in being able to construct the general case. Suppose that

$$T^{\mu\nu}{}_{\kappa\lambda} = U^{\mu}S^{\nu}{}_{\kappa\lambda}$$

and T is known to be a tensor and U is known to be a contravariant vector (i.e. a tensor of rank $\begin{pmatrix} 1 \\ 0 \end{pmatrix}$); then we claim that S is a $\begin{pmatrix} 1 \\ 2 \end{pmatrix}$ tensor. To do this we introduce an arbitrary contravariant vector η^κ and an arbitrary covariant vector ξ_ρ and multiply T above by $\xi_\mu\,\xi_\nu\,\eta^\kappa\,\eta^\lambda$ and sum (i.e. contract) over repeated indices:

$$T^{\mu\nu}{}_{\kappa\lambda}\,\xi_\mu\,\xi_\nu\,\eta^\kappa\eta^\lambda = (U^\mu\,\xi_\mu)S^\nu{}_{\kappa\lambda}\,\xi_\nu\,\eta^\kappa\eta^\lambda.$$

The left hand side is a scalar (since all indices are contracted), and so is $(U^\mu\,\xi_\mu)$. It follows that $S^\nu{}_{\kappa\lambda}\,\xi_\nu\,\eta^\kappa\eta^\lambda$ is also a scalar, i.e.

$$S'^\nu{}_{\kappa\lambda}\,\xi_\nu\,\eta'^\kappa\eta^\lambda = S^\nu{}_{\kappa\lambda}\,\xi_\nu\,\eta^\kappa\,\eta^\lambda.$$

However, since ξ and η are co- and contra-variant vectors this gives

$$S'^\nu{}_{\kappa\lambda}\frac{\partial x^\alpha}{\partial x'^\nu}\frac{\partial x'^\kappa}{\partial x^\rho}\frac{\partial x'^\lambda}{\partial x^\sigma}\,\xi_\alpha\,\eta^\rho\,\eta^\sigma = S^\nu{}_{\kappa\lambda}\,\xi_\nu\,\eta^\kappa\eta^\lambda = S^\alpha{}_{\rho\sigma}\,\xi_\alpha\,\eta^\rho\,\eta^\sigma,$$

where the second equality follows on relabelling the indices. But ξ and η are arbitrary vectors so this implies that

$$S'^\nu{}_{\kappa\lambda}\frac{\partial x^\alpha}{\partial x'^\nu}\frac{\partial x'^\kappa}{\partial x^\rho}\frac{\partial x'^\lambda}{\partial x^\sigma} = S^\alpha{}_{\rho\sigma}.$$

Now multiply by $\dfrac{\partial x'^\tau}{\partial x^\alpha}\dfrac{\partial x^\rho}{\partial x'^\zeta}\dfrac{\partial x^\sigma}{\partial x'^\mu}$ and use (3.20) to give

$$S'^\tau{}_{\zeta\mu} = \frac{\partial x'^\tau}{\partial x^\alpha}\frac{\partial x^\rho}{\partial x'^\zeta}\frac{\partial x^\sigma}{\partial x'^\mu}S^\alpha{}_{\rho\sigma},$$

which is the transformation law for a $\begin{pmatrix}1\\2\end{pmatrix}$ tensor.

3.5 Differential forms: Hodge duality

One-forms are dual to vectors, as we have seen; but they possess another sort of duality – they are dual to *lines*. The *integral* of a one-form over a line is a number (belonging to R^1). Consider for example the 1-form in R^3

$$\boldsymbol{\omega}_1 = a_1\,\mathbf{d}x + a_2\,\mathbf{d}y + a_3\,\mathbf{d}z. \tag{3.53}$$

The line integral of $\boldsymbol{\omega}_1$ over a line c_1 is

$$\int_{c_1}\omega_1 = \text{number}. \tag{3.54}$$

Let us call a line a 1-*chain*; then we have a duality between a 1-form and a 1-chain. This duality is a consequence of integration but follows from the previously noted duality between a 1-form and a vector, since integration is simply the summation along the path of the contraction of the form with the tangent vector at that point. Now the notion of a 1-chain is easily generalised: we may call an area a 2-chain c_2, a volume a 3-chain c_3, a 4-volume (say in space-time) a 4-chain c_4, and so on – and with this logic, c_0, a zero-chain, is a point:

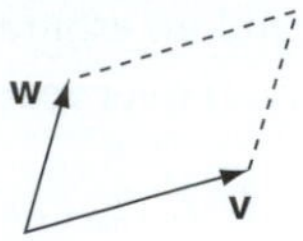

Fig. 3.7 The vectors **w** and **v** form a parallelogram.

$$\begin{aligned} c_0&: \text{0-chain}: && \text{point}\\ c_1&: \text{1-chain}: && \text{line}\\ c_2&: \text{2-chain}: && \text{area}\\ c_3&: \text{3-chain}: && \text{volume}\\ c_4&: \text{4-chain}: && \text{4-volume}\\ &&\text{etc.} \end{aligned} \tag{3.55}$$

Then we may define a 2-*form* $\boldsymbol{\omega}_2$ as dual to a 2-chain: when integrated over an area c_2 it gives a number:

$$\int_{c_2} \omega_2 = \text{number}. \tag{3.56}$$

We shall now see that a 2-form, defined in this way, is not simply a product of 1-forms, but an *antisymmetric* product. This is basically because c_2 itself is antisymmetric. Consider the area A of a parallelogram defined by the vectors **v** and **w** (see Fig. 3.7); it is given by

$$A = |\mathbf{v} \times \mathbf{w}|.$$

By convention $A > 0$, but $\mathbf{v} \times \mathbf{w} = -\ \mathbf{w} \times \mathbf{v}$, so A is antisymmetric under interchange of **v** and **w**. This notation is, however, unsatisfactory; it is the notation of 3-vectors, whereas we are concerned only with a 2-dimensional space. Taking **v** and **w** (locally) to be in the xy plane we have

$$\pm A = v_x w_y - v_y w_x = \begin{vmatrix} v_x & v_y \\ w_x & w_y \end{vmatrix} = \begin{vmatrix} v^1 & v^2 \\ w^1 & w^2 \end{vmatrix} \tag{3.57}$$

(we take $A > 0$). Now the components v^i may be written as

$$(3.58)\mathrm{v}^{\mathrm{i}} = \mathbf{dx}^{\mathrm{i}}(\mathbf{v})(3.58)$$

where $\mathbf{d}x^i$ is a 1-form and we are using the notation in (3.31). Then we define the 2-*form*[5]

$$\mathbf{d}x^i \wedge \mathbf{d}x^k = \mathbf{d}x^i \otimes \mathbf{d}x^k - \mathbf{d}x^k \otimes \mathbf{d}x^i. \tag{3.59}$$

This antisymmetric product is commonly called a 'wedge product'. It then follows that

[5] Some writers include a symmetry factor of ½ on the right hand side of this definition; see for example Choquet-Bruhat (1968), p. 56. Our definition is the same as that of Misner *et al.* (1973), p. 99, and Sexl & Urbantke (1983), p. 204.

$$\begin{aligned}\mathbf{d}x^i \wedge \mathbf{d}x^k\,(\mathbf{v},\mathbf{w}) &= \mathbf{d}x^i(\mathbf{v})\mathbf{d}x^k\,(\mathbf{w}) - \mathbf{d}x^k(\mathbf{v})\,\mathbf{d}x^i\,(\mathbf{w}) \\ &= v^i w^k - v^k w^i \\ &= \begin{vmatrix} v^i & v^k \\ w^i & w^k \end{vmatrix}.\end{aligned} \tag{3.60}$$

The quantity $\mathbf{d}x^i \wedge \mathbf{d}x^k$ is an *area 2-form*. A general 2-form, for example in R^3, in local (Cartesian) coordinates is (i, $k=1, 2, 3$)

$$\boldsymbol{\omega}_2 = a_{ik}\mathbf{d}x^i \wedge \mathbf{d}x^k \tag{3.61}$$

with

$$\int_{c_2} \omega_2 = \text{number}. \tag{3.62}$$

Since, from (3.59) we have

$$\mathbf{d}x^1 \wedge \mathbf{d}x^2 = -\,\mathbf{d}x^2 \wedge \mathbf{d}x^1, \quad \mathbf{d}x^1 \wedge \quad \mathbf{d}x^1 = 0, \text{ etc.} \tag{3.63}$$

then (3.61) gives

$$\boldsymbol{\omega}_2 = (a_{12} - a_{21})\mathbf{d}x^1 \wedge \mathbf{d}x^2 + (a_{23} - a_{32})\,\mathbf{d}x^2 \wedge \mathbf{d}x^3 + (a_{31} - a_{13})\mathbf{d}x^3 \wedge \mathbf{d}x^1,$$

so without loss of generality a_{ik} may be assumed to be antisymmetric in i and k – the symmetric part makes no contribution to $\boldsymbol{\omega}_2$. A general 2-form in R^3 may be written

$$\boldsymbol{\omega}_2 = P\,\mathbf{d}x^2 \wedge \mathbf{d}x^3 + Q\,\mathbf{d}x^3 \wedge \mathbf{d}x^1 + R\,\mathbf{d}x^1 \wedge \mathbf{d}x^2; \tag{3.64}$$

and we observe that there are *three basic 2-forms in* R^3. Similarly the *volume 3-form* is

$$\mathbf{d}x^1 \wedge \mathbf{d}x^2 \wedge \mathbf{d}x^3.$$

Clearly

$$\mathbf{d}x^i \wedge \mathbf{d}x^j \wedge \mathbf{d}x^k = \varepsilon^{ijk}\mathbf{d}x^1 \wedge \mathbf{d}x^2 \wedge \mathbf{d}x^3;$$

and there is only *one* basic 3-form in R^3. The generic 3-form is

$$\boldsymbol{\omega}_3 = F(x, y, z)\,\mathbf{d}x^1 \wedge \mathbf{d}x^2 \wedge \mathbf{d}x^3, \tag{3.65}$$

with

$$\int_{c_3} \omega_3 = \text{number}. \tag{3.66}$$

We may summarise the p-forms in R^3 in Table 3.1.

It should be clear that *there are no 4-forms or higher forms (5-forms, etc.) in* R^3. An example of a 4-form would be $F(x, y, z)\ \mathbf{d}x^1 \wedge \mathbf{d}x^2 \wedge \mathbf{d}x^3 \wedge \mathbf{d}x^2$, which is zero, by virtue of (3.63).

For completeness we also display p-forms in R^2 in Table 3.2.

Table 3.1 *p*-forms in R^3

p	p-form	Basis (local Cartesian)	Dimensionality
0	$f(x^1, x^2, x^3)$	1	1
1	$a_1\,\mathbf{d}x^1 + a_2\,\mathbf{d}x^2 + a_3\,\mathbf{d}x^3$	$\mathbf{d}x^1, \mathbf{d}x^2, \mathbf{d}x^3$	3
2	$P\,\mathbf{d}x^2 \wedge \mathbf{d}x^3 + Q\,\mathbf{d}x^3 \wedge \mathbf{d}x^1 + R\,\mathbf{d}x^1 \wedge \mathbf{d}x^2$	$\mathbf{d}x^2 \wedge \mathbf{d}x^3$, $\mathbf{d}x^3 \wedge \mathbf{d}x^1$, $\mathbf{d}x^1 \wedge \mathbf{d}x^2$	3
3	$F(x, y, z)\,\mathbf{d}x^1 \wedge \mathbf{d}x^2 \wedge \mathbf{d}x^3$	$\mathbf{d}x^1 \wedge \mathbf{d}x^2 \wedge \mathbf{d}x^3$	1

Table 3.2 *p*-forms in R^2

p	Basic p-forms	Dimensionality
0	1	1
1	$\mathbf{d}x, \mathbf{d}y$	2
2	$\mathbf{d}x \wedge \mathbf{d}y$	1

Table 3.3 *p*-forms in Minkowski space-time

p	Basic p-forms	Dimensionality
0	1	1
1	$\mathbf{d}t, \mathbf{d}x, \mathbf{d}y, \mathbf{d}z$	4
2	$\mathbf{d}t \wedge \mathbf{d}x, \mathbf{d}t \wedge \mathbf{d}y, \mathbf{d}t \wedge \mathbf{d}z, \mathbf{d}x \wedge \mathbf{d}y, \mathbf{d}y \wedge \mathbf{d}z, \mathbf{d}z \wedge \mathbf{d}x$	6
3	$\mathbf{d}t \wedge \mathbf{d}x \wedge \mathbf{d}y, \mathbf{d}t \wedge \mathbf{d}x \wedge \mathbf{d}z, \mathbf{d}t \wedge \mathbf{d}y \wedge \mathbf{d}z, \mathbf{d}x \wedge \mathbf{d}y \wedge \mathbf{d}z$	4
4	$\mathbf{d}t \wedge \mathbf{d}x \wedge \mathbf{d}y \wedge \mathbf{d}z$	1

It is clear that in R^2 there are no 3- or higher forms. Finally we display a table (3.3) of p-forms in Minkowski space-time ($n=4$). Again, it is clear that there are no 5- or higher forms in Minkowski space.

From the tables it is also clear that in an n-dimensional space, the number of independent p-forms is the same as the number of independent $(n-p)$-forms: for example in Minkowski space ($n=4$) the dimensionality of the basis of 1-forms is the same as that of 3-forms (both $=4$). This is in fact a general result, not confined to the cases we have considered, and it suggests the possibilty of a *mapping* between the two spaces of p-forms and $(n-p)$-forms. This mapping is effected by the *Hodge star operator* (or *Hodge* $*$ *operator*):

$$* \ (p\text{-form}) = (n-p)\text{-form}. \tag{3.67}$$

The general formula is the following:

$$* \ (\mathbf{d}x^{i_1} \wedge \mathbf{d}x^{i_2} \wedge \cdots \wedge \mathbf{d}x^{i_p}) = \frac{1}{(n-p)!}\varepsilon^{i_1 \dots i_p}{}_{i_{p+1} \dots i_n}\mathbf{d}x^{i_{p+1}} \wedge \cdots \wedge \mathbf{d}x^{i_n}. \tag{3.68}$$

The ε symbol above needs some explanation. For spaces with a *positive definite* metric no distinction needs to be made between upper and lower indices. In Minkowski space-time indices are lowered with the Minkowski metric tensor (2.18); they are also raised with it, since its inverse is itself. We work with the convention

$$\varepsilon^{0123} = 1 = -\varepsilon_{0123}. \tag{3.69}$$

So for $n=2$ we have

$$\begin{aligned} * \ 1 &= \frac{\varepsilon_{ij}}{2!} \mathbf{d}x^i \wedge \mathbf{d}x^j = \mathbf{d}x \wedge \mathbf{d}y, \\ * \ \mathbf{d}x &= \mathbf{d}x^1 = \varepsilon_{1i} \mathbf{d}x^i = \mathbf{d}y, \\ * \ \mathbf{d}y &= -\mathbf{d}x, \\ * \ \mathbf{d}x \wedge \mathbf{d}y &= \varepsilon_{12}\, 1 = 1. \end{aligned} \tag{3.70}$$

We may perform the $*$ operator a second time to give

$$\begin{aligned} &** 1 = 1, \\ &** \mathbf{d}x = -\mathbf{d}x, \quad ** \mathbf{d}y = -\mathbf{d}y, \\ &** \mathbf{d}x \wedge \mathbf{d}y = \mathbf{d}x \wedge \mathbf{d}y. \end{aligned} \tag{3.71}$$

For $n=3$ we have

$$\begin{aligned} &* \ 1 = \mathbf{d}x \wedge \mathbf{d}y \wedge \mathbf{d}z, \\ &* \ \mathbf{d}x = \mathbf{d}y \wedge \mathbf{d}z \quad \text{and cyclic perms,}^6 \\ &* \ \mathbf{d}x \wedge \mathbf{d}y = \mathbf{d}z \quad \text{and cyclic perms,} \\ &* \ \mathbf{d}x \wedge \mathbf{d}y \wedge \mathbf{d}z = 1; \end{aligned} \tag{3.72}$$

and hence

$$\begin{aligned} &** 1 = 1 \\ &** \mathbf{d}x = \mathbf{d}x, \quad ** \mathbf{d}y = \mathbf{d}y, \quad ** \mathbf{d}z = \mathbf{d}z, \\ &** \mathbf{d}x \wedge \mathbf{d}y \wedge \mathbf{d}z = \mathbf{d}x \wedge \mathbf{d}y \wedge \mathbf{d}z. \end{aligned} \tag{3.73}$$

Finally, for Minkowski space-time ($n=4$) we have

$$\begin{aligned} &* \ 1 = -\mathbf{d}t \wedge \mathbf{d}x \wedge \mathbf{d}y \wedge \mathbf{d}z, \\ &* \ \mathbf{d}t = \mathbf{d}x \wedge \mathbf{d}y \wedge \mathbf{d}z, \quad * \mathbf{d}x = \mathbf{d}t \wedge \mathbf{d}y \wedge \mathbf{d}z, \quad \text{etc.} \\ &* \ \mathbf{d}x \wedge \mathbf{d}y = -\mathbf{d}t \wedge \mathbf{d}z, \quad * \mathbf{d}t \wedge \mathbf{d}x = \mathbf{d}y \wedge \mathbf{d}z, \quad \text{etc.} \\ &* \ \mathbf{d}x \wedge \mathbf{d}y \wedge \mathbf{d}z = \mathbf{d}t, \quad * \mathbf{d}t \wedge \mathbf{d}y \wedge \mathbf{d}z = \mathbf{d}x, \quad \text{etc.} \\ &* \ \mathbf{d}t \wedge \mathbf{d}x \wedge \mathbf{d}y \wedge \mathbf{d}z = 1; \end{aligned} \tag{3.74}$$

and, as a consequence,

[6] i.e. $*\mathbf{d}y = \mathbf{d}z \wedge \mathbf{d}x$, $*\mathbf{d}z = \mathbf{d}x \wedge \mathbf{d}y$.

$$
\begin{aligned}
&** 1 = -1 \\
&** \mathbf{d}x = \mathbf{d}x, \quad \text{etc.} \\
&** \mathbf{d}x \wedge \mathbf{d}y = -\mathbf{d}x \wedge \mathbf{d}y, \quad \text{etc.} \\
&** \mathbf{d}x \wedge \mathbf{d}y \wedge \mathbf{d}z = \mathbf{d}x \wedge \mathbf{d}y \wedge \mathbf{d}z, \quad \text{etc.} \\
&** \mathbf{d}t \wedge \mathbf{d}x \wedge \mathbf{d}y \wedge \mathbf{d}z = -\mathbf{d}t \wedge \mathbf{d}x \wedge \mathbf{d}y \wedge \mathbf{d}z.
\end{aligned}
\tag{3.75}
$$

The general formula for a p-form in spaces with a positive definite metric (like R^2 and R^3) is

$$** \boldsymbol{\omega}_p = (-1)^{p(n-p)} \boldsymbol{\omega}_p; \tag{3.76}$$

in Minkowski space there is an extra minus sign: and since $n=4$ we find

$$** \boldsymbol{\omega}_p = (-1)^{p-1} \boldsymbol{\omega}_p, \tag{3.77}$$

in accordance with (3.75).

3.5.1 Remarks on the algebra of p-forms

Let us revert to considering 2-forms; and first show that if $\boldsymbol{\omega}$ and $\boldsymbol{\sigma}$ are both 1-forms

$$\boldsymbol{\omega} = a_i \boldsymbol{\theta}^i, \quad \boldsymbol{\sigma} = b_k \boldsymbol{\theta}^k,$$

then their wedge product is a 2-form:

$$
\begin{aligned}
\boldsymbol{\omega} \wedge \boldsymbol{\sigma} &= a_i b_k \boldsymbol{\theta}^i \wedge \boldsymbol{\theta}^k \\
&= 1/2(a_i b_k - a_k b_i) \boldsymbol{\theta}^i \wedge \boldsymbol{\theta}^k \\
&= c_{ik} \boldsymbol{\theta}^i \wedge \boldsymbol{\theta}^k
\end{aligned}
\tag{3.78}
$$

with

$$c_{ik} = -c_{ki} = 1/2(c_{ik} - c_{ki}) = c_{[ik]};$$

the coefficients c_{ik} are the components of an antisymmetric $\binom{0}{2}$ tensor. It is then clear that $\boldsymbol{\omega} \wedge \boldsymbol{\sigma}$ is a 2-form, and also that

$$\boldsymbol{\omega} \wedge \boldsymbol{\sigma} = -\boldsymbol{\sigma} \wedge \boldsymbol{\omega}. \tag{3.79}$$

Analogous relations hold for general wedge products. Let $\boldsymbol{\alpha}$ be a p-form and $\boldsymbol{\beta}$ a q-form, so that

$$\boldsymbol{\alpha} = a_{k_1 \ldots k_p} \boldsymbol{\theta}^{k_1} \wedge \cdots \wedge \boldsymbol{\theta}^{k_p} = a_{[k_1 \ldots k_p]} \boldsymbol{\theta}^{k_1} \wedge \cdots \wedge \boldsymbol{\theta}^{k_p} \tag{3.80}$$

and the coefficients $a_{[k_1 \ldots k_p]}$ are the components of a totally antisymmetric $\binom{0}{p}$ tensor. A similar formula holds for $\boldsymbol{\beta}$ and it then follows, by manipulations similar to those which lead to (3.79), that

$$\boldsymbol{\alpha} \wedge \boldsymbol{\beta} = (-1)^{pq} \boldsymbol{\beta} \wedge \boldsymbol{\alpha}. \tag{3.81}$$

3.5.2 A note on orientation

We saw above that the area A of a parallelogram defined by the vectors **v**, **w** is $\pm\begin{vmatrix} v_x & v_y \\ w_x & w_y \end{vmatrix}$ (and A is always taken to be positive). For simplicity take **v** and **w** to be at right angles, and let us assume that in some local Cartesian coordinate system **v** is in the $+x$ direction and **w** in the $+y$ direction; then

$$A = \begin{vmatrix} v_x & v_y \\ w_x & w_y \end{vmatrix} = \begin{vmatrix} v_x & 0 \\ 0 & w_y \end{vmatrix} > 0.$$

If, however, the x and y axes are interchanged (see Fig. 3.8), we find that

$$\begin{vmatrix} v_x & v_y \\ w_x & w_y \end{vmatrix} = \begin{vmatrix} 0 & v_y \\ w_x & 0 \end{vmatrix} < 0.$$

If a space has the property that it is possible to define $A > 0$ consistently over the whole space, the space is called *orientable*. Otherwise it is *non-orientable*. In an orientable space the existence of two distinguishable classes $A > 0$ and $A < 0$ allows a global distinction between right-handed and left-handed coordinate systems over the space, but this will not hold in a non-orientable one. An example of a non-orientable (2-dimensional) space is the *Möbius strip*, illustrated in Fig. 3.9. A coordinate system is set up at P, and it is seen that after transporting it round the band to Q the x and y axes have been interchanged, so a consistent definition of the sign of A over the surface is not possible. The *Möbius strip* is actually also an example of a *fibre bundle*. This is seen as follows: a cylinder is made by drawing a rectangle and joining together the edges marked with arrows, so that the arrows are aligned, as in Fig. 3.10(a). Coordinatising the rectangle, this means that the point $(1, y)$ becomes identified with the point $(-1, y)$. If, however, one of the arrows on the rectangle is inverted, then joining the edges in such a way that the arrows are still aligned results in a Möbius strip, as in Fig. 3.10(b). This corresponds to the identification of the points $(1, y)$ and $(-1, -y)$ in the original rectangle. Now compare the rectangles in (a) and (b). Moving from the point $(1, y)$, keeping y constant but with x decreasing, describes a journey on the cylinder where we eventually return to the original point – so that x has completed a circuit and y has remained unchanged. But on the Möbius strip after x has completed a circuit, from $x = 1$ to $x = -1$, y has changed. This means it is not possible to define a Cartesian coordinate system over the whole space; the space may be coordinatised by (x, y), but this does *not* represent a

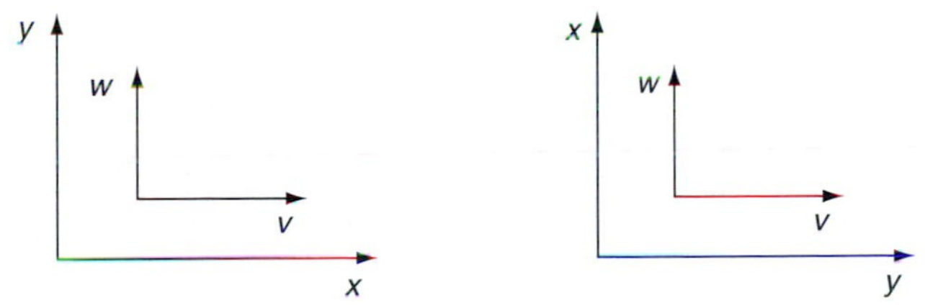

Fig. 3.8 (see text)

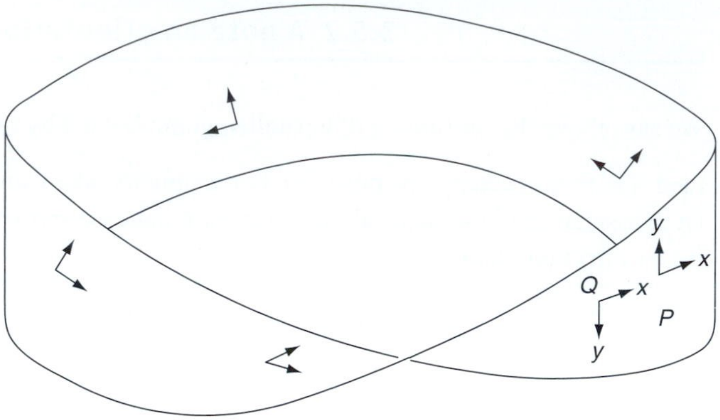

Fig. 3.9 Transport of coordinate axes (x, y) round the Möbius strip, from P to Q results in an interchange $x \leftrightarrow y$ (as may be seen by rotating the axes at Q by $\pi/2$).

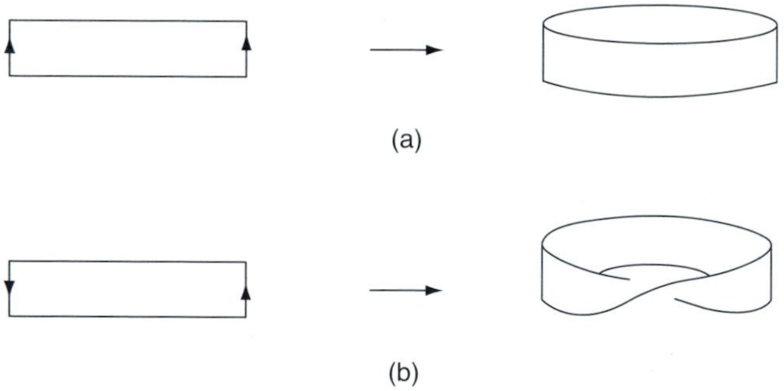

Fig. 3.10 Making (a) a cylinder, (b) a Möbius strip, from a rectangle by gluing edges together.

Cartesian product of x and y. Calling the x axis the *base space* and the y axis the *fibre*, a closed circuit in the base space results in a motion along the fibre. (The reader will recall that it was argued at the beginning of this chapter that space-time is a fibre bundle.)

3.6 Exterior derivative operator: generalised Stokes' theorem

We have seen that (in an n-dimensional space) we may introduce 1-forms, with a corresponding *line integral*, and 2-forms, with a corresponding *area integral* – as well as 3-forms, with a volume integral, and so on. Consider, however, Stokes' theorem

$$\int \mathbf{A} \cdot \mathrm{d}\mathbf{s} = \int \nabla \times \mathbf{A} \cdot \mathrm{d}\Sigma \,. \qquad (3.82)$$

The left hand side is a line integral – the integral of a 1-form, while the right hand side is an area integral – the integral of a 2-form. The very existence of a theorem like Stokes' theorem implies that there *must be a relation* between 1-forms and 2-forms. In a similar way, Gauss's theorem relates an integral over an area to one over a volume, implying a relation between 2-forms and 3-forms. We now investigate this relation and find a beautiful generalisation of Stokes' theorem, applicable to general forms, which yields both Stokes' theorem and Gauss's theorem as special cases.

The key to finding the relation is to define an *exterior derivative operator* **d** which converts a p-form into a $(p+1)$-form. Let $\boldsymbol{\omega}$ be the p-form

$$\boldsymbol{\omega} = a_{1,\dots,p}(x)\,\mathbf{d}x^1 \wedge \cdots \wedge \mathbf{d}x^p. \tag{3.83}$$

then

$$\mathbf{d}\boldsymbol{\omega} = \frac{\partial a_{i_1\cdots i_p}}{\partial x^k}\mathbf{d}x^k \wedge \mathbf{d}x^{i_1} \wedge \cdots \wedge \mathbf{d}x^{i_p}. \tag{3.84}$$

Clearly $\mathbf{d}\boldsymbol{\omega}$ is a $(p+1)$-form. For example let $\boldsymbol{\theta}$ be a 1-form in R^3

$$\boldsymbol{\theta} = a_x\,\mathbf{d}x + a_y\,\mathbf{d}y + a_z\,\mathbf{d}z, \tag{3.85}$$

then

$$\begin{aligned}\mathbf{d}\boldsymbol{\theta} &= \frac{\partial a_x}{\partial y}\mathbf{d}y \wedge \mathbf{d}x + \frac{\partial a_x}{\partial z}\mathbf{d}z \wedge \mathbf{d}x + \frac{\partial a_y}{\partial x}\mathbf{d}x \wedge \mathbf{d}y \\ &\quad + \frac{\partial a_y}{\partial z}\mathbf{d}z \wedge \mathbf{d}y + \frac{\partial a_z}{\partial x}\mathbf{d}x \wedge \mathbf{d}z + \frac{\partial a_z}{\partial y}\mathbf{d}y \wedge \mathbf{d}z \\ &= \left(\frac{\partial a_y}{\partial x} - \frac{\partial a_x}{\partial y}\right)\mathbf{d}x \wedge \mathbf{d}y + \left(\frac{\partial a_z}{\partial y} - \frac{\partial a_y}{\partial z}\right)\mathbf{d}y \wedge \mathbf{d}z \\ &\quad + \left(\frac{\partial a_x}{\partial z} - \frac{\partial a_z}{\partial x}\right)\mathbf{d}z \wedge \mathbf{d}x,\end{aligned} \tag{3.86}$$

which is evidently a 2-form. Note that the coefficients of the three basis forms are respectively the z, x and y components of $\nabla \times \mathbf{a}$. Similarly let $\boldsymbol{\omega}$ be the 2-form (in R^3)

$$\boldsymbol{\omega} = b_x\,\mathbf{d}y \wedge \mathbf{d}z + b_y\,\mathbf{d}z \wedge \mathbf{d}x + b_z\,\mathbf{d}x \wedge \mathbf{d}y, \tag{3.87}$$

then

$$\mathbf{d}\boldsymbol{\omega} = \left(\frac{\partial b_x}{\partial x} + \frac{\partial b_y}{\partial y} + \frac{\partial b_z}{\partial z}\right)\mathbf{d}x \wedge \mathbf{d}y \wedge \mathbf{d}z\,, \tag{3.88}$$

a 3-form. Its coefficient is $\nabla \cdot \mathbf{b}$. The exterior derivative operator **d** operating on a 1-form gives a 2-form, and on a 2-form gives a 3-form; it combines, as is clear from the manipulations above, the operations of differentiation and antisymmetrisation. These features now give an interesting and surprising result, and to illustrate it let us calculate $\mathbf{d}^2\boldsymbol{\theta} = \mathbf{d}(\mathbf{d}\boldsymbol{\theta})$, which should, according to the logic above, be a 3-form.

$$
\begin{aligned}
\mathbf{d}^2\boldsymbol{\theta} &= \mathbf{d}\left[\left(\frac{\partial a_y}{\partial x} - \frac{\partial a_x}{\partial y}\right)\mathbf{d}x \wedge \mathbf{d}y + \left(\frac{\partial a_z}{\partial y} - \frac{\partial a_y}{\partial z}\right)\mathbf{d}y \wedge \mathbf{d}z + \left(\frac{\partial a_x}{\partial z} - \frac{\partial a_z}{\partial x}\right)\mathbf{d}z \wedge \mathbf{d}x\right] \\
&= \left\{\frac{\partial}{\partial z}\left(\frac{\partial a_y}{\partial x} - \frac{\partial a_x}{\partial y}\right) + \frac{\partial}{\partial x}\left(\frac{\partial a_z}{\partial y} - \frac{\partial a_y}{\partial z}\right) + \frac{\partial}{\partial y}\left(\frac{\partial a_x}{\partial z} - \frac{\partial a_z}{\partial x}\right)\right\}\mathbf{d}x \wedge \mathbf{d}y \wedge \mathbf{d}z \\
&= (\nabla \cdot \nabla \times \mathbf{a})\,\mathbf{d}x \wedge \mathbf{d}y \wedge \mathbf{d}z \\
&= 0,
\end{aligned}
\tag{3.89}
$$

since $\nabla \cdot \nabla \times \mathbf{a} = \text{div curl } \mathbf{a} = 0$ for any well-behaved vector field $\mathbf{a}$. That is, $\mathbf{d}^2$ acting on a general 1-form gives a vanishing 3-form.

This, in fact, is a general result; for example let $f(x, y, z)$ be a 0-form (a function) then $\mathbf{d}f$ is a 1-form

$$
\mathbf{d}f = \frac{\partial f}{\partial x}\mathbf{d}x + \frac{\partial f}{\partial y}\mathbf{d}y + \frac{\partial f}{\partial z}\mathbf{d}z,
$$

and

$$
\begin{aligned}
\mathbf{d}^2 f &= \mathbf{d}\left\{\frac{\partial f}{\partial x}\mathbf{d}x + \frac{\partial f}{\partial y}\mathbf{d}y + \frac{\partial f}{\partial z}\mathbf{d}z\right\} \\
&= (\nabla \times \nabla f)_z\,\mathbf{d}x \wedge \mathbf{d}y + (\nabla \times \nabla f)_x\,\mathbf{d}y \wedge \mathbf{d}z + (\nabla \times \nabla f)_y\,\mathbf{d}z \wedge \mathbf{d}x \\
&= 0,
\end{aligned}
$$

since $\nabla \times \nabla f = \text{curl grad } f = 0$ for any well-behaved function f. The general result,

$$
\mathbf{d}^2 = 0, \tag{3.90}
$$

is known as the *Poincaré lemma*; $\mathbf{d}^2$, acting on any differential form, gives zero.[7] The following is a useful identity: if $\boldsymbol{\omega}$ is a p-form and $\boldsymbol{\theta}$ a q-form, then

$$
\mathbf{d}(\boldsymbol{\omega} \wedge \boldsymbol{\theta}) = \mathbf{d}\omega \wedge \boldsymbol{\theta} + (-1)^p\,\boldsymbol{\omega} \wedge \mathbf{d}\theta. \tag{3.91}
$$

Proof: Put

$$
\boldsymbol{\omega} = a_{1\ldots p}\,\mathbf{d}x^{i_1} \wedge \cdots \wedge \mathbf{d}x^{i_p}, \quad \boldsymbol{\theta} = b_{1\ldots q}\,\mathbf{d}x^{k_1} \wedge \cdots \wedge \mathbf{d}x^{k_q}
$$

then

$$
\begin{aligned}
\mathbf{d}(\boldsymbol{\omega} \wedge \boldsymbol{\theta}) &= \mathbf{d}(a_{1\ldots p}\,b_{1\ldots q}\,\mathbf{d}x^{i_1} \wedge \cdots \wedge \mathbf{d}x^{i_p} \wedge \mathbf{d}x^{k_1} \wedge \cdots \wedge \mathbf{d}x^{k_q} \\
&= \{(\mathbf{d}a_{1\ldots p})\,b_{1\ldots q} + a_{1\ldots p}\,(\mathbf{d}b_{1\ldots q})\} \wedge \mathbf{d}x^{i_1} \wedge \cdots \wedge \mathbf{d}x^{i_p} \wedge \mathbf{d}x^{k_1} \wedge \cdots \wedge \mathbf{d}x^{k_q} \\
&= (\mathbf{d}a_{1\ldots p} \wedge \mathbf{d}x^{i_1} \wedge \cdots \wedge \mathbf{d}x^{i_p}) \wedge (b_{1\ldots q}\,\mathbf{d}x^{k_1} \wedge \cdots \wedge \mathbf{d}x^{k_q}) \\
&\qquad + (-1)^p (a_{1\ldots p}\,\mathbf{d}x^{i_1} \wedge \cdots \wedge \mathbf{d}x^{i_p}) \wedge (\mathbf{d}b_{1\ldots q} \wedge \mathbf{d}x^{k_1} \wedge \cdots \wedge \mathbf{d}x^{k_q}) \\
&= \mathbf{d}\omega \wedge \boldsymbol{\theta} + (-1)^p \boldsymbol{\omega} \wedge \mathbf{d}\boldsymbol{\theta}.
\end{aligned}
$$

□

[7] For a general proof, see for example Schutz (1980), p. 140.

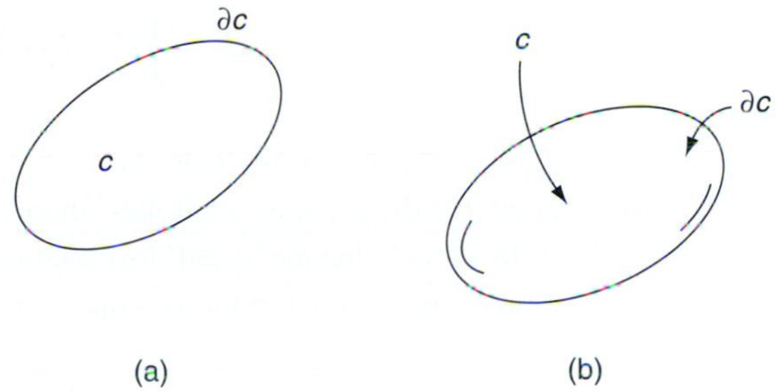

Fig. 3.11 (a) An area c is bounded by the closed line ∂c; (b) a volume c is bounded by the (closed) surface ∂c.

3.6.1 Generalised Stokes' theorem

Let $\boldsymbol{\omega}$ be a p-form, and let c be a $(p+1)$-chain. Define ∂, the *boundary operator* on chains, so that ∂c is the boundary of c ; ∂c is a p-chain. Two simple examples are drawn in Fig. 3.11; in (a) c is an area (2-chain) which is bounded by the closed line ∂c , a 1-chain. In (b) c is a volume (3-chain) bounded by the surface ∂c, a (closed) 2-chain. Note that in both these cases the boundary is itself *closed*; ∂c has *no* boundary, or

$$\partial\,(\partial c) = \partial^2 c = 0.$$

This is a general result for p-chains:

$$\partial^2 = 0\,, \tag{3.92}$$

and may be understood as being a result 'dual' to the Poincaré lemma $\mathbf{d}^2 = 0$, (3.90) above. The boundary operator ∂ is dual to the exterior derivative operator $\mathbf{d}$.

Having defined the boundary operator we are now in a position to state the generalised Stokes' theorem, which is

$$\int_{\partial c} \boldsymbol{\omega} = \int_c \mathbf{d}\boldsymbol{\omega}. \tag{3.93}$$

Stokes' theorem holds in any space, but to illustrate it let us work in R^3; and first consider the case $p=1$. Then $\boldsymbol{\omega}$ is a 1-form, of the type (3.85), and c is an area, with boundary ∂c, as in Fig. 3.11(a). The 2-form $\mathbf{d}\boldsymbol{\omega}$ is given by (3.86), where, as remarked already, the coefficients are the components of $\nabla \times \mathbf{a}$. Then (3.93) gives

$$\int_{\partial c} a.\mathrm{d}l = \int_c (\nabla \times a).n\,\mathrm{d}\Sigma \tag{3.94}$$

where $\mathrm{d}\Sigma$ is an element of surface area, with unit normal $\mathbf{n}$. This is clearly Stokes' theorem. As a second example take the case $p=2$, so $\boldsymbol{\omega}$ is a 2-form, and therefore of the form (3.87); $\mathbf{d}\boldsymbol{\omega}$ is the 3-form given by (3.88). The 3-chain c is a volume V with boundary $\partial c = \partial V$ (Fig. 3.11(b)) and (3.92) then gives

$$\int_{\partial V} b \cdot n \, d\Sigma = \int_V (\nabla \cdot b) \, dV. \tag{3.95}$$

The reader will recognise this as *Gauss's theorem*.

It will be appreciated that the generalised Stokes' theorem is a neat and powerful theorem. The reader will doubtless recall that the 'usual' formulation of Stokes' and Gauss's theorems requires the stipulation of 'directional' notions – the normal **n** is an *outward*, not an inward, normal; and in Stokes' theorem the path round the closed boundary is taken in an *anticlockwise* sense. These notions are however automatically encoded in the present formulation based on the exterior derivative operator, which, as we have seen, antisymmetrises and differentiates at the same time.

3.6.2 Closed and exact forms

It may be helpful to make some more general remarks about differential forms and the spaces in which they are defined. First, if $\boldsymbol{\omega}_p$ is a p-form in some space and $\boldsymbol{\omega}_{p-1}$ is a $(p-1)$-form, we can make the following definitions. If $\mathbf{d}\boldsymbol{\omega}_p = 0$, $\boldsymbol{\omega}_p$ is called *closed*

$$\mathbf{d}\boldsymbol{\omega}_p = 0; \quad \boldsymbol{\omega}_p \text{ closed.} \tag{3.96}$$

If, on the other hand, $\boldsymbol{\omega}_p = \mathbf{d}\boldsymbol{\omega}_{p-1}$ for some $(p-1)$-form $\boldsymbol{\omega}_{p-1}$, $\boldsymbol{\omega}_p$ is called *exact*,

$$\boldsymbol{\omega}_p = \mathbf{d}\boldsymbol{\omega}_{p-1}; \quad \boldsymbol{\omega}_p \text{ exact.} \tag{3.97}$$

Because of the Poincaré lemma (3.90), all exact forms are closed. But are all closed forms exact? For example, if $\mathbf{B} = \nabla \times \mathbf{A}$, then $\nabla \cdot \mathbf{B} = 0$; but if $\nabla \cdot \mathbf{B} = 0$, does it follow that $\mathbf{B} = \nabla \times \mathbf{A}$ for some $\mathbf{A}$? The answer is 'yes' if the space is R^n, which is topologically trivial; but it is not 'yes' in general. Because of the duality between forms and chains the question may be reformulated as a question about chains, and hence as a question about the *space itself*, rather than the forms defined in it. So we make the following definition: let c_p be a p-chain. If $\partial c_p = 0$, c_p is *closed*

$$\partial c_p = 0; \quad c_p \text{ closed.} \tag{3.98}$$

If, however, $c_p = \partial c_{p+1}$ for some c_{p+1}, then c_p is the *boundary* of a $(p+1)$-chain:

$$c_p = \partial c_{p+1}; \quad c_p \text{ boundary.} \tag{3.99}$$

Then because $\partial^2 = 0$ ((3.92) above) it follows that all chains which are boundaries are also closed – boundaries themselves have no boundary, or, as Misner *et al.* (1973) put it, 'the boundary of a boundary is zero'. But the question remains, if c_p is a boundary, is it also closed? – does $\partial c_p = 0$ imply $c_p = \partial c_{p+1}$? This is the dual of the question, 'are all closed forms exact?', but it has the advantage that the situation is much easier to visualise. We first of all may state that in R^n, the answer is 'yes' – all chains which are closed – which have no boundary – are themselves the boundaries of other chains. For example, in R^3, a circle S^1 is the boundary of the area it encloses and a spherical surface S^2 is the boundary of the volume it encloses. In spaces with a non-trivial topology, however, we may get more interesting

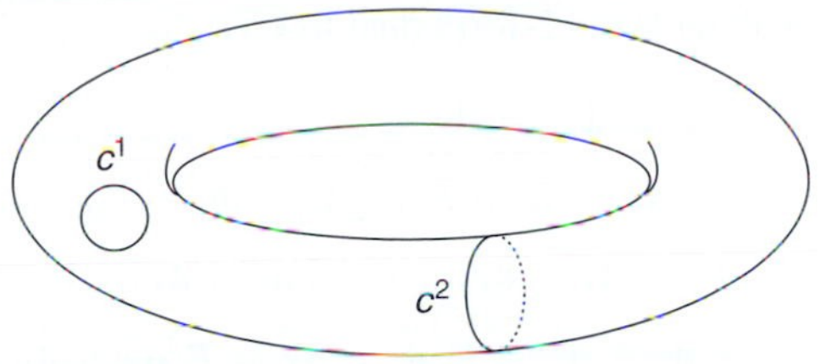

Fig. 3.12 Circles c^1 and c^2 on the surface of a torus.

answers. Consider for example a torus ('doughnut') T^2 shown in Fig. 3.12. This is a 2-dimensional space – it is the *surface only* that is being considered. Now consider the 1-chains c^1 and c^2 on T^2. They are both closed: $\partial c^1 = 0$, $\partial c^2 = 0$. However, c^1 is the boundary of an area (a 2-chain) a^1, $c^1 = \partial a^1$; whereas c^2 is *not* the boundary of any area a^2 defined on T^2, $c^2 \neq \partial a^2$. By virtue of duality this means we can state that there are 2-forms $\boldsymbol{\omega}$ defined on T^2 which are closed but not exact; $\mathbf{d}\boldsymbol{\omega} = 0$, but $\boldsymbol{\omega} \neq \mathbf{d}\boldsymbol{\theta}$, where $\boldsymbol{\theta}$ is a 1-form. These considerations mark the introduction of the subjects of homology, cohomology, de Rham's theorem, etc.; see 'Further reading' for other references to these topics.

3.7 Maxwell's equations and differential forms

We now consider the formulation of Maxwell's equations (in Minkowski space) using differential forms. In this section for simplicity we use units in which $c = 1$, so that $x^0 = t$. The basic 1-, 2-, 3- and 4-forms in Minkowski space are shown above in Table 3.3. We may therefore define a 1-form $\mathbf{A}$:

$$\mathbf{A} = A_0\,\mathbf{d}x^0 + A_1\,\mathbf{d}x^1 + A_2\,\mathbf{d}x^2 + A_3\,\mathbf{d}x^3 = A_\mu\,\mathbf{d}x^\mu\,, \tag{3.100}$$

where (see (2.85) and (2.86)) $A_0 = -\phi$ (scalar potential) and $A_i = A^i$ (vector potential); and a 2-form $\mathbf{F}$:

$$\begin{aligned}\mathbf{F} &= {}^1\!/\!_2 F_{\mu\nu}\,\mathbf{d}x^\mu \wedge \mathbf{d}x^\nu \\ &= -F_{0i}\,\mathbf{d}x^i \wedge \mathbf{d}x^0 + F_{ij}\,\mathbf{d}x^i \wedge \mathbf{d}x^j \\ &= E_i\,\mathbf{d}x^i \wedge \mathbf{d}t + B_x\,\mathbf{d}y \wedge \mathbf{d}z + B_y\,\mathbf{d}z \wedge \mathbf{d}x + B_z\,\mathbf{d}x \wedge \mathbf{d}y \\ &= (E_x\,\mathbf{d}x + E_y\,\mathbf{d}y + E_z\,\mathbf{d}z) \wedge \mathbf{d}t + B_x\,\mathbf{d}y \wedge \mathbf{d}z + B_y\,\mathbf{d}z \wedge \mathbf{d}x + B_z\,\mathbf{d}x \wedge \mathbf{d}y.\end{aligned} \tag{3.101}$$

It is then straightforward to calculate the 3-form $\mathbf{dF}$:

$$\begin{aligned}\mathbf{dF} = &\left(\frac{\partial E_y}{\partial x} - \frac{\partial E_x}{\partial y}\right)\mathbf{d}x \wedge \mathbf{d}y \wedge \mathbf{d}t + \left(\frac{\partial E_z}{\partial y} - \frac{\partial E_y}{\partial z}\right)\mathbf{d}y \wedge \mathbf{d}z \wedge \mathbf{d}t \\ &+ \left(\frac{\partial E_x}{\partial z} - \frac{\partial E_z}{\partial x}\right)\mathbf{d}z \wedge \mathbf{d}x \wedge \mathbf{d}t + \nabla\cdot\mathbf{B}\,\mathbf{d}x \wedge \mathbf{d}y \wedge \mathbf{d}z + \frac{\partial Bx}{\partial t}\mathbf{d}t \wedge \mathbf{d}y \wedge \mathbf{d}z \\ &+ \frac{\partial By}{\partial t}\mathbf{d}t \wedge \mathbf{d}z \wedge \mathbf{d}x + \frac{\partial Bz}{\partial t}\mathbf{d}t \wedge \mathbf{d}x \wedge \mathbf{d}y.\end{aligned} \tag{3.102}$$

We also calculate the 2-form dual to $\mathbf{F}$:

$$\begin{aligned}
{*\mathbf{F}} &= {}^1\!/\!_2 F_{\mu\nu} * \mathbf{d}x^{\mu} \wedge \mathbf{d}x^{\nu} \\
&= F_{10} * \mathbf{d}x \wedge \mathbf{d}t + F_{20} * \mathbf{d}y \wedge \mathbf{d}t + F_{30} * \mathbf{d}z \wedge \mathbf{d}t \\
&\quad + F_{12} * \mathbf{d}x \wedge \mathbf{d}y + F_{31} * \mathbf{d}z \wedge \mathbf{d}x + F_{23} * \mathbf{d}y \wedge \mathbf{d}z \\
&= -E_x\, \mathbf{d}y \wedge \mathbf{d}z - E_y\, \mathbf{d}z \wedge \mathbf{d}x - E_z\, \mathbf{d}x \wedge \mathbf{d}y + (B_z\, \mathbf{d}z + B_y\, \mathbf{d}y + B_x\, \mathbf{d}x) \wedge \mathbf{d}t
\end{aligned} \tag{3.103}$$

as well as its exterior derivative (a 3-form):

$$\begin{aligned}
\mathbf{d} * \mathbf{F} = &- \nabla \cdot \mathbf{E}\, \mathbf{d}x \wedge \mathbf{d}y \wedge \mathbf{d}z - \frac{\partial E_x}{\partial t} \mathbf{d}t \wedge \mathbf{d}y \wedge \mathbf{d}z - \frac{\partial E_y}{\partial t} \mathbf{d}t \wedge \mathbf{d}z \wedge \mathbf{d}x \\
&- \frac{\partial E_z}{\partial t} \mathbf{d}t \wedge \mathbf{d}x \wedge \mathbf{d}y + \left(\frac{\partial B_z}{\partial x} - \frac{\partial B_x}{\partial z}\right) \mathbf{d}x \wedge \mathbf{d}z \wedge \mathbf{d}t \\
&+ \left(\frac{\partial B_y}{\partial x} - \frac{\partial B_x}{\partial y}\right) \mathbf{d}x \wedge \mathbf{d}y \wedge \mathbf{d}t + \left(\frac{\partial B_z}{\partial y} - \frac{\partial B_y}{\partial z}\right) \mathbf{d}y \wedge \mathbf{d}z \wedge \mathbf{d}t.
\end{aligned} \tag{3.104}$$

We may also write down the current 3-form $\mathbf{j} = j_\mu * \mathbf{d}x^\mu$. With $j_\mu = (-\rho, \mathbf{j})$ (see (2.104)) we have

$$\begin{aligned}
\mathbf{j} &= j_0 * \mathbf{d}x^0 + j_i * \mathbf{d}x^i \\
&= -\rho\, \mathbf{d}x \wedge \mathbf{d}y \wedge \mathbf{d}z + j_x\, \mathbf{d}y \wedge \mathbf{d}z \wedge \mathbf{d}t + j_y\, \mathbf{d}z \wedge \mathbf{d}x \wedge \mathbf{d}t + j_z\, \mathbf{d}x \wedge \mathbf{d}y \wedge \mathbf{d}t.
\end{aligned} \tag{3.105}$$

Now consider the equation

$$\mathbf{d} * \mathbf{F} = \mathbf{j}. \tag{3.106}$$

Comparing the coefficients of $\mathbf{d}x \wedge \mathbf{d}y \wedge \mathbf{d}z$ gives $\nabla \cdot \mathbf{E} = \rho$; and inspection of the coefficients of $\mathbf{d}y \wedge \mathbf{d}z \wedge \mathbf{d}t$ yields the x component of $\nabla \times B - \dfrac{\partial \mathbf{E}}{\partial t} = \mathbf{j}$. From (2.103) these are the inhomogeneous Maxwell's equations

$$\partial_\nu F^{\mu\nu} = j^\mu. \tag{3.107}$$

The homogeneous Maxwell's equations are

$$\mathbf{dF} = 0, \tag{3.108}$$

since, from (3.101) the coefficient of $\mathbf{d}x \wedge \mathbf{d}y \wedge \mathbf{d}z$ gives $\nabla \cdot \mathbf{B} = 0$; and the coefficient of $\mathbf{d}x \wedge \mathbf{d}y \wedge \mathbf{d}t$ gives the z component of $\nabla \times E + \dfrac{\partial \mathbf{B}}{\partial t} = 0$. Together, these are the homogeneous equations – see (2.103) and (2.105).

Note that the equation $\mathbf{dF} = 0$ implies, in a topologically trivial space like Minkowski space-time, that $\mathbf{F} = \mathbf{dA}$ – that is, if $\mathbf{F}$ is closed, it is exact – and in component language this is simply

$$F_{\mu\nu} = \partial_\mu A_\nu - \partial_\nu A_\mu, \tag{3.109}$$

as in (2.91); $F_{\mu\nu}$ is the 4-dimensional curl of A_μ.

3.8 Metric tensor

We have so far been concerned with spaces in which there are points and a notion of 'nearness' of points, so that we can take derivatives; and we can also define vectors and differential forms. But one notion which has been missing is that of the *length* of a vector, or (equivalently) the *distance* between two points. These are additional concepts, and a new entity is needed to define them. This entity is called the *metric* or the *metric tensor.* A general space endowed with a metric is called a *Riemannian space*.

Let us begin by considering the simple case of R^2, in which we may define the vectors **v** and **u** . Their Cartesian components may be written

$$\mathbf{v} = (v_x,\, v_y) = (v^1, v^2); \quad \mathbf{u} = (u_x, u_y) = (u^1, u^2);$$

or we may write, in polar coordinates

$$\mathbf{v} = (v_r,\, v_\theta) = (v^1, v^2),$$

and so on. As is well known, from the elementary theory of vectors we define

$$v^2 = \mathbf{v}\cdot\mathbf{v} = |\mathbf{v}|^2 = v_x^2 + v_y^2 \tag{3.110}$$

as the (magnitude)^2 of the vector (vector field) **v** at a particular point. This is a *scalar* quantity. Similarly **u** · **u** and **v** · **u** are scalars. So the situation is that we have a vector **v** (or two vectors **v** and **u**) and we now want to introduce a scalar quantity, its 'length' (or the 'scalar product' of two vectors), *without* talking about 1-forms. To do this we introduce a new geometric object. We write, in some coordinate system,

$$\mathbf{v}\cdot\mathbf{v} = v^2 = g_{ik}v^i v^k = g_{11}(v^1)^2 + (g_{12} + g_{21})v^1 v^2 + g_{22}(v^2)^2 \tag{3.111}$$

and

$$\mathbf{v}\cdot\mathbf{u} = g_{ik}v^i u^k = g_{11}v^1 u^1 + g_{12}v^1 u^2 + g_{21}v^2 u^1 + g_{22}v^2 u^2, \tag{3.112}$$

where g_{ik} are the components of the *metric tensor* in this coordinate system. So in Cartesian coordinates, we have, comparing (3.110) and (3.111), $g_{11}=g_{22}=1$, $g_{12}=g_{21}=0$; or, in matrix form

$$[\text{Cartesian coordinates}] \quad g_{ik} = \begin{pmatrix} 1 & 0 \\ 0 & 1 \end{pmatrix}. \tag{3.113}$$

In addition, the metric tensor is used to express the *distance* ds between the points (x^1, x^2) and $(x^1 + dx^1, x^2 + dx^2)$:

$$ds^2 = g_{ik}\, dx^i\, dx^k, \tag{3.114}$$

where of course the summation convention is being used (as in (3.112)). The notation indicates that under a change of coordinates $x^i \to x'^i$ we have

$$g_{ij} \to g'_{ij} = \frac{\partial x^m}{\partial x'^i} \frac{\partial x^n}{\partial x'^j} g_{mn} \tag{3.115}$$

so that $\mathbf{v} \cdot \mathbf{v}$, $\mathbf{v} \cdot \mathbf{u}$ and $\mathrm{d}s^2$ are indeed scalars. This suggests that g_{ij} transforms as a $\begin{pmatrix} 0 \\ 2 \end{pmatrix}$ tensor, a rank 2 covariant tensor, a (Cartesian) product of two 1-forms (which is not a 2-form!). Then in polar coordinates $x^1 = r$, $x^2 = \theta$, using (3.113) and writing $(x^1, x^2) = (r, \theta)$ we have for example

$$g_{22} = \frac{\partial x^m}{\partial \theta} \frac{\partial x^n}{\partial \theta} g_{mn} = \left(\frac{\partial x}{\partial \theta}\right)^2 g_{11} + \left(\frac{\partial y}{\partial \theta}\right)^2 g_{22} = r^2 \cos^2\theta + r^2 \sin^2\theta = r^2;$$

and similar elementary calculations lead to (dropping the primes)

$$[\text{polar coordinates}] \quad g_{ij} = \begin{pmatrix} 1 & 0 \\ 0 & r^2 \end{pmatrix} \tag{3.116}$$

and hence, from (3.114) $\mathrm{d}s^2 = \mathrm{d}r^2 + r^2\,\mathrm{d}\theta^2$, as expected.

Since the metric tensor 'maps' two vectors into a scalar, it can be represented as the product of two 1-forms, and we may write

$$\mathbf{g}(\mathbf{u}, \mathbf{v}) = \mathbf{u} \cdot \mathbf{v}. \tag{3.117}$$

Writing the basis 1-forms as $\mathbf{d}x^\mu$ and the basis vectors as $\mathbf{e}_\nu$ with $\langle \mathbf{d}x^\mu, \mathbf{e}_\nu \rangle = \delta^\mu{}_\nu$ (cf. (3.20)), so that $\mathbf{u} = u^\mu \mathbf{e}_\mu$, we put

$$\mathbf{g} = g_{\mu\nu}\, \mathbf{d}x^\mu \otimes \mathbf{d}x^\nu. \tag{3.118}$$

Then

$$\mathbf{g}(\mathbf{u}, \mathbf{v}) = g_{\mu\nu} \langle \mathbf{d}x^\mu, u^\lambda \mathbf{e}_\lambda \rangle \langle \mathbf{d}x^\nu, v^\kappa \mathbf{e}_\kappa \rangle = g_{\mu\nu} u^\mu v^\nu = \mathbf{u} \cdot \mathbf{v}. \tag{3.119}$$

Note that

$$\mathbf{g}(\mathbf{e}_\mu, \mathbf{e}_\nu) = g_{\lambda\kappa} \langle \mathbf{d}x^\lambda, \mathbf{e}_\mu \rangle \langle \mathbf{d}x^\kappa, \mathbf{e}_\nu \rangle = g_{\mu\nu}. \tag{3.120}$$

Now consider the formula

$$\mathrm{d}s^2 = g_{\mu\nu}\, \mathrm{d}x^\mu\, \mathrm{d}x^\nu. \tag{3.121}$$

Here $\mathrm{d}x^\mu$ is the old-fashioned infinitesimal separation between the two points x^μ and $x^\mu + \mathrm{d}x^\mu$; it is *not* a 1-form. The mathematical language in Equations (3.118) and (3.121) may be different, but the ultimate content is the same. The line element $\mathrm{d}s^2$ represents the squared length of an infinitesimal displacement '$\mathrm{d}x^\mu$' in an unspecified direction; this is the content of (3.121). Using our upgraded mathematical machinery, we may now express the displacement $\mathrm{d}x^\mu$ by the vector $\mathbf{\Delta} = \mathrm{d}x^\mu\, \mathbf{e}_\mu$. Then, from (3.119), the metric $\mathbf{g}$ contracted against the vectorial displacement $\mathbf{\Delta}$ gives

$$\mathbf{g}(\mathbf{\Delta}, \mathbf{\Delta}) = g_{\mu\nu}\, \mathrm{d}x^\mu\, \mathrm{d}x^\nu = \mathrm{d}s^2. \tag{3.122}$$

Thus the content of the two equations is the same; only the language differs slightly. It should be noted that the symbol '$\otimes$' is often omitted, and then (3.118) is written

$$\mathbf{g} = g_{\mu\nu}\, \mathbf{d}x^{\mu}\, \mathbf{d}x^{\nu}. \tag{3.123}$$

In fact, verging on sloppiness, the above equation is sometimes written[8]

$$\mathbf{g} = \mathbf{ds}^2 = \mathrm{d}s^2 = g_{\mu\nu}\, \mathbf{d}x^{\mu}\, \mathbf{d}x^{\nu}. \tag{3.124}$$

3.8.1 Holonomic and anholonomic (coordinate and non-coordinate) bases

We continue our discussion of holonomic and anholonomic bases in Section 3.2 above by incorporating the metric tensor. It is most straightforward to illustrate the ideas by considering the simple case of E^3, Euclidean 3-dimensional space, in Cartesian and spherical polar coordinates. The general form for the metric will be

$$\mathbf{g} = g_{ik}\, \boldsymbol{\theta}^i \otimes \boldsymbol{\theta}^k. \tag{3.125}$$

In Cartesian coordinates the basis one-forms are $\boldsymbol{\theta}^i = \mathbf{d}x^i$, and with

$$\boldsymbol{\theta}^1 = \mathbf{d}x^1 = \mathbf{d}x, \quad \boldsymbol{\theta}^2 = \mathbf{d}x^2 = \mathbf{d}y, \quad \boldsymbol{\theta}^3 = \mathbf{d}x^3 = \mathbf{d}z, \tag{3.126}$$

and

$$g_{ik} = \begin{pmatrix} 1 & 0 & 0 \\ 0 & 1 & 0 \\ 0 & 0 & 1 \end{pmatrix} = \delta_{ik}, \tag{3.127}$$

we find

$$\mathbf{g} = \mathbf{d}s^2 = \mathbf{d}x \otimes \mathbf{d}x + \mathbf{d}y \otimes \mathbf{d}y + \mathbf{d}z \otimes \mathbf{d}z\,. \tag{3.128}$$

The vectors dual to the basis 1-forms (3.126) are

$$\mathbf{e}_1 = \frac{\partial}{\partial x}, \quad \mathbf{e}_2 = \frac{\partial}{\partial y}, \quad \mathbf{e}_3 = \frac{\partial}{\partial z}, \tag{3.129}$$

since they clearly obey

$$\langle \boldsymbol{\omega}^m, \mathbf{e}_n \rangle = \delta^m{}_n\,. \tag{3.130}$$

Now let us turn to spherical polars, in which

$$\mathrm{d}s^2 = \mathrm{d}r^2 + r^2\, \mathrm{d}\theta^2 + r^2 \sin^2\theta\, \mathrm{d}\phi^2.$$

In the language of forms this becomes

$$\mathbf{d}s^2 = \mathbf{d}r \otimes \mathbf{d}r + r^2\, \mathbf{d}\theta \otimes \mathbf{d}\theta + r^2 \sin^2\theta\, \mathbf{d}\phi \otimes \mathbf{d}\phi\,. \tag{3.131}$$

[8] The above remarks owe much to Misner *et al.* (1973), Box 3.2.

We now have a choice. In a *holonomic* (or *coordinate*) basis we may choose the basis 1-forms as

$$\boldsymbol{\theta}^1 = \mathbf{d}r, \quad \boldsymbol{\theta}^2 = \mathbf{d}\theta, \quad \boldsymbol{\theta}^3 = \mathbf{d}\phi \tag{3.132}$$

and then

$$g_{ik} = \begin{pmatrix} 1 & 0 & 0 \\ 0 & r^2 & 0 \\ 0 & 0 & r^2 \sin^2\theta \end{pmatrix}. \tag{3.133}$$

The vectors dual to (3.132) are

$$\mathbf{e}_1 = \frac{\partial}{\partial r}, \quad \mathbf{e}_2 = \frac{\partial}{\partial \theta}, \quad \mathbf{e}_3 = \frac{\partial}{\partial \phi}. \tag{3.134}$$

In an *anholonomic* (non-coordinate) basis, on the other hand, we may choose the basis 1-forms to be

$$\boldsymbol{\theta}^1 = \mathbf{d}r, \quad \boldsymbol{\theta}^2 = r\,\mathbf{d}\theta, \quad \boldsymbol{\theta}^3 = r \sin\theta\,\mathbf{d}\phi \tag{3.135}$$

with

$$g_{ik} = \begin{pmatrix} 1 & 0 & 0 \\ 0 & 1 & 0 \\ 0 & 0 & 1 \end{pmatrix}. \tag{3.136}$$

Because g_{ik} is the unit matrix bases of this type are commonly called *orthonormal* bases. The vectors dual to (3.135) are

$$\mathbf{e}_1 = \frac{\partial}{\partial r}, \quad \mathbf{e}_2 = \frac{1}{r}\frac{\partial}{\partial \theta}, \quad \mathbf{e}_3 = \frac{1}{r \sin\theta}\frac{\partial}{\partial \phi}. \tag{3.137}$$

A characteristic of a non-coordinate basis is that the basis vectors typically do not commute. For example, from (3.137) we have

$$[\mathbf{e}_1, \mathbf{e}_2] f(r, \theta, \phi) = \frac{\partial}{\partial r}\left(\frac{1}{r}\frac{\partial f}{\partial \theta}\right) - \frac{1}{r}\frac{\partial}{\partial \theta}\left(\frac{\partial f}{\partial r}\right) = -\frac{1}{r^2}\frac{\partial f}{\partial r} = -\frac{1}{r}\,\mathbf{e}_2 f,$$

$$[\mathbf{e}_1, \mathbf{e}_2] = -\frac{1}{r}\,\mathbf{e}_2, \tag{3.138}$$

as in (3.15) above. In a holonomic basis, on the other hand, the basis vectors commute. In both holonomic and anholonomic bases we have

$$\langle \boldsymbol{\omega}^\mu, \mathbf{e}_\nu \rangle = \delta^\mu{}_\nu. \tag{3.139}$$

The elements $g_{\mu\nu}$ are the components of the metric tensor in some coordinate system, and may, for many purposes, be regarded as being defined by Equation (3.121)

$$\mathrm{d}s^2 = g_{\mu\nu}\,\mathrm{d}x^\mu\,\mathrm{d}x^\nu. \tag{3.140}$$

Let us now make some observations about $g_{\mu\nu}$. First, we may obviously write

$$g_{\mu\nu} \equiv {}^1\!/\!_2(g_{\mu\nu} + g_{\nu\mu}) + {}^1\!/\!_2(g_{\mu\nu} - g_{\nu\mu});$$

the first term of which is symmetric, and the second antisymmetric, under the interchange $\mu \leftrightarrow \nu$. The antisymmetric combination makes no contribution to ds^2 in (3.140), so we may, without loss of generality, assume that $g_{\mu\nu}$ itself is symmetric:

$$g_{\mu\nu} = g_{\nu\mu}. \tag{3.141}$$

Second, let us observe that, because of the way in which the metric is introduced – see Equation (3.118) – it has the effect of associating with any vector a one-form. Translated into more old-fashioned language, it has the effect of associating a contravariant vector V^μ with a covariant one V_μ; the metric tensor $g_{\mu\nu}$ may be used to 'lower the index' on V^μ, and to *define*

$$V_\mu = g_{\mu\nu} V^\nu. \tag{3.142}$$

Clearly this may repeated any number of times, so upper indices in any tensor may be lowered:

$$T_{\lambda\ldots\kappa}{}^{\mu\ldots\nu} = g_{\lambda\rho} \ldots g_{\kappa\sigma} T^{\rho\ldots\sigma\mu\ldots\nu}. \tag{3.143}$$

Third, we may define $g^{\mu\nu}$, with upper indices, as the inverse of $g_{\mu\nu}$:

$$g^{\mu\nu} g_{\nu\rho} = \delta^\mu{}_\rho; \tag{3.144}$$

and it is clear that $g^{\mu\nu}$ may be used to *raise* indices:

$$U^\mu = g^{\mu\nu} U_\nu. \tag{3.145}$$

It is a trivial exercise to check that, by virtue of (3.144), if an index of some tensor is raised, and then lowered again, the same tensor is recovered.

3.8.2 Tensor densities: volume elements

We are concerned with the way in which the components of a geometric object transform on a change of coordinate system $x \to x'$. For example the components of a $\begin{pmatrix}1\\1\end{pmatrix}$ tensor, $T^\mu{}_\nu$, transform according to

$$T'^\mu{}_\nu = \frac{\partial x'^\mu}{\partial x^\rho} \frac{\partial x^\sigma}{\partial x'^\nu} T^\rho{}_\sigma. \tag{3.146}$$

We now wish to generalise this. Define

$$D = \det \frac{\partial x^\mu}{\partial x'^\nu}. \tag{3.147}$$

For example, in four dimensions

$$D = \begin{vmatrix} \frac{\partial x^0}{\partial x'^0} & \frac{\partial x^0}{\partial x'^1} & \frac{\partial x^0}{\partial x'^2} & \frac{\partial x^0}{\partial x'^3} \\ \frac{\partial x^1}{\partial x'^0} & , & , & , \\ , & , & , & , \\ \frac{\partial x^3}{\partial x'^0} & , & , & \frac{\partial x^3}{\partial x'^3} \end{vmatrix}. \tag{3.148}$$

We then define a *tensor density of weight w*, denoted $S^{\mu}{}_{\nu}$, (taking again as example the case of a $\binom{1}{1}$ tensor) as an object transforming according to

$$S'^{\mu}{}_{\nu} = D^{w} \frac{\partial x'^{\mu}}{\partial x^{\rho}} \frac{\partial x^{\sigma}}{\partial x'^{\nu}} S^{\rho}{}_{\sigma}. \tag{3.149}$$

Similar definitions hold for tensors of arbitrary rank.[9]

Now consider the totally antisymmetric Levi-Cività symbol (in four dimensions, but the generalisation to other dimensions is obvious)

$$\varepsilon^{\kappa\lambda\mu\nu} = \begin{cases} +1 \text{ if } (\kappa\lambda\mu\nu) \text{ is an even permutation of } (0123), \\ -1 \text{ if } (\kappa\lambda\mu\nu) \text{ is an odd permutation of } (0123), \\ 0 \quad \text{otherwise.} \end{cases} \tag{3.150}$$

This definition is to hold in *all* coordinate systems: it follows that $\varepsilon^{\kappa\lambda\mu\nu}$ is a tensor density of weight $w=1$. To see this consider a *tensor* $T_{\kappa\lambda\mu\nu}$ which is completely antisymmetric:

$$T_{\kappa\lambda\mu\nu} = T_{[\kappa\lambda\mu\nu]}.$$

Then

$$T'_{0123} = \frac{\partial x^{\kappa}}{\partial x'^0} \frac{\partial x^{\lambda}}{\partial x'^1} \frac{\partial x^{\mu}}{\partial x'^2} \frac{\partial x^{\nu}}{\partial x'^3} T_{\kappa\lambda\mu\nu}.$$

Every term in the sum on the right hand side above is proportional to T_{0123}; in fact

$$T_{0'1'2'3'} = D T_{0123}$$

(note again that $T_{\kappa\lambda\mu\nu}$ is a tensor, *not* a tensor density). Now raise the indices on *T*. It is easily seen that $T^{\kappa\lambda\mu\nu}$ is also completely antisymmetric in its indices. However,

$$T'^{0123} = \frac{\partial x'^0}{\partial x^{\kappa}} \frac{\partial x'^1}{\partial x^{\lambda}} \frac{\partial x'^2}{\partial x^{\mu}} \frac{\partial x'^3}{\partial x^{\nu}} T^{\kappa\lambda\mu\nu} = \det \frac{\partial x'}{\partial x} . T^{0123} = \frac{1}{D} T^{0123}.$$

(Note: $T^{\kappa\lambda\mu\nu}$ is also a *tensor*.) Hence $T_{\kappa\lambda\mu\nu}$ and $T^{\kappa\lambda\mu\nu}$ have different values in different coordinate systems; whereas the Levi-Cività symbol must possess the same value in all systems. This will follow if we declare $\varepsilon^{\kappa\lambda\mu\nu}$ to be a tensor density of weight $w=1$; for then

[9] This definition differs from that given by Weinberg (1972) (p. 99); Weinberg would define (3.149) as being the transformation law for a tensor density of weight $-w$. Our definition coincides with, for example, that of Papapetrou (1974), p. 11.

$$\varepsilon'^{\kappa\lambda\mu\nu} = D.\frac{\partial x'^{\kappa}}{\partial x^{\rho}}\frac{\partial x'^{\lambda}}{\partial x^{\sigma}}\frac{\partial x'^{\mu}}{\partial x^{\tau}}\frac{\partial x'^{\nu}}{\partial x^{\zeta}}\varepsilon^{\rho\sigma\tau\zeta} = D.\det\frac{\partial x'}{\partial x}.\varepsilon^{\kappa\lambda\mu\nu} = \varepsilon^{\kappa\lambda\mu\nu}.$$

Hence $\varepsilon'^{0123} = \varepsilon^{0123} = 1$, and so on.

Note that $g = \det g_{\mu\nu}$ is a scalar density of weight $w = 2$; we have

$$g_{\mu'\nu'} = \frac{\partial x^{\rho}}{\partial x^{\mu'}}\frac{\partial x^{\sigma}}{\partial x^{\nu'}}g_{\rho\sigma},$$

hence

$$g' = \left(\det\frac{\partial x}{\partial x'}\right)^2 g = D^2 g,$$

or

$$\sqrt{g'} = D\sqrt{g}; \tag{3.151}$$

or, in spaces where $g < 0$ (for example spaces with the signature of Minkowski space)

$$\sqrt{-g'} = D\sqrt{-g}. \tag{3.152}$$

An important application of tensor densities is the notion of *volume elements*. Let us show that in an n-dimensional space with basis 1-forms $\boldsymbol{\theta}^1, \ldots, \boldsymbol{\theta}^n$, the volume element (an n-form) is

$$\boldsymbol{\eta} = \sqrt{\pm g}\,\boldsymbol{\theta}^1 \wedge \ldots \wedge \boldsymbol{\theta}^n. \tag{3.153}$$

We begin by specialising to 4-dimensional space-time (which, however, is easily generalisable to n dimensions). Let us lower the indices on $\varepsilon^{\kappa\lambda\mu\nu}$.

$$\varepsilon_{\kappa\lambda\mu\nu} = g_{\kappa\rho}\,g_{\lambda\sigma}\,g_{\mu\tau}\,g_{\nu\zeta}\,\varepsilon^{\rho\sigma\tau\zeta} = g\,\varepsilon^{\kappa\lambda\mu\nu}.$$

Then, defining

$$\mathrm{d}\tau = \mathrm{d}^4x = \frac{1}{4!}\varepsilon_{\kappa\lambda\mu\nu}\,\mathrm{d}x^{\kappa}\,\mathrm{d}x^{\lambda}\,\mathrm{d}x^{\mu}\,\mathrm{d}x^{\nu} \tag{3.154}$$

we have, under $x \to x'$,

$$\begin{aligned}\mathrm{d}\tau \to \mathrm{d}\tau' &= \frac{1}{4!}\varepsilon'_{\kappa\lambda\mu\nu}\frac{\partial x'^{\kappa}}{\partial x^{\rho}}\frac{\partial x'^{\lambda}}{\partial x^{\sigma}}\frac{\partial x'^{\mu}}{\partial x^{\tau}}\frac{\partial x'^{\nu}}{\partial x^{\zeta}}\,\mathrm{d}x^{\rho}\,\mathrm{d}x^{\sigma}\,\mathrm{d}x^{\tau}\,\mathrm{d}x^{\zeta}\\ &= \frac{1}{4!}\det\frac{\partial x'}{\partial x}\,\varepsilon_{\rho\sigma\tau\zeta}\,\mathrm{d}x^{\rho}\,\mathrm{d}x^{\sigma}\,\mathrm{d}x^{\tau}\,\mathrm{d}x^{\zeta}\\ &= \frac{1}{D}\,\mathrm{d}\tau.\end{aligned} \tag{3.155}$$

Hence, in a space with negative signature, (3.152) and (3.155) give

$$\sqrt{-g'}\,\mathrm{d}\tau' = \sqrt{-g}\,\mathrm{d}\tau, \tag{3.156}$$

and it follows that $\boldsymbol{\eta}$ in (3.153) above is the invariant volume element.

As a simple illustration consider E^3 in spherical polar coordinates. In the holonomic basis the basis 1-forms are given by (3.132) and the metric by (3.133), giving

$$g = r^4 \sin^2\theta, \quad \sqrt{g} = r^2 \sin\theta$$

and

$$\boldsymbol{\eta} = r^2 \sin\theta \, \mathbf{d}r \wedge \mathbf{d}\theta \wedge \mathbf{d}\phi. \tag{3.157}$$

In the anholonomic system the basis 1-forms are given by (3.135) and the metric by (3.136), giving $g = 1$ and

$$\boldsymbol{\eta} = \mathbf{d}r \wedge (r\,\mathbf{d}\theta) \wedge (r \sin\theta\,\mathbf{d}\phi) = r^2 \sin\theta\,\mathbf{d}r \wedge \mathbf{d}\theta \wedge \mathbf{d}\phi,$$

as in the holonomic system – and as expected.

3.9 Absolute differentiation: connection forms

In the previous sections we have set up the language of vectors, forms and tensors. Our next task is to differentiate them. Suppose a vector has components V^μ in some coordinate system, then the derivative with respect to x^ν is commonly written

$$\frac{\partial V^\mu}{\partial x^\nu} = V^\mu{}_{,\nu}\,. \tag{3.158}$$

This is an object with two indices and one might be inclined to think it is a $\binom{1}{1}$ tensor, but that is wrong. In coordinate systems other than Cartesian ones the coordinate *axes themselves* move around when one moves to a different point in the space, and this must be taken into account for a proper consideration of differentiation. (We must not forget that the field equations of gravitation are, in the non-relativistic case – and therefore also in the relativistic case – second order differential equations; see Equations (1.5) and (1.6). It is therefore crucial to arrive at a correct treatment of differentiation – ultimately differentiation in a curved space.)

We begin by considering the simple case of a vector field in a 2-dimensional (locally) flat space E^2. In Cartesian coordinates we write

$$\mathbf{V} = V^x\,\mathbf{e}_x + V^y\,\mathbf{e}_y\,,$$

and in polar coordinates

$$\mathbf{V} = V^r\,\mathbf{e}_r + V^\phi\,\mathbf{e}_\phi\,.$$

The basis vectors are sketched in Fig. 3.13. In general we write

$$\mathbf{V} = V^i\,\mathbf{e}_i\;(i = 1, 2). \tag{3.159}$$

As noted above, $\mathbf{V}$ is itself invariant under coordinate transformations $x \to x'$. In a similar way we may write a 1-form $\boldsymbol{\omega}$ as

$$\boldsymbol{\omega} = \omega_i\,\boldsymbol{\theta}^i, \tag{3.160}$$

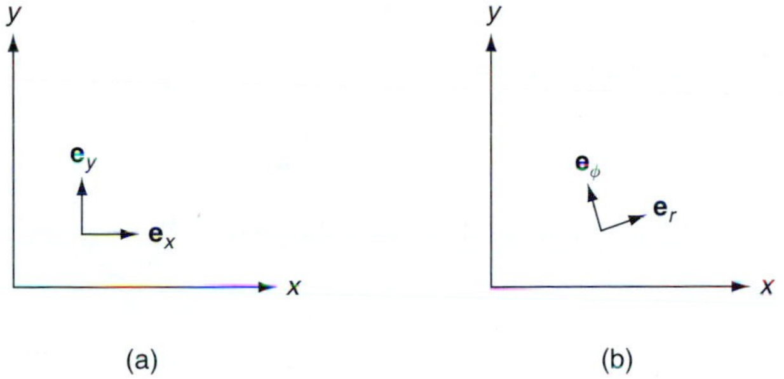

Fig. 3.13 Basis vectors in the plane: (a) $\mathbf{e}_x$ and $\mathbf{e}_y$ (b) $\mathbf{e}_r$ and $\mathbf{e}_\phi$.

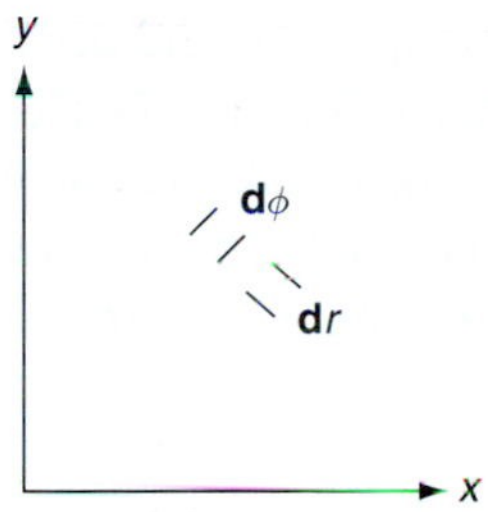

Fig. 3.14 The differential forms $\mathbf{d}r$ and $\mathbf{d}\phi$.

where $\boldsymbol{\theta}^i$ are the basis 1-forms, for example

$$\boldsymbol{\theta}^i = (\mathbf{d}x, \mathbf{d}y) \ \text{ or } \ (\mathbf{d}r, \mathbf{d}\phi)$$

in a holonomic basis, or

$$\boldsymbol{\theta}^i = (\mathbf{d}r, r\,\mathbf{d}\phi)$$

in an anholonomic one. The forms $\mathbf{d}r$ and $\mathbf{d}\phi$ are sketched in Fig. 3.14. Like $\mathbf{V}$, $\boldsymbol{\omega}$ is coordinate-invariant. Now consider the case of a *constant* vector field, $\mathbf{V}=\text{const}$. 'Constant' means that the vector has the same magnitude and direction at all points; and this means that its *Cartesian* coordinates are constants, V^x, $V^y=\text{const}$. The basis vectors $\mathbf{e}_x$, $\mathbf{e}_y$ are constant over the plane, but it is clear that $\mathbf{e}_r$, $\mathbf{e}_\phi$ are not – their orientation will clearly change as they move over the plane, as shown in Fig. 3.15. As a consequence, even for constant $\mathbf{V}$, V^r and V^ϕ are not constant; they are the 'projections' of a constant vector field over changing basis directions. In general we may write

$$\nabla_i\mathbf{V} = \nabla_i(V^m\,\mathbf{e}_m) = (\partial_i V^m)\,\mathbf{e}_m + V^m(\partial_i\mathbf{e}_m) = V^m{}_{,i}\,\mathbf{e}_m + V^m\,\mathbf{e}_{m,\,i}\,. \tag{3.161}$$

We call this the *absolute* derivative of $\mathbf{V}$. In Cartesian coordinates $\mathbf{e}_{m,i}=0$ so a constant $\mathbf{V}$ would imply $\partial_i V^m=0$; $\partial V^x/\partial x=0$ and so on. In the general case this is not true, however, and a constant $\mathbf{V}$ gives

$$0 = \nabla_i\mathbf{V} = V^m{}_{,i}\,\mathbf{e}_m + V^m\,\mathbf{e}_{m,\,i} \tag{3.162}$$

and $\mathbf{e}_{m,i}\neq 0$ (see Fig. 3.15); equally, of course, $V^m{}_{,i}\neq 0$ ($m=r$, ϕ).

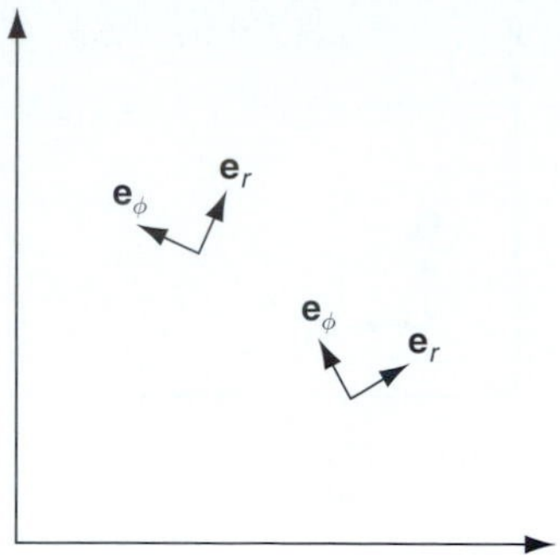

Fig. 3.15 The basis vectors $\mathbf{e}_r$ and $\mathbf{e}_\phi$ change orientation as they move over the plane.

Equation (3.162) represents the derivative of $\mathbf{V}$ in a particular direction x^i, but this suggests that we may make an 'improvement' to it by multiplying by the basis 1-form $\boldsymbol{\theta}^i$ (and summing over i). Recall that the 1-form $\mathbf{d}f=(\partial_i f)\boldsymbol{\theta}^i$ is the derivative of f in an *unspecified* direction, while $\partial_i f$ is its derivative in the direction x^i (in a coordinate basis). In a similar way, therefore, we define

$$\nabla \mathbf{V} = (\nabla_i \mathbf{V})\boldsymbol{\theta}^i = (V^m{}_{,i}\,\mathbf{e}_m + V^m\,\mathbf{e}_{m,i})\,\boldsymbol{\theta}^i. \tag{3.163}$$

This represents the derivative of $\mathbf{V}$ in an *unspecified* direction. We may describe it as a 'vector-valued one-form' (to use the term employed by Misner *et al.* (1973), page 349); alternatively, it is a $\begin{pmatrix}1\\1\end{pmatrix}$ tensor.

We may actually calculate the quantities $\mathbf{e}_{m,i}$, that is $\mathbf{e}_{r,r}$, $\mathbf{e}_{r,\phi}$, $\mathbf{e}_{\phi,r}$ and $\mathbf{e}_{\phi,\phi}$. We have $x=r\cos\phi$, $y=r\sin\phi$, so in the holonomic basis $\mathbf{e}_r=\dfrac{\partial}{\partial r}$, $\mathbf{e}_\phi=\dfrac{\partial}{\partial \phi}$ and

$$\mathbf{e}_r = \frac{\partial x}{\partial r}\,\mathbf{e}_x + \frac{\partial y}{\partial r}\,\mathbf{e}_y = (\cos\phi)\mathbf{e}_x + (\sin\phi)\mathbf{e}_y;$$

similarly

$$\mathbf{e}_\phi = -(r\sin\phi)\mathbf{e}_x + (r\cos\phi)\mathbf{e}_y.$$

(Note, in passing, that $|\mathbf{e}_x|=|\mathbf{e}_y|=1$, hence $|\mathbf{e}_r|=1$, $|\mathbf{e}_\phi|=r$; $\mathbf{e}_\phi$ is not a unit vector.) Then

$$\frac{\partial}{\partial r}\,\mathbf{e}_r = \mathbf{e}_{r,r} = \frac{\partial}{\partial r}\,(\cos\phi\,\mathbf{e}_x + \sin\phi\,\mathbf{e}_y) = 0;$$
$$\frac{\partial}{\partial \phi}\,\mathbf{e}_r = \mathbf{e}_{r,\phi} = \frac{\partial}{\partial \phi}\,(\cos\phi\,\mathbf{e}_x + \sin\phi\,\mathbf{e}_y) = -\sin\phi\,\mathbf{e}_x + \cos\phi\,\mathbf{e}_y = \frac{1}{r}\,\mathbf{e}_\phi.$$

Similarly

$$\mathbf{e}_{\phi,r} = \frac{1}{r}\,\mathbf{e}_\phi, \quad \mathbf{e}_{\phi,\phi} = -r\,\mathbf{e}_r.$$

We see that $\mathbf{e}_{m,i}$ are *vectors*. They may therefore be expressed in terms of the basis vectors $\mathbf{e}_r$, $\mathbf{e}_\phi$, so we put

$$\mathbf{e}_{m,i} = \Gamma^k{}_{m\,i}\,\mathbf{e}_k. \tag{3.164}$$

The coefficients $\Gamma^k{}_{mi}$ are called the *connection coefficients* or *Christoffel symbols*. So in E^2 in polar coordinates, with $x^1 = r$, $x^2 = \phi$, and in the holonomic basis $\mathbf{e}_r = \frac{\partial}{\partial r}, \mathbf{e}_\phi = \frac{\partial}{\partial \phi}$, the above equations give

$$\begin{aligned} \Gamma^1{}_{11} &= \Gamma^2{}_{11} = \Gamma^1{}_{12} = \Gamma^1{}_{21}\,\Gamma^2{}_{22} = 0; \\ \Gamma^2{}_{12} &= \Gamma^2{}_{21} = \frac{1}{r}, \quad \Gamma^1{}_{22} = -r. \end{aligned} \tag{3.165}$$

In the *anholonomic basis* $\mathbf{e}_r = \frac{\partial}{\partial r}, \mathbf{e}_\phi = \frac{1}{r}\frac{\partial}{\partial \phi}$, on the other hand, we have $\mathbf{e}_r = \cos\phi\ \boldsymbol{e}_x +$ $\sin\phi\ \boldsymbol{e}_y$ and $\mathbf{e}_\phi = -\sin\phi\ \mathbf{e}_x + \cos\phi\ \mathbf{e}_y$, from which it is easy to find $\mathbf{e}_{r,r} = 0$, $\mathbf{e}_{r,\phi} = \mathbf{e}_\phi$, $\mathbf{e}_{\phi,r} = 0$, $\mathbf{e}_{\phi,\phi} = \frac{1}{r}\frac{\partial}{\partial\phi}(\mathbf{e}_\phi) = -\mathbf{e}_r$. We adopt the notation that in an anholonomic basis the connection coefficients are written $\gamma^i{}_{kl}$ rather than $\Gamma^i{}_{kl}$ so we have, here

$$\begin{aligned} &\gamma^1{}_{11} = \gamma^2{}_{11} = \gamma^1{}_{12} = \gamma^1{}_{21} = \gamma^2{}_{21} = \gamma^2{}_{22} = 0; \\ &\gamma^2{}_{12} = \frac{1}{r}, \quad \gamma^1{}_{22} = -\frac{1}{r}. \end{aligned} \tag{3.166}$$

Note that the connection coefficients are different in holonomic and anholonomic bases; and in particular – and this turns out to be a general result – that in a holonomic basis they are *symmetric* in the lower indices

$$\Gamma^i{}_{km} = \Gamma^i{}_{mk} \text{ (holonomic basis)}; \tag{3.167}$$

a result which does not in general hold in an anholonomic one, and is evident in (3.166).

Returning to the general case, combining Equations (3.163) and (3.164) gives

$$\begin{aligned} \nabla\mathbf{V} &= (\nabla_\mu \mathbf{V})\,\boldsymbol{\theta}^\mu \\ &= (V^\nu{}_{,\mu}\mathbf{e}_\nu + \Gamma^\nu{}_{\lambda\mu}V^\lambda\mathbf{e}_\nu)\,\boldsymbol{\theta}^\mu \\ &= (V^\nu{}_{,\mu} + \Gamma^\nu{}_{\lambda\mu}V^\lambda)\,\mathbf{e}_\nu\,\boldsymbol{\theta}^\mu \\ &= (V^\nu{}_{,\mu} + \Gamma^\nu{}_{\lambda\mu}V^\lambda)\,\mathbf{e}_\nu \otimes \boldsymbol{\theta}^\mu, \end{aligned} \tag{3.168}$$

where in the third line the suffices have been relabelled, and in the last line the symbol $\otimes$ has been inserted for correctness (though we shall generally omit this symbol). A further definition is commonly made; we put

$$V^\mu{}_{;\nu} = V^\mu{}_{,\nu} + \Gamma^\mu{}_{\kappa\nu}V^\kappa \tag{3.169}$$

so that (3.168) may be written

$$\nabla\mathbf{V} = V^\mu{}_{;\nu}\,\mathbf{e}_\mu\,\boldsymbol{\theta}^\nu. \tag{3.170}$$

This is known as the *absolute* or *covariant* derivative of a vector **V**. In a basis given by $\mathbf{e}_\mu$ and $\boldsymbol{\theta}^\nu$ the components of the covariant derivative are $V^\mu{}_{;\nu}$ given by (3.169). Yet another definition is coined:

$$\boldsymbol{\omega}^\kappa{}_\lambda = \Gamma^\kappa{}_{\lambda\mu}\,\boldsymbol{\theta}^\mu \tag{3.171}$$

is called the *connection one-form*. It is clear that (3.168) may be written as

$$\nabla \mathbf{V} = (\mathbf{d}V^\mu + \boldsymbol{\omega}^\mu{}_\kappa V^\kappa)\,\mathbf{e}_\mu, \tag{3.172}$$

in view of (3.171) and the fact that the 1-form $\mathbf{d}V^\mu$ may be written $V^\mu{}_{,\nu}\,\boldsymbol{\theta}^\nu$.

Note that the above equations imply that

$$\nabla \mathbf{e}_\mu = \boldsymbol{\omega}^\kappa{}_\mu\,\mathbf{e}_\kappa \tag{3.173}$$

since this will yield

$$\nabla_\mu \mathbf{e}_\nu = \langle \nabla \mathbf{e}_\nu, \mathbf{e}_\mu\rangle = \langle \boldsymbol{\omega}^\kappa{}_\nu\,\mathbf{e}_\kappa, \mathbf{e}_\mu\rangle = \Gamma^\kappa{}_{\nu\lambda}\,\mathbf{e}_\kappa\,\langle \boldsymbol{\theta}^\lambda, \mathbf{e}_\mu\rangle = \Gamma^\kappa{}_{\nu\mu}\,\mathbf{e}_\kappa$$

and then

$$\begin{aligned}\nabla_\mu \mathbf{V} &= \nabla_\mu(V^\lambda\,\mathbf{e}_\lambda) = (\partial_\mu V^\lambda)\mathbf{e}_\lambda + V^\lambda(\Gamma^\kappa{}_{\lambda\mu}\,\mathbf{e}_\kappa)\\ &= (V^\nu{}_{,\mu} + \Gamma^\nu{}_{\kappa\mu}\,V^\kappa)\,\mathbf{e}_\nu = V^\nu{}_{;\mu}\,\mathbf{e}_\nu = \langle \nabla \mathbf{V}, \mathbf{e}_\mu\rangle.\end{aligned}$$

This is how differentiation is to be performed on vectors, or, in slightly more old-fashioned language, on contravariant vectors – Equation (3.169) gives the 'covariant derivative of a contravariant vector'. How is differentiation to be performed on 1-forms (or equivalently on covariant vectors)? This is not difficult to deduce, since it follows from the duality of forms and vectors.

To begin, let us note that for 0-forms (scalar functions $f(x^\mu)$) ∇f simply becomes $\mathbf{d}f$:

$$\nabla f = \mathbf{d}f = \partial_\mu f\,\mathbf{d}x^\mu \tag{3.174}$$

where $\mathbf{d}x^\mu$ is the basis of 1-forms in a holonomic system. This gives

$$\nabla_\mu f = \partial_\mu f = f_{,\mu}. \tag{3.175}$$

Turning to 1-forms, what is $\nabla \mathbf{d}x^\mu$? Use Equation (3.173) and the relation (3.27) in the form $\langle \mathbf{d}x^\mu, \mathbf{e}_\nu\rangle = \delta^\mu{}_\nu$ to give (assuming the Leibnitz rule)

$$\nabla\langle \mathbf{d}x^\mu, \mathbf{e}_\nu\rangle = 0 = \langle \nabla\,\mathbf{d}x^\mu, \mathbf{e}_\nu\rangle + \langle \mathbf{d}x^\mu, \nabla \mathbf{e}_\nu\rangle,$$

hence

$$\langle \nabla\,\mathbf{d}x^\mu, \mathbf{e}_\nu\rangle = -\langle \mathbf{d}x^\mu, \boldsymbol{\omega}^\lambda{}_\nu\,\mathbf{e}_\lambda\rangle = -\,\boldsymbol{\omega}^\lambda{}_\nu\langle \mathbf{d}x^\mu, \mathbf{e}_\lambda\rangle = -\boldsymbol{\omega}^\mu{}_\nu. \tag{3.176}$$

Now put

$$\nabla\,\mathbf{d}x^\mu = \tilde{\boldsymbol{\omega}}^\mu_\nu \otimes \mathbf{d}x^\nu, \tag{3.177}$$

a tensor product of 1-forms; that is, a $\begin{pmatrix}0\\2\end{pmatrix}$ tensor, not a 2-form. Then

$$\langle \nabla\,\mathbf{d}x^\mu, \mathbf{e}_\lambda\rangle = \tilde{\boldsymbol{\omega}}^\mu{}_\nu\langle \mathbf{d}x^\nu, \mathbf{e}_\lambda\rangle = \tilde{\boldsymbol{\omega}}^\mu{}_\lambda, \tag{3.178}$$

so from (3.176), (3.178)

$$\tilde{\boldsymbol{\omega}}^\mu{}_\nu = -\boldsymbol{\omega}^\mu{}_\nu. \tag{3.179}$$

Then if $\tilde{\boldsymbol{\omega}}^\mu{}_\nu = \tilde{\Gamma}^\mu{}_{\nu\lambda}\,\mathbf{d}x^\lambda$ it follows that

$$\tilde{\Gamma}^\mu{}_{\nu\lambda} = -\,\Gamma^\mu{}_{\nu\lambda}. \tag{3.180}$$

In conclusion, then we have

$$\nabla \mathbf{d}x^\mu = -\boldsymbol{\omega}^\mu{}_\nu \otimes \mathbf{d}x^\nu = -\Gamma^\mu{}_{\nu\lambda}\,\mathbf{d}x^\lambda \otimes \mathbf{d}x^\nu. \tag{3.181}$$

Operating on 1-forms ∇ is equivalent to $\mathbf{d}$ so we have, for $\boldsymbol{\theta}^\mu$, the *Cartan structure equation*

$$\nabla\boldsymbol{\theta}^\mu = \mathbf{d}\boldsymbol{\theta}^\mu = -\boldsymbol{\omega}^\mu{}_\nu \wedge \boldsymbol{\theta}^\nu. \tag{3.182}$$

Having found the operation of ∇ on the basis 1-forms, it is straightforward to calculate the effect of ∇ on a general 1-form:

$$\begin{aligned}\nabla \mathbf{A} &= \nabla(A_\mu\,\mathbf{d}x^\mu) = \mathbf{d}A_\mu \otimes \mathbf{d}x^\mu - A_\mu\,\boldsymbol{\omega}^\mu{}_\nu \otimes \mathbf{d}x^\nu\\ &= A_{\mu,\nu}\,\mathbf{d}x^\nu \otimes \mathbf{d}x^\mu - \Gamma^\lambda{}_{\nu\mu}\,A_\lambda\,\mathbf{d}x^\mu \otimes \mathbf{d}x^\nu\\ &= (A_{\mu,\nu} - \Gamma^\lambda{}_{\mu\nu}A_\lambda)\,\mathbf{d}x^\mu \otimes dx^\nu.\end{aligned}$$

Or, with

$$\nabla\mathbf{A} = A_{\mu;\nu}\,\mathbf{d}x^\mu \otimes \mathbf{d}x^\nu \tag{3.183}$$

we have

$$A_{\mu;\nu} = A_{\mu,\nu} - \Gamma^\lambda{}_{\mu\nu}\,A_\lambda, \tag{3.184}$$

which is the formula for the absolute or covariant derivative of a covariant vector.

3.9.1 Tensors

What is the general formula for the absolute derivative of a $\begin{pmatrix}p\\q\end{pmatrix}$ tensor? Consider for definiteness a $\begin{pmatrix}2\\0\end{pmatrix}$ tensor

$$\mathbf{t} = t^{\mu\nu}\,\mathbf{e}_\mu \otimes \mathbf{e}_\nu. \tag{3.185}$$

Then

$$\begin{aligned}\nabla_\lambda \mathbf{t} &= \nabla_\lambda(t^{\mu\nu}\,\mathbf{e}_\mu \otimes \mathbf{e}_\nu)\\ &= (\partial_\lambda t^{\mu\nu})\,\mathbf{e}_\mu \otimes \mathbf{e}_\nu + t^{\mu\nu}(\nabla_\lambda\,\mathbf{e}_\mu) \otimes \mathbf{e}_\nu + t^{\mu\nu}\mathbf{e}_\mu \otimes (\nabla_\lambda \mathbf{e}_\nu)\\ &= (\partial_\lambda t^{\mu\nu} + \Gamma^\mu{}_{\rho\lambda}\,t^{\rho\nu} + \Gamma^\nu{}_{\rho\lambda}t^{\mu\rho})\,\mathbf{e}_\mu \otimes \mathbf{e}_\nu\\ &\equiv t^{\mu\nu}{}_{;\lambda}\,\mathbf{e}_\mu \otimes \mathbf{e}_\nu,\end{aligned}$$

so that

$$t^{\mu\nu}{}_{;\lambda} = t^{\mu\nu}{}_{,\lambda} + \Gamma^\mu{}_{\rho\lambda}\,t^{\rho\nu} + \Gamma^\nu{}_{\rho\lambda}\,t^{\mu\rho}. \tag{3.186}$$

It should be clear that the general formula is

$$\begin{aligned}T^{\kappa\lambda\ldots}{}_{\mu\nu\ldots;\rho} = T^{\kappa\lambda\ldots}{}_{\mu\nu\ldots,\rho}\\ &+\Gamma^\kappa{}_{\sigma\rho}\,T^{\sigma\lambda\ldots}{}_{\mu\nu\ldots} + (\text{similar term for each upper index})\\ &-\Gamma^\sigma{}_{\mu\rho}\,T^{\kappa\lambda\ldots}{}_{\sigma\nu\ldots} - (\text{similar term for each lower index}).\end{aligned} \tag{3.187}$$

Now we have that $\mathbf{A} = A_\mu \, \mathbf{d}x^\mu$ is a 1-form, or a $\begin{pmatrix} 0 \\ 1 \end{pmatrix}$ tensor, while $\mathbf{V} = V^\nu \, \mathbf{e}_\nu$ is a vector, or a $\begin{pmatrix} 1 \\ 0 \end{pmatrix}$ tensor, so that $\nabla\mathbf{V}$ is a $\begin{pmatrix} 1 \\ 1 \end{pmatrix}$ tensor, and $\nabla\mathbf{A}$ a $\begin{pmatrix} 0 \\ 2 \end{pmatrix}$ tensor. The *metric tensor* **g**, however, relates vectors and 1-forms, so is in effect a $\begin{pmatrix} 0 \\ 2 \end{pmatrix}$ tensor – see Equation (3.118). Suppose that

$$\mathbf{A} = \mathbf{g}(\mathbf{V}, \ldots), \quad \text{i.e. } A_\mu = g_{\mu\nu} V^\nu \tag{3.188}$$

and that

$$\nabla\mathbf{A} = \mathbf{g}(\nabla\mathbf{V}, \ldots), \quad \text{i.e. } A_{\mu;\rho} = g_{\mu\nu} V^\nu{}_{;\rho}. \tag{3.189}$$

We also have, however (by Leibnitz's rule),

$$\nabla\mathbf{A} = \nabla[\mathbf{g}(\mathbf{V}, \ldots)] = \nabla\mathbf{g}(\mathbf{V}, \ldots) + \mathbf{g}(\nabla\mathbf{V}, \ldots) \tag{3.190}$$

and comparison with (3.189) gives, since **V** is an arbitrary vector,

$$\nabla\mathbf{g} = 0 \Rightarrow \mathbf{d}g_{\mu\nu} = \boldsymbol{\omega}_{\mu\nu} + \boldsymbol{\omega}_{\nu\mu}, \tag{3.191}$$

which is called the *metric compatibility condition*. In index notation the equations above become

$$A_{\mu;\kappa} = (g_{\mu\rho} V^\rho)_{;\kappa} = g_{\mu\rho;\kappa} V^\rho + g_{\mu\rho} V^\rho{}_{;\kappa}$$

and comparison with (3.31) gives $g_{\mu\rho;\kappa} V^\rho = 0$; or, since V^μ is an arbitrary vector,

$$g_{\mu\rho;\kappa} = 0, \tag{3.192}$$

which is the component version of the metric compatibility condition.

This condition actually enables us to find an expression for the connection coefficients $\Gamma^\mu{}_{\nu\rho}$ in terms of the metric tensor and its (ordinary) derivatives. Working in a holonomic basis, in which $\Gamma^\mu{}_{\nu\rho} = \Gamma^\mu{}_{\rho\nu}$, equation (3.192) reads

$$g_{\mu\rho,\kappa} - \Gamma^\sigma{}_{\mu\kappa} g_{\sigma\rho} - \Gamma^\sigma{}_{\rho\kappa} g_{\mu\sigma} = 0. \tag{i}$$

Rewriting this equation after a cyclic permutation of the indices $\mu \to \rho \to \kappa \to \mu$ gives

$$g_{\rho\kappa,\mu} - \Gamma^\sigma{}_{\rho\mu} g_{\sigma\kappa} - \Gamma^\sigma{}_{\kappa\mu} g_{\rho\sigma} = 0, \tag{ii}$$

and after a further cyclic permutation

$$g_{\kappa\mu,\rho} - \Gamma^\sigma{}_{\kappa\rho} g_{\sigma\mu} - \Gamma^\sigma{}_{\mu\rho} g_{\kappa\sigma} = 0. \tag{iii}$$

Now take the combination (i) + (ii) – (iii). The symmetry of the connection coefficients in the lower indices means that four of the six terms involving Γ cancel and we find

$$g_{\mu\rho,\kappa} + g_{\rho\kappa,\mu} - g_{\kappa\mu,\rho} = 2\Gamma^\sigma{}_{\mu\kappa} g_{\rho\sigma}. \tag{3.193}$$

Multiplying this equation by $g^{\rho\nu}$ gives a $\delta^\nu{}_\sigma$ on the right hand side, and hence

$$\Gamma^{\nu}{}_{\mu\kappa} = {}^1\!/\!_2\, g^{\nu\rho}\,(g_{\mu\rho,\kappa} + g_{\kappa\rho,\mu} - g_{\mu\kappa,\rho}). \tag{3.194}$$

This is the promised formula for the connection coefficients. Its use will be illustrated below.

3.10 Parallel transport

The treatment of absolute ('covariant') differentiation above is rather formal and also differs from the historical approach, which was more pictorial and in which the notion of *parallel transport* played an important part. We now give an outline of the older conception of covariant differentiation, which has the virtue of visualisability. This will involve some repetition of results already obtained, but also gives a more rounded understanding. Consider the derivative of a vector $\frac{\partial V_\mu}{\partial x^\nu}$. We denote it, as in (3.158) above, by

$$\frac{\partial V_\mu}{\partial x^\nu} \equiv \partial_\nu V_\mu \equiv V_{\mu,\,\nu}. \tag{3.158}$$

This object has two indices, which might seduce us into believing it is a rank $\binom{0}{2}$ tensor. Is it? Under the coordinate transformation $x^\mu \to x'^\mu$ we have, from (3.38),

$$V_\mu \to V'_\mu(x) = \frac{\partial x^\rho}{\partial x'^\mu}\,(x)\, V_\rho(x),$$

hence

$$\frac{\partial V'_\mu}{\partial x'^\nu} = \frac{\partial}{\partial x'^\nu}\left\{\frac{\partial x^\rho}{\partial x'^\mu}\, V_\rho\right\} = \frac{\partial x^\rho}{\partial x'^\mu}\,\frac{\partial x^\sigma}{\partial x'^\nu}\,\frac{\partial V_\rho}{\partial x^\sigma} + \frac{\partial^2 x^\rho}{\partial x'^\nu \partial x'^\mu}\, V_\rho;$$

or, using the notation (3.158),

$$V'_{\mu,\,\nu} = \frac{\partial x^\rho}{\partial x'^\mu}\,\frac{\partial x^\sigma}{\partial x'^\nu}\, V_{\rho,\sigma} + \frac{\partial^2 x^\rho}{\partial x'^\nu \partial x'^\mu}\, V_\rho. \tag{3.195}$$

This is *not* the transformation law for a tensor. The first term on the right hand side is tensorial, but the second term spoils the tensorial character: the differential of a vector is *not* a tensor. This is clearly seen to be a consequence of the fact that the transformation 'matrix' $\frac{\partial x^\rho}{\partial x'^\mu}$ is, for non-linear transformations, a function of the coordinate x, so the second derivative is non-vanishing. The point is that a tensor is a geometric object whose existence is independent of coordinate systems. If it is non-zero in one coordinate system it is non-zero in all. For *linear* transformations (e.g. $x'^\mu = x^\mu + \alpha^\mu{}_\nu x^\nu$, $\alpha = \text{const}$) the second term in (3.195) vanishes so if $V_{\mu,\nu}$ is non-zero in one coordinate system it is indeed non-zero in all. But for *non-linear* transformations this is not true ; $V_{\mu,\nu}$ could be zero in one coordinate system but not in another one. And recall that non-linear transformations (for example accelerations) play a central role in General Relativity because they mimic gravitational fields.

This non-tensorial nature of the derivative $V_{\mu,\nu}$ is a fundamental problem. How do we solve it? How may we define a genuine tensor which is related to the derivative of a vector?

We may identify the source of the problem by looking at the very definition of differentiation: we have a vector field $V(x)$, which has the value $V(x)$ at the point x, and the value $V(x+\mathrm{d}x) = V + \mathrm{d}V$ at the point $x + \mathrm{d}x$, as shown in Fig. 3.16. Then $\mathrm{d}V$ is the difference between two vectors at *different points*, and this is *not* a vector – it was noted above that the difference between two vectors at the *same* point is a vector, since the transformation coefficients $\frac{\partial x'^\mu}{\partial x^\lambda}(x)$ are the same, but this condition clearly ceases to hold now. Schematically we may write

$$\frac{\partial V}{\partial x} = \lim \frac{\mathrm{d}V}{\mathrm{d}x}, \tag{3.196}$$

the limit being taken as $\mathrm{d}x \to 0$. Then $\mathrm{d}V$ is not a vector, but $\mathrm{d}x$ *is* a vector, so the ratio of them is *not* a tensor, exactly as in (3.195) above. We may give a simple illustration of this. Consider a *constant* vector in the plane (a flat space). This is illustrated in Fig. 3.17 by a vector **V** at one point and a vector **W** at another point, and **V** and **W** are parallel, **V** || **W**. In Cartesian coordinates this means that $V_x = W_x$, $V_y = W_y$, so $V_i = W_i$, or

$$(\text{Cartesian coordinates}) \quad (V_i - W_i) = 0. \tag{3.197}$$

In polar coordinates, on the other hand, as is clear from Fig. 3.17, $V_r \neq W_r$, $V_\theta \neq W_\theta$, hence $V_i \neq W_i$, or

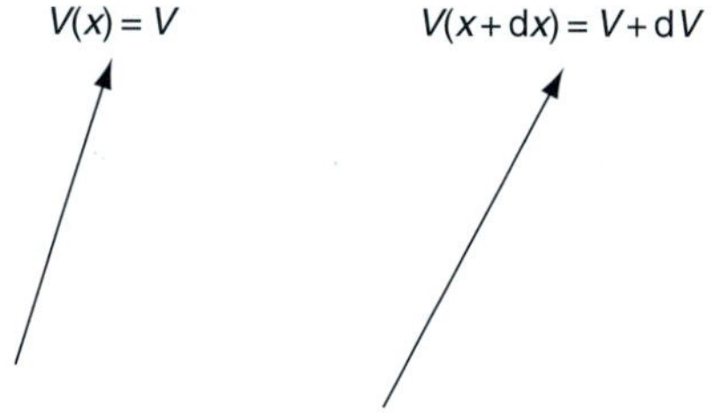

Fig. 3.16 A vector field $V(x)$ at the points x and $x + \mathrm{d}x$.

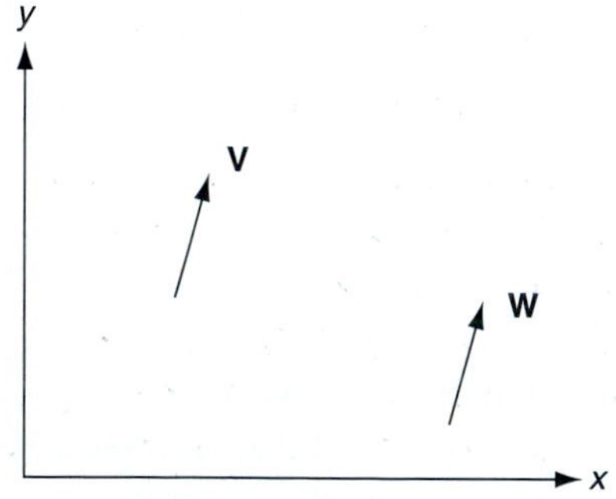

Fig. 3.17 The vectors **V** and **W** are parallel; their Cartesian components are equal but their polar coordinates are clearly different.

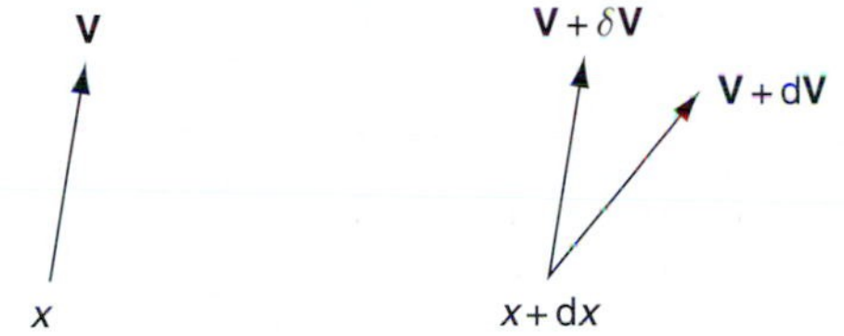

Fig. 3.18 **Two vector fields are defined at the point $x + dx$; $V + \delta V$ is parallel to V.**

$$\text{(polar coordinatess)} \quad (V_i - W_i) \neq 0. \tag{3.198}$$

We see that the difference between the vectors **V** and **W** is zero in one system but non-zero in another. It therefore cannot be a vector, which by definition has the *homogeneous* transformation law (3.38): if all components of **V** are zero in one frame, they are zero in all frames. We see the force of the observation in (3.196); dV, being the difference between two vectors at two different points, is not a vector.

To construct a derivative which is a genuine tensor we must take the difference between two vectors at the *same* point. So at the point $x + dx$ we consider two vectors; one is the vector $V + dV$, as in Fig. 3.16. The other is a vector constructed at $x + dx$ which is *parallel* to V at x. It is denoted $V + \delta V$; this is shown in Fig. 3.18. The vectors V and $V + \delta V$ are like the vectors V and W in Fig. 3.17, and $V + \delta V$ is said to be obtained by *parallel transport* from x to $x + dx$. We have then

$$\begin{aligned} &V^\mu(x) + dV^\mu = V^\mu(x + dx) \quad \text{vector field at } x + dx, \\ &V^\mu(x) + \delta V^\mu \quad \text{vector at } x + dx \textit{ parallel transported} \text{ from } x. \end{aligned} \tag{3.199}$$

$V + \delta V$ is *by definition* a vector at the point $x + dx$ which has the same *Cartesian components* as V does at x. So in a Cartesian coordinate system $\delta V^\mu = 0$, but it is not zero in general. The quantity δV is not a vector, but $V^\mu(x) + \delta V^\mu$ *is* a vector at $x + dx$. We now have two vectors at $x + dx$, as in (3.199), and the difference between them is a genuine vector, which we denote DV and define thus:

$$DV^\mu = (V^\mu + dV^\mu) - (V^\mu \delta V^\mu) = dV^\mu - \delta V^\mu. \tag{3.200}$$

As a conceptual scheme this is fine, but how do we write δV^μ? It is highly reasonable to suppose that it is proportional to dx and also to V, so we write

$$\delta V^\mu = -\Gamma^\mu{}_{\lambda\nu} V^\lambda dx^\nu, \tag{3.201}$$

and the coefficient $\Gamma^\mu{}_{\lambda\nu}$ is the *Christoffel symbol*, or *connection coefficient*, already introduced. (The term 'connection coefficient' comes about because this quantity *connects* the value of a vector field at one point with the value at another. It amounts to an additional structure possessed by the space.) Equations (3.200) and (3.201) give

$$DV^\mu = dV^\mu + \Gamma^\mu{}_{\lambda\nu} V^\lambda dx^\nu$$

so the 'true' derivative – the *tensorial* derivative – of V is

$$\frac{\mathrm{D}V^{\mu}}{\mathrm{d}x^{\nu}} = \frac{\partial V^{\mu}}{\partial x^{\nu}} + \Gamma^{\mu}{}_{\lambda\nu}\, V^{\lambda}. \tag{3.202a}$$

In an extension of the notation in Equation (3.158) this equation is commonly written

$$V^{\mu}{}_{;\nu} = V^{\mu}{}_{,\nu} + \Gamma^{\mu}{}_{\lambda\nu}\, V^{\lambda}, \tag{3.202b}$$

as in (3.169) above. This is the *covariant derivative* of a contravariant vector. To find the covariant derivative of a covariant vector W_{μ} note simply that under parallel transport the scalar product $V^{\mu} W_{\mu}$ must be unchanged,

$$\delta\,(V^{\mu} W_{\mu}) = 0,$$

hence $(\delta V^{\mu})W_{\mu} + V^{\mu}(\delta W_{\mu}) = 0$, and therefore

$$V^{\mu}\,\delta W_{\mu} = -\; W_{\mu}\,\delta V^{\mu} = \Gamma^{\mu}{}_{\lambda\nu}\, W_{\mu}\, V^{\lambda}\, \mathrm{d}x^{\nu} = \Gamma^{\lambda}{}_{\mu\nu}\, W_{\lambda}\, \mathrm{d}x^{\nu}\, V^{\mu},$$

where Equation (3.201) has been used, and in the last step a relabelling $\lambda \leftrightarrow \mu$. The vector field V^{μ} is however arbitrary, so this implies that

$$\delta W_{\mu} = \Gamma^{\lambda}{}_{\mu\nu}\, W_{\lambda}\, \mathrm{d}x^{\nu}.$$

This equation parallels Equation (3.201) for a contravariant vector. The covariant derivative of W_{μ} follows from an analogous argument to that involving V^{μ}, so that finally

$$\frac{\mathrm{D}W_{\mu}}{\mathrm{d}x^{\nu}} = \frac{\partial W_{\mu}}{\partial x^{\nu}} - \Gamma^{\lambda}{}_{\mu\nu}\, W_{\lambda}, \tag{3.203a}$$

or, in alternative notation,

$$W_{\mu;\nu} = W_{\mu,\nu} - \Gamma^{\lambda}{}_{\mu\nu}\, W_{\lambda}, \tag{3.203b}$$

as in (3.184) above.

As we have already noted, the connection coefficients $\Gamma^{\lambda}{}_{\mu\nu}$ do not transform as the components of a tensor under general coordinate transformations. Their transformation law is shown below – Equation (3.207) – but from that law it will be seen that the quantity

$$T^{\rho}{}_{\mu\nu} = \Gamma^{\rho}{}_{\mu\nu} - \Gamma^{\rho}{}_{\nu\mu}, \tag{3.204}$$

does indeed transform as a tensor, since the final term in (3.207), which is symmetric in μ and ν, will drop away in the transformation law of $T^{\rho}{}_{\mu\nu}$, leaving only the homogeneous term on the right hand side, and thus demonstrating the tensorial nature. The quantity $T^{\rho}{}_{\mu\nu}$ is called the *torsion tensor.* In General Relativity it is generally assumed that space-time is *torsion-free*, so the connection coefficients are symmetric in their lower indices

$$\Gamma^{\rho}{}_{\mu\nu} = \Gamma^{\rho}{}_{\nu\mu}, \tag{3.205}$$

in a coordinate basis.

3.11 Some relations involving connection coefficients

In the expressions for covariant derivatives, (3.202) and (3.203), we have seen that the left hand sides, $V^{\mu}{}_{;\nu}$ and $W_{\mu;\nu}$, are tensors, while the ordinary derivatives, $V^{\mu}{}_{,\nu}$ and $W_{\mu,\nu}$ are not. This is equivalent to the observation that $\Gamma^{\lambda}{}_{\mu\nu}$ is not a tensor, as we have seen from its definition. We may ask: how does $\Gamma^{\lambda}{}_{\mu\nu}$ transform under a coordinate transformation?

Consider Equation (3.203b). Both $W_{\mu;\nu}$ and W_{μ} are tensors; under the transformation $x^{\mu} \to x'^{\mu}$ we have

$$W'_{\mu;\nu} = \frac{\partial x^{\rho}}{\partial x'^{\mu}} \frac{\partial x^{\sigma}}{\partial x'^{\nu}} W_{\rho;\sigma}; \quad W'_{\lambda} = \frac{\partial x^{\tau}}{\partial x'^{\lambda}} W_{\tau}.$$

In the x' coordinate system we will have

$$W'_{\mu;\nu} = W'_{\mu,\nu} - \Gamma'^{\lambda}_{\mu\nu} W'_{\lambda},$$

where

$$W'_{\mu,\nu} = \frac{\partial}{\partial x'^{\nu}} \left\{ \frac{\partial x^{\tau}}{\partial x'^{\mu}} W_{\tau} \right\} = \frac{\partial^2 x^{\tau}}{\partial x'^{\nu} \partial x'^{\mu}} W_{\tau} + \frac{\partial x^{\tau}}{\partial x'^{\mu}} W_{\tau,\rho} \frac{\partial x^{\rho}}{\partial x'^{\nu}}.$$

Hence

$$\frac{\partial x^{\rho}}{\partial x'^{\mu}} \frac{\partial x^{\sigma}}{\partial x'^{\nu}} W_{\rho;\sigma} = \frac{\partial^2 x^{\tau}}{\partial x'^{\nu} \partial x'^{\mu}} W_{\tau} + \frac{\partial x^{\tau}}{\partial x'^{\mu}} \frac{\partial x^{\rho}}{\partial x'^{\nu}} W_{\tau,\rho} - \Gamma'^{\lambda}{}_{\mu\nu} \frac{\partial x^{\tau}}{\partial x'^{\lambda}} W_{\tau}. \tag{3.206}$$

Substituting for $W_{\tau,\rho}$ from (3.203b) gives

$$\frac{\partial x^{\rho}}{\partial x'^{\mu}} \frac{\partial x^{\sigma}}{\partial x'^{\nu}} W_{\rho;\sigma} = \frac{\partial^2 x^{\iota}}{\partial x'^{\nu} \partial x'^{\mu}} W_{\tau} + \frac{\partial x^{\tau}}{\partial x'^{\mu}} \frac{\partial x^{\rho}}{\partial x'^{\nu}} W_{\tau;\rho} + \frac{\partial x^{\tau}}{\partial x'^{\mu}} \frac{\partial x^{\rho}}{\partial x'^{\nu}} \Gamma^{\alpha}{}_{\tau\rho} W_{\alpha}$$
$$- \Gamma'^{\lambda}_{\mu\nu} \frac{\partial x^{\tau}}{\partial x'^{\lambda}} W_{\tau}.$$

The left hand side of this equation is the same as the second term on the right, so with a bit of relabelling we have

$$\Gamma'^{\lambda}{}_{\mu\nu} \frac{\partial x^{\tau}}{\partial x'^{\lambda}} W_{\tau} = \left(\frac{\partial x^{\sigma}}{\partial x'^{\mu}} \frac{\partial x^{\rho}}{\partial x'^{\nu}} \Gamma^{\tau}{}_{\sigma\rho} + \frac{\partial^2 x^{\tau}}{\partial x'^{\mu} \partial x'^{\nu}} \right) W_{\tau}.$$

Since W_{τ} is an arbitrary vector this is an identity in its coefficients, and on multiplying by $\frac{\partial x'^{\kappa}}{\partial x^{\tau}}$ we find

$$\Gamma'^{\kappa}{}_{\mu\nu} = \frac{\partial x'^{\kappa}}{\partial x^{\tau}} \frac{\partial x^{\sigma}}{\partial x'^{\mu}} \frac{\partial x^{\rho}}{\partial x'^{\nu}} \Gamma^{\tau}{}_{\sigma\rho} + \frac{\partial x'^{\kappa}}{\partial x^{\tau}} \frac{\partial^2 x^{\tau}}{\partial x'^{\nu} \partial x'^{\mu}}. \tag{3.207}$$

This is the transformation law for the connection coefficients $\Gamma^{\lambda}{}_{\mu\nu}$. The first term on the right of (3.207) is tensorial (homogeneous), while the second term spoils the tensorial nature – entirely as expected.

It is also true, and is not difficult to show, that an equivalent transformation law is (Problem 3.2)

$$\Gamma'^{\lambda}{}_{\mu\nu} = \frac{\partial x'^{\lambda}}{\partial x^{\tau}}\frac{\partial x^{\sigma}}{\partial x'^{\mu}}\frac{\partial x^{\rho}}{\partial x'^{\nu}}\Gamma^{\tau}{}_{\sigma\rho} - \frac{\partial x^{\sigma}}{\partial x'^{\mu}}\frac{\partial x^{\rho}}{\partial x'^{\nu}}\frac{\partial^2 x'^{\lambda}}{\partial x^{\rho}\partial x^{\sigma}}. \tag{3.208}$$

For future purposes it is useful to find an expression for the contracted quantity $\Gamma^{\mu}{}_{\kappa\mu}$. The definition

$$\Gamma^{\nu}{}_{\mu\kappa} = {}^1\!/\!_2\, g^{\nu\rho}(g_{\mu\rho,\kappa} + g_{\kappa\rho,\mu} - g_{\mu\kappa,\rho}) \tag{3.194}$$

involves $g^{\nu\rho}$, the 'inverse' of $g_{\nu\rho}$. Put

$$\Delta^{\nu\rho} = \text{minor of } g_{\nu\rho},$$

then

$$g^{\nu\rho} = \frac{1}{g}\Delta^{\nu\rho}$$

where g is the determinant of $g_{\nu\rho}$. On differentiation,

$$\mathrm{d}g = (\mathrm{d}g_{\nu\rho})\,\Delta^{\nu\rho} = (\mathrm{d}g_{\nu\rho})\,g\,g^{\nu\rho}. \tag{3.209}$$

In words, this equation means: dg is the differential of the determinant g made up from the components of the metric tensor $g_{\nu\rho}$. It can be obtained by taking the differential of each component of $g_{\nu\rho}$ and multiplying it by its coefficient in the determinant, i.e. by the minor $\Delta^{\nu\rho}$. This equation implies that

$$\partial_\lambda g = g\,g^{\mu\nu}(\partial_\lambda g_{\mu\nu}). \tag{3.210}$$

Now

$$\partial_\lambda \ln g = \frac{1}{g}\partial_\lambda g$$

and hence

$$\partial_\lambda \ln\sqrt{g} = \frac{1}{\sqrt{g}}(\partial_\lambda\sqrt{g}) = \frac{1}{2g}\,\partial_\lambda g;$$

and equally, when the metric is of negative signature,

$$\partial_\lambda \ln\sqrt{-g} = \frac{1}{2g}\partial_\lambda g. \tag{3.211}$$

From (3.194) we have

$$\Gamma^{\mu}{}_{\lambda\mu} = {}^1\!/\!_2\; g^{\mu\nu}\partial_\lambda\, g_{\mu\nu},$$

and (3.210) and (3.211) then give

$$\Gamma^{\mu}{}_{\lambda\mu} = \frac{1}{2g}\frac{\partial g}{\partial x^{\lambda}} = \frac{\partial(\ln\sqrt{-g})}{\partial x^{\lambda}}. \tag{3.212}$$

3.11.1 Derivatives of scalar and tensor densities

Tensor densities were defined in (3.149) above. Here we shall only be concerned with densities of weight $w = 1$, so a scalar density Q transforms under $x^\mu \to x'^\mu$ as

$$Q \to Q' = QD, \quad D = \det \frac{\partial x}{\partial x'}, \tag{3.213}$$

and a tensor density, for example $T_{\mu\nu}$, as

$$T_{\mu\nu} \to T'_{\mu\nu} = D \frac{\partial x^\kappa}{\partial x'^\mu} \frac{\partial x^\lambda}{\partial x'^\nu} T_{\kappa\lambda}. \tag{3.214}$$

The derivatives of these quantities do not transform in the same way, so as usual we want to construct 'covariant' derivatives. Differentiating (3.213) with respect to x'^μ gives

$$\frac{\partial Q'}{\partial x'^\mu} = \frac{\partial Q}{\partial x^\nu} \frac{\partial x^\nu}{\partial x'^\mu} D + Q \frac{\partial D}{\partial x'^\mu}. \tag{3.215}$$

It is the second term – and also the presence of D in the first term – which prevents $Q_{,\mu}$ from being a covariant vector. To proceed, we find an expression for $\partial D/\partial x'^\mu$ in the last term. Recall that for any matrix $a^\kappa{}_\lambda$ with determinant a we have

$$a = a^\kappa{}_\lambda \Delta^\lambda{}_\kappa \tag{3.216}$$

where $\Delta^\lambda{}_\kappa$ is the minor of $a^\kappa{}_\lambda$. Then (in analogy with (3.209) above)

$$\frac{\partial a}{\partial x'^\mu} = \frac{\partial}{\partial x'^\mu} (a^\kappa{}_\lambda) \cdot \Delta^\lambda{}_\kappa.$$

Now put

$$a^\kappa_\lambda = \frac{\partial x^\kappa}{\partial x'^\lambda} \Rightarrow a = D, \tag{3.217}$$

and then

$$\frac{\partial D}{\partial x'^\mu} = \frac{\partial^2 x^\kappa}{\partial x'^\lambda \partial x'^\mu} \Delta^\lambda{}_\kappa. \tag{3.218}$$

We now need to find an expression for $\Delta^\lambda{}_\kappa$: note that, as well as (3.216) we also have

$$a^\kappa{}_\lambda \Delta^\lambda{}_\mu = \delta^\kappa{}_\mu a.$$

Then, from (3.217)

$$\frac{\partial x^\kappa}{\partial x'^\lambda} \Delta^\lambda{}_\mu = \delta^\kappa{}_\mu D.$$

Multiplying by $\partial x'^\nu/\partial x^\kappa$ gives

$$\Delta^\nu{}_\mu = \frac{\partial x'^\nu}{\partial x^\mu} D,$$

which, when substituted into (3.218) gives

$$\frac{\partial D}{\partial x'^{\mu}} = \frac{\partial^2 x^{\kappa}}{\partial x'^{\lambda}\,\partial x'^{\mu}}\frac{\partial x'^{\lambda}}{\partial x^{\kappa}} D. \tag{3.219}$$

It is this term which appears in (3.215). It is convenient, however, to find another expression for it. Putting $\kappa = \nu$ in (3.207) gives

$$\Gamma'^{\kappa}{}_{\mu\kappa} = \frac{\partial x^{\sigma}}{\partial x'^{\mu}}\Gamma^{\rho}{}_{\sigma\rho} + \frac{\partial x'^{\kappa}}{\partial x^{\tau}}\frac{\partial^2 x^{\tau}}{\partial x'^{\kappa}\,\partial x'^{\mu}}$$

or

$$\frac{\partial x'^{\kappa}}{\partial x^{\tau}}\frac{\partial^2 x^{\tau}}{\partial x'^{\kappa}\,\partial x'^{\mu}} = \Gamma'^{\kappa}{}_{\mu\kappa} - \frac{\partial x^{\sigma}}{\partial x'^{\mu}}\Gamma^{\rho}{}_{\sigma\rho}$$

so (3.219) becomes

$$\frac{\partial D}{\partial x'^{\mu}} = \left\{\Gamma'^{\kappa}{}_{\mu\kappa} - \frac{\partial x^{\sigma}}{\partial x'^{\mu}}\Gamma^{\rho}{}_{\sigma\rho}\right\} D. \tag{3.220}$$

Substituting this into (3.215) gives

$$\begin{aligned}\frac{\partial Q'}{\partial x'^{\mu}} &= \frac{\partial Q}{\partial x^{\nu}}\frac{\partial x^{\nu}}{\partial x'^{\mu}} D + Q\left\{\Gamma'^{\kappa}{}_{\mu\kappa} - \frac{\partial x^{\sigma}}{\partial x'^{\mu}}\Gamma^{\rho}{}_{\sigma\rho}\right\} D \\ &= \frac{\partial Q}{\partial x^{\nu}}\frac{\partial x^{\nu}}{\partial x'^{\mu}} D + Q'\Gamma'^{\kappa}{}_{\mu\kappa} - Q\frac{\partial x^{\sigma}}{\partial x'^{\mu}}\Gamma^{\rho}{}_{\sigma\rho} D\end{aligned}$$

or

$$\frac{\partial Q'}{\partial x'^{\mu}} - Q'\Gamma'^{\kappa}{}_{\mu\kappa} = \frac{\partial x^{\nu}}{\partial x'^{\mu}}\left\{\frac{\partial Q}{\partial x^{\nu}} - Q\Gamma^{\kappa}{}_{\kappa\nu}\right\} D. \tag{3.221}$$

Hence $\dfrac{\partial Q'}{\partial x'^{\nu}} - Q\Gamma^{\kappa}{}_{\kappa\nu}$ is a *covariant vector density*. We may call it the covariant derivative of a scalar density:

$$Q_{;\nu} = Q_{,\nu} - \Gamma^{\kappa}{}_{\kappa\nu}\, Q, \tag{3.222}$$

and then (3.221) is

$$Q'_{;\mu} = \frac{\partial x^{\nu}}{\partial x'^{\mu}} Q_{;\nu}\, D. \tag{3.223}$$

This definition of a covariant vector density parallels that of a scalar density, Equation (3.149) above.

The definition of the covariant derivative of a tensor density follows the same lines: if $T_{\mu\nu}$ is the density in (3.214) above, its covariant derivative is

$$T_{\mu\nu;\rho} = T_{\mu\nu,\rho} - \Gamma^{\kappa}{}_{\mu\rho}T_{\kappa\nu} - \Gamma^{\kappa}{}_{\nu\rho}T_{\mu\kappa} - \Gamma^{\alpha}{}_{\alpha\rho}T_{\mu\nu}, \tag{3.224}$$

with analogous definitions for tensor densities of any rank: the crucial last term is always the same.

3.11.2 Note on torsion and curvature

We have been concerned with the absolute derivative operator ∇. Its component in a direction $\mathbf{e}_\mu$ is ∇_μ. Now ordinary derivative operators commute: $\frac{\partial^2 f}{\partial x^\mu \partial x^\nu} = \frac{\partial^2 f}{\partial x^\nu \partial x^\mu}$; or

$$[\partial_\mu, \partial_\nu] f = 0,$$

where f is any scalar field. Do absolute derivative operators commute? There are two separate questions: whether the commutators acting on a *scalar* field, or on a *vector* field, give zero:

$$(1)\ \nabla_\mu \nabla_\nu f \stackrel{?}{=} \nabla_\nu \nabla_\mu f, \tag{3.225}$$

$$(2)\ \nabla_\mu \nabla_\nu \mathbf{V} \stackrel{?}{=} \nabla_\nu \nabla_\mu \mathbf{V}. \tag{3.226}$$

The difference between ordinary and absolute derivative operators is that the latter depend also on the frame vectors $\mathbf{e}_\mu$, so it is to be expected that the above questions are actually questions about the space being considered. The answers to the questions are, in fact, that the commutator acting on a scalar field always vanishes,

$$(1)\ [\nabla_\mu, \nabla_\nu] f = 0 \text{ always} \tag{3.227}$$

whereas if the commutator acting on a vector field vanishes, the space has *no curvature*, i.e. is *flat*:

$$(2)\ [\nabla_\mu, \nabla_\nu]\, \mathbf{V} = 0 \Rightarrow \text{space has } \textit{no curvature} \text{ (is flat)}. \tag{3.228}$$

The whole spirit of General Relativity, of course, is that space-time is *curved*, so $[\nabla_\mu, \nabla_\nu]\, \mathbf{V} \neq 0$; more will be said about this below. However, in conventional General Relativity (Einstein's theory) space-time is supposed to have zero torsion. There are in existence theories of space-time with torsion, and these have particular consequences for particles with spin, for example spin ½ particles in a gravitational field (described by the Dirac equation) – see Section 11.4 below for some more information on this. Theories with torsion were also used in some early versions of unified field theories, for example the Einstein–Strauss theory.[10]

These theories lie outside 'mainstream' General Relativity, so let us here investigate the consequences of *zero torsion*. The torsion tensor is given by (3.204) above and a space-time with zero torsion is therefore characterised by

$$\Gamma^\rho{}_{\mu\nu} = \Gamma^\rho{}_{\nu\mu}, \tag{3.229}$$

as in (3.205); the connection coefficients, calculated in a coordinate basis, are symmetric in their lower indices.

[10] Einstein & Straus (1946); see also Schrödinger (1985).

In terms of forms the no torsion condition is

$$\boldsymbol{\omega}^{\mu}{}_{\nu} \wedge \mathbf{d}x^{\nu} = 0, \tag{3.230}$$

since this gives (see (3.171) and put $\boldsymbol{\theta}^{\nu} = \mathbf{d}x^{\nu}$)

$$\boldsymbol{\omega}^{\mu}{}_{\nu} \wedge \mathbf{d}x^{\nu} = \Gamma^{\mu}{}_{\nu\kappa}\, \mathbf{d}x^{\kappa} \wedge \mathbf{d}x^{\nu} = 0,$$

and since the wedge product is antisymmetric in ν and κ this implies that $\Gamma^{\mu}{}_{\nu\kappa}$ is symmetric in its lower indices, which is condition (3.229).

In the general case, in which the basis $\boldsymbol{\theta}^{\mu}$ is not necessarily holonomic, we define the *torsion two-form*

$$\boldsymbol{\Sigma}^{\mu} = \mathbf{d}\boldsymbol{\theta}^{\mu} + \boldsymbol{\omega}^{\mu}{}_{\kappa} \wedge \boldsymbol{\theta}^{\kappa}. \tag{3.231}$$

Space is torsion-free if $\boldsymbol{\Sigma}^{\mu} = 0$. This is the content of (3.182), which in a coordinate basis $\boldsymbol{\theta}^{\mu} = \mathbf{d}x^{\mu}$ gives $\mathbf{d}\boldsymbol{\theta}^{\mu} = 0$ and (3.231) becomes the condition (3.230).

3.12 Examples

Let us illustrate the methods above by considering the simple cases of the Euclidean plane E^2 and the 2-sphere S^2 – flat and curved 2-dimensional spaces. The equations we shall need are (3.125) and (3.124), involving the metric tensor

$$\mathbf{g} = g_{\mu\nu}\boldsymbol{\theta}^{\mu} \otimes \boldsymbol{\theta}^{\nu}, \tag{3.125}$$

$$\mathrm{d}s^2 = g_{\mu\nu}\, \mathbf{d}x^{\mu}\, \mathbf{d}x^{\nu}, \tag{3.124}$$

and, for the connection 1-forms, (3.173) and (3.182),

$$\nabla \mathbf{e}_{\mu} = \boldsymbol{\omega}^{\kappa}{}_{\mu}\, \mathbf{e}_{\kappa}, \tag{3.173}$$

$$\mathbf{d}\,\boldsymbol{\theta}^{\mu} = -\boldsymbol{\omega}^{\mu}{}_{\nu} \wedge \boldsymbol{\theta}^{\nu}, \tag{3.182}$$

the last of which is the *zero torsion condition*. The connection coefficients $\Gamma^{\kappa}{}_{\lambda\mu}$ are defined by (3.171)

$$\boldsymbol{\omega}^{\kappa}{}_{\lambda} = \Gamma^{\kappa}{}_{\lambda\mu}\, \boldsymbol{\theta}^{\mu}, \quad \boldsymbol{\omega}^{\kappa}{}_{\lambda} = \gamma^{\kappa}{}_{\lambda\mu}\, \boldsymbol{\theta}^{\mu}, \tag{3.171}$$

for coordinate and non-coordinate bases respectively. In a *coordinate basis* these are given by (3.194)

$$\Gamma^{\nu}{}_{\mu\kappa} = {}^1\!/\!{}_2 g^{\nu\rho}(g_{\mu\rho,\kappa} + g_{\kappa\rho,\mu} - g_{\mu\kappa,\rho}). \tag{3.194}$$

The metric compatibility condition is (3.191)

$$\nabla \mathbf{g} = 0. \tag{3.191}$$

From (3.125) we have

$$\nabla \mathbf{g} = (\mathbf{d}g_{\mu\nu})\, \boldsymbol{\theta}^\mu \boldsymbol{\theta}^\nu + g_{\mu\nu}(-\boldsymbol{\omega}^\mu{}_\kappa \boldsymbol{\theta}^\kappa)\, \boldsymbol{\theta}^\nu + g_{\mu\nu}\boldsymbol{\theta}^\mu(-\boldsymbol{\omega}^\nu{}_\kappa \boldsymbol{\theta}^\kappa)$$
$$= (\mathbf{d}g_{\mu\nu} - \boldsymbol{\omega}^\kappa{}_\mu g_{\kappa\nu} - \boldsymbol{\omega}^\kappa{}_\nu g_{\mu\kappa})\, \boldsymbol{\theta}^\mu \boldsymbol{\theta}^\nu,$$

so (3.191) gives

$$\mathbf{d}\, g_{\mu\nu} = \boldsymbol{\omega}_{\mu\nu} + \boldsymbol{\omega}_{\nu\mu}, \tag{3.232}$$

where

$$\boldsymbol{\omega}_{\mu\nu} \equiv g_{\mu\kappa} \boldsymbol{\omega}^\kappa{}_\nu. \tag{3.233}$$

3.12.1 Plane E^2: coordinate basis

In polar coordinates

$$ds^2 = dr^2 + r^2\, d\phi^2 \tag{3.234}$$

and the coordinate basis 1-forms are

$$\boldsymbol{\theta}^1 = \mathbf{d}r, \quad \boldsymbol{\theta}^2 = \mathbf{d}\phi, \tag{3.235}$$

with corresponding basis vectors

$$\mathbf{e}_1 = \frac{\partial}{\partial r}, \quad \mathbf{e}_2 = \frac{\partial}{\partial \phi}.$$

The metric tensor has components $g_{11}=1$, $g_{22}=r^2$, $g_{12}=g_{21}=0$, or in matrix form (using Latin rather than Greek indices in this 2-dimensional example)

$$g_{ik} = \begin{pmatrix} 1 & 0 \\ 0 & r^2 \end{pmatrix} \quad g^{ik} = \begin{pmatrix} 1 & 0 \\ 0 & \frac{1}{r^2} \end{pmatrix}. \tag{3.236}$$

Equation (3.232), the metric compatibility condition, yields

$$\mu = 1, \quad \nu = 2: \quad \boldsymbol{\omega}_{12} = -\boldsymbol{\omega}_{21}; \tag{A}$$

$$\mu = \nu = 1: \quad \boldsymbol{\omega}_{11} = 0 \Rightarrow \boldsymbol{\omega}^1{}_1 = 0; \tag{B}$$

$$\mu = \nu = 2: \quad \boldsymbol{\omega}_{22} = r\,\mathbf{d}r = r\boldsymbol{\theta}^1 \Rightarrow \boldsymbol{\omega}^2{}_2 = \frac{1}{r}\boldsymbol{\theta}^1. \tag{C}$$

Equations (B) and (C) give (with (3.171))

$$\Gamma^1{}_{11} = \Gamma^1{}_{12} = 0; \quad \Gamma^2{}_{21} = \frac{1}{\mathrm{r}}, \quad \Gamma^2{}_{22} = 0. \tag{3.237}$$

The zero torsion condition (3.182) gives:

$$\mu = 1: \quad \boldsymbol{\omega}^1{}_\kappa \wedge \boldsymbol{\theta}^\kappa = 0$$
$$\therefore \boldsymbol{\omega}^1{}_2 \wedge \boldsymbol{\theta} = 0 \Rightarrow \boldsymbol{\omega}^1{}_2 \propto \boldsymbol{\theta}^2 = \mathbf{d}\phi. \tag{D}$$

$$\mu = 2: \quad \boldsymbol{\omega}^2{}_1 \wedge \mathbf{d}r + \boldsymbol{\omega}^2{}_2 \wedge \mathbf{d}\phi = 0. \tag{E}$$

By virtue of (C), Equation (E) gives

$$\boldsymbol{\omega}^2{}_1 \wedge \mathbf{d}r = -\frac{1}{r}\mathbf{d}r \wedge \mathbf{d}\phi = \frac{1}{r}\,\mathbf{d}\phi \wedge \mathbf{d}r \Rightarrow \boldsymbol{\omega}^2{}_1 = \frac{1}{r}\mathbf{d}\phi = \frac{1}{r}\boldsymbol{\theta}^2. \tag{F}$$

(F) clearly gives

$$\Gamma^2{}_{11} = 0, \quad \Gamma^2{}_{12} = \frac{1}{r} = \Gamma^2{}_{21}. \tag{3.238}$$

(F) also gives $\boldsymbol{\omega}_{21} = g_{22}\boldsymbol{\omega}^2{}_1 = r\,\mathbf{d}\phi$, whence (A) gives $\boldsymbol{\omega}_{12} = -\,r\,\mathbf{d}\phi$ and hence

$$\boldsymbol{\omega}^1{}_2 = g^{11}\,\boldsymbol{\omega}_{12} = -r\,\mathbf{d}\phi = -r\,\boldsymbol{\theta}^2,$$

which yields

$$\Gamma^1{}_{21} = 0 = \Gamma^1{}_{12}, \quad \Gamma^1{}_{22} = -r. \tag{3.239}$$

Equations (3.237)–(3.239) agree with (3.165) above; and also with (3.194): for example

$$\Gamma^1{}_{22} = {}^1\!/\!{}_2\, g^{1k}(g_{2k,2} + g_{2k,2} - g_{22,k}) = {}^1\!/\!{}_2\, g^{11}(-g_{22,1}) = -{}^1\!/\!{}_2\frac{\partial}{\partial r}(r^2) = -r.$$

It is left as a (useful) exercise to check the other connection coefficients.

So far we have worked in the polar coordinate system. Let us briefly dispose of the plane in Cartesian coordinates. Here $ds^2 = dx^2 + dy^2$ so taking $x^1 = x$, $x^2 = y$ the metric tensor is

$$g_{ik} = \begin{pmatrix} 1 & 0 \\ 0 & 1 \end{pmatrix}, \quad g^{ik} = \begin{pmatrix} 1 & 0 \\ 0 & 1 \end{pmatrix};$$

and since its values are *constants*, the connection coefficients, given by (3.194), which depend on the derivatives of the metric tensor, will all be zero:

$$\Gamma^i{}_{km} = 0.$$

The same conclusion follows from the equations involving the connection forms. Equation (3.232) clearly gives

$$\boldsymbol{\omega}_{11} = \boldsymbol{\omega}_{22} = 0, \qquad \boldsymbol{\omega}_{12} = -\boldsymbol{\omega}_{21}.$$

With $\boldsymbol{\theta}^1 = \mathbf{d}x$, $\boldsymbol{\theta}^2 = \mathbf{d}y$, Equation (3.227) with $\mu = 1$ gives

$$\boldsymbol{\omega}^1{}_2 \wedge \boldsymbol{\theta}^2 = 0,$$

implying that $\boldsymbol{\omega}^1{}_2 \propto \mathbf{d}y$ and hence $\boldsymbol{\omega}^2{}_1 \propto \mathbf{d}y$. With $\mu = 2$, however, Equation (3.182) gives

$$\boldsymbol{\omega}^2{}_1 \wedge \boldsymbol{\theta}^2 = 0,$$

and hence $\boldsymbol{\omega}^2{}_1 \sim \mathbf{d}x$, in contradiction with the above. We conclude that $\boldsymbol{\omega}^2{}_1 = 0$; and thence that all the connection forms – and therefore all the connection coefficients – vanish.

Let us now calculate the connection coefficients in an *orthonormal* basis. We still have the line element (3.234) but choose the basis 1-forms to be

$$\boldsymbol{\theta}^1 = \mathbf{d}r, \quad \boldsymbol{\theta}^2 = r\,\mathbf{d}\phi, \tag{3.240}$$

so that the metric tensor is

$$g_{ik} = \begin{pmatrix} 1 & 0 \\ 0 & 1 \end{pmatrix} = g^{ik}; \tag{3.241}$$

an orthonormal basis. Then $\mathbf{d}g_{ik}=0$ and, from (3.228) $\boldsymbol{\omega}_{ik}=-\boldsymbol{\omega}_{ki}$, so that $\boldsymbol{\omega}_{11}=\boldsymbol{\omega}_{22}=0=\boldsymbol{\omega}^1{}_1=\boldsymbol{\omega}^2{}_2$, and $\boldsymbol{\omega}_{12}=-\ \boldsymbol{\omega}_{21}$ is the only connection form to be found. The connection coefficients, denoted $\gamma^i{}_{km}$ in an anholonomic basis, are defined by (3.171)

$$\boldsymbol{\omega}^i{}_k = \gamma^i{}_{km}\boldsymbol{\theta}^m,$$

and we then have

$$\gamma^1{}_{11} = \gamma^1{}_{12} = 0, \quad \gamma^2{}_{21} = \gamma^2{}_{22} = 0. \tag{3.242}$$

The zero torsion condition $\mathbf{d}\boldsymbol{\theta}^i + \boldsymbol{\omega}^i{}_k \wedge \boldsymbol{\theta}^k = 0$ gives

$$\begin{aligned} i = 1:&\quad \boldsymbol{\omega}^1{}_2 \wedge r\,\mathbf{d}\phi = 0 \Rightarrow \boldsymbol{\omega}^1{}_2 \sim \mathbf{d}\phi, \\ i = 2:&\quad \mathbf{d}r \wedge \mathbf{d}\phi + \boldsymbol{\omega}^2{}_1 \wedge \mathbf{d}r = 0 \Rightarrow \boldsymbol{\omega}^2{}_1 = \mathbf{d}\phi = \frac{1}{r}\boldsymbol{\theta}^2, \\ &\quad \boldsymbol{\omega}^1{}_2 = -\mathbf{d}\phi = -\frac{1}{r}\boldsymbol{\theta}^2 \end{aligned}$$

and hence

$$\gamma^2{}_{11} = 0, \quad \gamma^2{}_{12} = \frac{1}{r}, \quad \gamma^1{}_{21} = 0, \quad \gamma^1{}_{22} = -\frac{1}{r}. \tag{3.243}$$

Equations (3.242), (3.243) are in agreement with (3.166). Note, comparing these with Equations (3.238)–(3.239), that $\gamma^i{}_{km} \neq \Gamma^i{}_{km}$; the connection coefficients are different in holonomic and anholonomic systems. And in particular $\Gamma^i{}_{km}$ is symmetric, $\Gamma^i{}_{km}=\Gamma^i{}_{mk}$, whereas $\gamma^i{}_{km}$ is not: for example $\gamma^2{}_{12}\neq 0$, but $\gamma^2{}_{21}=0$.

3.12.2 Sphere S^2

The metric for a 2-sphere of radius a is

$$\mathrm{d}s^2 = a^2(\mathrm{d}\theta^2 + \sin^2\theta\,\mathrm{d}\phi^2). \tag{3.244}$$

In a coordinate basis $\boldsymbol{\theta}^1=\mathbf{d}\theta$, $\boldsymbol{\theta}^2=\mathbf{d}\phi$ and the metric tensor and its inverse are

$$g_{ik} = \begin{pmatrix} a^2 & 0 \\ 0 & a^2\sin^2\theta \end{pmatrix}, \quad g^{ik} = \begin{pmatrix} \frac{1}{a^2} & 0 \\ 0 & \frac{1}{a^2\sin^2\theta} \end{pmatrix}.$$

Now consider the zero torsion equation (3.182),

$$\mathbf{d}\,\boldsymbol{\theta}^i + \boldsymbol{\omega}^i{}_k \wedge \boldsymbol{\theta}^k = 0.$$

It yields

$$i = 1: \quad \boldsymbol{\omega}^1{}_1 \wedge \mathbf{d}\theta + \boldsymbol{\omega}^1{}_2 \wedge \mathbf{d}\phi = 0, \tag{A}$$

$$i = 2: \quad \boldsymbol{\omega}^2{}_1 \wedge \mathbf{d}\theta + \boldsymbol{\omega}^2{}_2 \wedge \mathbf{d}\phi = 0. \tag{B}$$

The metric compatibility condition

$$\mathbf{d}g_{ik} = \boldsymbol{\omega}_{ik} + \boldsymbol{\omega}_{ki}$$

yields

$$i = k = 1: \quad \boldsymbol{\omega}_{11} = 0, \tag{C}$$

$$i = 1,\ k = 2: \quad \boldsymbol{\omega}_{21} = -\boldsymbol{\omega}_{12}, \tag{D}$$

$$i = k = 2: \quad \boldsymbol{\omega}_{22} = a^2 \sin\theta\cos\theta\ \mathbf{d}\theta \Rightarrow \boldsymbol{\omega}^2{}_2 = \cot\theta\ \mathbf{d}\theta = \cot\theta\ \boldsymbol{\theta}^1. \tag{E}$$

Equation (C) gives

$$\Gamma^1{}_{11} = \Gamma^1{}_{12} = 0, \tag{3.245}$$

while (E) gives

$$\Gamma^2{}_{21} = \cot\theta, \quad \Gamma^2{}_{22} = 0. \tag{3.246}$$

(B) and (E) give $\boldsymbol{\omega}^2{}_1 = \cot\theta\ \mathbf{d}\phi = \cot\theta\ \boldsymbol{\theta}^2$, hence

$$\Gamma^2{}_{11} = 0, \quad \Gamma^2{}_{12} = \cot\theta, \tag{3.247}$$

On the other hand, from $\boldsymbol{\omega}^2{}_1 = \cot\theta\ \mathbf{d}\phi$ we deduce that $\boldsymbol{\omega}^1{}_2 = -\sin\theta\cos\theta\ \mathbf{d}\phi = -\sin\theta\cos\theta\ \boldsymbol{\theta}^2$, and hence

$$\Gamma^1{}_{21} = 0,\ \Gamma^1{}_{22} = -\sin\theta\cos\theta. \tag{3.248}$$

These values of the connection coefficients agree with Equation (3.194); for example

$$\Gamma^2{}_{22} = \tfrac{1}{2}g^{1m}(2g_{2m,2} - g_{22,m}) = -\tfrac{1}{2}g^{11}\,g_{22,1}$$
$$= -\frac{1}{2a^2}\frac{\partial}{\partial\theta}(a^2\sin^2\theta) = -\sin\theta\cos\theta.$$

Now perform the calculation in an orthonormal basis. We have

$$ds^2 = (\boldsymbol{\theta}^1)^2 + (\boldsymbol{\theta}^2)^2$$

with

$$\boldsymbol{\theta}^1 = a\,\mathbf{d}\theta, \quad \boldsymbol{\theta}^2 = a\sin\theta\,\mathbf{d}\phi, \quad g_{ij} = \begin{pmatrix} 1 & 0 \\ 0 & 1 \end{pmatrix}.$$

Equation (3.182) gives

$$i=1:\quad \boldsymbol{\omega}^1{}_2\wedge a\sin\theta\,\mathbf{d}\phi=0\Rightarrow\boldsymbol{\omega}^1{}_2\sim\mathbf{d}\phi,$$
$$i=2:\quad a\cos\theta\,\mathbf{d}\theta\wedge\mathbf{d}\phi+\boldsymbol{\omega}^2{}_1\wedge(a\,\mathbf{d}\theta)=0\Rightarrow\boldsymbol{\omega}^2{}_1=\cos\theta\,\mathbf{d}\phi;\tag{3.248a}$$

and these two equations are consistent. Then (3.171) gives

$$\gamma^2{}_{11}=\frac{1}{a}\cot\theta=-\gamma^1{}_{22},\quad\text{others}=0.\tag{3.249}$$

Again, as in the case of the plane, the connection coefficients differ in coordinate and non-coordinate bases. The general formula for them in a coordinate basis is (3.194). Is there a general formula in a non-coordinate basis? Indeed there is, and we now derive it.

3.13 General formula for connection coefficients

The zero torsion condition is $\mathbf{d}\boldsymbol{\theta}^\kappa=-\boldsymbol{\omega}^\kappa{}_\lambda\wedge\boldsymbol{\theta}^\lambda$. This is a 2-form, and the basis of 2-forms is $\boldsymbol{\theta}^\mu\wedge\boldsymbol{\theta}^\nu$ so we may put

$$\mathbf{d}\boldsymbol{\theta}^\kappa=-{}^1\!/\!_2C^\kappa{}_{\mu\nu}\boldsymbol{\theta}^\mu\wedge\boldsymbol{\theta}^\nu\tag{3.250}$$

where $C^\kappa{}_{\mu\nu}=-C^\kappa{}_{\nu\mu}$ are some coefficients. Then

$$\begin{aligned}{}^1\!/\!_2\,C^\kappa{}_{\mu\nu}\boldsymbol{\theta}^\mu\wedge\boldsymbol{\theta}^\nu&=\boldsymbol{\omega}^\kappa{}_\lambda\wedge\boldsymbol{\theta}^\lambda=\gamma^\kappa{}_{\lambda\mu}\boldsymbol{\theta}^\mu\wedge\boldsymbol{\theta}^\lambda\\&={}^1\!/\!_2(\gamma^\kappa{}_{\lambda\mu}-\gamma^\kappa{}_{\mu\lambda})\,\boldsymbol{\theta}^\mu\wedge\boldsymbol{\theta}^\lambda\\&={}^1\!/\!_2(\gamma^\kappa{}_{\nu\mu}-\gamma^\kappa{}_{\mu\nu})\,\boldsymbol{\theta}^\mu\wedge\boldsymbol{\theta}^\nu,\end{aligned}$$

on relabelling, from which

$$C^\kappa{}_{\mu\nu}=\gamma^\kappa{}_{\nu\mu}-\gamma^\kappa{}_{\mu\nu}.\tag{3.251}$$

Now let us lower the indices. Define

$$C_{\kappa\lambda\mu}\equiv g_{\kappa\rho}\,C^\rho{}_{\lambda\mu},\quad\gamma_{\kappa\lambda\mu}\equiv g_{\kappa\rho}\,\gamma^\rho{}_{\lambda\mu}.\tag{3.252}$$

Then (3.251) is

$$C_{\kappa\lambda\mu}=\gamma_{\kappa\mu\lambda}-\gamma_{\kappa\lambda\mu}.\tag{a}$$

Now cyclically permute the indices $\kappa\to\lambda\to\mu\to\kappa$ to give

$$C_{\lambda\mu\kappa}=\gamma_{\lambda\kappa\mu}-\gamma_{\lambda\mu\kappa},\tag{b}$$

and permuting them once more gives

$$C_{\mu\kappa\lambda}=\gamma_{\mu\lambda\kappa}-\gamma_{\mu\kappa\lambda}.\tag{c}$$

Taking Equations (a) + (b) – (c) we find

$$\begin{aligned}C_{\kappa\lambda\mu}+C_{\lambda\mu\kappa}-C_{\mu\kappa\lambda}=&(\gamma_{\kappa\mu\lambda}+\gamma_{\mu\kappa\lambda})-(\gamma_{\lambda\mu\kappa}+\gamma_{\mu\lambda\kappa})\\&+(\gamma_{\lambda\kappa\mu}+\gamma_{\kappa\kappa\mu})-2\gamma_{\kappa\lambda\mu}.\end{aligned}\tag{3.253}$$

Now use the metric compatibility condition

$$\mathbf{d}g_{\mu\nu} = \boldsymbol{\omega}_{\mu\nu} + \boldsymbol{\omega}_{\nu\mu}, \tag{3.254}$$

the left hand side vanishing, of course, in an orthonormal system. In a coordinate basis

$$\mathbf{d}g_{\mu\nu} = \partial_\lambda g_{\mu\nu}\,\mathbf{d}x^\lambda = g_{\mu\nu,\lambda}\,\mathbf{d}x^\lambda, \tag{3.255}$$

while in a non-coordinate basis

$$\mathbf{d}g_{\mu\nu} = e_\lambda(g_{\mu\nu})\,\boldsymbol{\theta}^\lambda; \tag{3.256}$$

for example in the non-coordinate basis (3.240) above, $e_2 = \frac{1}{r}\frac{\partial}{\partial\phi}$. Then in the general case

$$e_\lambda(g_{\mu\nu})\,\boldsymbol{\theta}^\lambda = \boldsymbol{\omega}_{\mu\nu} + \boldsymbol{\omega}_{\nu\mu} = (\gamma_{\mu\nu\lambda} + \gamma_{\nu\mu\lambda})\,\boldsymbol{\theta}^\lambda,$$

so that

$$e_\lambda(g_{\mu\nu}) = \gamma_{\mu\nu\lambda} + \gamma_{\nu\mu\lambda}. \tag{3.257}$$

Substituting this equation into (3.253) (more precisely, the three terms in brackets on the right) gives

$$\gamma_{\mu\nu\lambda} = \tfrac{1}{2}\left[e_\lambda(g_{\mu\nu}) + e_\nu(g_{\lambda\mu}) - e_\mu(e_{\lambda\nu}) - C_{\mu\nu\lambda} - C_{\nu\lambda\mu} + C_{\lambda\mu\nu}\right]. \tag{3.258}$$

This is the general formula. In a *coordinate* (holonomic) basis $e_\lambda(g_{\mu\nu}) \to g_{\mu\nu,\lambda}$ and $C_{\lambda\mu\nu} = 0$ (since $\mathbf{d}\boldsymbol{\theta}^\mu = 0$), so we recover (with $\gamma \to \Gamma$ and the first index lowered – see (3.252)) the formula (3.194). In an *orthonormal* basis ($g_{\mu\nu} = \delta_{\mu\nu}$, up to signs ± for Lorentzian signature), however, $e_\lambda(g_{\mu\nu}) = 0$ so

$$\gamma_{\mu\nu\lambda} = -\tfrac{1}{2}[C_{\mu\nu\lambda} + C_{\nu\lambda\mu} - C_{\lambda\mu\nu}]. \tag{3.259}$$

We illustrate this by returning to the calculation above, of E^2 in an anholonomic basis. The metric and basis forms are given by (3.234) and (3.240) and $e_i\,(g_{km}) = 0$. The constants C_{ikm} are found from (3.250)

$$\mathbf{d}\boldsymbol{\theta}^i = -\tfrac{1}{2}C^i{}_{km}\boldsymbol{\theta}^k \wedge \boldsymbol{\theta}^m.$$

This gives, with (3.240) and $i = 1$:

$$\mathbf{d}\boldsymbol{\theta}^1 = 0 = -\tfrac{1}{2}C^1{}_{km}\boldsymbol{\theta}^k \wedge \boldsymbol{\theta}^m = -C^1{}_{12}\,\boldsymbol{\theta}^1 \wedge \boldsymbol{\theta}^2,$$

hence

$$C^1{}_{12} = C^1{}_{21} = 0 \Rightarrow C_{112} = C_{121} = 0; \tag{3.260}$$

and with $i = 2$:

$$\mathbf{d}\boldsymbol{\theta}^2 = \mathbf{d}r \wedge \mathbf{d}\phi = -\tfrac{1}{2}C^2{}_{km}\boldsymbol{\theta}^k \wedge \boldsymbol{\theta}^m = -C^2{}_{12}\boldsymbol{\theta}^1 \wedge \boldsymbol{\theta}^2 = -r\,C^2{}_{12}\,\mathbf{d}r \wedge \mathbf{d}\phi,$$

hence

$$C^2{}_{12} = -\frac{1}{r}, \quad C^2{}_{21} = \frac{1}{r} \Rightarrow C_{212} = -\frac{1}{r}, \quad C_{221} = \frac{1}{r}. \tag{3.261}$$

Then (3.259) gives

$$\gamma_{122} = -\frac{1}{2}(C_{122} + C_{221} - C_{212}) = -\frac{1}{r},$$

(since $C_{122} \equiv 0$; any $C_{\lambda\kappa\mu} = 0$ if $\kappa = \mu$) and

$$\gamma_{212} = -\frac{1}{2}(C_{212} + C_{122} - C_{221}) = \frac{1}{r}.$$

The remaining γ_{ikm} are all zero. For example

$$\gamma_{211} = -\frac{1}{2}(C_{211} + C_{112} - C_{121}) = 0.$$

We then conclude that the only non-zero components of $\gamma^i{}_{km}$ are

$$\gamma^1{}_{22} = -\frac{1}{r}, \quad \gamma^2{}_{12} = \frac{1}{r},$$

as was found in (3.243) above. It is not an accident that the quantities $C^i{}_{jk}$ appearing in (3.250) have the same form as the *structure constants* (3.15) in the Lie algebra of the basis vectors. In fact they are the same quantities, so that corresponding to the formula (3.250),

$$\mathbf{d\theta}^\kappa = -\frac{1}{2}C^k{}_{\mu\nu}\,\boldsymbol{\theta}^\mu \wedge \boldsymbol{\theta}^\nu,$$

we also have a corresponding formula in the dual vector space

$$[\mathbf{e}_\kappa, \mathbf{e}_\lambda] = C^\mu{}_{\kappa\lambda}\,\mathbf{e}_\mu. \tag{3.262}$$

For example, for E^2 in the anholonomic basis above

$$[\mathbf{e}_1, \mathbf{e}_2] = \left[\frac{\partial}{\partial r}, \frac{1}{r}\frac{\partial}{\partial \phi}\right] = -\frac{1}{r^2}\frac{\partial}{\partial \phi} = -\frac{1}{r}\,\mathbf{e}_2,$$

and hence

$$C^2{}_{12} = -\frac{1}{r}, \quad C^2{}_{21} = \frac{1}{r}, \quad \text{others} = 0,$$

i.e.

$$C_{212} = -\frac{1}{r}, \quad C_{221} = \frac{1}{r}, \quad \text{others} = 0,$$

as in (3.261) above. This is an alternative, and sometimes rather quicker, method of finding $C_{\kappa\lambda\mu}$. Clearly, in a holonomic basis $e_\kappa = \partial/\partial x^\kappa$, so all $C^\kappa{}_{\lambda\mu} = 0$, as noted before. The proof of (3.262) is long and slightly involved and the interested reader is referred to 'Further reading'.

In this chapter we have devoted much attention to connection coefficients. We have seen that a flat space, for example E^2, allows a Cartesian coordinate system to be set up throughout, and in this case the connection coefficients are all zero, but in any other coordinate system they will not all vanish. In a curved space, on the other hand, for example S^2, the connection coefficients are not all zero. So if we are presented with a metric tensor for some space and calculate the connection coefficients, to find that some of them are not zero, we *shall not know* whether the space is flat or curved. This is basically because the connection coefficients are *not tensors*: if the space happens to be flat they will all vanish in a Cartesian coordinate system, but it does not follow that they will vanish in another coordinate system.

It is important in General Relativity to know whether a space-time is flat or curved, because this corresponds to the presence or absence of a gravitational field. To tell if a space is really curved we need first to define a *test* of curvature, and then from this to arrive at a *tensor* – a 'curvature tensor'. If this is non-zero in one coordinate system it is non-zero in all, so the space is curved. This is the subject of the next chapter.

Further reading

Good general references for this chapter, written in a style relatively accessible to physicists, are Misner *et al.* (1973), Frankel (1979) and Schutz (1980). More sophisticated accounts are to be found in a large number of texts, each with its own emphasis and point of view (and some written for a purely mathematical readership): Choquet-Bruhat (1968), Hawking & Ellis (1973), Eguchi *et al.* (1980), Choquet-Bruhat *et al.* (1982), Crampin & Pirani (1986), De Felice & Clarke (1990), Nakahara (1990), Martin (1991), Frankel (1997) and Cartan (2001).

A very enlightening account of the modelling of space-time as a differentiable manifold is to be found in Kopczyński & Trautman (1992); see also Schröder (2002). A good, and for many years the standard, reference on differential forms for physicists is Flanders (1989). Westenholtz (1978) also provides a useful and quite complete account. Accounts written for more mathematically inclined readers are Cartan (1971) and Schreiber (1977). Introductions to homology, cohomology and de Rham's theorem may be found in Misner (1964), Choquet-Bruhat (1968), Flanders (1989) and Frankel (1997). An old but highly authoritative and very readable account of parallel transport is to be found in Levi-Cività (1927), and a nice potted history appears in Martin (1991), Section 6.3.

Problems

3.1 By differentiating the equation

$$g^{ij}g_{jk} = \delta^i{}_k$$

with respect to x^l, show that

$$\frac{\partial g^{im}}{\partial x^l} = -g^{mk}\,g^{ij}\,\frac{\partial g_{jk}}{\partial x_l}$$

and hence that

$$\frac{\partial g^{im}}{\partial x^l} + g^{ij}\Gamma^m{}_{jl} + g^{mj}\Gamma^i{}_{jl} = 0.$$

Hence show that $g^{ij}{}_{;k} = 0$.

3.2 By differentiating the equation

$$\frac{\partial x'^{\lambda}}{\partial x^{\rho}} \frac{\partial x^{\rho}}{\partial x'^{\nu}} = \delta^{\lambda}{}_{\nu}$$

with respect to x'^{μ}, and using Equation (3.207), show that

$$\Gamma'^{\lambda}{}_{\mu\nu} = \frac{\partial x'^{\lambda}}{\partial x^{\rho}} \frac{\partial x^{\sigma}}{\partial x'^{\mu}} \frac{\partial x^{\tau}}{\partial x'^{\nu}} \Gamma^{\rho}{}_{\sigma\tau} - \frac{\partial x^{\sigma}}{\partial x'^{\mu}} \frac{\partial x^{\rho}}{\partial x'^{\nu}} \frac{\partial^2 x'^{\lambda}}{\partial x^{\rho} \partial x^{\sigma}}.$$

3.3 Calculate the connection coefficients $\Gamma^{i}{}_{jk}$ in E^3 (Euclidean 3-dimensional space) in spherical polar coordinates.

3.4 Consider the sphere S^2 in the orthonormal basis $\boldsymbol{\theta}^1 = a\,\mathbf{d}\theta$, $\boldsymbol{\theta}^2 = a \sin\theta\,\mathbf{d}\phi$. Calculate the structure constants $C^{i}{}_{jk}$ from the Lie algebra (3.262) generated by the vectors $\mathbf{e}^i$ dual to $\boldsymbol{\theta}^i$ and thence show that $\boldsymbol{\omega}^2{}_1 = \cos\theta\,\mathbf{d}\phi$, as in Section 3.12.2.

4 Differential geometry II: geodesics and curvature

We continue our account of differential geometry by discussing the topics of geodesics and curvature.

4.1 Autoparallel curves and geodesics

In the last chapter we defined the notion of the absolute derivative of a vector in an unspecified direction

$$\nabla \mathbf{V} = V^{\kappa}{}_{;\lambda}\, \mathrm{d}x^{\lambda} \otimes \mathbf{e}_{\kappa} \tag{4.1}$$

where $V^{\kappa}{}_{;\lambda} = V^{\kappa}{}_{,\lambda} + \Gamma^{\kappa}{}_{\mu\lambda}\, V^{\mu}$ (in a holonomic basis). The derivative of $\mathbf{V}$ in the direction $\mathbf{U}$ is

$$\nabla_{\mathbf{U}} \mathbf{V} = \langle \nabla \mathbf{V}, \mathbf{U} \rangle = \mathrm{V}^{\kappa}{}_{;\lambda} U^{\lambda}\, \mathbf{e}_{\kappa}. \tag{4.2}$$

If $\mathbf{U} = \mathbf{e}_{\mu}$ we may write this as

$$\nabla_{\mu} \mathbf{V} = V^{k}{}_{;\mu}\, \mathbf{e}_{\kappa}. \tag{4.3}$$

We now use this to define the notion of *parallel transport* along a curve – this is a refinement of what was discussed in Section 3.10 above. Let C be some curve in R^n. We may parametrise this curve by λ, so the curve is the mapping $R \to R^n$, $\lambda \to \mathrm{C}(\lambda)$. Let $\mathbf{t}$ be the tangent vector to the curve at some point with coordinates x^{μ} (see Fig. 4.1):

$$\mathbf{t} = t^{\mu} \mathbf{e}_{\mu}, \quad t^{\mu} = \frac{\mathrm{d}x^{\mu}}{\mathrm{d}\lambda}. \tag{4.4}$$

If $\mathbf{V}$ is a vector field defined in R^n, and therefore in particular at each point on C, we define it to be *parallel transported* along C if

$$\nabla_{\mathbf{t}} \mathbf{V} = t^{\kappa}\, V^{\lambda}{}_{;\kappa}\, \mathbf{e}_{\lambda} = t^{\kappa} (V^{\lambda}{}_{,\kappa} + \Gamma^{\lambda}{}_{\mu\kappa} V^{\mu})\, \mathbf{e}_{\lambda} = 0. \tag{4.5}$$

Consider for simplicity flat space. The statement that two vectors (of equal magnitude, let us say) are 'parallel' means that their *Cartesian components* are equal. If, then, a vector (field) $\mathbf{V}$ is defined to be parallel along a curve, its Cartesian components are unchanged, so $\frac{\mathrm{d}V^x}{\mathrm{d}\lambda} = \frac{\mathrm{d}V^y}{\mathrm{d}\lambda} = 0$, but $\frac{\mathrm{d}}{\mathrm{d}\lambda} = t^{\mu} \partial_{\mu}$, hence $\frac{\partial V^x}{\partial x} = \frac{\partial V^x}{\partial y} = \cdots = 0$, and the first term, $V^{\lambda}{}_{,\kappa}$, in the bracket in (4.5) vanishes. But in a Cartesian system all $\Gamma^{\lambda}{}_{\mu\kappa} = 0$, hence $\nabla_{\mathbf{t}} \mathbf{V} = 0$ in Cartesian coordinates. But this is a tensor equation, and so defines – at least in flat

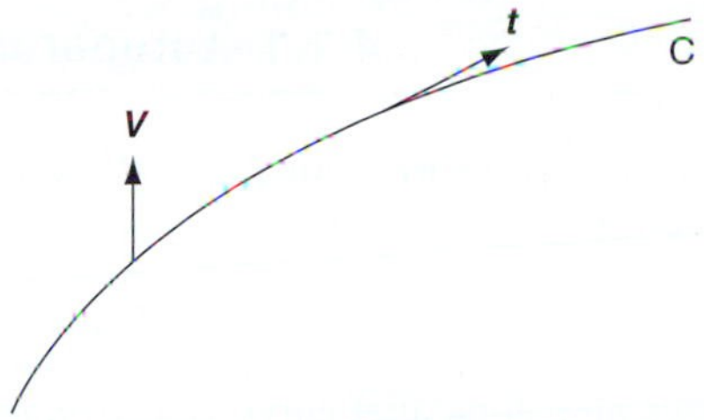

Fig. 4.1 A curve C on which is defined a vector field **V**, and **t** is a tangent vector at any point.

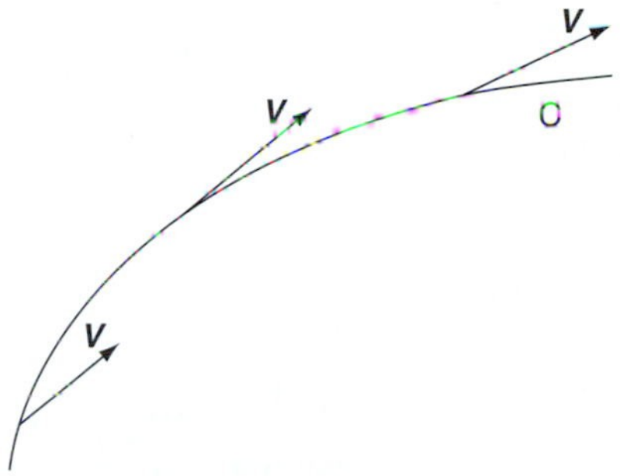

Fig. 4.2 The vector field **V** is parallel to itself along the curve C; it is a tangent to C at one point but not at others, hence tangent vectors are not parallel transported along the curve.

spaces – parallelism (parallel transport) in any coordinate system. Note, by the way, that the *polar* components of two parallel vectors are of course *not* equal, but equally the connection coefficients $\Gamma^{\lambda}{}_{\mu\kappa}$ are not all zero in a non-Cartesian coordinate system – the *absolute derivative* will still vanish. Equation (4.5) defines parallelism along a curve.

It is convenient to note a couple of points. First, returning to the case of flat space briefly, if C is an *arbitary* curve, tangent vectors are *not* parallel-transported into tangent vectors – Fig. 4.2 illustrates this simple point. This continues to hold in general (curved) spaces, so a special subset of curves exists, which have the property that tangent vectors *are* parallel transported into tangent vectors. These are called *autoparallel curves*, and are treated below. The second remark is that in the definition (4.5) the vector field **t** is defined *only along C*, not over the whole space. To be more precise over this, we need to establish the existence of a vector field **Y** (say), which when restricted to C gives the tangent vector **t**, and then show that $\nabla_{\mathbf{Y}}\mathbf{V}$ is independent of **Y**, apart from its restriction to C(λ). This is discussed, for example, in Helgason (1978).

It follows from the definition above that the *scalar product* (inner product) of two vectors **U**, **V** is preserved under parallel transport,

$$\nabla(\mathbf{U}\cdot\mathbf{V}) = 0. \qquad (4.6)$$

This follows from the definition $\mathbf{U}\cdot\mathbf{V} = \mathbf{g}(\mathbf{U}, \mathbf{V}) = g_{\mu\nu}U^{\mu}V^{\nu}$, where

$$\mathbf{g} = g_{\kappa\lambda}\theta^{\kappa}\otimes\theta^{\lambda}, \text{ hence } \nabla\mathbf{g} = g_{\kappa\lambda;\mu}\theta^{\kappa}\otimes\theta^{\lambda}\otimes\theta^{\mu} = 0,$$

whence (4.6) follows.

4.1.1 Autoparallel curves

A curve C is *defined* to be autoparallel if the tangent vector $\boldsymbol{t}$ is transported along it by parallel transport

$$\nabla_{\mathbf{t}}\mathbf{t} = 0. \tag{4.7}$$

In a flat space an autoparallel curve is a *straight line* – it is only along straight lines that tangent vectors at different points are parallel to each other – see Fig. 4.1 above. In curved spaces there are no straight lines, so we may describe an autoparallel curve as the *straightest* path (between two points). From (4.5)

$$\nabla_{\mathbf{t}}\mathbf{t} = t^\nu t^\mu{}_{;\nu}\,\mathbf{e}_\mu, \tag{4.8}$$

so (4.7) gives $t^\nu\, t^\mu{}_{;\nu}=0$, or

$$t^\nu(t^\mu{}_{,\nu} + \Gamma^\mu{}_{\rho\nu} t^\rho) = 0,$$

which, with $t^\mu = \dfrac{\mathrm{d}x^\mu}{\mathrm{d}\lambda}$ gives

$$\frac{\mathrm{d}x^\nu}{\mathrm{d}\lambda}\left[\frac{\partial}{\partial x^\nu}\left(\frac{\mathrm{d}x^\mu}{\mathrm{d}\lambda}\right) + \Gamma^\mu{}_{\rho\nu}\frac{\mathrm{d}x^\rho}{\mathrm{d}\lambda}\right] = 0,$$

that is,

$$\frac{\mathrm{d}^2x^\mu}{\mathrm{d}\lambda^2} + \Gamma^\mu{}_{\nu\rho}\frac{\mathrm{d}x^\nu}{\mathrm{d}\lambda}\frac{\mathrm{d}x^\rho}{\mathrm{d}\lambda} = 0. \tag{4.9}$$

This is the equation for an autoparallel curve. Our next task is to show that in a *metric* space it is also the equation for a *geodesic*.

4.1.2 Geodesics

In a flat space, the straight line joining two points has two properties: (i) it is the 'straightest' path – tangent vectors at different points are parallel, (ii) it is the *shortest* path between the two points. This second property of course involves the notion of distance, i.e. $g_{\mu\nu}$. In a general space the straightest path becomes an autoparallel curve (and exists even in an affine space), while the shortest path becomes a *geodesic*, whose equation we shall now derive.

The length of arc between the two points x^μ and $x^\mu + \mathrm{d}x^\mu$ is $\mathrm{d}s$, where

$$\mathrm{d}s^2 = g_{\mu\nu}\,\mathrm{d}x^\mu\,\mathrm{d}x^\nu,$$

so the length of arc between P and Q (see Fig. 4.3) is

$$s = \int_P^Q \sqrt{g_{\mu\nu}(x)\,\mathrm{d}x^\mu\,\mathrm{d}x^\nu}\,. \tag{4.10}$$

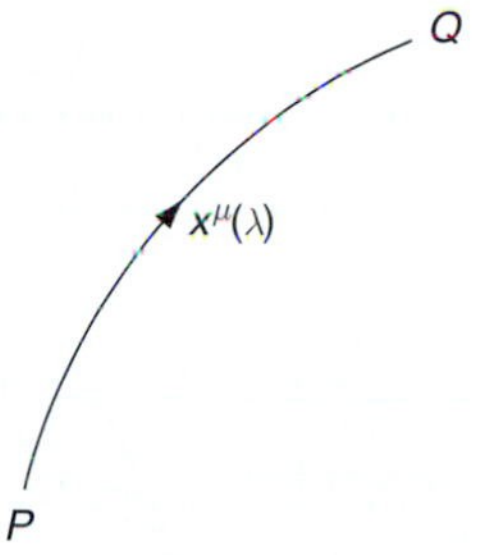

Fig. 4.3 **A parametrised curve between points *P* and *Q*.**

Let us parametrise the curve by λ, $x^\mu(\lambda)$, with the end points $P(\lambda_1)$, $Q(\lambda_2)$. For example, in space-time λ could be proper time. Putting $\dot{x}^\mu = \dfrac{dx^\mu}{d\lambda}$, we have

$$s = \int_{\lambda_1}^{\lambda_2} \sqrt{g_{\mu\nu}(x)\dot{x}^\mu \dot{x}^\nu \, d\lambda}\,, \tag{4.11}$$

or

$$s = \int_{\lambda_1}^{\lambda_2} \sqrt{L}\, d\lambda, \quad L = L(x^\mu, \dot{x}^\mu) = g_{\mu\nu}(x)\dot{x}^\mu \dot{x}^\nu, \tag{4.12}$$

and L may be termed a 'Lagrangian'.

Now suppose the curve is deformed, so that $x^\mu(\lambda) \to x^\mu(\lambda) + \delta x^\mu(\lambda)$ (see Fig. 4.4), but with $\delta x^\mu(\lambda_1) = \delta x^\mu(\lambda_2) = 0$, so that the end-points are fixed.

A geodesic is a path of *extremal* length (in general minimal), so under this deformation or variation we would have $\delta \int ds = 0$. It is more convenient, however, to take instead $\delta \int ds^2 = 0$, which is, with L in (4.12)

$$\delta \int L\, d\lambda = 0. \tag{4.13}$$

then

$$\int \left(\frac{\partial L}{\partial \dot{x}^\mu} \delta \dot{x}^\mu + \frac{\partial L}{\partial x^\mu} \delta x^\mu \right) d\lambda = 0, \tag{4.14}$$

where $\delta \dot{x}^\mu = \dfrac{d}{d\lambda} \delta x^\mu$. Now note that the following integral I is zero:

$$I = \int_P^Q \frac{d}{d\lambda} \left(\frac{\partial L}{\partial \dot{x}^\mu} \delta x^\mu \right) d\lambda = \frac{\partial L}{\partial \dot{x}^\mu} \delta x^\mu |^Q{}_P = 0.$$

On the other hand

$$I = \int \left\{ \frac{d}{d\lambda} \left(\frac{\partial L}{\partial \dot{x}^\mu} \right) \delta x^\mu + \frac{\partial L}{\partial \dot{x}^\mu} \delta \dot{x}^\mu \right\} d\lambda;$$

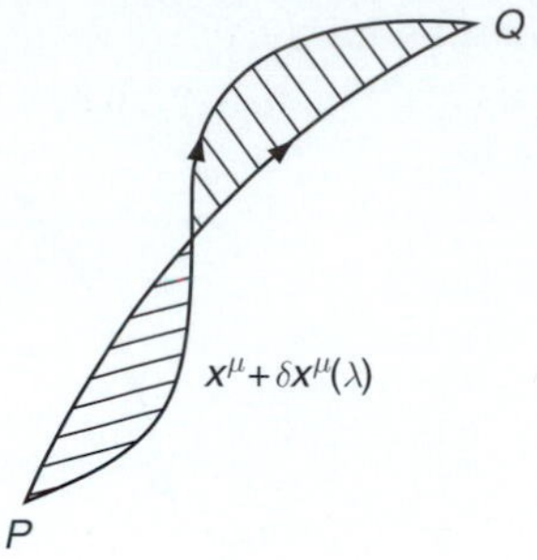

Fig. 4.4 **A curve deformed, but with the end-points *P* and *Q* fixed.**

hence

$$\int \frac{\partial L}{\partial \dot{x}^\mu}\,\delta\dot{x}^\mu\,\mathrm{d}\lambda = -\int \frac{\mathrm{d}}{\mathrm{d}\lambda}\left(\frac{\partial L}{\partial \dot{x}^\mu}\right)\delta x^\mu\,\mathrm{d}\lambda. \tag{4.15}$$

Substituting this in (4.14) gives

$$\int L^{-1/2}\left\{\frac{\partial L}{\partial x^\mu} - \frac{\mathrm{d}}{\mathrm{d}\lambda}\left(\frac{\partial L}{\partial \dot{x}^\mu}\right)\right\}\delta x^\mu\,\mathrm{d}\lambda = 0,$$

and since δx^μ is arbitrary this implies

$$\frac{\partial L}{\partial x^\mu} - \frac{\mathrm{d}}{\mathrm{d}\lambda}\left(\frac{\partial L}{\partial \dot{x}^\mu}\right) = 0. \tag{4.16}$$

This is the *Euler–Lagrange equation* for the geodesic, where

$$L(x,\dot{x}) = g_{\kappa\lambda}(x)\,\dot{x}^\kappa\,\dot{x}^\lambda. \tag{4.17}$$

It is clear that

$$\frac{\partial L}{\partial x^\mu} = g_{\kappa\lambda,\mu}\,\dot{x}^\kappa\,\dot{x}^\lambda, \quad \frac{\partial L}{\partial \dot{x}^\mu} = 2g_{\mu\kappa}\,\dot{x}^\kappa$$

and hence

$$\frac{\mathrm{d}}{\mathrm{d}\lambda}\left(\frac{\partial L}{\partial \dot{x}^\mu}\right) = 2\frac{\mathrm{d}}{\mathrm{d}\lambda}(g_{\mu\kappa}\dot{x}^\kappa) = 2g_{\mu\kappa,\lambda}\,\dot{x}^\kappa\,\dot{x}^\lambda + 2\,g_{\mu\kappa}\,\ddot{x}^\kappa,$$

so that (4.16) gives

$$g_{\kappa\lambda,\mu}\dot{x}^\kappa\,\dot{x}^\lambda - 2g_{\mu\kappa,\lambda}\dot{x}^\kappa\,\dot{x}^\lambda - 2\,g_{\mu\kappa}\ddot{x}^\kappa = 0.$$

Multiplying by $g^{\mu\rho}$ this gives

$$\ddot{x}^\rho + g^{\mu\rho}g_{\mu\kappa,\lambda}\,\dot{x}^\kappa\,\dot{x}^\lambda - {}^1\!/\!{}_2\,g^{\mu\rho}g_{\kappa\lambda,\mu}\,\dot{x}^\kappa\,\dot{x}^\lambda = 0.$$

The $g_{\mu\kappa,\lambda}$ in the second term may, however, be replaced by $^1/_2\,(g_{\mu\kappa,\lambda}+g_{\mu\lambda,\kappa})$ since $\dot{x}^\kappa\dot{x}^\lambda$ is symmetric under $\kappa \leftrightarrow \lambda$ interchange, so its coefficient may be replaced by its symmetric part only. This then gives

$$\ddot{x}_\rho + {}^1/_2 g^{\rho\mu}(g_{\mu\kappa,\lambda} + g_{\mu\lambda,\kappa} - g_{\kappa\lambda,\mu})\dot{x}^\kappa\dot{x}^\lambda = 0;$$

or, in view of (3.194),

$$\ddot{x}_\rho + \Gamma^\rho{}_{\kappa\lambda}\,\dot{x}^\kappa\dot{x}^\lambda = 0. \tag{4.18}$$

This is the equation for a geodesic, and we see it is the same as (4.9), the equation for an autoparallel curve. Hence it is true, in a general space as well as in a flat one, that *the straightest path between two points is also the shortest one*. We now consider some examples of geodesics.

4.1.3 Examples

The first example is almost trivial – the plane in Cartesian coordinates. Here $ds^2 = dx^2 + dy^2$, $g_{11} = g_{22} = 0$, $g_{12} = g_{21} = 0$, so all $\Gamma^i{}_{km} = 0$ and the geodesic equations become (with s the parameter along the path)

$$\frac{d^2x}{ds^2} = \frac{d^2y}{ds^2} = 0,$$

whose solutions are $x = as + c_1$, $y = bs + c_2$ (a, b, c_1, c_2 constants), hence $y = mx + c$, a straight line.

For a second example consider the plane in polar coordinates. Here $ds^2 = dr^2 + r^2\,d\phi^2$, and the connection coefficients are given by (3.237), (3.239) (with, of course, $x^1 = r$, $x^2 = \phi$), so the geodesic equations for $\rho = 1, 2$ are

$$\frac{d^2r}{ds^2} - r\left(\frac{d\phi}{ds}\right)^2 = 0, \tag{i}$$

$$\frac{d^2\phi}{ds^2} + \frac{2}{r}\frac{dr}{ds}\frac{d\phi}{ds} = 0. \tag{ii}$$

We must verify that these equations describe straight lines. If $\frac{d\phi}{ds} = 0$, then $\phi = \text{const}$ and from (i) $r = as + b$; this is a straight line through the origin. Otherwise, if $\frac{d\phi}{ds} = \phi' \neq 0$, divide (ii) by ϕ':

$$\frac{1}{\phi'}\frac{d\phi'}{ds} + \frac{2}{r}\frac{dr}{ds} = 0.$$

On integration this yields

$$\ell n\,|\phi'| + \ell n\, r^2 = 0 \Rightarrow r^2\phi' = r^2\frac{d\phi}{ds} = h = \text{const.} \tag{iii}$$

Dividing the equation for the metric by ds^2 and using (iii) gives

$$1 = \left(\frac{dr}{ds}\right)^2 + r^2\left(\frac{d\phi}{ds}\right)^2 = \left(\frac{dr}{ds}\right)^2 + \frac{h^2}{r^2}$$

and hence

$$\frac{dr}{ds} = \pm\frac{1}{r}(r^2 - h^2)^{1/2}. \tag{iv}$$

On the other hand (iii) gives

$$\frac{d\phi}{ds} = \frac{h}{r^2}. \tag{v}$$

Dividing equations (iv) and (v) gives

$$\frac{d\phi}{dr} = \pm\frac{h}{r(r^2 - h^2)^{1/2}} = \pm\frac{d}{dr}\left(\cos^{-1}\frac{h}{r}\right)$$

and hence

$$\phi - \phi_0 = \pm\cos^{-1}\frac{h}{r} \Rightarrow \frac{h}{r} = \cos(\phi - \phi_0),$$

which is the equation of a straight line in polar coordinates – see for example Fig. 4.5.

Our third example of a geodesic is the sphere S^2. We shall prove that a geodesic on a sphere is a great circle, that is, the intersection of a plane through the origin of the sphere and the spherical surface. The metric is

$$ds^2 = a^2\,d\theta^2 + a^2\sin^2\theta\,d\phi^2,$$

and with $x^1 = \theta$, $x^2 = \phi$ we have, from (3.245)–(3.248)

$$\Gamma^2{}_{21} = \Gamma^2{}_{12} = \cot\theta, \quad \Gamma^1{}_{22} = -\sin\theta\cos\theta.$$

The geodesic equation with $\mu = 2$ is then

$$\ddot{\phi} + 2\cot\theta\,\dot{\phi}\dot{\theta} = 0,$$

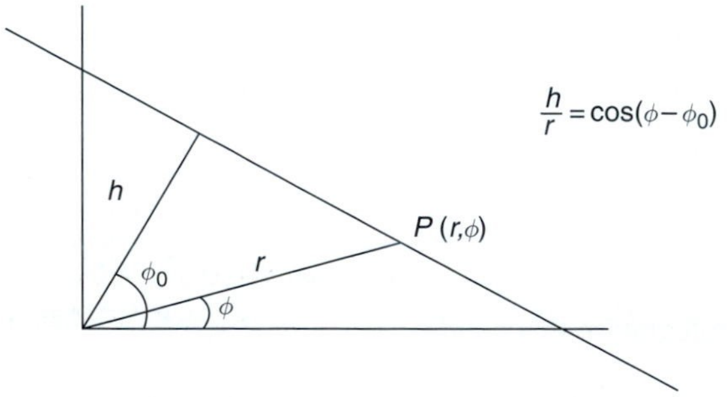

Fig. 4.5 A straight line in polar coordinates.

(with $\dot{\phi} = \frac{\mathrm{d}\phi}{\mathrm{d}s}$, etc.) On integration this gives

$$\sin^2\theta . \dot{\phi} = h = \text{const} \Rightarrow \dot{\phi} = \frac{h}{\sin^2\theta}. \qquad \text{(a)}$$

From the equation for the metric we have

$$1 = a^2\dot{\theta}^2 + a^2 \sin^2\theta \, \dot{\phi}^2,$$

which, on substituting from (a) gives

$$\dot{\theta} = \frac{\left(\sin^2\theta - a^2h^2\right)^{1/2}}{a \sin\theta}. \qquad \text{(b)}$$

Dividing (a) and (b) gives

$$\frac{\dot{\phi}}{\dot{\theta}} = \frac{\mathrm{d}\phi}{\mathrm{d}\theta} = \frac{h \operatorname{cosec}^2\theta}{\left(\dfrac{1}{a^2 - h^2 \operatorname{cosec}^2\theta}\right)^{1/2}},$$

which on integration gives

$$\phi - \phi_0 = -\cos^{-1}\left(\frac{h \cot\theta}{\sqrt{\dfrac{1}{a^2} - h^2}}\right),$$

i.e.

$$\cos(\phi - \phi_0) + \frac{h \cot\theta}{\sqrt{\dfrac{1}{a^2} - h^2}} = 0;$$

which is an equation of the form

$$A \cos\phi + B \sin\phi + C \cot\theta = 0,$$

which, in turn, is of the form $\alpha x + \beta y + \gamma z = 0$, where

$$x = A \sin\theta \cos\phi, \quad y = A \sin\theta \sin\phi, \quad z = A \cos\theta.$$

It thus represents a plane section of the sphere through the origin, i.e. a great circle. This is what we set out to prove.

4.2 Geodesic coordinates

Consider a point P on a geodesic. A nearby point has the coordinates $x^\mu(s)$, where s is the arc length from P (see Fig. 4.6). Now expand $x^\mu(s)$ about $x^\mu{}_P$ by Taylor's theorem:

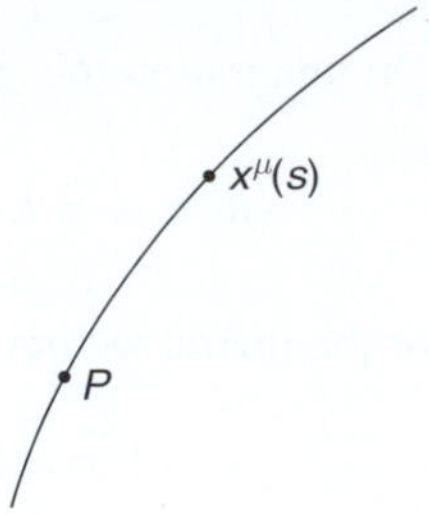

Fig. 4.6 Points on a geodesic; s is the arc length from the point P.

$$x^\mu(s) = x^\mu{}_P + \dot{x}^\mu{}_P s + {}^1\!/\!_2 \ddot{x}^\mu{}_P s^2 + \frac{1}{6}\dddot{x}^\mu{}_P s^3 + O(s^4). \tag{4.19}$$

The point $x^\mu(s)$, is, however, on the geodesic, so

$$\ddot{x}^\mu + \Gamma^\mu{}_{\nu\kappa}\dot{x}^\nu\dot{x}^\kappa = 0 \tag{4.20}$$

hence

$$\ddot{x}^\mu{}_P = -(\Gamma^\mu{}_{\nu\kappa})_P v^\nu v^\kappa, \quad v^\nu = \dot{x}^\nu{}_P.$$

To find $\dddot{x}^\mu{}_P$, differentiate (4.20):

$$\dddot{x}^\mu + \dot{\Gamma}^\mu{}_{\nu\kappa}\dot{x}^\nu\dot{x}^\kappa + 2\Gamma^\mu{}_{\nu\kappa}\ddot{x}^\nu\dot{x}^\kappa = 0.$$

Now putting $\dot{\Gamma}^\mu{}_{\nu\kappa} = \Gamma^\mu{}_{\nu\kappa,\lambda}\dot{x}^\lambda$, substituting from (4.20) for $\ddot{x}^\nu$ and relabelling gives

$$\begin{aligned}\dddot{x}^\mu &= (-\Gamma^\mu{}_{\kappa\lambda,\rho} + 2\Gamma^\mu{}_{\alpha\kappa}\Gamma^\alpha{}_{\rho\lambda})\dot{x}^\kappa\dot{x}^\lambda\dot{x}^\rho \\ &= \mathrm{Sym}[-\Gamma^\mu{}_{\kappa\lambda,\,\rho} + 2\Gamma^\mu{}_{\alpha\kappa}\Gamma^\alpha{}_{\rho\lambda}]\dot{x}^\kappa\dot{x}^\lambda\dot{x}^\rho \\ &\equiv \Lambda^\mu{}_{\kappa\lambda\rho}\dot{x}^\kappa\dot{x}^\lambda\dot{x}^\rho,\end{aligned} \tag{4.21}$$

where 'Sym' means taking the part of the object in brackets which is symmetric with respect to the indices κ, λ and ρ. We then have

$$\dddot{x}^\mu{}_p = (\Lambda^\mu{}_{\kappa\lambda\rho})_P v^\kappa v^\lambda v^\rho$$

and hence, substituting in (4.19),

$$x^\mu(s) = x^\mu{}_P + v^\mu{}_s - {}^1\!/\!_2(\Gamma^\mu{}_{\nu\kappa})_P v^\nu v^\kappa s^2 + \frac{1}{6}(\Lambda^\mu{}_{\kappa\lambda\rho})_P v^\kappa v^\lambda v^\rho s^3 + O(s^4), \tag{4.22}$$

with $v^\mu = \dot{x}^\mu{}_P$. Having obtained this expansion we now define *geodesic coordinates* y^μ by the equation

$$x^\mu = x^\mu{}_P + y^\mu - {}^1\!/\!_2(\Gamma^\mu{}_{\nu\kappa})_P v^\nu v^\kappa s^2 + \frac{1}{6}(\Lambda^\mu{}_{\kappa\lambda\rho})_P v^\kappa v^\lambda v^\rho s^3 + O(s^4),$$

so that

$$v^\mu s - y^\mu + O(s^4) = 0,$$

or

$$y^\mu = v^\mu s + O(s^4), \tag{4.23}$$

so that $\dot{y}^\mu{}_P = v^\mu = \dot{x}^\mu{}_P$. Comparing Equations (4.21) and (4.22) we see that, in geodesic coordinates,

$$(\Gamma^\mu{}_{\nu\kappa})_P = 0, \tag{4.24}$$

and that $(\Lambda^\mu{}_{\kappa\,\lambda\,\rho})_P = 0$, or equivalently that

$$(\Gamma^\mu{}_{\kappa\lambda,\rho})_P + (\Gamma^\mu{}_{\rho\kappa,\lambda})_P + (\Gamma^\mu{}_{\lambda\rho,\kappa})_P = 0. \tag{4.25}$$

The condition (4.24) requires $(g_{\mu\nu,\kappa})_P = 0$, but (4.25) is too weak to require that the first derivative of the connection coefficients vanish, or (what is the same) that the second derivative of the metric tensor vanish at *P*. We therefore have, in geodesic coordinates

$$(g_{\mu\nu,\kappa})_P = 0, \;\; \text{but} \;\; (g_{\mu\nu,\kappa\rho})_P \neq 0, \tag{4.26}$$

or, equivalently,

$$(\Gamma^\mu{}_{\nu\kappa})_P = 0, \;\; \text{but} \;\; (\Gamma^\mu{}_{\nu\kappa,\,\rho})_P \neq 0. \tag{4.27}$$

Thus the first derivative of the metric tensor vanishes at *P*, but not nearby. This implies that, at *P*, $g_{\mu\nu} = \text{const}$, corresponding to Minkowski (pseudo-Euclidean) space, and therefore to a *locally inertial frame*. This is the significance of geodesic coordinates; they determine a coordinate system which is locally Minkowski.

4.3 Curvature

Everyone would consider themselves to have an intuitive understanding of curvature – for example, that of a curved surface, but the task of the mathematician is to formulate a *definition* of it. This is done by pursuing the considerations already begun in Section 3.10 on parallelism and parallel transport. Consider the question; when are two vectors at two distinct points *parallel*? In Euclidean space two vectors are said to be parallel if their Cartesian components are the same. In curved spaces, what do we say? Consider the sketches in Fig. 4.7. In (a), a flat space, it is clear whether the vectors **U** and **V** are parallel or not. (As drawn, they are.) In (b) the space is a cylinder, which may be unrolled to make it into a plane. It then becomes case (a), so that it is clear whether **U** and **V** are parallel. A cylinder is said to be a ‘developable surface’. A cone is also developable. In some sense they are not ‘truly’ curved, since they may be unrolled to be made flat. A sphere S^2, on the other hand, is not developable – it cannot be made flat without stretching or tearing. Then the question, whether **U** is parallel to **V**, now becomes a matter of definition. Figure 4.7(c) shows both **U** and **V** as east-pointing (tangent) vectors on the equator, at opposite sides of the sphere. The equator is a geodesic and we saw above that under parallel transport a tangent

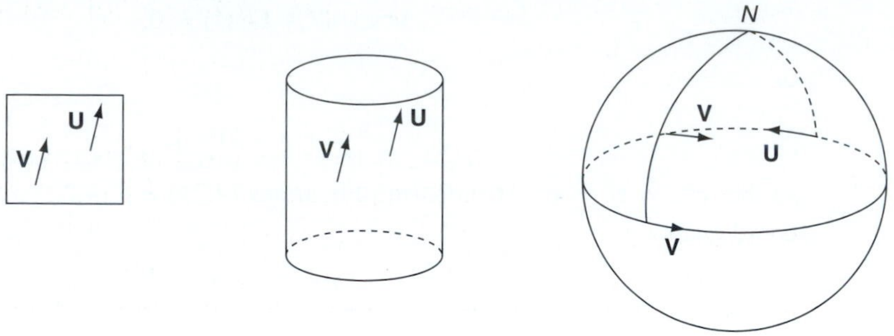

Fig. 4.7 Vectors **V** and **U** (a) in a flat space, (b) on the surface of a cylinder, (c) on the surface of a sphere.

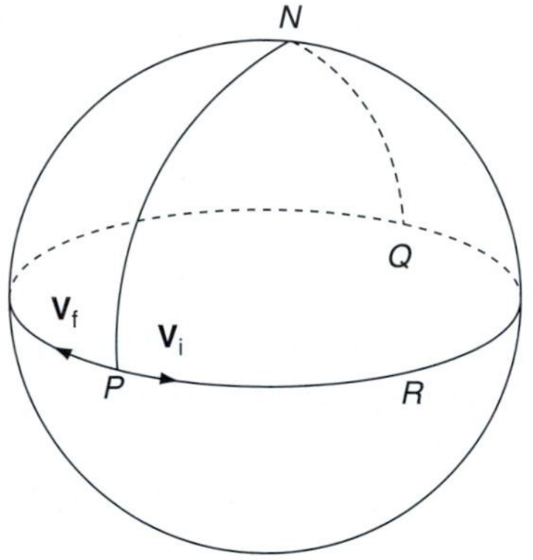

Fig. 4.8 An vector $\mathbf{V}_i$ is parallel transported along the path *PNQRP*, finishing as the vector $\mathbf{V}_f$.

vector to a geodesic remains a tangent vector, so under parallel transport along the equator **V** becomes the same as **U**, so they are parallel by definition.

Consider, however, an alternative route over which to transport **V** to the point where **U** is, that is the route over the north pole N. Since **V** is initially perpendicular to this path, it remains perpendicular to it (since the angle between a vector and the tangent vector to the geodesic is preserved under parallel transport), so when **V** arrives at the equator on the far side of the sphere it is pointing due west. Are we then obliged to say that **V** is still parallel to **U**? Is there a contradiction? There is no contradiction. What is happening is precisely an indication that the surface is *curved*, and that on a curved surface *parallel transport is not integrable*.

Consider for example (Fig. 4.8) the parallel transport of **V** round the closed path *PNQRP*. The vector starts out as $\mathbf{V}_i$ at P (east pointing) and when it arrives at Q, via N, it is west pointing. It remains west pointing along the equator from Q to P, so that the final specimen $\mathbf{V}_f$ has been rotated through π from its orientation as $\mathbf{V}_i$: $\mathbf{V}_f \neq \mathbf{V}_i$. (Actually, the angle of rotation, π in this case, is also the solid angle subtended by the area enclosed by the path at the origin; this is a general result.) Parallelism, defined in this way, is not integrable. Instead,

$\delta\mathbf{V} = \mathbf{V}_f - \mathbf{V}_i$ gives a measure of the *curvature* of the path enclosed. We now investigate this matter more thoroughly.

4.3.1 Round trips by parallel transport

Consider, as shown in Fig. 4.9, the general case of the parallel transport of the vector V^μ round the closed loop *ABCDA* on a (2-dimensional) surface, where the lines *AB* etc. are not necessarily geodesics. Under parallel transport in the direction **U** we have

$$\nabla_{\mathbf{U}}\mathbf{V} = V^\kappa{}_{;\lambda}\, U^\lambda\, \mathbf{e}_\kappa = (V^\kappa{}_{,\lambda} + \Gamma^\kappa{}_{\mu\lambda}\, V^\mu)\, U^\lambda\, \mathbf{e}_\kappa = 0,$$

hence, with $U^\lambda = \delta x^\lambda$,

$$\delta V^\kappa = -\Gamma^\kappa{}_{\lambda\mu}\, V^\lambda \delta x^\mu. \tag{4.28}$$

So, going from *A* to *B*

$$V^\kappa(B) - [V^\kappa(A)]_\mathrm{i} = -\int_{x^2=b} \Gamma^\kappa{}_{\lambda\mu}\, V^\lambda\, \mathrm{d}x^\mu = -\int_{x^2=b} \Gamma^\kappa{}_{\lambda 1} V^\lambda\, \mathrm{d}x^1,$$

where the subscript i denotes initial. Similarly, over the paths *B* to *C*, *C* to *D* and finally *D* back to *A*,

$$V^\kappa(C) - V^\kappa(B)] = -\int_{x^1=a+\delta a} \Gamma^\kappa{}_{\lambda 2} V^\lambda\, \mathrm{d}x^2,$$

$$V^\kappa(D) - V^\kappa(C) = +\int_{x^2=b+\delta b} \Gamma^\kappa{}_{\lambda 1}\, V^\lambda\, \mathrm{d}x^1,$$

$$[V^\kappa(A)]_\mathrm{f} - V^\kappa\,(D) = +\int_{x^1=a} \Gamma^\kappa{}_{\lambda 2}\, V^\lambda\, \mathrm{d}x^2,$$

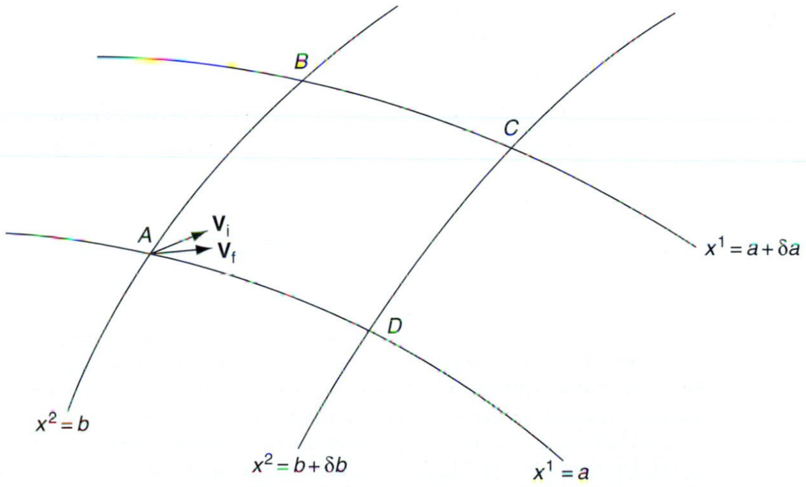

Fig. 4.9 **Parallel transport of a vector along the closed path *ABCDA*.**

where f denotes final. Adding these four equations gives

$$
\begin{aligned}
&[V^{\kappa}(A)]_{\mathrm{f}} - [V^{\kappa}(A)]_{\mathrm{i}} \\
&\quad = \left\{ -\int_{x^1=a+\delta a} + \int_{x^1=a} \right\} \Gamma^{\kappa}{}_{\lambda 2}\, V^{\lambda}\, \mathrm{d}x^2 + \left\{ \int_{x^2=b+\delta b} - \int_{x^2=b} \right\} \Gamma^{\kappa}{}_{\lambda 1}\, V^{\lambda}\, \mathrm{d}x^1 \\
&\quad \approx \int (-\delta a)\partial_1 (\Gamma^{\kappa}{}_{\lambda 2}\, V^{\lambda})\, \mathrm{d}x^2 + \int (\delta b)\partial_2 (\Gamma^{\kappa}{}_{\lambda 1}\, V^{\lambda})\, \mathrm{d}x^1 \\
&\quad \approx \delta a\, \delta b \left\{ -\partial_1 (\Gamma^{\kappa}{}_{\lambda 2} V^{\lambda}) + \partial_2 (\Gamma^{\kappa}{}_{\lambda 1}\, V^{\lambda}) \right\} \\
&\quad = \delta a\, \delta b \left\{ (-\Gamma^{\kappa}{}_{\lambda 2,1} + \Gamma^{\kappa}{}_{\lambda 1,2}))\, V^{\lambda} - \Gamma^{\kappa}{}_{\lambda 2}(V^{\lambda}{}_{,1}) + \Gamma^{\kappa}{}_{\lambda 1}(V^{\lambda}{}_{,2}) \right\}.
\end{aligned}
$$

The last two terms, involving the derivatives of V^{λ}, may be approximated by substituting from (4.28), giving

$$\Delta V^{\kappa} \approx \delta a\, \delta b\, [\Gamma^{\kappa}{}_{\lambda 1,2} - \Gamma^{\kappa}{}_{\lambda 2,1)} + \Gamma^{\kappa}{}_{\mu 2}\, \Gamma^{\mu}{}_{\lambda 1} - \Gamma^{\kappa}{}_{\mu 1}\, \Gamma^{\mu}{}_{\lambda 2}]\, V^{\lambda}. \tag{4.29}$$

Writing $\delta a\, \delta b$ as $\delta x^1\, \delta x^2$ and observing that the quantity in square brackets is antisymmetric under $1 \leftrightarrow 2$, (4.29) may be written more generally as

$$\Delta V^{\kappa} = {}^{1}\!/_{2}\, \delta x^{\mu}\, \delta x^{\nu}\, [\Gamma^{\kappa}{}_{\lambda \mu,\nu} - \Gamma^{\kappa}{}_{\lambda \nu,\mu} + \Gamma^{\kappa}{}_{\rho \nu}\, \Gamma^{\rho}{}_{\lambda \mu} - \Gamma^{\kappa}{}_{\rho \mu}\, \Gamma^{\rho}{}_{\lambda \nu}]\, V^{\lambda};$$

or

$$\Delta V^{\kappa} = {}^{1}\!/_{2}\, R^{\kappa}_{\ \lambda \mu \nu}\, V^{\lambda}\, \Delta A^{\nu \mu}, \tag{4.30}$$

where $\Delta A^{\nu\mu}$ is the area enclosed by the path and

$$R^{\kappa}{}_{\lambda \mu \nu} = \Gamma^{\kappa}{}_{\lambda \nu,\mu} - \Gamma^{\kappa}{}_{\lambda \mu,\nu} + \Gamma^{\kappa}{}_{\rho \mu}\, \Gamma^{\rho}{}_{\lambda \nu} - \Gamma^{\kappa}{}_{\rho \nu}\, \Gamma^{\rho}{}_{\lambda \mu} \tag{4.31}$$

is the *Riemann–Christoffel* (or simply *Riemann*) *curvature tensor.*[1] The fact that (unlike the connection coefficients) $R^{\kappa}{}_{\lambda \mu \nu}$ is actually a *tensor* follows from the quotient theorem: every other term in (4.30) is a tensor, so $R^{\kappa}{}_{\lambda \mu \nu}$ also is. It will be appreciated that this tensor is a property of the manifold itself, since it depends only on the connection coefficients and their derivatives. It tells us whether a space is curved or not, and since the connection coefficients depend only on the metric tensor $g_{\mu\nu}$ and its first order derivatives, the Riemann tensor also depends only on the metric tensor and its (first and second order) derivatives. Given the metric, the curvature follows.[2]

It need hardly be emphasised that this point is a high point in our study of gravity, since we saw in Chapter 1 that in Einstein's view gravity was to be described as a curvature of space-time. We now know how to describe that curvature, so a very significant mile-stone has been reached. There is, however, still some work required to appreciate exactly how curvature is to be incorporated into Einstein's theory. This will emerge in due course; but it is

[1] Some books define the Riemann–Christoffel tensor with an overall minus sign relative to this one; unfortunately there is no general agreement. Our convention agrees with that of Landau & Lifshitz (1971), Misner *et al.* (1973), Stephani (1982, 2004), and Rindler (2001). It differs by a minus sign from that of Weinberg (1972).

[2] It should be noted that this reasoning concerns a path on a surface, implicitly embedded in R^3, a flat space of one higher dimension. In considering space-time, however, we would need to talk about a surface embedded in a 4-dimensional (or, in general, an n-dimensional, space, spanned by geodesics. This can indeed be done; the student is referred to the literature for details.

nevertheless worth reflecting on the significance of the point we have reached. After rather a long slog we have reached something like the summit ridge of a mountain. At present the view is not all that great but as we wander along the ridge the clouds will disperse and we shall be rewarded with more spectacular views.

Before broaching the next topic it is convenient to note that all the above manipulations were carried out in a holonomic (coordinate) basis. In an anholonomic basis the formula for the Riemann tensor is

$$R^{\kappa}{}_{\lambda\mu\nu} = \gamma^{\kappa}{}_{\lambda\nu,\mu} - \gamma^{\kappa}{}_{\lambda\mu,\nu} + \gamma^{\kappa}{}_{\rho\mu}\gamma^{\rho}{}_{\lambda\nu} - \gamma^{\kappa}{}_{\rho\nu}\gamma^{\rho}{}_{\lambda\mu} - C^{\rho}{}_{\mu\nu}\gamma^{\kappa}{}_{\rho\lambda}. \tag{4.32}$$

This formula will be derived below (Eq. (4.50)).

4.4 Symmetries of the Riemann tensor

We shall now study some properties of the Riemann tensor, given by (4.30), as defining curvature in *space-time*. It should therefore be emphasised once more that the results above, derived in the 2-dimensional case, also hold in n dimensions. Since the Riemann tensor is of rank 4, it has $4^4 = 256$ components in 4-dimensional space-time, but not all of them are independent. For example, it follows from the definition (4.31) that

$$R^{\kappa}{}_{\lambda\mu\nu} = -R^{\kappa}{}_{\lambda\nu\mu}$$

so that of the 16 combinations of the last two indices, 4 are zero and only 6 are independent. The curvature tensor possesses other symmetries in addition, but to examine them more fully it is more convenient firstly to consider the completely covariant $\left(\text{rank}\begin{pmatrix}0\\4\end{pmatrix}\right)$ tensor $R_{\kappa\lambda\mu\nu} = g_{\kappa\rho}R^{\rho}{}_{\lambda\mu\nu}$, and secondly to work in the geodesic coordinate system at the given point. From (4.27) this means that, in this system

$$R^{\kappa}{}_{\lambda\mu\nu} = \Gamma^{\kappa}{}_{\lambda\nu,\mu} - \Gamma^{\kappa}{}_{\lambda\mu,\nu}.$$

From (3.194) and (4.26) we have, in addition,

$$\Gamma^{\kappa}{}_{\lambda\nu,\mu} = {}^{1}\!/\!{}_{2}\, g^{\kappa\rho}(g_{\rho\lambda,\nu\mu} + g_{\rho\nu,\lambda\mu} - g_{\lambda\nu,\rho\mu}),$$

so that finally

$$R_{\kappa\lambda\mu\nu} = {}^{1}\!/\!{}_{2}\,(g_{\kappa\mu,\lambda\nu} - g_{\kappa\nu,\lambda\mu} + g_{\lambda\nu,\kappa\mu} - g_{\lambda\mu,\kappa\nu}). \tag{4.33}$$

This has the immediately verifiable properties

$$\begin{aligned}
R_{\kappa\lambda\mu\nu} &= -R_{\kappa\lambda\nu\mu} && \text{(i)}\\
R_{\kappa\lambda\mu\nu} &= -R_{\lambda\kappa\mu\nu} && \text{(ii)}\\
R_{\kappa\lambda\mu\nu} &= +R_{\mu\nu\kappa\lambda} && \text{(iii)}\\
3R_{\kappa[\lambda\mu\nu]} &= R_{\kappa\lambda\mu\nu} + R_{\kappa\mu\nu\lambda} + R_{\kappa\nu\lambda\mu} = 0. && \text{(iv)}
\end{aligned} \tag{4.34}$$

How many independent components does this leave $R_{\kappa\lambda\mu\nu}$ with (in 4-dimensional space-time)? Condition (i) means that antisymmetry in the last two indices implies that there are only $4 \times 3/2 = 6$ independent combinations of these. Condition (ii) means that the same is true of the first pair of indices. Then writing

$$R_{\kappa\lambda\mu\nu} = R_{AB}$$

where A refers to the first pair and B to the second pair of indices, condition (iii) means that $R_{AB} = R_{BA}$, so we have what we could regard as a *symmetric* 6×6 matrix, which will therefore have $6 \times 7/2 = 21$ linearly independent components. Thus of the 256 components of $R_{\kappa\lambda\mu\nu}$, conditions (i)–(iii) imply that only 21 are independent (and many, of course, are zero). Equation (iv) imposes only *one* extra constraint; it is easy to check that $R_{\lambda[\kappa\mu\nu]} = -R_{\kappa[\lambda\mu\nu]}$ and similarly for the other indices (up to an overall sign), so no new information is added to (iv). We conclude that $R_{\kappa\lambda\mu\nu}$ has only 20 (linearly) independent components.[3] (Although this conclusion was reached by making use of the geodesic coordinate system, it will be appreciated that the result holds in general.)

4.5 Ricci tensor and curvature scalar

Two other quantities related to the Riemann tensor play an important part in General Relativity. The *Ricci tensor* $R_{\mu\nu}$ is defined by

$$R_{\mu\nu} = R^{\rho}{}_{\mu\rho\nu} = g^{\rho\sigma} R_{\sigma\mu\rho\nu}; \tag{4.35}$$

it amounts to taking the 'trace' of the Riemann tensor on the first and third indices. It is a $\binom{0}{2}$ tensor which is also *symmetric*

$$R_{\nu\mu} = g^{\rho\sigma} R_{\sigma\nu\rho\mu} = g^{\sigma\rho} R_{\sigma\mu\rho\nu} = R_{\mu\nu}, \tag{4.36}$$

where (4.34 iii) above has been used (as well as the symmetry of the metric tensor). The Ricci tensor $R_{\mu\nu}$ therefore has 10 independent components in space-time.

The *curvature scalar* or *Ricci scalar* R is the contraction of the Ricci tensor with the (contravariant) metric tensor:

$$R = g^{\mu\nu} R_{\mu\nu}. \tag{4.37}$$

It is obviously a scalar, and therefore the same in all coordinate systems – unlike $R_{\mu\nu}$ and $R^{\kappa}{}_{\lambda\mu\nu}$. We now give some 2-dimensional examples of the calculation of curvature.

4.5.1 Plane and sphere: holonomic basis

As usual, in these 2-dimensional examples we use Latin indices. $R^{i}{}_{klm}$ has only one component (see Footnote 3 above), which we may take to be $R^{1}{}_{212}$. The Ricci tensor has

[3] In n dimensions the number of independent components of the Riemann tensor is $n^2(n^2 - 1)/12$; see for example Weinberg (1972), pp. 142–3.

three components: R_{11}, $R_{12}=R_{21}$, R_{22}. We illustrate the foregoing formulae by calculating $R^i{}_{klm}$, R_{ik} and R in the cases of E^2 and S^2, in holonomic and anholonomic bases. We shall indeed find that E^2 is flat, and S^2 is curved.

We begin with the plane in polar coordinates, $ds^2=dr^2+r^2\,d\theta^2$, in the coordinate basis $x^1=r$, $x^2=\phi$ (alternatively $\mathbf{\theta}^1=\mathbf{d}r$, $\mathbf{\theta}^2=\mathbf{d}\phi$), and from Equations (3.237)–(3.239) we have, for the connection coefficients

$$\Gamma^1{}_{11}=\Gamma^1{}_{12}\Gamma^1{}_{21}=\Gamma^2{}_{11}=\Gamma^2{}_{22}=0;\quad \Gamma^2{}_{12}=\Gamma^2{}_{21}=\frac{1}{r},\quad \Gamma^1{}_{22}=-r.$$

Hence, from (4.31)

$$\begin{aligned} R^1{}_{212} &= \Gamma^1{}_{22,1}-\Gamma^1{}_{21,2}+\Gamma^1{}_{i1}\Gamma^i{}_{22}-\Gamma^1{}_{i2}\Gamma^i{}_{21} \\ &= \frac{\partial}{\partial r}(-r)-\Gamma^1{}_{21}\Gamma^2{}_{21}=-1-(-1)=0. \end{aligned} \tag{4.38}$$

Hence the space is flat, and of course

$$R_{ik}=0,\quad R=0.$$

The *sphere* S^2 has metric $ds^2=a^2\,(d\theta^2+\sin^2\theta\,d\phi^2)$ and in the holonomic basis $\mathbf{\theta}^1=\mathbf{d}\theta$, $\mathbf{\theta}^2=\mathbf{d}\phi$ the connection coefficients are, from (3.245)–(3.248),

$$\begin{aligned} &\Gamma^1{}_{11}=\Gamma^1{}_{12}=\Gamma^1{}_{21}=0,\quad \Gamma^1{}_{22}=-\sin\theta\cos\theta; \\ &\Gamma^2{}_{11}=\Gamma^2{}_{22}=0,\quad \Gamma^2{}_{12}=\Gamma^2{}_{21}=\cot\theta \end{aligned}$$

then

$$\begin{aligned} R^1{}_{212} &= \Gamma^1{}_{22,1}-\Gamma^1{}_{21,2}+\Gamma^1{}_{i1}\Gamma^i{}_{22}-\Gamma^1{}_{i2}\Gamma^i{}_{21} \\ &= \frac{\partial}{\partial\theta}(-\sin\theta\cos\theta)-\Gamma^1{}_{22}\Gamma^2{}_{21}=\sin^2\theta. \end{aligned} \tag{4.39}$$

This is non-zero, hence the space is curved. At the north and south poles, $\sin^2\theta=0$, but this is not an intrinsic feature; it depends on the coordinate system, and since the sphere is a homogeneous space the north and south poles may be chosen anywhere.

The components of the Ricci tensor are

$$\begin{aligned} R_{11}=R^i{}_{1i1}=R^2{}_{121}=g^{22}R_{2121}=g^{22}g_{11}R^1{}_{212}=\frac{1}{a^2\sin^2\theta}\cdot a^2\cdot\sin^2\theta=1; \\ R_{22}=R^i_{2i2}=R^1{}_{212}=\sin^2\theta;\quad R_{12}=R^i_{1i2}=0; \end{aligned} \tag{4.40}$$

and the curvature scalar is

$$R=g^{11}R_{11}+g^{22}R_{22}=\frac{1}{a^2}+\frac{1}{a^2}+\frac{2}{a^2}. \tag{4.41}$$

It is this last quantity which gives a measure of the curvature of the space, which is independent of the coordinate system. Shortly below we shall calculate the above tensors for S^2 in an anholonomic system; we expect that the Riemann and Ricci tensors will be different from (4.38) and (4.39), but the curvature scalar should again be $2/a^2$. It is worth remarking that the curvature scalar decreases as the radius a increases – in accord with intuition.

4.5.2 Plane and sphere: anholonomic basis

Now return to the plane E^2, this time using the anholonomic basis (3.240). The connection coefficients, from (3.242), (3.243) are

$$\gamma^1{}_{11} = \gamma^1{}_{12} = \gamma^1{}_{21} = 0, \quad \gamma^1{}_{22} = -\frac{1}{r},$$
$$\gamma^2{}_{11} = \gamma^2{}_{21} = \gamma^2{}_{22} = 0, \quad \gamma^2{}_{12} = \frac{1}{r}.$$

Hence, from (4.32),

$$\begin{aligned} R^1{}_{212} &= \gamma^1{}_{22,1} - \gamma^1{}_{21,2} + \gamma^1{}_{i1}\gamma^i{}_{22} - \gamma^1{}_{i2}\gamma^i{}_{21} - C^k{}_{12}\gamma^1{}_{k2} \\ &= \frac{\partial}{\partial r}\left(-\frac{1}{r}\right) - C^2{}_{12}\gamma^1{}_{22} = 0, \end{aligned} \tag{4.42}$$

where (3.261) has been used. We recover the result that the space is flat.

Finally we return to the sphere S^2, this time using an anholonomic basis. From (3.249) we have

$$\gamma^1{}_{11} = \gamma^1{}_{21} = \gamma^2{}_{22} = \gamma^2{}_{21} = \gamma^2{}_{11} = 0, \quad \gamma^2{}_{12} = -\gamma^1{}_{22} = \frac{1}{a}\cot\theta,$$

and hence

$$\begin{aligned} R^1{}_{212} &= \gamma^1{}_{22,1} - \gamma^1{}_{21,2} + \gamma^1{}_{i1}\gamma^i{}_{22} - \gamma^1{}_{i2}\gamma^i{}_{21} - C^k{}_{12}\gamma^1{}_{k2} \\ &= -\frac{1}{a^2}\frac{\partial}{\partial\theta}(\cot\theta) - C^2{}_{12}\gamma^1{}_{22} = \frac{1}{a^2}\operatorname{cosec}^2\theta - C^2{}_{12}\gamma^1{}_{22}. \end{aligned}$$

$C^2{}_{12}$ may be calculated from the Lie algebra (3.262). With

$$\mathbf{e}_1 = \frac{1}{a}\frac{\partial}{\partial\theta}, \quad \mathbf{e}_2 = \frac{1}{a\sin\theta}\frac{\partial}{\partial\phi}$$

we have

$$[\mathbf{e}_1, \mathbf{e}_2] = -\frac{\cot\theta}{a\sin\theta}\frac{\partial}{\partial\phi} = -\frac{\cot\theta}{a}\mathbf{e}_2,$$

hence $C^2{}_{12} = -\dfrac{\cot\theta}{a}$ and

$$R^1{}_{212} = \frac{1}{a^2}\operatorname{cosec}^2\theta - \frac{\cot^2\theta}{a^2} = \frac{1}{a^2}. \tag{4.43}$$

Then the components of the Ricci tensor are

$$R_{11} = R^2{}_{121} = g^{22}g_{11}R^1{}_{212} = \frac{1}{a^2}; \quad R_{22} = R^1{}_{212} = \frac{1}{a^2}; \quad R_{12} = 0,$$

and the curvature scalar is

$$R = g^{11} R_{11} + g^{22} R_{22} = \frac{2}{a^2}, \tag{4.44}$$

as in (4.41) above, and as expected.

4.6 Curvature 2-form

We have defined the Riemann tensor above, and argued that, by virtue of the quotient theorem, it really is a tensor. In this section we present an alternative and more formal argument for its tensorial nature. This is couched in the language of differential forms. We introduce a 2-form, the *curvature 2-form*, which we shall demonstrate to have true tensorial properties, and whose components in a general basis are precisely those of the Riemann curvature tensor. We shall illustrate the usefulness of the curvature 2-form by calculating the curvature of a 2-sphere. This calculation turns out to be considerably shorter than the equivalent one which involves finding the components of the Riemann tensor.

The curvature 2-form (really a matrix, or a tensor, of 2-forms) is denoted $\mathbf{\Omega}^{\mu}{}_{\nu}$ and *defined* as

$$\mathbf{\Omega}^{\mu}{}_{\nu} = \mathbf{d}\boldsymbol{\omega}^{\mu}{}_{\nu} + \boldsymbol{\omega}^{\mu}{}_{\lambda} \wedge \boldsymbol{\omega}^{\mu}{}_{\nu}. \tag{4.45}$$

(It might be remarked that this is similar in form to the definition of the torsion 2-form, Equation (3.231).) We shall first prove that $\mathbf{\Omega}^{\mu}{}_{\nu}$ is a tensor. We have first to find the transformation law for $\boldsymbol{\omega}^{\mu}{}_{\nu}$. Transforming from coordinates x^{μ} to x'^{μ} let us *define*

$$p^{\mu'}{}_{\nu} = \frac{\partial x'\,\mu}{\partial x^{\nu}}, \quad p^{\nu}{}_{\kappa'} = \frac{\partial x^{\nu}}{\partial x'^{\kappa}}, \tag{4.46}$$

transferring the prime from x to the index concerned. This notational change makes the tensorial manipulations slightly more transparent. Then we have for the basis vectors

$$\mathbf{e}_{\mu'} = p^{\nu}{}_{\mu'}\,\mathbf{e}_{\nu}$$

and

$$\begin{aligned}
\nabla\,\mathbf{e}_{\mu'} &= \nabla(p^{\nu}{}_{\mu'}\,\mathbf{e}_{\nu}) = (\mathbf{d}p^{\nu}{}_{\mu'})\,\mathbf{e}_{\nu} + p^{\nu}{}_{\mu'}\,\nabla\,\mathbf{e}_{\nu} \\
&= (\mathbf{d}p^{\nu}{}_{\mu'})\,\mathbf{e}_{\nu} + p^{\nu}{}_{\mu'}\,\boldsymbol{\omega}^{\kappa}{}_{\nu}\,\mathbf{e}_{\kappa} \\
&= (\mathbf{d}p^{\kappa}{}_{\mu'} + p^{\nu}{}_{\mu'}\,\boldsymbol{\omega}^{\kappa}{}_{\nu})\,\mathbf{e}_{\kappa} \\
&= (\mathbf{d}p^{\kappa}{}_{\mu'} + p^{\nu}{}_{\mu'}\,\boldsymbol{\omega}^{\kappa}{}_{\nu})\,p^{\lambda'}{}_{\kappa}\,\mathbf{e}_{\lambda'}.
\end{aligned}$$

On the other hand

$$\nabla\,\mathbf{e}_{\mu'} = \boldsymbol{\omega}^{\nu'}{}_{\mu'}\,\mathbf{e}_{\nu'}\,;$$

hence

$$\boldsymbol{\omega}^{\nu'}{}_{\mu'} = p^{\nu'}{}_{\kappa}\,p^{\lambda}{}_{\mu'}\,\boldsymbol{\omega}^{\kappa}{}_{\gamma} + p^{\nu'}{}_{\kappa}\,\mathbf{d}p^{\kappa}{}_{\mu'}. \tag{4.47}$$

Then

$$\begin{aligned}
\boldsymbol{\Omega}^{\mu'}{}_{\nu'} &= \mathbf{d}\boldsymbol{\omega}^{\mu'}{}_{\nu'} + \boldsymbol{\omega}^{\mu'}{}_{\kappa'} \wedge \boldsymbol{\omega}^{\kappa'}{}_{\nu'} \\
&= \mathbf{d}\,[p^{\mu'}{}_{\lambda} p^{\kappa}{}_{\nu'}\, \boldsymbol{\omega}^{\lambda}{}_{\kappa} + p^{\mu'}{}_{\lambda}\, \mathbf{d}p^{\lambda}{}_{\nu'}] \\
&\quad + (p^{\mu'}{}_{\lambda} p^{\sigma}{}_{\kappa'}\, \boldsymbol{\omega}^{\lambda}{}_{\sigma} + p^{\mu'}{}_{\lambda}\, \mathbf{d}P^{\lambda}{}_{k'}) \wedge (p^{\kappa'}{}_{\chi} p^{\rho}_{\nu'}\, \boldsymbol{\omega}^{\chi}{}_{\rho} + p^{\kappa'}{}_{\chi}\, \mathbf{d}p^{\chi}{}_{\nu'}) \\
&= [(\mathbf{d}\,p^{\mu'}{}_{\lambda})\, p^{\kappa}{}_{\nu'} + p^{\mu'}{}_{\lambda}\, (\mathbf{d}p^{\kappa}{}_{\nu'})] \wedge \boldsymbol{\omega}^{\lambda}{}_{\kappa} \\
&\quad + p^{\mu'}{}_{\lambda} p^{\kappa}{}_{\nu'}\, \mathbf{d}\boldsymbol{\omega}^{\lambda}{}_{\kappa} + \mathbf{d}p^{\mu'}{}_{\lambda} \wedge \mathbf{d}p^{\lambda}{}_{\nu'} \\
&\quad + \left(p^{\mu'}{}_{\lambda} p^{\sigma}{}_{\kappa'}\, \boldsymbol{\omega}^{\lambda}{}_{\sigma} - p^{\lambda}{}_{\kappa'}\, \mathbf{d}p^{\mu'}{}_{\lambda}\right) \wedge \left(p^{\kappa'}{}_{\chi} p^{\rho}_{\nu'}\, \boldsymbol{\omega}^{\chi}_{\rho} + p^{\kappa'}{}_{\chi}\, \mathbf{d}p^{\chi}{}_{\nu'}\right) \cdot \\
&= [(\mathbf{d}p^{\mu'}{}_{\lambda})\, p^{\kappa}{}_{\nu'} + p^{\mu'}{}_{\lambda}(\mathbf{d}p^{\kappa}{}_{\nu'})] \wedge \boldsymbol{\omega}^{\lambda}{}_{\kappa} \\
&\quad + p^{\mu'}{}_{\lambda} p^{\kappa}{}_{\nu'}\, \mathbf{d}\boldsymbol{\omega}^{\lambda}{}_{\kappa} + \mathbf{d}p^{\mu'}{}_{\lambda} \wedge \mathbf{d}p^{\lambda}{}_{\nu'} \\
&\quad + p^{\mu'}{}_{\lambda} p^{\rho}_{\nu'}\, \boldsymbol{\omega}^{\lambda}{}_{\sigma} \wedge \boldsymbol{\omega}^{\sigma}_{\rho} - p^{\rho}_{\nu'}\, \mathbf{d}p^{\mu'}{}_{\lambda} \wedge \boldsymbol{\omega}^{\lambda}{}_{\rho} \\
&\quad + p^{\mu'}{}_{\lambda}\, \boldsymbol{\omega}^{\lambda}{}_{\sigma} \wedge \mathbf{d}p^{\sigma}{}_{\nu'} \; - \; \mathbf{d}p^{\mu'}{}_{\lambda} \wedge \mathbf{d}p^{\lambda}{}_{\nu'}.
\end{aligned}$$

There are eight terms in the above final expression. Numbering them 1 to 8, we see that terms 1 and 6 cancel, as also do terms 2 and 7, and terms 4 and 8. This leaves terms 3 and 5, giving

$$\boldsymbol{\Omega}^{\mu'}{}_{\nu'} = p^{\mu'}{}_{\lambda} p^{\kappa}{}_{\nu'}\, (\mathbf{d}\boldsymbol{\omega}^{\lambda}{}_{\kappa} + \boldsymbol{\omega}^{\lambda}{}_{\sigma} \wedge \boldsymbol{\omega}^{\sigma}{}_{\kappa}) = p^{\mu'}{}_{\lambda} p^{k}{}_{\nu'}\, \boldsymbol{\Omega}^{\lambda}{}_{\kappa}; \tag{4.48}$$

which completes the proof that $\boldsymbol{\Omega}^{\mu}{}_{\nu}$ is indeed a tensorial 2-form. It may therefore be expanded in terms of the basis 1-forms $\boldsymbol{\theta}^{\mu}$:

$$\boldsymbol{\Omega}^{\mu}{}_{\nu} = \tfrac{1}{2} R^{\mu}{}_{\nu\rho\sigma}\, \boldsymbol{\theta}^{\rho} \wedge \boldsymbol{\theta}^{\sigma}. \tag{4.49}$$

We shall now show that the quantity $R^{\mu}{}_{\nu\rho\sigma}$ is indeed the Riemann–Christoffel curvature tensor. We have, from (4.45)

$$\begin{aligned}
\boldsymbol{\Omega}^{\mu}{}_{\nu} &= \mathbf{d}\boldsymbol{\omega}^{\mu}{}_{\nu} + \boldsymbol{\omega}^{\mu}{}_{\lambda} \wedge \boldsymbol{\omega}^{\lambda}{}_{\nu} \\
&= \mathbf{d}(\gamma^{\mu}{}_{\nu\rho} \boldsymbol{\theta}^{\rho}) + \gamma^{\mu}{}_{\lambda\rho} \gamma^{\lambda}{}_{\nu\sigma}\, \boldsymbol{\theta}^{\rho} \wedge \boldsymbol{\theta}^{\sigma} \\
&= (\mathbf{d}\gamma^{\mu}{}_{\nu\rho}) \wedge \boldsymbol{\theta}^{\rho} + \gamma^{\mu}{}_{\nu\kappa}\, \mathbf{d}\boldsymbol{\theta}^{k} + \gamma^{\mu}{}_{\lambda\rho} \gamma^{\lambda}{}_{\nu\sigma}\, \boldsymbol{\theta}^{\rho} \wedge \boldsymbol{\theta}^{\sigma} \\
&= \gamma^{\mu}{}_{\nu\rho,\sigma}\, \boldsymbol{\theta}^{\sigma} \wedge \boldsymbol{\theta}^{\rho} - \tfrac{1}{2} \gamma^{\mu}{}_{\nu\kappa} C^{\kappa}{}_{\rho\sigma}\, \boldsymbol{\theta}^{\rho} \wedge \boldsymbol{\theta}^{\sigma} + \gamma^{\mu}{}_{\lambda\rho} \gamma^{\lambda}{}_{\nu\sigma}\, \boldsymbol{\theta}^{\rho} \wedge \boldsymbol{\theta}^{\sigma}
\end{aligned}$$

(in this last line $\gamma^{\mu}{}_{\nu\rho,\sigma}$ should strictly be written as $e_{\sigma}(\gamma^{\mu}{}_{\nu\rho})$; and Equation (3.171) has been used in the second term)

$$= \tfrac{1}{2} [\gamma^{\mu}{}_{\nu\rho,\sigma} - \gamma^{\mu}{}_{\nu\sigma,\rho} + \gamma^{\mu}{}_{\lambda\rho} \gamma^{\lambda}{}_{\nu\sigma} - \gamma^{\mu}{}_{\lambda\sigma} \gamma^{\lambda}{}_{\nu\rho} - \gamma^{\mu}{}_{\nu\kappa} C^{\kappa}{}_{\rho\sigma}]\, \boldsymbol{\theta}^{\rho} \wedge \boldsymbol{\theta}^{\sigma}. \tag{4.50}$$

Equation (4.50) is of the form of (4.49) with

$$R^{\mu}{}_{\nu\rho\sigma} = \gamma^{\mu}{}_{\nu\rho,\sigma} - \gamma^{\mu}{}_{\nu\sigma,\rho} + \gamma^{\mu}{}_{\lambda\rho} \gamma^{\lambda}{}_{\nu\sigma} - \gamma^{\mu}{}_{\lambda\sigma} \gamma^{\lambda}{}_{\nu\rho} - \gamma^{\mu}{}_{\nu\kappa} C^{\kappa}{}_{\rho\sigma}, \tag{4.51}$$

as in Equation (4.32) above – which we have now proved. In a holonomic basis $C^{\kappa}{}_{\rho\sigma} = 0$ and $\gamma^{\kappa}{}_{\mu\nu} \to \Gamma^{\kappa}{}_{\mu\nu}$ and we recover (4.31).

Let us illustrate this procedure for S^2. In a particular anholonomic basis the non-zero connection 1-forms are (see (3.248a)) $\boldsymbol{\omega}^{1}{}_{2} = -\cos\theta\, \mathbf{d}\phi$, so

$$\begin{aligned}
\boldsymbol{\Omega}^{1}{}_{2} &= \mathbf{d}\boldsymbol{\omega}^{1}{}_{2} + \boldsymbol{\omega}^{1}{}_{i} \wedge \boldsymbol{\omega}^{i}{}_{2} = \mathbf{d}\boldsymbol{\omega}^{1}{}_{2} = \sin\theta\, \mathbf{d}\theta \wedge \mathbf{d}\phi \\
&= \frac{1}{a^2}\, \boldsymbol{\theta}^{1} \wedge \boldsymbol{\theta}^{2} = \tfrac{1}{2} R^{1}{}_{2ik}\, \boldsymbol{\theta}^{i} \wedge \boldsymbol{\theta}^{k}
\end{aligned}$$

and hence

$$R^1{}_{212} = \frac{1}{a^2} \Rightarrow R = \frac{2}{a^2}$$

as in (4.41) above. I hope the reader will notice that the calculation of curvature is much simpler using differential forms, in this case the curvature 2-form. Apart from the greater clarity that forms shed on the conceptual nature of the calculations, there is no doubt that they make calculations much simpler.

Let us finally investigate the connection between the curvature of a space and the non-commutativity of absolute derivatives when acting on vector fields. First define the *curvature operator* ρ $(\mathbf{U}, \mathbf{V})$:

$$\rho\,(\mathbf{U}, \mathbf{V}) = \nabla_{\mathbf{U}} \nabla_{\mathbf{V}} - \nabla_{\mathbf{V}} \nabla_{\mathbf{U}} - \nabla_{[\mathbf{U},\mathbf{V}]}. \tag{4.52}$$

We recall that if $\mathbf{U}$ and $\mathbf{V}$ are vectors, then so is $[\mathbf{U},\mathbf{V}]$. In a holonomic basis $[\mathbf{U},\mathbf{V}]=0$ if $\mathbf{U}$, $\mathbf{V}$ are both basis vectors, so the last term above disappears. It is clear that ρ is antisymmetric in its arguments,

$$\rho\,(\mathbf{U}, \mathbf{V}) = -\rho\,(\mathbf{V}, \mathbf{U}). \tag{4.53}$$

In the basis $\mathbf{e}_\mu$, with

$$[\mathbf{e}_\mu, \mathbf{e}_\nu] = C^\lambda{}_{\mu\nu}\, \mathbf{e}_\lambda,$$

we have

$$\rho\,(\mathbf{e}_\mu, \mathbf{e}_\nu) = \nabla_{e_\mu} \nabla_{e_\nu} - \nabla_{e_\nu} \nabla_{e_\mu} - \nabla_{C^\lambda{}_{\mu\nu} e_\lambda} \equiv \nabla_\mu \nabla_\nu - \nabla_\nu \nabla_\mu - C^\lambda{}_{\mu\nu} \nabla_\lambda. \tag{4.54}$$

Since $\boldsymbol{\nabla}_\mu$ acts on a vector to give a vector, then $\rho(\mathbf{e}_\mu, \mathbf{e}_\nu)$ also operates on one vector to give another. We have

$$\nabla_\nu\, \mathbf{e}_\kappa = \gamma^\rho{}_{\kappa\nu}\, \mathbf{e}_\rho \tag{4.55}$$

hence

$$\nabla_\mu \nabla_\nu \mathbf{e}_\kappa = \nabla_\mu\,(\gamma^\rho{}_{\kappa\nu}\, \mathbf{e}_\rho) = \gamma^\rho{}_{\kappa\nu,\mu}\, \mathbf{e}_\rho + \gamma^\rho{}_{\kappa\nu} \gamma^\sigma{}_{\rho\mu}\, \mathbf{e}_\sigma$$

(where, strictly, the first term should be $e_\mu(\gamma^\rho{}_{\kappa\nu})\mathbf{e}_\rho$), and so

$$\begin{aligned}
&(\boldsymbol{\nabla}_\mu \boldsymbol{\nabla}_\nu - \boldsymbol{\nabla}_\nu \boldsymbol{\nabla}_\mu - C^\lambda{}_{\mu\nu} \boldsymbol{\nabla}_\lambda)\, \mathbf{e}_\kappa \\
&= (\gamma^\rho{}_{\kappa\nu,\mu} - \gamma^\rho{}_{\kappa\mu,\nu} + \gamma^\rho{}_{\kappa\nu} \gamma^\sigma{}_{\rho\mu} + \gamma^\rho{}_{\kappa\mu} \gamma^\sigma{}_{\rho\nu} - C^\lambda{}_{\mu\nu} \gamma^\rho{}_{\kappa\lambda})\, \mathbf{e}_\rho \\
&= R^\rho{}_{\kappa\mu\nu} \mathbf{e}_\rho,
\end{aligned} \tag{4.56}$$

with $R^\rho{}_{\kappa\mu\nu}$ as in (4.51) above. In particular, in a holonomic system

$$\rho\,(\mathbf{e}_\mu, \mathbf{e}_\nu)\, \mathbf{e}_\kappa = [\boldsymbol{\nabla}_\mu, \boldsymbol{\nabla}_\nu]\, \mathbf{e}_\kappa = R^\rho{}_{\kappa\mu\nu}\, \mathbf{e}_\rho, \tag{4.57}$$

with $R^\rho{}_{\kappa\mu\nu}$ as in (4.31) above. We then clearly see the connection between the non-commutativity of the absolute derivative in a particular space and the curvature of the space. It follows straightforwardly from (4.55) that (Problem 4.4)

$$V^{\lambda}{}_{;\mu;\nu} - V^{\lambda}{}_{;\nu;\mu} = -R^{\lambda}{}_{\rho\mu\nu} V^{\rho}, \tag{4.58}$$

$$V^{\lambda}{}_{;\mu;\nu} - V_{\lambda;\nu;\mu} = +R^{\rho}{}_{\lambda\mu\nu} V_{\rho}. \tag{4.59}$$

Finally, it is worth noting, for the record, that in theories *with torsion*, the right hand sides of the equations above become modified to include torsion terms.[4]

4.7 Geodesic deviation

Geodesic deviation is a manifestation of curvature. Let me begin by giving a physical motivation for introducing it. In General Relativity a particle under the influence of no force other than gravity follows a geodesic. Because of the Equivalence Principle, however, the acceleration of one freely falling body has no significance in General Relativity – see Section 1.2 above. An observer moving with the body also follows a geodesic, and the particle appears (and therefore *is*) at rest. What *is* observable, however, is the relative acceleration of particles on neighbouring geodesics. An example was given in Section 1.2, where observers (particles) in a freely falling lift on the Earth approach one another, in fact accelerate towards one another. This is the so-called 'tidal effect' of the Earth's gravitational field. We shall now investigate the relative acceleration of neighbouring geodesics geometrically and shall find that this depends on the (Riemannian) *curvature* of the space, thus giving another connection between a realistic gravitational field (for example that of the Earth) and the curvature of space (or space-time).

Consider, then, a one-parameter family of geodesics $x^{\mu}(s, v)$, where v labels the geodesic and s the arc length along a given geodesic. This is illustrated in Fig. 4.10. Let $\mathbf{t}$ be the tangent to a geodesic

$$t^{\mu} = \frac{\partial x^{\mu}}{\partial s}. \tag{4.60}$$

The geodesic equation is $\nabla_{\mathbf{t}}\mathbf{t} = 0$ (autoparallelism), i.e.

$$\nabla_{\mathbf{t}}\mathbf{t} = t^{\mu}{}_{;\nu}\, t^{\nu}\, \mathbf{e}_{\mu} = 0. \tag{4.61}$$

We label points on neighbouring geodesics by s and connect these points, labelled by v and $v + \delta v$, by an infinitesimal vector $\boldsymbol{\eta}$:

$$\eta^{\mu} = \frac{\partial x^{\mu}}{\partial v}\, \delta v\,. \tag{4.62}$$

Now let us calculate $\nabla_{\mathbf{t}}\boldsymbol{\eta}$:

[4] See, for example, de Felice and Clarke (1990), p. 105.

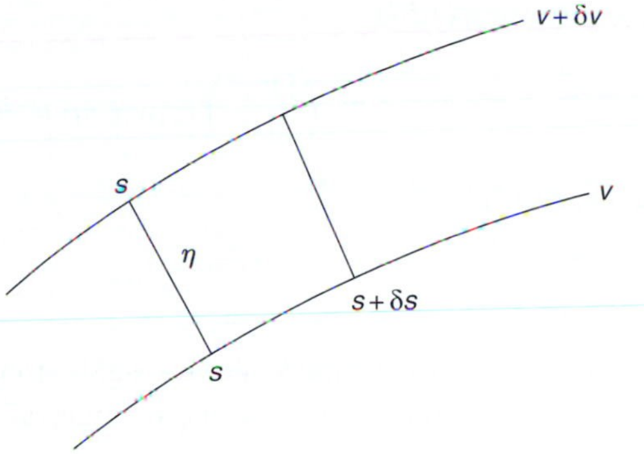

Fig. 4.10 Two geodesics; v labels the geodesic and s the arc length along a given geodesic.

$$\begin{aligned}\nabla_{\mathbf{t}}\boldsymbol{\eta} = \eta^{\mu}{}_{;\nu}\, t^{\nu}\mathbf{e}_{\mu} &= \left\{\frac{\partial\eta^{\mu}}{\partial x^{\nu}} + \Gamma^{\mu}{}_{\rho\nu}\eta^{\rho}\right\}\frac{\partial x^{\nu}}{\partial s}\,\mathbf{e}_{\mu}\\ &= \left\{\frac{\partial\eta^{\mu}}{\partial s} + \Gamma^{\mu}{}_{\rho\nu}\frac{\partial x^{\rho}}{\partial v}\frac{\partial x^{\nu}}{\partial s}\,\delta v\right\}\mathbf{e}_{\mu}\\ &= \left\{\frac{\partial^2 x^{\mu}}{\partial s\,\partial v} + \Gamma^{\mu}_{\rho\nu}\frac{\partial x^{\rho}}{\partial v}\frac{\partial x^{\nu}}{\partial s}\right\}\delta v\,\mathbf{e}_{\mu}. \end{aligned}\tag{4.63}$$

Similarly, let us calculate $\nabla_{\boldsymbol{\eta}}\mathbf{t}$:

$$\begin{aligned}\nabla_{\boldsymbol{\eta}}\mathbf{t} = t^{\mu}{}_{;\nu}\,\eta^{\nu}\,\mathbf{e}_{\mu} &= \left\{\frac{\partial t^{\mu}}{\partial x^{\nu}} + \Gamma^{\mu}{}_{\rho\nu}t^{\rho}\right\}\frac{\partial x^{\nu}}{\partial v}\,\delta v\,\mathbf{e}_{\mu}\\ &= \left\{\frac{\partial t^{\mu}}{\partial s} + \Gamma^{\mu}{}_{\rho\nu}\frac{\partial x^{\rho}}{\partial s}\frac{\partial x^{\nu}}{\partial v}\right\}\delta v\,\mathbf{e}_{\mu}\\ &= \left\{\frac{\partial^2 x^{\mu}}{\partial s\,\partial v} + \Gamma^{\mu}{}_{\rho\nu}\frac{\partial x^{\rho}}{\partial s}\frac{\partial x^{\nu}}{\partial v}\right\}\delta v\,\mathbf{e}_{\mu}\\ &= \nabla_{\mathbf{t}}\boldsymbol{\eta}, \end{aligned}\tag{4.64}$$

because of the commutativity of second order partial derivatives and the symmetry of $\Gamma^{\mu}{}_{\rho\nu}$. We want to calculate the relative acceleration of two geodesics, that is,

$$\nabla_{\mathbf{t}}\,(\nabla_{\mathbf{t}}\boldsymbol{\eta}) = \nabla_{\mathbf{t}}\,\nabla_{\mathbf{t}}\boldsymbol{\eta}.$$

The geodesic equation $\nabla_{\mathbf{t}}\mathbf{t}=0$ implies that $\nabla_{\boldsymbol{\eta}}\,\nabla_{\mathbf{t}}\mathbf{t}=0$, which is the same as

$$\{\nabla_{\mathbf{t}}\,\nabla_{\boldsymbol{\eta}} + [\nabla_{\boldsymbol{\eta}},\,\nabla_{\mathbf{t}}]\}\,\mathbf{t}=0\tag{4.65}$$

and hence, using (4.64) and (4.65)

$$\nabla_{\mathbf{t}}\nabla_{\mathbf{t}}\,\boldsymbol{\eta} = -\,[\nabla_{\boldsymbol{\eta}},\,\nabla_{\mathbf{t}}]\,\mathbf{t} = [\nabla_{\mathbf{t}},\nabla_{\boldsymbol{\eta}}]\,\mathbf{t} = \rho\,(\mathbf{t},\,\boldsymbol{\eta})\,\mathbf{t}.$$

Inserting components, this equation is

$$\nabla_{\mathbf{t}}\nabla_{\mathbf{t}}\,(\eta^{\mu}\,\mathbf{e}_{\mu}) = \rho\,(t^{\lambda}\,\mathbf{e}_{\lambda},\,\eta^{\kappa}\,\mathbf{e}_{\kappa})\,t^{\mu}\,\mathbf{e}_{\mu} = t^{\lambda}\,\eta^{\kappa}\,t^{\mu}\,\rho\,(\mathbf{e}_{\lambda},\,\mathbf{e}_{\kappa})\,\mathbf{e}_{\mu};$$

which gives, using (4.57)

$$\left(\frac{\mathrm{D}^2\eta^\sigma}{\mathrm{d}s^2}\right)\mathbf{e}_\sigma = t^\lambda\,\eta^\kappa\,t^\mu\,R^\sigma{}_{\mu\lambda\kappa}\,\mathbf{e}_\sigma$$

and hence, finally,

$$\frac{\mathrm{D}^2\eta^\sigma}{\mathrm{d}s^2} = R^\sigma{}_{\mu\lambda\kappa}\,t^\mu\,t^\lambda\,\eta^\kappa. \tag{4.66}$$

This is the relative acceleration of two neighbouring geodesics, written in terms of the absolute derivative (denoted D, in the notation of (3.202a)). We have reverted to the notation of ordinary, rather than partial, derivatives, to emphasise the fact that we can forget about the one-parameter family of curves; these equations apply to neighbouring geodesics. Hence the deviation of geodesics from straight lines (in which the relative acceleration is zero) does indicate the presence of curvature; and hence the 'tidal effect' in the lift in Einstein's thought-experiment is in turn an indication of a curved space-time.

4.8 Bianchi identities

The Bianchi identities are interesting in their own right, but for our purposes they are introduced here only because of their usefulness in the context of Einstein's equations for the gravitational field. They are identities involving the covariant derivatives of the Riemann tensor; we therefore start by taking the (exterior) derivative of the curvature 2-form.

The curvature 2-form is, from Equation (4.45)

$$\mathbf{\Omega}^\mu{}_\nu = \mathbf{d}\boldsymbol{\omega}^\mu{}_\nu + \boldsymbol{\omega}^\mu{}_\lambda \wedge \boldsymbol{\omega}^\lambda{}_\nu. \tag{4.67}$$

Hence (recall that $\mathbf{d}^2 = 0$, and Equation (3.90))

$$\mathbf{d}\mathbf{\Omega}^\mu{}_\nu = \mathbf{d}\boldsymbol{\omega}^\mu{}_\lambda \wedge \boldsymbol{\omega}^\lambda{}_\nu - \boldsymbol{\omega}^\mu{}_\lambda \wedge \mathbf{d}\boldsymbol{\omega}^\lambda{}_\nu. \tag{4.68}$$

On the other hand,

$$\begin{aligned}
\boldsymbol{\omega}^\mu{}_\lambda \wedge \mathbf{\Omega}^\lambda{}_\nu &= \boldsymbol{\omega}^\mu{}_\lambda \wedge \mathbf{d}\boldsymbol{\omega}^\lambda{}_\nu + \boldsymbol{\omega}^\mu{}_\lambda \wedge \boldsymbol{\omega}^\lambda{}_\kappa \wedge \boldsymbol{\omega}^\kappa{}_\nu,\\
\mathbf{\Omega}^\mu{}_\lambda \wedge \boldsymbol{\omega}^\lambda{}_\nu &= \mathbf{d}\boldsymbol{\omega}^\mu{}_\lambda \wedge \boldsymbol{\omega}^\lambda{}_\nu + \boldsymbol{\omega}^\mu{}_\kappa \wedge \boldsymbol{\omega}^\kappa{}_\lambda \wedge \boldsymbol{\omega}^\lambda{}_\nu.
\end{aligned}$$

The last terms in the above equations are identical (κ and λ are both dummy indices), so (4.68) gives

$$\mathbf{d}\mathbf{\Omega}^\mu{}_\nu = \mathbf{\Omega}^\mu{}_\lambda \wedge \boldsymbol{\omega}^\lambda{}_\nu - \boldsymbol{\omega}^\mu{}_\lambda \wedge \mathbf{\Omega}^\lambda{}_\nu. \tag{4.69}$$

These are the Bianchi identities written in terms of the curvature 2-form. To find their expression in terms of the Riemann tensor, take an arbitrary point P in the manifold and choose geodesic local coordinates there, in which (see (4.27)) $\Gamma^\mu{}_{\nu\lambda} = 0$, and hence $\boldsymbol{\omega}^\mu{}_\nu = 0$, so (4.67) gives

$$\mathbf{d\Omega}^{\mu}{}_{\nu} = 0 \quad \text{(geodesic coordinates)}. \tag{4.70}$$

In a holonomic basis we have, from (4.40),

$$\mathbf{\Omega}^{\mu}{}_{\nu} = {}^1\!/\!_2 R^{\mu}{}_{\nu\rho\sigma}\, \mathbf{d}x^{\rho} \wedge \mathbf{d}x^{\sigma},$$

hence

$$\mathbf{d\Omega}^{\mu}{}_{\nu} = {}^1\!/\!_2 R^{\mu}{}_{\nu\rho\sigma,\lambda}\, \mathbf{d}x^{\lambda} \wedge \mathbf{d}x^{\rho} \wedge \mathbf{d}x^{\sigma} \tag{4.71}$$

with the consequence that

$$R^{\mu}{}_{\nu\rho\sigma,\lambda} + R^{\mu}{}_{\nu\sigma\lambda,\rho} + R^{\mu}{}_{\nu\lambda\rho,\sigma} = 0, \tag{4.72}$$

(the symmetry in the three indices in (4.72) cancelling the antisymmetry in the wedge products of (4.71), thus yielding (4.70).) This equation holds in a geodesic coordinate system, in which covariant and ordinary derivatives are the same (the connection coefficients vanishing); so in a geodesic coordinate system (4.72) may be replaced by

$$R^{\mu}{}_{\nu\rho\sigma;\lambda} + R^{\mu}{}_{\nu\sigma\lambda;\rho} + R^{\mu}{}_{\nu\lambda\rho;\sigma} = 0. \tag{4.73}$$

These, however, are *tensor* equations, so hold in any coordinate system. They are the *Bianchi identities*. They, like the other results in differential geometry that we have obtained, will find their application to General Relativity in the succeeding chapters.

Further reading

See the 'Further reading' for Chapter 3.

Problems

4.1 Show that the 2-dimensional space with metric

$$ds^2 = y\, dx^2 + x\, dy^2$$

is curved, and that the curvature scalar is

$$R = \frac{1}{2xy}\left(\frac{1}{x} + \frac{1}{y}\right).$$

4.2 A Riemannian space is said to be of constant curvature if the metric and Riemann tensors are such that

$$R_{hijk} = K(g_{hj}\, g_{ik} - g_{hk}\, g_{ij})$$

where K is a numerical constant. If the metric of such a (3-dimensional) space is

$$ds^2 = dr^2 + f^2(r)(d\theta^2 + \sin^2\theta\, d\phi^2),$$

with $x^1 = r$, $x^2 = \theta$, $x^3 = \phi$, show that the three independent non-zero components of R_{hijk} are R_{1212}, R_{1313} and R_{2323}. Calculate R_{1212} and show that K may be $1, 0$ or -1 according as $f = \sin r$, r or $\sinh r$ (up to a multiplicative constant), provided that $f = 0$ at $r = 0$.

(You may assume that the only non-zero Christoffel symbols are

$\Gamma^1{}_{22}$, $\Gamma^1{}_{33}$, $\Gamma^2{}_{21} = \Gamma^2{}_{12}$ and $\Gamma^3{}_{31} = \Gamma^3{}_{13}$.)

4.3 A Riemannian 4-space has metric

$$ds^2 = e^{2\sigma}[(dx^1)^2 + (dx^2)^2 + (dx^3)^2 + (dx^4)^2]$$

where $\sigma = \sigma(x^1, x^2, x^3, x^4)$. If t^μ is the unit tangent to a geodesic prove that, along the geodesic,

$$\frac{dt^\mu}{\delta\sigma} + 2(\sigma_\nu t^\nu)\, t^\mu = \sigma^\mu$$

where $\sigma_\mu = \dfrac{\partial\sigma}{\partial x^\mu}$.

4.4 Given that $[\nabla_\mu, \nabla_\nu]\, \mathbf{e}_\kappa = R^\rho{}_{\kappa\mu\nu}\, \mathbf{e}_\rho$, show that for a vector $\mathbf{V} = V^\lambda \mathbf{e}_\lambda$,

$$V^\lambda{}_{;\mu;\nu} - V^\lambda{}_{;\nu;\mu} = -R^\lambda{}_{\rho\mu\nu} V^\rho.$$

Show also that the covariant components obey

$$V_{\lambda;\mu;\nu} - V_{\lambda;\nu;\mu} = +R^\rho_{\lambda\mu\nu} V_\rho .$$

4.5 Prove that the Bianchi identity (4.64) implies that

$$R^\mu{}_{\rho;\mu} = \tfrac{1}{2} R_{,\rho}.$$

5 Einstein field equations, the Schwarzschild solution and experimental tests of General Relativity

Important milestones in the early history of General Relativity were the Einstein field equations, Schwarzschild's solution to them and the observational consequences of this solution. The Schwarzschild solution[1] describes the space-time in the vicinity of a static, spherically symmetric mass, like the Sun, and the observational tests of this solution include the precession of the perihelion of planetary orbits – in particular the orbit of Mercury – and the bending of light in a gravitational field. A more recent test is the so-called radar echo delay of a signal sent from one planet (Earth) and reflected back from another one. An additional test of General Relativity, which depends only on the Equivalence Principle and not on the field equations, is the gravitational red-shift of light. The successful passing of these tests established General Relativity as the 'correct' theory of gravity. A feature of the Schwarzschild solution, not emphasised in the early days but given great prominence since, is the presence of the 'Schwarzschild' radius, which is the signature for the phenomenon of black holes. These matters are the concerns of this chapter. We begin with a comparison of the geodesic equation and the Newtonian limit of a weak, static gravitational field.

5.1 Newtonian limit

Consider the case of a weak, static field (such as, to a good approximation, that of the Sun) and a particle moving slowly in it ($v \ll c$). With $x^0 = ct, x^1 = x, x^2 = y, x^3 = z$, an *inertial* frame is one in which the metric tensor is

$$g_{\mu\nu} = \eta_{\mu\nu} = \mathrm{diag}\,(-1, 1, 1, 1),$$

so that $\mathrm{d}s^2 = -c^2\,\mathrm{d}t^2 + \mathrm{d}x^2 + \mathrm{d}y^2 + \mathrm{d}z^2$. A *weak* field is one for which

$$g_{\mu\nu} = \eta_{\mu\nu} + h_{\mu\nu} \tag{5.1}$$

with $|h_{\mu\nu}| \ll 1$; each element of $g_{\mu\nu}$ is close to its inertial value. Non-relativistic motion, on the other hand, implies that $\tau \approx t$, $\dfrac{\mathrm{d}x^0}{\mathrm{d}\tau} \approx c$, $\dfrac{\mathrm{d}x^i}{\mathrm{d}\tau} \approx v^i \ll c$, so the geodesic Equation (4.18) with $\rho = i$ becomes

$$\frac{1}{c^2}\,\frac{\mathrm{d}^2 x^i}{\mathrm{d}t^2} + \Gamma^i{}_{00} = 0 \tag{5.2}$$

[1] Schwarzschild (1916a).

since the terms in $\frac{dx^i}{d\tau}$ are neglected. (We may write this equation as

$$\frac{d^2x^i}{dt^2} = a^i = -c^2\,\Gamma^i{}_{00}:$$

then the right hand side represents the 'gravitational force', which gives the particle its acceleration. It is better, however, to discard the use of the term 'force' and accustom ourselves to thinking instead of gravity implying a curved space-time, in which particles move *freely*.) The connection coefficient in (5.2) is

$$\begin{aligned}\Gamma^i{}_{00} &= {}^1\!/\!_2\, g^{i\sigma}(2\,g_{0\sigma,0} - g_{00,\sigma}) = -{}^1\!/\!_2\, g^{ik} g_{00,k} \\ &\approx -{}^1\!/\!_2\,\eta^{ik}\frac{\partial h_{00}}{\partial x^k} = -{}^1\!/\!_2\nabla_i\, h_{00},\end{aligned} \tag{5.3}$$

where the second and third equalities follow because the field is static, the fourth because we substitute (5.1), noting that η is constant, but keeping only first order terms in η, and the final equality since $\eta^{ik} = \delta^{ik}$ for spacelike indices. Putting (5.3) into (5.2) gives

$$\frac{d^2x^i}{dt^2} = \frac{c^2}{2}\,\nabla_i\, h_{00}. \tag{5.4}$$

This is to be compared with Newton's equation (see (1.1)–(1.4))

$$\frac{d^2\mathbf{x}}{dt^2} = \mathbf{g} = -\nabla\phi \tag{5.5}$$

where ϕ is the gravitational potential. Comparison of (5.4) and (5.5) gives

$$h_{00} = -\frac{2\phi}{c^2}$$

and hence

$$g_{00} = -\left(1 + \frac{2\phi}{c^2}\right). \tag{5.6}$$

We have found one component of the metric tensor $g_{\mu\nu}$! Actually, this is all we can find by comparing Einstein's theory with Newton's, since Newton's theory describes gravitation by means of one function only, the scalar field ϕ. (This naturally raises the question, how do we find the other components of $g_{\mu\nu}$?)

At a distance r from a gravitating body of mass M, we have $\phi = -\frac{MG}{r}$, so $g_{00} = -\left(1 - \frac{2GM}{rc^2}\right)$ and

$$ds^2 = -\left(1 - \frac{2GM}{rc^2}\right)c^2\,dt^2 + \cdots. \tag{5.7}$$

The 'time-time' component of the metric tensor has been found. This is actually sufficient to derive the gravitational red-shift, which is therefore a consequence of the Equivalence Principle alone, since the above reasoning relies only on this. We now turn our attention to the field equations, which will yield the other components of $g_{\mu\nu}$ (answering the question above).

5.2 Einstein field equations

The question to be answered is: given a particular matter distribution, what are the $g_{\mu\nu}$? There are two considerations to be borne in mind:

(i) the equations we find must be tensor equations, and the tensors we have to hand, to describe the gravitational field, are $R^{\kappa}{}_{\lambda\mu\nu}$, $R_{\mu\nu}$ and R;
(ii) in a larger context we may summarise the present situation in Table 5.1.

The field equations we are looking for are the relativistic generalisations of Laplace's and Poisson's equations, in the cases of a vacuum and matter, respectively; and they should therefore yield these equations in the non-relativistic limit. Laplace's and Poisson's equations are second order differential equations, and we have seen that in the weak field limit ϕ becomes essentially g_{00}, so the relativistic field equations should be second order differential equations in $g_{\mu\nu}$; and, indeed, the tensors above, $R^{\kappa}{}_{\lambda\mu\nu}$, $R_{\mu\nu}$ and R, all contain second order derivatives of $g_{\mu\nu}$, since $R^{\kappa}{}_{\lambda\mu\nu}$ involves derivatives of $\Gamma^{\kappa}{}_{\lambda\mu}$ and $\Gamma^{\kappa}{}_{\lambda\mu}$ involves derivatives of $g_{\mu\nu}$. Apart from these guidelines, finding the field equations is a matter of guess-work. Let us begin with the vacuum field equations.

5.2.1 Vacuum field equations

The right hand side will clearly be zero, since there is no matter. The equation $R^{\kappa}{}_{\lambda\mu\nu}=0$ is no good, though, because that would imply that outside a massive body space-time is flat and there is no gravitational field. So let us consider the equations

$$R_{\mu\nu} = 0. \tag{5.8}$$

These are 10 equations for 10 unknowns $g_{\mu\nu}$. They do *not* imply that $R^{\kappa}{}_{\lambda\mu\nu}=0$, since as we saw in Chapter 4 the Riemann tensor has 20 independent components whereas the Ricci tensor only has 10, so that Equations (5.8) would constitute 10 relations among the 20 independent components of the Riemann tensor. The Equations (5.8) are in fact the vacuum Einstein field equations. Their general solution is unknown, but particular solutions are known, corresponding to cases with particular symmetry, for example the Schwarzschild

Table 5.1

	Non-relativistic	Relativistic
Equations of motion	Newton: $\frac{d^2x^i}{dt^2} = -\nabla_i\phi$	Geodesic equation: $\frac{d^2x^\mu}{ds^2} + \Gamma^\mu{}_{\nu\rho}\frac{dx^\nu}{ds}\frac{dx^\rho}{ds} = 0$
Field equations	Laplace: $\nabla^2\phi=0$ vacuum Poisson: $\nabla^2\phi=4\pi\rho G$ matter	Einstein field equations

solution. We shall show below that the non-relativistic limit of (5.8) gives Laplace's equation.

As far as the matter field equations are concerned, it is clear that the right hand side must be a tensor describing the matter distribution, which in the non-relativistic limit reduces essentially to ρ, as in Poisson's equation. This tensor is called the energy-momentum tensor.

5.2.2 Energy-momentum tensor

The energy-momentum tensor (or stress-energy-momentum tensor, or stress-energy tensor), denoted $T_{\mu\nu}$, is a rank 2 tensor. It is not to be confused with the energy-momentum 4-vector (a rank 1 tensor). In this section we exhibit $T_{\mu\nu}$ for *dust*, that is, incoherent matter whose particles do not interact. The definition is

$$T^{\mu\nu} = \rho_0 \, u^\mu \, u^\nu \tag{5.9}$$

or more properly

$$T^{\mu\nu}(x) = \rho_0(x) \, u^\mu(x) \, u^\nu(x), \tag{5.10}$$

where ρ_0 is the proper density of matter, that is the density moving with the flow, and u^μ is its 4-velocity

$$u^\mu = \frac{1}{c} \frac{\mathrm{d}x^\mu}{\mathrm{d}\tau}. \tag{5.11}$$

We then have

$$\mathrm{d}s^2 = -c^2 \, \mathrm{d}\tau^2 = -c^2 \, \mathrm{d}t^2 + \mathrm{d}x^2 + \mathrm{d}y^2 + \mathrm{d}z^2 = -c^2 \, \mathrm{d}t^2 \left(1 - \frac{v^2}{c^2}\right)$$

and hence

$$\frac{\mathrm{d}\tau}{\mathrm{d}t} = \left(1 - \frac{v^2}{c^2}\right)^{1/2} = \frac{1}{\gamma},$$

so that

$$T^{00} = \rho_0 \left(\frac{\mathrm{d}t}{\mathrm{d}\tau}\right)^2 = \gamma^2 \rho_0 = \rho. \tag{5.12}$$

Here ρ is the mass density in a moving frame. The two factors of γ are easy to understand: in Special Relativity the mass (strictly, mass-energy) of moving material increases over its rest value by a factor γ, and the volume decreases by the same factor. So $c^2 T^{00}$ is the relativistic energy density of matter. Similarly

$$T^{0i} = \rho_0 \, u^0 u^i = \rho_0 \frac{1}{c^2} \frac{\mathrm{d}x^0}{\mathrm{d}\tau} \frac{\mathrm{d}x^i}{\mathrm{d}\tau} = \gamma^2 \rho_0 \frac{v^i}{c} = \rho \frac{v^i}{c}, \tag{5.13}$$

where $v^i = \dfrac{\mathrm{d}x^i}{\mathrm{d}t}$. And finally

$$T^{ik} = \rho_0 \frac{1}{c^2} \frac{\mathrm{d}x^i}{\mathrm{d}\tau} \frac{\mathrm{d}x^k}{\mathrm{d}\tau} = \gamma^2 \rho_0 \frac{v^i v^k}{c^2} = \rho \frac{v^i v^k}{c^2}. \tag{5.14}$$

Putting these components together we may display $T^{\mu\nu}$ as a matrix

$$T^{\mu\nu} = \rho \begin{pmatrix} 1 & \frac{v_x}{c} & \frac{v_y}{c} & \frac{v_z}{c} \\ \frac{v_x}{c} & \frac{v_x{}^2}{c^2} & \frac{v_x v_y}{c^2} & \frac{v_x v_z}{c^2} \\ \frac{v_y}{c} & \frac{v_y v_x}{c^2} & \frac{v_y{}^2}{c^2} & \frac{v_y v_z}{c^2} \\ \frac{v_z}{c} & \frac{v_z v_x}{c^2} & \frac{v_z v_y}{c^2} & \frac{v_z{}^2}{c^2} \end{pmatrix}. \tag{5.15}$$

In the non-relativistic approximation the dominant component of $T^{\mu\nu}$ is $T^{00} \approx \rho \approx \rho_0$, the density of matter in space.

Energy and momentum are of course conserved quantities, so we expect $T^{\mu\nu}$ to obey a conservation law. We need to find the form of this law, whatever it is, in General Relativity. In Special Relativity it is

$$T^{\mu\nu}{}_{,\nu} = 0. \tag{5.16}$$

To see that this is indeed a conservation law, put $\mu = 0$:

$$\begin{aligned} T^{00}{}_{,0} + T^{0i}{}_{,i} &= 0, \\ \frac{1}{c}\frac{\partial \rho}{\partial t} + \frac{1}{c}\frac{\partial}{\partial x^i}(\rho v^i) &= 0, \\ \frac{\partial \rho}{\partial t} + \nabla \cdot (\rho \mathbf{v}) &= 0. \end{aligned} \tag{5.17}$$

This expresses the law of conservation of energy (strictly mass-energy). The reader might find it useful to be reminded of a similar law in electrodynamics, that of the conservation of charge Q. Suppose the charge inside a bounded volume V is Q, then if ρ is the *charge* density (see Fig 5.1),

$$Q = \int \rho \, \mathrm{d}V.$$

In any particular frame there are also moving charges so there is also a current density

$$\mathbf{j} = \rho \mathbf{v}$$

where $\mathbf{v}$ is the velocity of the moving charge. Charge will then flow out of (or into) the bounding volume V through its surface S, so that the total charge enclosed in V will in general change. In a time δt the change in the charge enclosed is

$$\delta Q_1 = \frac{\partial Q}{\partial t}\,\delta t = \left\{ \int \frac{\partial \rho}{\partial t}\,\mathrm{d}V \right\} \delta t.$$

The charge escaping through the surface is, on the other hand,

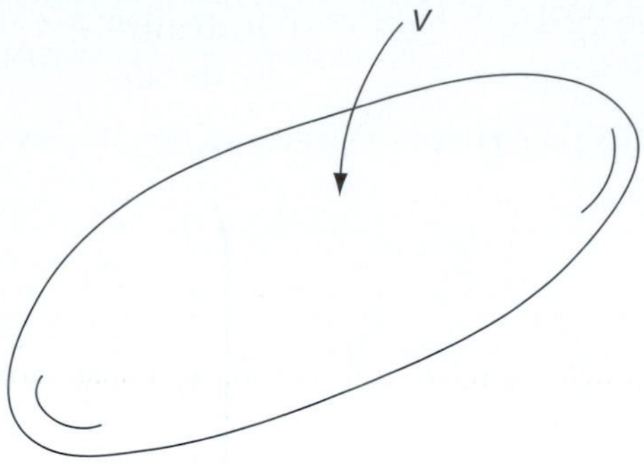

Fig. 5.1 A volume V containing a total charge Q, with charge density ρ.

$$\delta Q_2 = \left\{ \int \mathbf{j} \cdot \mathrm{d}\mathbf{S} \right\} \delta t = \left\{ \int \nabla \cdot \mathbf{j} \, \mathrm{d}V \right\} \delta t,$$

where the first integral is taken over the closed surface and the second equality follows from Gauss's theorem. The law of charge conservation is $\delta Q_1 + \delta Q_2 = 0$, i.e.

$$\int \left(\frac{\partial \rho}{\partial t} + \nabla \cdot \mathbf{j} \right) \mathrm{d}V = 0,$$

which implies that, in differential form,

$$\frac{\partial \rho}{\partial t} + \nabla \cdot (\rho \, \mathbf{v}) = 0$$

where ρ is the *charge* density and $\mathbf{v}$ its velocity at any point. This equation is of the same form as (5.17) above, which therefore expresses the conservation of energy, or more strictly, mass-energy.

Now put $\mu = i$ in Equation (5.16). This gives, successively,

$$T^{i0}{}_{,0} + T^{ik}{}_{,k} = 0,$$

$$\frac{1}{c} \frac{\partial}{\partial t} \left(\frac{1}{c} \rho v^i \right) + \frac{1}{c^2} \frac{\partial}{\partial x^i} (\rho v^i v^k) = 0,$$

$$\frac{\partial}{\partial t} (\rho v^i) + \nabla \cdot (\rho v^i \mathbf{v}) = 0. \tag{5.18}$$

This is of the same form as (5.17) above, except that ρ, the energy density, has been replaced by ρv^i, the momentum density, so (5.18) expresses conservation of momentum (density). We have therefore shown that (5.16) expresses conservation of energy-momentum in Special Relativity. In General Relativity the conservation law follows by replacing the ordinary derivative by a covariant one:

$$T^{\mu\nu}{}_{;\nu} = 0. \tag{5.19}$$

Recall once more that the form of the energy-momentum tensor we have been considering is appropriate only for dust particles, on which no forces act: the particles do not interact.

A more general stress tensor describes a fluid, in which there is pressure (exerting a force and therefore an acceleration). There is also a stress tensor for the electromagnetic field, which is important in cosmology. This will be considered later.

5.2.3 Matter field equations

We claim that the relativistic version of Lapace's (vacuum) equation is $R_{\mu\nu}=0$, Equation (4.8). We want the relativistic version of Poisson's equation, and it is clear that ρ should be replaced by $T^{\mu\nu}$. We might guess that the desired equation is

$$R_{\mu\nu} = \frac{8\pi G}{c^2}\, T_{\mu\nu}\ (?) \tag{5.20}$$

Both sides of the equation are rank $\begin{pmatrix} 0 \\ 2 \end{pmatrix}$ tensors. (It is obvious that having defined $R_{\mu\nu}$ and $T_{\mu\nu}$, we can raise and lower indices at will. The only requirement is that in a given equation the tensorial type is the same throughout.) The constant $\frac{8\pi G}{c^2}$ will be justified later. Is the above equation credible? It is not, because $T^{\mu\nu}{}_{;\nu}=0$ but $R^{\mu\nu}{}_{;\nu}\neq 0$. This latter fact follows from the Bianchi identities (4.73):

$$R^{\mu}{}_{\nu\rho\sigma;\lambda} + R^{\mu}{}_{\nu\rho\lambda;\rho} + R^{\mu}{}_{\nu\lambda\rho;\,\sigma} = 0.$$

Putting $\mu=\rho$ and contracting gives

$$R_{\nu\sigma;\lambda} + R^{\mu}{}_{\nu\sigma\lambda;\mu} - R_{\nu\lambda;\sigma} = 0,$$

where the minus sign in the last term follows from the antisymmetry under $\lambda \leftrightarrow \rho$. On multiplying by $g^{\nu\sigma}$ we find, successively

$$\begin{aligned} R_{;\lambda} - R^{\mu}{}_{\lambda;\mu} - R^{\sigma}{}_{\lambda;\sigma} &= 0, \\ \delta^{\rho}{}_{\lambda} R_{;\rho} - 2\, R^{\rho}{}_{\lambda;\rho} &= 0, \\ (\delta^{\rho}{}_{\lambda} R - 2\, R^{\rho}{}_{\lambda})_{;\rho} &= 0, \\ (R^{\rho\lambda} - {}^1\!/\!{}_2 g^{\rho\lambda} R)_{;\rho} &= 0; \end{aligned} \tag{5.21}$$

or

$$G^{\mu\nu}{}_{;\nu} = 0, \tag{5.22}$$

where

$$G^{\mu\nu} = R^{\mu\nu} - {}^1\!/\!{}_2\, g^{\mu\nu} R \tag{5.23}$$

is called the *Einstein tensor*. It is not the Ricci tensor, which has vanishing covariant divergence, but the Einstein tensor. Then, instead of (5.20) we propose

$$R_{\mu\nu} - {}^1\!/\!{}_2 g_{\mu\nu} R = \frac{8\pi G}{c^2} T_{\mu\nu} \tag{5.24}$$

Bridge over the river Tisza in Szeged. Hungarian graffiti artists are clearly well educated but seem to use unusual units for the gravitational constant. I am grateful to Dr Geretovszkyné Varjú Katalin for sending this picture.

for the relativistic matter field equations. Note that in the *absence* of matter, $T_{\mu\nu}=0$, (5.24) gives

$$R_{\mu\nu} - {}^1\!/_2\, g_{\mu\nu} R = 0,$$

which on multiplying by $g^{\mu\nu}$ gives $R - 2R = 0$, (since $g^{\mu\nu} g_{\mu\nu} = 4$), hence $R = 0$, and

$$R_{\mu\nu} = 0,$$

which are indeed the vacuum field equations (5.8).

We now have to show that the weak field non-relativistic limit of (5.24) gives Poisson's equation. First multiply (5.24) by $g^{\mu\nu}$:

$$R - 2R = \frac{8\pi G}{c^2} T \Rightarrow R = -\frac{8\pi G}{c^2} T$$

where $T = g^{\mu\nu} T_{\mu\nu}$. Hence the field equations (5.24) may be rewritten as

$$R_{\mu\nu} = \frac{8\pi G}{c^2} (T_{\mu\nu} - {}^1\!/_2\, g_{\mu\nu} T). \tag{5.25}$$

Neglecting terms in $\frac{v^2}{c^2}$ and $\rho \frac{v}{c}$, the energy momentum tensor becomes, from (5.15),

$$T^{\mu\nu} = \begin{pmatrix} \rho & 0 & 0 & 0 \\ 0 & 0 & 0 & 0 \\ 0 & 0 & 0 & 0 \\ 0 & 0 & 0 & 0 \end{pmatrix}.$$

In the weak field approximation (5.1) above $g_{\mu\nu} = \eta_{\mu\nu} + h_{\mu\nu}$. To lowest order in h this implies that

$$g^{\mu\nu} = \eta^{\mu\nu} - h^{\mu\nu}, \tag{5.26}$$

since then

$$g^{\mu\kappa}g_{\kappa\nu} = (\eta^{\mu\kappa} - h^{\mu\kappa})(\eta_{\kappa\nu} + h_{\kappa\nu}) = \delta^\mu{}_\nu - h^\mu{}_\nu + h^\mu{}_\nu + \mathrm{O}(h^2) = \delta^\mu{}_\nu$$

to lowest order (*linear approximation*), as required. Then, also to lowest order $T = -\rho$ and

$$T^{\mu\nu} - {}^1\!/\!{}_2\, g^{\mu\nu}T = \frac{\rho}{2}\begin{pmatrix} 1 & 0 & 0 & 0 \\ 0 & 1 & 0 & 0 \\ 0 & 0 & 1 & 0 \\ 0 & 0 & 0 & 1 \end{pmatrix} = \frac{\rho}{2}\delta^{\mu\nu}; \tag{5.27}$$

and also

$$T_{\mu\nu} - {}^1\!/\!{}_2\, g_{\mu\nu}T \approx \eta_{\mu\rho}\,\eta_{\nu\kappa}(T^{\rho\kappa} - {}^1\!/\!{}_2\, T g^{\rho\kappa}) = \frac{\rho}{2}\,\delta_{\mu\nu}. \tag{5.28}$$

Now let us calculate $R_{\mu\nu}$. From (4.31) and (4.35) we have

$$R_{\mu\nu} = \Gamma^\kappa{}_{\mu\nu,\kappa} - \Gamma^\kappa{}_{\mu\kappa,\nu} + \Gamma^\kappa{}_{\rho\kappa}\Gamma^\rho{}_{\mu\nu} - \Gamma^\kappa{}_{\rho\nu}\Gamma^\rho{}_{\mu\kappa}. \tag{5.29}$$

To lowest (first) order in h the connection coefficients are

$$\Gamma^\kappa{}_{\mu\nu} = {}^1\!/\!{}_2\, g^{\kappa\sigma}(g_{\sigma\mu,\nu} + g_{\sigma\nu,\mu} - g_{\mu\nu,\sigma}) = {}^1\!/\!{}_2\, \eta^{\kappa\sigma}(h_{\sigma\mu,\nu} + h_{\sigma\nu,\mu} - h_{\mu\nu,\sigma})$$

so the third and fourth terms on the right hand side of (5.29) are of order h^2 and can be ignored in the linear approximation. We then have, as may easily be verified,

$$R_{\mu\nu} = {}^1\!/\!{}_2\, \eta^{\kappa\sigma}(h_{\sigma\nu,\mu\kappa} + h_{\mu\kappa,\sigma\nu} - h_{\mu\nu,\sigma\kappa} - h_{\sigma\kappa,\mu\nu}).$$

Thus, in the static approximation

$$\begin{aligned} R_{00} &= {}^1\!/\!{}_2\, \eta^{\kappa\sigma}(h_{\sigma 0,\mu 0} + h_{0\kappa,\sigma 0} - h_{00,\sigma\kappa} - h_{\sigma\kappa,00}) \\ &= -{}^1\!/\!{}_2\, \eta^{\kappa\sigma} h_{00,\sigma\kappa} = -{}^1\!/\!{}_2\left(-\frac{1}{c^2}\frac{\partial^2}{\partial t^2} + \nabla^2\right)h_{00} \\ &= -{}^1\!/\!{}_2\, \nabla^2 h_{00}. \end{aligned}$$

From above $h_{00} = -\frac{2\phi}{c^2}$ and the above equation gives $R_{00} = \frac{1}{c^2}\nabla^2\phi$. The field equation (5.25) with $\mu = \nu = 0$ then gives (see (5.28))

$$\frac{1}{c^2}\,\nabla^2\phi = \frac{8\pi G}{c^2}\,\frac{\rho}{2} \Rightarrow \nabla^2\phi = 4\pi G\rho,$$

which is Poisson's equation (1.6).

Before leaving this section it might be useful to reflect once more on Table 5.1, showing the equations of motion for test particles, and the gravitational field equations. It is clear that given a particular distribution of (heavy) matter, the curvature of space-time (that is, essentially, $g_{\mu\nu}$) is fixed by the field equations; and therefore the motion of test bodies is also fixed – they simply follow a geodesic in the curved space. Wheeler summarises this in one of his famous aphorisms: 'matter tells space how to curve, and space tells matter how to move'. This might sound circular, but it isn't because in the first statement the amount of matter involved is large – large enough to cause significant curvature in space-time, whereas the 'matter' in the second phrase is simply a test body, which is small enough not to give rise to any gravitational field of its own. It will be appreciated, however, that when the mass of a test body becomes large enough to have an effect on the spatial geometry, then the problem becomes complicated and non-linear. In fact the gravitational field equations themselves are non-linear and the only method of procedure is one based on an approximation scheme – a 'post-Newtonian' approximation (General Relativity being a post-Newtonian theory).

Einstein, Infeld and Hoffmann did a significant amount of work on this problem, which has become known as the *problem of motion*. They came to the interesting conclusion that the equations of motion in General Relativity were actually consequences of the field equations themselves; a situation which, for example, does not hold in electrodynamics. This general topic of investigation was of particular significance to Einstein since he took the view that a point mass was a singularity in the gravitational field, and this would link very intimately questions about the field and its equations with those of how a point mass moves in the field. From the point of view of modern particle physics, however, this point of view would seem too simple, since it appears that there are no massive particles in nature which do not *also* carry other field quantities. The electron carries charge, the neutrino weak isospin and weak hypercharge, the quarks colour; so these particles, on Einstein's view, are also singularities in the electromagnetic, weak isospin and weak hypercharge, and colour fields. Perseverance with Einstein's question would take us into the territory of unified field theories. These theories belong, however, to the quantum regime, whereas General Relativity is, at the level we are at present concerned, a classical theory. Einstein's question then enters difficult and unknown territory.

5.3 Schwarzschild solution

We are concerned with a solution of the vacuum field equations $R_{\mu\nu}=0$, in the case corresponding to the Solar System; that is, to a very good approximation, the field produced by a static spherically symmetric body at rest. The static condition means that $g_{\mu\nu}$ is independent of x^0; and in addition $\mathrm{d}s^2$ is invariant under $x^0 \to -x^0$ (time reversal), so there must be no terms involving $\mathrm{d}x^i\,\mathrm{d}x^0$ in the expansion of $\mathrm{d}s^2$. This means that $g_{i0}=g_{0i}=0$. With these conditions, the most general form of the space-time line element compatible with spherical symmetry is

$$ds^2 = -U(r)\,c^2\,dt^2 + V(r)\,dr^2 + W(r)\,r^2\,(d\theta^2 + \sin^2\theta\,d\phi^2). \tag{5.30}$$

By virtue of the special symmetry conditions obtaining here, the ten components of the metric tensor, in general each dependent on all the x^μ, reduce to only three functions, and these are functions of r alone. The field equations are second order differential equations for these functions. In fact the three functions may be reduced to two. Recall that r is only a radial *parameter*, not an actual distance, so it may be replaced by any function of r. Put $Wr^2 = \hat{r}^2$. Then $\hat{r} = \sqrt{W}r$ and $\dfrac{d\hat{r}}{dr} = \sqrt{W}\left(1 + \dfrac{r}{2W}\dfrac{dW}{dr}\right)$, so

$$V\,dr^2 = \frac{V}{W}\left(1 + \frac{r}{2W}\frac{dW}{dr}\right)^{-2} d\hat{r}^2 \equiv \hat{V}\,d\hat{r}^2.$$

We may write, at the same time, $U(r) = \hat{U}(\hat{r})$. The effect of all this is to replace r by $\hat{r}$, to replace U and V by corresponding functions with hats, but W is replaced by unity in (5.30). Then removing the hats, and putting, at the same time,

$$U(r) = e^{2\nu(r)}, \quad V(r) = e^{2\lambda(r)},$$

the line element becomes

$$ds^2 = -e^{2\nu}\,c^2\,dt^2 + e^{2\lambda}\,dr^2 + r^2(d\theta^2 + \sin^2\theta\,d\phi^2). \tag{5.31}$$

The metric tensor, in covariant and contravariant versions, is then

$$g_{\mu\nu} = \begin{pmatrix} -e^{2\nu} & 0 & 0 & 0 \\ 0 & e^{2\lambda} & 0 & 0 \\ 0 & 0 & r^2 & 0 \\ 0 & 0 & 0 & r^2\sin^2\theta \end{pmatrix}, \quad g^{\mu\nu} = \begin{pmatrix} -e^{-2\nu} & 0 & 0 & 0 \\ 0 & e^{-2\lambda} & 0 & 0 \\ 0 & 0 & \dfrac{1}{r^2} & 0 \\ 0 & 0 & 0 & \dfrac{1}{r^2\sin^2\theta} \end{pmatrix}. \tag{5.32}$$

The two unknown functions $\nu(r)$ and $\lambda(r)$ will now be found from the vacuum field equations. We first find the connection coefficients: for example

$$\begin{aligned}\Gamma^1{}_{00} &= \tfrac{1}{2}g^{1\sigma}(2g_{0\sigma,0} - g_{00,\sigma}) = -\tfrac{1}{2}g^{11}g_{00,1} = -\tfrac{1}{2}e^{-2\lambda}\frac{\partial}{\partial r}(-e^{2\nu}) \\ &= \nu'\,e^{2\nu-2\lambda},\end{aligned} \tag{5.33}$$

where $\nu' = \dfrac{d\nu}{dr}$. Similarly the other connection coefficients turn out to be

$$\begin{aligned}&\Gamma^0{}_{10} = \Gamma^0{}_{01} = \nu', \\ &\Gamma^1{}_{11} = \lambda', \quad \Gamma^1{}_{22} = -r\,e^{-2\lambda}, \quad \Gamma^1{}_{33} = -r\sin^2\theta\,e^{-2\lambda}, \\ &\Gamma^2{}_{12} = \Gamma^2{}_{21} = \Gamma^3{}_{13} = \Gamma^3{}_{31} = \frac{1}{r}, \quad \Gamma^2{}_{33} = -\sin\theta\cos\theta, \\ &\Gamma^3{}_{23} = \Gamma^3{}_{32} = \cot\theta, \quad \text{others} = 0.\end{aligned} \tag{5.34}$$

These are now to be substituted in the vacuum equations

$$R_{\mu\nu} = \Gamma^{\kappa}{}_{\mu\nu,\kappa} - \Gamma^{\kappa}{}_{\mu\kappa,\nu} + \Gamma^{\kappa}{}_{\rho\kappa}\Gamma^{\rho}{}_{\mu\nu} - \Gamma^{\kappa}{}_{\rho\nu}\Gamma^{\rho}{}_{\mu\kappa} = 0.$$

We have, consulting (5.33), (5.34) and remembering the static condition,

$$\begin{aligned} R_{00} &= \Gamma^{\kappa}{}_{00,\kappa} + \Gamma^{\kappa}{}_{0\kappa,0} + \Gamma^{\kappa}{}_{\rho\kappa}\Gamma^{\rho}{}_{00} - \Gamma^{\kappa}{}_{\rho 0}\Gamma^{\rho}{}_{0\kappa} \\ &= \Gamma^{1}{}_{00,1} + \Gamma^{\kappa}{}_{1\kappa}\Gamma^{1}{}_{00} - (\Gamma^{1}{}_{00}\Gamma^{0}{}_{01} + \Gamma^{0}{}_{10}\Gamma^{1}{}_{00}) \\ &= \frac{\partial}{\partial r}(\nu' e^{2\nu-2\lambda}) + \nu' e^{2\nu-2\lambda}\left(\nu' + \lambda' + \frac{2}{r}\right) - 2(\nu')^2 e^{2\nu-2\lambda} \\ &= e^{2\nu-2\lambda}\left(\nu'' + \nu'^2 - \nu'\lambda' + \frac{2\nu'}{r}\right). \end{aligned}$$

Then the field equations give

$$R_{00} = \left(\nu'' + \nu'^2 - \nu'\lambda' + \frac{2\nu'}{r}\right) e^{2\nu-2\lambda} = 0. \tag{A}$$

Similarly,

$$R_{11} = -\nu'' + \nu'\lambda' + \frac{2\lambda'}{r} - \nu'^2 = 0, \tag{B}$$

$$R_{22} = (-1 - r\nu' + r\lambda')\, e^{-2\lambda} + 1 = 0, \tag{C}$$

$$\begin{aligned} R_{33} &= R_{22}\sin^2\theta, \\ R_{\mu\nu} &= 0 \quad (\mu \neq \nu). \end{aligned} \tag{5.35}$$

These are three independent equations for the functions $\nu(r)$ and $\lambda(r)$. The factor $e^{2\nu-2\lambda}$ in Equation (A) is non-zero, so Equations (A) and (B) give, on adding

$$\lambda' + \nu' = 0 \Rightarrow \lambda(r) + \nu(r) = \text{const.}$$

As $r \to \infty$, however, the metric (5.32) must approach the Minkowski metric $\eta_{\mu\nu}$, so $\lambda, \nu \to 0$, hence the constant above is zero and

$$\lambda(r) = -\nu(r).$$

Then Equation (C) gives

$$(1 + 2r\nu')e^{2\nu} = 1 \Rightarrow (r\,e^{2\nu})' = 1 \Rightarrow r\,e^{2\nu} = r - 2m,$$

where $2m$ is a constant of integration. That is,

$$e^{2\nu} = 1 - \frac{2m}{r},$$

which on comparison with (5.32) gives

$$g_{00} = -\left(1 - \frac{2m}{r}\right), \quad g_{11} = \frac{1}{1 - \dfrac{2m}{r}}.$$

In the weak field approximation, however, $g_{00} = -\left(1 - \frac{2GM}{rc^2}\right)$, so we can identify the constant of integration as $m = \frac{GM}{c^2}$ and finally

$$ds^2 = -\left(1 - \frac{2GM}{rc^2}\right)c^2\,dt^2 + \frac{dr^2}{1 - \frac{2GM}{rc^2}} + r^2(d\theta^2 + \sin^2\theta\,d\phi^2) \tag{5.36}$$

or, equivalently,

$$ds^2 = -\left(1 - \frac{2m}{r}\right)c^2\,dt^2 + \left(1 - \frac{2m}{r}\right)^{-1} dr^2 + r^2(d\theta^2 + \sin^2\theta\,d\phi^2),$$

with

$$m = \frac{MG}{c^2}. \tag{5.37}$$

This is the Schwarzschild solution. It is *exact* (and, as a consequence, the weak field approximation found above is now seen also to be exact – an exact solution of the field equations). As $r \to \infty$ Schwarzschild space-time approaches Minkowski space-time, as desired. The solution holds for the space-time *outside* a body of mass M. We shall see that it gives small corrections to the Newtonian predictions for the motions of light and the planets. It is worth noting that, for this vacuum solution, M is simply the total mass of the 'gravitating' body; the actual *distribution* of matter inside the body is irrelevant. This is a feature shared with Newtonian physics, where for example the gravitational potential due to a mass M depends only on the mass, not on the distribution of matter inside it.

5.3.1 Apparent 'singularity' at $r = 2m$

A glance at (5.36) above shows that g_{11}, the coefficient of dr^2, becomes singular as $r \to \frac{2GM}{c^2}$. What are the consequences of this? Does it give trouble? We shall see that the consequences are very far-reaching, giving rise to black holes, which have to be investigated carefully. Here, however, we may note that in the Solar System – which is what is relevant for many of the tests of General Relativity – this 'singularity' gives no trouble. Let me make the following observations:

(i) For the Sun, $M = 1.99 \times 10^{30}$ kg, so

$$\frac{2GM}{c^2} = \frac{2 \times 1.99 \times 10^{30} \times 6.67 \times 10^{-11}}{9 \times 10^{16}} = 2.95\ \text{km}.$$

This quantity is called the Schwarzschild radius r_S, so for the Sun

$$r_S = \frac{2GM}{c^2} = 2m = 2.95\ \text{km}; \tag{5.38}$$

and we may write the Schwarzschild metric as

$$ds^2 = -\left(1 - \frac{r_S}{r}\right)c^2\,dt^2 + \frac{dr^2}{1 - \frac{r_S}{r}} + r^2(d\theta^2 + \sin^2\theta\,d\phi^2). \tag{5.39}$$

The question then is: do we have trouble when $r \to r_S$? The answer is No, since the Schwarzschild surface (a spherical surface) $r = r_S$ is *inside* the Sun, whose radius is $R = 6.96 \times 10^5$ km. So at $r = r_S$ the Schwarzschild solution above does *not hold*, since it is a solution to the *vacuum* field equations. The region in which General Relativity will be tested in the Solar System is that for which $r > R$, hence $\frac{r_S}{r} < 4.2 \times 10^{-6}$, and Schwarzschild space-time differs only very slightly from Minkowski space-time.

(ii) We now know, however, that when stars become sufficiently old they collapse, and very heavy stars may collapse to $r < r_S$, a radius smaller than their Schwarzschild radius. In that case the surface $r = r_S$ is in the 'vacuum', outside the star, and the problems related to the Schwarzschild surface become real, and will be dealt with in Chapter 7, where we shall see that light cannot escape from a star whose radius is less than $2GM/c^2$. We may note here, however, that the 'singularity' $r = 2m$ is more like a coordinate singularity than an actual singularity in the geometry. By analogy, at the north and south poles on a sphere $g_{22} = \sin^2\theta$, $g^{22} = \frac{1}{\sin^2\theta} \to \infty$ as $\theta \to 0, \pi$. The (contravariant components of the) metric tensor may become singular, but nothing unusual appears at the poles; a sphere is, after all, a homogeneous space, and no one point is different from any other point.

(iii) It is interesting to note that the Schwarzschild radius first appeared in the work of Laplace in 1798 (the fourth year of the French Republic). Consider a particle of mass m escaping from the gravitational pull of a body (planet, star) of mass M and radius R. Using a Newtonian argument, the work done to escape is

$$\int_R^\infty F\,dr = mMG \int_R^\infty \frac{dr}{r^2} = \frac{mMG}{R}.$$

Supposing light to consist of particles of mass m travelling at speed $v = c$ with kinetic energy $\frac{1}{2}mc^2$(!), we may then deduce that the light will not escape if $\frac{mMG}{R} > \frac{1}{2}mc^2$, i.e. if $R < \frac{2GM}{c^2}$, precisely the Schwarzschild radius! If the radius of a star is smaller than its Schwarzschild radius, it will not shine. It is amusing and odd that despite the gross errors in this derivation, Laplace reached the same conclusion as that mentioned above concerning black holes. We may take the argument a little further. If the density of a (spherical) star is ρ, the above condition – that the star will not release light – becomes

$$R^2 > \frac{3c^2}{8\pi G\rho};$$

so if a star (of given density) is big enough, it will not shine. As Laplace remarked, 'il est donc possible que les plus grands corps de l'univers, soient par cela même, invisibles.' (it is therefore possible that the largest objects in the universe are for that very reason invisible.)

5.3.2 Isotropic form of the Schwarzschild solution

For some purposes it is useful to express the Schwarzschild metric in a form that makes explicit its isotropic character. We introduce a new radial coordinate

$$\rho = {}^1\!/_2\left(r - m + \sqrt{r^2 - 2mr}\right),$$

or equivalently

$$r = \rho\left(1 + \frac{m}{2\rho}\right)^2.$$

It is then easy to see that

$$1 - \frac{2m}{r} = \frac{\left(1 - \dfrac{m}{2\rho}\right)^2}{\left(1 + \dfrac{m}{2\rho}\right)^2}$$

and that

$$\left(1 - \frac{2m}{r}\right)^{-1} \mathrm{d}r^2 = \left(1 + \frac{m}{2\rho}\right)^4 \mathrm{d}\rho^2,$$

so the Schwarzschild line element becomes

$$\begin{aligned} \mathrm{d}s^2 &= -\frac{\left(1 - \dfrac{m}{2\rho}\right)^2}{\left(1 + \dfrac{m}{2\rho}\right)^2} c^2\,\mathrm{d}t^2 + \left(1 + \frac{m}{2\rho}\right)^4 \left(\mathrm{d}\rho^2 + \rho^2\,\mathrm{d}\theta^2 + \rho^2 \sin^2\theta\,\mathrm{d}\phi^2\right) \\ &= -\frac{\left(1 - \dfrac{m}{2\rho}\right)^2}{\left(1 + \dfrac{m}{2\rho}\right)^2} c^2\,\mathrm{d}t^2 + \left(1 + \frac{m}{2\rho}\right)^4 \left(\mathrm{d}x^2 + \mathrm{d}y^2 + \mathrm{d}z^2\right), \end{aligned} \tag{5.40}$$

where the variables x, y and z are defined by $x = \rho \sin\theta \cos\phi$, $y = \rho \sin\theta \sin\phi$, $z = \rho \cos\theta$. This is the so-called isotropic form of the metric in Schwarzschild space-time.

5.4 Time dependence and spherical symmetry: Birkhoff's theorem

Now consider a slightly more general problem; that of a spherically symmetric but time-dependent gravitational field, satisfying the vacuum field equations. For example we might have a radially pulsating star (the pulsations compatible with spherical symmetry). Then the most general form of the metric is

$$ds^2 = -P(r,t)\, c^2\, dt^2 + Q(r,t)\, dr^2 + 2R(r,t)\, dr\, dt + S(r,t)\, r^2 (d\theta^2 + \sin^2\theta\, d\phi^2).$$

As before, we can put $\hat{r} = \sqrt{S(r,t)}r$, so the final term above is $\hat{r}^2(d\theta^2 + \sin^2\theta\, d\phi^2)$, and P, Q and R become replaced by new functions. Then with simple relabelling we get $(d\Omega^2 = d\theta^2 + \sin^2\theta\, d\phi^2)$

$$ds^2 = -P(r,t)\, c^2\, dt^2 + Q(r,t)\, dr^2 + 2R(r,t)\, dr\, dt + r^2\, d\Omega^2. \tag{5.41}$$

Now we can find a function $f(r, t)$ such that

$$f(r,t)\ [P(r,t)\, c\, dt - R(r,t)\, dr]$$

is a *perfect differential*, $c\, dF(r, t)$

$$f(r,t)\ [P(r,t)\, c\, dt - R(r,t)\, dr] = c\, dF(r,t).$$

The condition for this is

$$\frac{\partial}{\partial r}[f(r,t)\, P(r,t)] = -\frac{\partial}{\partial t}[f(r,t)\, R(r,t)].$$

This is regarded as an equation for $f(r, t)$, which may be put in the form

$$\frac{\partial f}{\partial t} = \frac{1}{R}\left\{\frac{\partial f}{\partial r}P + f\frac{\partial P}{\partial r} + f\frac{\partial R}{\partial t}\right\}.$$

Then, given $f(r, t_0)$ for all r at $t = t_0$, we can find $f(r, t_0 + dt)$. It then follows that

$$-\frac{1}{f^2}\frac{1}{P}\, c^2\, dF^2 = -Pc^2\, dt^2 + 2Rc\, dt\, dr - \frac{R^2}{P} dr^2$$

and hence that

$$ds^2 = -\frac{1}{f^2}\frac{1}{P}c^2\, dF^2 + \left(Q + \frac{R^2}{P}\right) dr^2 + r^2\, d\Omega^2. \tag{5.42}$$

Comparing this with (5.41) we note that the term in $dr\, dt$ has disappeared. We may now *define* $F(r, t)$ to be a new time parameter t'

$$t' = F(r,t);$$

then dropping the prime we may write (5.42) as

$$ds^2 = -U(r,t)\, c^2\, dt^2 + V(r,t)\, dr^2 + r^2\, d\Omega^2.$$

This is similar to (5.30) above, except that U and V are now functions of t as well as of r. Analogously to (5.31), then, we may put

$$ds^2 = -e^{2\nu(r,t)}\, c^2\, dt^2 + e^{2\lambda(r,t)}\, dr^2 + r^2 (d\theta^2 + \sin^2\theta\, d\phi^2). \tag{5.43}$$

We remind ourselves that this is the general form of the metric in a situation with spherical symmetry but time dependence also. As before, we now calculate the connection coefficients $\Gamma^{\mu}{}_{\nu\kappa}$ and impose the vacuum field equations to find $\nu(r, t)$ and $\lambda(r, t)$.

In addition to the Christoffel symbols found above, it turns out that the following are non-zero:

$$\Gamma^{0}{}_{00} = \frac{1}{c}\frac{\partial \nu}{\partial t} = \frac{\dot{\nu}}{c}, \quad \Gamma^{1}{}_{01} = \Gamma^{1}{}_{10} = \frac{\dot{\lambda}}{c}, \quad \Gamma^{1}{}_{11} = \frac{\dot{\lambda}}{c}\mathrm{e}^{2(\lambda-\nu)}.$$

As for the Ricci tensor, R_{22} and R_{33} are unchanged. R_{00} and R_{11} acquire the following extra terms:

$$R_{00} = \cdots + \frac{1}{c^2}\left(\dot{\lambda}\dot{\nu} - \ddot{\lambda} - \dot{\lambda}^2\right), \quad R_{11} = \cdots + \frac{1}{c^2}\left(\ddot{\lambda} + \dot{\lambda}^2 - \dot{\lambda}\dot{\nu}\right)\mathrm{e}^{2(\lambda-\nu)}.$$

In addition, there is one non-diagonal component of $R_{\mu\nu}$:

$$R_{01} = \frac{2}{rc}\dot{\lambda}.$$

Hence the condition $R_{01} = 0$ implies that $\dot{\lambda} = 0$, so λ depends only on r; and all the extra terms in R_{00} and R_{11} therefore vanish. The solution to the field equations is therefore found from (A), (B) and (C) above (after (5.34)), except that $\nu(r, t)$ is now a function of r and t. The equation $\lambda' + \nu' = 0$ then gives, on integration,

$$\lambda(r) + \nu(r, t) = \eta(t),$$

a function of t only, and (C) gives

$$(1 - 2r\lambda')\,\mathrm{e}^{-2\lambda(r)} = 1 \Rightarrow \mathrm{e}^{-2\lambda} = 1 - \frac{2m}{r}.$$

Since $\lambda(r)$ is independent of t, so is m; $m = \text{const}$, as in the static case. Then the solution is

$$\mathrm{d}s^2 = -\mathrm{e}^{2\eta(t)}\left(1 - \frac{2m}{r}\right)c^2\,\mathrm{d}t^2 + \left(1 - \frac{2m}{r}\right)^{-1}\mathrm{d}r^2 + r^2(\mathrm{d}\theta^2 + \sin^2\theta\,\mathrm{d}\phi^2)$$

with $m = \frac{MG}{c^2}$. This differs from the Schwarzschild solution by the factor $\mathrm{e}^{2\eta(t)}$ in the first term. We can, however, redefine the time coordinate: put

$$t' = \int^{t} \mathrm{e}^{\eta(t)}\,\mathrm{d}t,$$

then the first term becomes $-(1 - \frac{2m}{r})c^2\,\mathrm{d}t'^2$, and we recover the Schwarzschild metric again. We have therefore shown that *any spherically symmetric solution of the field equations is necessarily static*. This is Birkhoff's theorem. It follows that a star pulsating radially has the same external field as a star at rest – an interesting result! In other words, a radially pulsating star emits no gravitational radiation. We shall indeed see below that gravitational radiation is quadrupolar in nature; a star must oscillate in a quadrupolar manner to emit radiation – and this mode of oscillation does not possess spherical symmetry.

5.5 Gravitational red-shift

The Schwarzschild metric tensor

$$g_{\mu\nu} = \mathrm{diag}\left[-\left(1-\frac{2GM}{rc^2}\right), \left(1-\frac{2GM}{rc^2}\right)^{-1}, r^2, r^2\sin^2\theta\right] \tag{5.44}$$

is constant in time, independent of $x^0 = ct$. The parameter t is therefore called *world time*. (Its choice is not unique; it may for instance be multiplied by an arbitrary constant.)

Consider the lapse of *proper time* τ between two events at a *fixed spatial point* in Schwarzschild space-time:

$$\mathrm{d}s^2 = -c^2\,\mathrm{d}\tau^2 = -g_{00}\,c^2\,\mathrm{d}t^2, \quad g_{00}(r) = \left(1-\frac{2GM}{rc^2}\right),$$

hence

$$\mathrm{d}\tau = \sqrt{g_{00}}\,\mathrm{d}t. \tag{5.45}$$

The element of proper time $\mathrm{d}\tau$ is measured by a clock at the particular point, while the element of world time $\mathrm{d}t$ is fixed for the whole manifold. From (5.44) we see that $g_{00} < 1$, so $\mathrm{d}\tau < \mathrm{d}t$; in an easy slogan, clocks go slower in a gravitational field. In fact *time itself* goes slower in a gravitational field. Consider two equal, originally synchronous, clocks at some point in a field, both in an inertial frame. Now move one of them to another point, in a gravitational field, for a certain time and then bring it back to rejoin the first clock. The lapses of proper time of the clocks will not be the same; the one which has spent time in the gravitational field will be behind ('younger'). By the Equivalence Principle this is actually analogous to the twin paradox; the twin who has undergone an acceleration is younger; and an acceleration is equivalent to a gravitational field.

How are we to observe this distortion of time by a gravitational field? Suppose a particular physical process takes a certain *time* to occur, and this time is measured by a clock. The process might be a nuclear decay half-life, or it might be light emitted in a particular atomic transition, which has a particular frequency (an inverse time). We might consider measuring this characteristic time or frequency at different points in a gravitational field (where g_{00} differs) and looking for a variation. This however will not work; there will be no effect, because while the gravitational field affects the physical process it *also* affects the measuring apparatus (the clock, or the frequency measuring device), so the recorded result will always be the same.

To get an observable effect we must compare time dilation effects at two different points in the gravitational field. Let us observe, at r_1, light coming from an atomic transition at r_2. The wavelength of the light corresponds to a definite $\mathrm{d}s$, or a definite proper time $\mathrm{d}s \sim c\,\mathrm{d}\tau$. Figure 5.2 shows the emission of two successive wave crests travelling from r_2 to r_1. Let the world-time interval between the emission of these crests at r_2 be Δt_2; this is then *also* the world-time interval between the *reception* of the crests at r_1, since the two crests of light take the same time to travel.

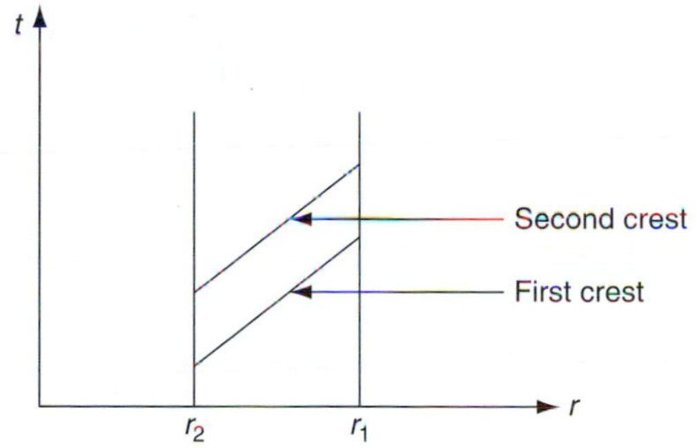

Fig. 5.2 Emission of successive crests of light, from r_2 to r_1.

At $r = r_2$, $d\tau_2$ = proper time interval between successive crests = period of light = (frequency)$^{-1}$, so we may write.

$$d\tau_2 = \frac{1}{\nu_2} = \frac{\lambda_2}{c} = \Delta t_2 \sqrt{g_{00}(r_2)}.$$

Also we have

$$d\tau_1 = \frac{1}{\nu_1} = \frac{\lambda_1}{c} = \Delta t_2 \sqrt{g_{00}(r_1)},$$

where the reader will note the same Δt_2 in both expressions, as explained above. Dividing, we have

$$\frac{\nu_1}{\nu_2} = \sqrt{\frac{g_{00}(r_2)}{g_{00}(r_1)}} \text{ with } g_{00}(r) = 1 - \frac{2GM}{rc_2}. \tag{5.46}$$

The meaning of the symbols is:

ν_1 = frequency of light emitted at r_2 and received at r_1,
ν_2 = frequency of light emitted and received at r_2
  = frequency of light emitted and received at r_1,

as per the comments above. In the case of light from the Sun received on Earth we put $r = r_2$ = surface of Sun, $r = r_1$ = surface of Earth, so

ν_1 = frequency of light emitted on Sun, received on Earth,
ν_2 = frequency of light emitted and received on Sun
  = frequency of light emitted and received on Earth, in the laboratory.

Since in the two cases $r = r_1$ and $r = r_2$ we have $\dfrac{2GM}{rc^2} \ll 1$, the expression in (5.46) may be evaluated by a binomial expansion to first order, giving

$$\frac{\nu_1}{\nu_2} = \left(1 - \frac{2GM}{r_2c^2}\right)^{1/2} \left(1 - \frac{2GM}{r_1c^2}\right)^{-1/2}$$
$$\approx \left(1 - \frac{GM}{r_2c^2}\right)\left(1 + \frac{GM}{r_1c^2}\right) \approx 1 + \frac{GM}{c^2}\left(\frac{1}{r_1} - \frac{1}{r_2}\right). \tag{5.47}$$

Since $r_1 > r_2$, then $\nu_1 < \nu_2$ – the light is *red-shifted*. For the Earth–Sun system,

$$r_2 = R_S = \text{solar radius} = 6.96 \times 10^8 \text{m}$$
$$r_1 = R = \text{Earth–Sun distance} = 1.5 \times 10^{11} \text{m} \gg R_S,$$

so to leading order we have

$$\frac{\nu_1}{\nu_2} = 1 - \frac{GM_S}{R_Sc^2} \approx 1 - 2.12 \times 10^{-6}, \tag{5.48}$$

or if $\nu_1 = \nu_2 + \Delta\nu$, ($\Delta\nu$ = frequency shift)

$$\frac{\Delta\nu}{\nu} = -2.12 \times 10^{-6}. \tag{5.49}$$

This is the predicted magnitude of the frequency shift of light from the Sun. Spectral lines from the Sun are compared with the 'same' lines observed in the laboratory. The effect is actually rather small, and difficult to detect, because broadening of spectral lines at $T = 3000\,\text{K}$ and Doppler shifts due to convection currents in the solar atmosphere both tend to mask the effect being looked for. Nevertheless Brault (1963) gives a value 1.05 ± 0.05 times the predicted one.

The effect is larger in white dwarfs, which have a similar mass to the Sun's, but a radius smaller by a factor between 10 and 100, so $\frac{\Delta\nu}{\nu}$ is greater by a similar factor. The problems caused by Doppler broadening and so on are less severe, but there is a problem in estimating the mass of white dwarfs. Nevertheless, for 40 Eridani B the predicted and observed red-shifts are[2]

$$\text{predicted } \frac{\Delta\nu}{\nu} = -(5.7 \pm 1) \times 10^{-5}; \quad \text{observed } \frac{\Delta\nu}{\nu} = -(7 \pm 1) \times 10^{-5}.$$

A more precise test for the gravitational frequency shift, however, is not astronomical but terrestrial, and is actually a blue-shift rather than a red-shift. It is very small indeed in magnitude, but can be measured very precisely. It was performed by Pound and Rebka in 1960.[3] A gamma ray from a 14.4 keV atomic transition in ^{57}Fe *falls* vertically in the Earth's gravitational field through a distance of 22.6 metres. From the formulae above, with

$$g_{00}(r) = -\left(1 + \frac{2\phi(r)}{c^2}\right) \tag{5.50}$$

we have, since $\phi/c^2 \ll 1$,

[2] Popper (1954).
[3] Pound & Rebka (1960).

$$\frac{\nu_1}{\nu_2} = \sqrt{\frac{g_{00}(r_2)}{g_{00}(r_1)}} = 1 + \frac{1}{c^2}(\phi(r_2) - \phi(r_1)). \tag{5.51}$$

Light travels from r_2 to r_1: ν_2 is the frequency measured in a laboratory at r_1, and ν_1 is the frequency of the radiation emitted at r_2 and received at r_1. Then, with $\nu_1 = \nu_2 + \Delta\nu$, we have

$$\frac{\Delta\nu}{\nu} = \frac{1}{c^2}(\phi(r_2) - \phi(r_1)), \tag{5.52}$$

with $\Delta\nu < 0$ giving a blue-shift and $\Delta\nu > 0$ a red-shift. With M_E the mass of the earth and R_E its radius we have $\phi(r) = -\frac{GM_E}{r}$, $r_1 = R_E$, $r_2 = R_E + z$, where z is the height of the tower down which the gamma ray falls. Hence

$$\frac{\Delta\nu}{\nu} = \frac{GM_E}{c^2}\left(\frac{1}{R_E} - \frac{1}{R_E + z}\right) \approx \frac{GM_E z}{c^2 {R_E}^2} = \frac{gz}{c^2}$$

where g is the acceleration due to gravity at the Earth's surface = 9.78 m s^{-2}. With z = 22.6 m this gives

$$\frac{\Delta\nu}{\nu} = 2.46 \times 10^{-15}, \tag{5.53}$$

a very small *blue-shift*. The 14.4 keV line in ^{57}Fe has a width Γ with $\Gamma/\nu = 1.13 \times 10^{-12}$. The width is actually small, due to the Mossbauer effect, yet nevertheless Γ/ν is 460 times the value of $\Delta\nu/\nu$ above and, on the face of it, it would seem impossible to detect this blue-shift. Pound and Rebka, however, had the clever idea of superimposing onto this gravitational frequency shift a *Doppler* frequency shift, by moving the gamma ray source up and down with a velocity $v_0 \cos \omega t$. In the experiment the gamma ray is detected (at the bottom of the tower) by resonant absorption in ^{57}Fe, and the effect of moving the detector is to change the counting rate in such a way that the *gravitational* frequency shift (which is much smaller) can actually be observed.[4] The experimental result was

$$\left(\frac{\Delta\nu}{\nu}\right)_{\text{exp}} = (2.57 \pm 0.26) \times 10^{-15}, \tag{5.54}$$

in striking confirmation of the prediction (5.53). This result has since been improved.[5]

The most precise test of the gravitational frequency shift to date, however, involves a rather different type of experiment, using a hydrogen maser in a rocket at an altitude of 10 000 km and comparing its frequency to that of a similar clock on the ground. The experiment was performed in 1976 by Vessot and Levine, who found agreement between the observed and predicted frequency shifts at the 70×10^{-6} level.[6]

Let me conclude this section with a couple of remarks. First, tests of the gravitational red-shift (or blue-shift) are in essence tests of the Equivalence Principle. They do not involve

[4] For details the reader is referred to a good account in Weinberg (1972), pp. 82–83.

[5] Pound & Snider (1964).

[6] Vessot & Levine (1979), Vessot *et al.* (1980). For more details see Will (1993, 2001).

Einstein's field equations, so are not tests of General Relativity in its full form. Second, the above equations may actually be derived by treating light as a stream of photons with energy $E = h\nu$, which decreases as the light 'climbs out of' the gravitational field. This type of derivation, presented in some books, is not exactly wrong, but it should be clear that if one takes it seriously, then what is being tested is *also* quantum theory. The approach above has nothing to do with quantum theory. Light is simply described by a frequency, an inverse time. Elsewhere in General Relativity (for example the bending of its path in a gravitational field) light is described by a null geodesic.[7]

5.6 Geodesics in Schwarzschild space-time

We are interested in the motion of planets and of light in the gravitational field of the Sun, i.e. of test bodies in Schwarzschild space-time. This motion is given by the geodesic equation

$$\frac{d^2x^\mu}{ds^2} + \Gamma^\mu{}_{\nu\rho}\frac{dx^\nu}{ds}\frac{dx^\rho}{ds} = 0. \tag{5.55}$$

The connection coefficients $\Gamma^\mu{}_{\nu\rho}$ have already been found in (5.33) and (5.34) above. With $\lambda = -\nu$ and $e^{2\nu} = 1 - \frac{2m}{r}$ we then have

$$\begin{aligned}
&\Gamma^0{}_{10} = \Gamma^0{}_{01} = \frac{m}{r(r-2m)},\\
&\Gamma^1{}_{11} = -\frac{m}{r(r-2m)}, \quad \Gamma^1{}_{22} = -(r-2m), \quad \Gamma^1{}_{33} = -(r-2m)\sin\theta,\\
&\Gamma^2{}_{12} = \Gamma^2{}_{21} = \Gamma^3{}_{13} = \Gamma^3{}_{31} = \frac{1}{r}, \quad \Gamma^2{}_{33} = -\sin\theta\cos\theta,\\
&\Gamma^3{}_{23} = \Gamma^3{}_{32} = \cot\theta, \quad \text{others} = 0.
\end{aligned} \tag{5.56}$$

With $x^0 = ct$, $\dot{t} = \frac{dt}{d\tau}$ etc., the geodesic equation with $\mu = 0$ is

$$\ddot{t} + \frac{2m}{r(r-2m)}\dot{t}\dot{r} = 0$$

or

$$\frac{d}{ds}\left[\left(1 - \frac{2m}{r}\right)\dot{t}\right] = 0,$$

which is integrated to give

$$\left(1 - \frac{2m}{r}\right)\dot{t} = b = \text{const.} \tag{5.57}$$

The $\mu = 2$ geodesic equation is $\ddot{\theta} + \Gamma^2{}_{\kappa\lambda}\dot{x}^\kappa\dot{x}^\lambda = 0$, which with (5.56) gives

[7] For more elaboration of this point see Okun *et al.* (2000).

$$\ddot{\theta} + \frac{2}{r}\dot{r}\dot{\theta} - \sin\theta\cos\theta\,\dot{\phi}^2 = 0. \tag{5.58}$$

The $\mu = 3$ equation likewise becomes

$$\ddot{\phi} + \frac{2}{r}\dot{r}\dot{\phi} + 2\cot\theta\dot{\theta}\dot{\phi} = 0. \tag{5.59}$$

Instead of the $\mu = 1$ geodesic equation take the line element

$$\mathrm{d}s^2 = -\left(1 - \frac{2m}{r}\right)c^2\,\mathrm{d}t^2 + \left(1 - \frac{2m}{r}\right)^{-1}\mathrm{d}r^2 + r^2(\mathrm{d}\theta^2 + \sin^2\theta\,\mathrm{d}\phi^2) \tag{5.60}$$

and divide it by $\mathrm{d}s^2 = -c^2\,\mathrm{d}\tau^2$ to give

$$1 = \left(1 - \frac{2m}{r}\right)\dot{t}^2 - \left(1 - \frac{2m}{r}\right)^{-1}\frac{\dot{r}^2}{c^2} - \frac{r^2}{c^2}(\dot{\theta}^2 + \sin^2\theta\,\dot{\phi}^2). \tag{5.61}$$

The advantage of writing this equation is that for light $\mathrm{d}s^2 = 0$ so the left hand side of (5.61) becomes 0 instead of 1. It is then possible to treat the paths of light and planets in a similar way.

Now consider a geodesic passing through a point P on the 'equator' $\theta = \pi/2$, tangent to the equatorial plane $\dot{\theta} = 0$. Then (5.58) gives $\ddot{\theta} = 0$, so $\dot{\theta}$ is *always* zero, hence θ is always $\pi/2$, and *planar motion is allowed* – in General Relativity, as in Newtonian physics. Then (5.59) gives

$$\ddot{\phi} + \frac{2}{r}\dot{r}\dot{\phi} = 0$$

or

$$\frac{\mathrm{d}}{\mathrm{d}s}(r^2\dot{\phi}) = 0$$

or, on integration,

$$r^2\dot{\phi} = a = \text{const.} \tag{5.62}$$

Now substitute (5.57) and (5.62) into (5.61), noting that

$$\dot{r} = \frac{\mathrm{d}r}{\mathrm{d}\phi}\dot{\phi} = \frac{\mathrm{d}r}{\mathrm{d}\phi}\cdot\frac{a}{r^2};$$

then

$$1 = \left(1 - \frac{2m}{r}\right)^{-1}b^2 - \frac{1}{c^2}\left(1 - \frac{2m}{r}\right)^{-1}a^2\frac{1}{r^4}\left(\frac{\mathrm{d}\phi}{\mathrm{d}r}\right)^2 - \frac{a^2}{c^2r^2}.$$

Then noting that $\frac{1}{r^4}\left(\frac{\mathrm{d}\phi}{\mathrm{d}r}\right)^2 = \left[\frac{\mathrm{d}}{\mathrm{d}\phi}\left(\frac{1}{r}\right)\right]^2$, and multiplying by $\left(1 - \frac{2m}{r}\right)/a^2$ gives

$$\left[\frac{\mathrm{d}}{\mathrm{d}\phi}\left(\frac{1}{r}\right)\right]^2 + \frac{1}{r^2} = \frac{c^2(b^2 - 1)}{a^2} + \frac{2mc^2}{ra^2} + \frac{2m}{r^3}. \tag{5.63}$$

Differentiating with respect to ϕ gives

$$2\frac{\mathrm{d}}{\mathrm{d}\phi}\left(\frac{1}{r}\right)\frac{\mathrm{d}^2}{\mathrm{d}\phi^2}\left(\frac{1}{r}\right)+\frac{2}{r}\frac{\mathrm{d}}{\mathrm{d}\phi}\left(\frac{1}{r}\right)=\frac{2mc^2}{a^2}\frac{\mathrm{d}}{\mathrm{d}\phi}\left(\frac{1}{r}\right)+\frac{6m}{r^2}\frac{\mathrm{d}}{\mathrm{d}\phi}\left(\frac{1}{r}\right).$$

Rejecting the solution $\frac{\mathrm{d}}{\mathrm{d}\phi}\left(\frac{1}{r}\right)=0$, which is a circle, we find

$$\frac{\mathrm{d}^2}{\mathrm{d}\phi^2}\left(\frac{1}{r}\right)+\frac{1}{r}=\frac{mc^2}{a^2}+\frac{3m}{r^2}, \tag{5.64}$$

which is the differential equation for the orbit. In the case of light the left hand side of (5.61) is 0 instead of 1, and (5.64) becomes replaced by

$$\frac{\mathrm{d}^2}{\mathrm{d}\phi^2}\left(\frac{1}{r}\right)+\frac{1}{r}=\frac{3m}{r^2}. \tag{5.65}$$

In these equations it will be recalled that $m=\frac{GM}{c^2}$, M being the mass of the gravitating body.

5.7 Precession of planetary orbits

In Newtonian theory the differential equation for a planet moving in a non-cicular orbit is

$$\frac{\mathrm{d}^2u}{\mathrm{d}\phi^2}+u=\frac{1}{p} \tag{5.66}$$

where $u=1/r$ and $p=a^2/GM=a^2/mc^2$ (see Problem 5.3). The quantity p, the semi-latus rectum of the ellipse, is given by

$$p=a_0(1-e^2)$$

where a_0 is the semi major axis and e the eccentricity. The general solution to Equation (5.66) is

$$u=\frac{1}{r}=\frac{1}{p}[1+e\cos(\phi-\alpha)],$$

where e and α are constants of integration. We may choose e to be non-negative ($e\geq 0$), since the cosine term changes sign when $\alpha\to\alpha+\pi$. In addition a suitable rotation of the coordinate system allows us to choose $\alpha=0$, so the solution above becomes

$$u=\frac{1}{r}=\frac{1}{p}(1+e\cos\phi),\quad e\geq 0. \tag{5.67}$$

This is the equation for an ellipse with eccentricity e. In the case $e=0$ the ellipse becomes a circle. The particle's closest approach is at $\phi=0$: this is called the perihelion if M is the Sun (and the perigee if M is the Earth).

The general relativistic equation for the orbit, (5.64), has a correction term added to the Newtonian equation (5.66). The relative magnitude of this term, in the case of Mercury, where $r\sim 6\times 10^7$ km, is

$$\frac{3m}{r^2}\Big/\frac{1}{r}=\frac{3m}{r}\sim 10^{-7}.$$

For the last term in (5.64) we then substitute (5.67) and neglect the term in e^2, giving

$$\begin{aligned}\frac{\mathrm{d}^2}{\mathrm{d}\phi^2}\left(\frac{1}{r}\right)+\frac{1}{r}&=\frac{mc^2}{a^2}+\frac{3m^3c^4}{a^4}(1+2e\cos\phi)\\&\approx\frac{mc^2}{a^2}+\frac{6m^3c^4}{a^4}\,e\cos\phi.\end{aligned}\tag{5.68}$$

The solution to (5.68) is the solution to (5.66) (which is (5.67)) plus the solution to

$$\frac{\mathrm{d}^2}{\mathrm{d}\phi^2}\left(\frac{1}{r}\right)+\frac{1}{r}=\frac{6m^3c^4}{a^4}\,e\cos\phi,$$

that is,

$$\frac{1}{r}=\frac{mc^2}{a^2}(1+e\cos\phi)+\frac{6em^3c^4}{a^4}\,\phi\sin\phi$$

(where it may be noted that the last term is *non-periodic*)

$$\begin{aligned}&=\frac{mc^2}{a^2}\left(1+e\cos\phi+\frac{3em^2c^2}{a^2}\phi\sin\phi\right)\\&=\frac{mc^2}{a^2}\left[1+e\cos\left\{\phi\left(1-\frac{3m^2c^2}{a^2}\right)\right\}\right]+\mathrm{O}\left(\frac{m}{r}\right)^2.\end{aligned}$$

The perihelion occurs when $1/r$ is a maximum, that is when $\cos\left[\phi\left(1-\frac{3m^2c^2}{a^2}\right)\right]=1$, hence

$$\begin{aligned}\phi&=0,\ \frac{2\pi}{1-\frac{3m^2c^2}{a^2}},\ \frac{4\pi}{1-\frac{3m^2c^2}{a^2}},\ \ldots\\&\approx 0,\ 2\pi+\frac{6\pi m^2c^2}{a^2},\ \ldots\equiv 0,\ 2\pi+\delta\phi,\ \ldots\end{aligned}$$

where $\delta\phi$ is the *precession* of the ellipse in one revolution:

$$\delta\phi=\frac{6\pi m^2c^2}{a^2}=\frac{6\pi m}{p}=\frac{6\pi MG}{c^2a_0(1-e^2)}\approx 0.1''\tag{5.69}$$

where M is the mass of the Sun and a_0 the semi-major axis of Mercury $\approx 5.8\times 10^{10}$ m and $e = 0.21$. This is only a tiny amount, but the precession is cumulative and in one hundred *Earth* years

$$\delta\phi_{100}=43.03''.\tag{5.70}$$

An exaggerated picture of a precessing elliptical orbit is shown in Fig. 5.3.

The total observed precession of Mercury is $5600.73\pm0.41''$ per century. The Newtonian-calculated precession, caused by the motion of the other planets (of which Venus, Earth and

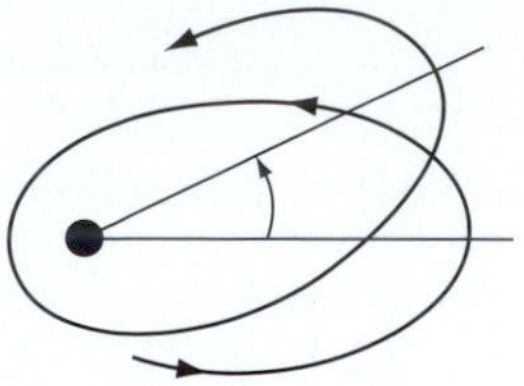

Fig. 5.3 A precessing elliptical orbit.

Jupiter give the largest contributions) is $5557.62 \pm 0.4''$.[8] The discrepancy between these figures is 43.11 ± 0.4. This is consistent with (5.70) and the effect is therefore well explained by General Relativity.

5.8 Deflection of light

We use Equations (5.57)–(5.59), but (5.61) has zero on the left hand side when light, rather than a massive test body, is being considered. Then the equation for the 'orbit' is (5.65):

$$\frac{d^2}{d\phi^2}\left(\frac{1}{r}\right) + \frac{1}{r} = \frac{3m}{r^2}. \tag{5.65}$$

where, as before, $3m/r^2 \gg 1/r$. The solution to $\frac{d^2}{d\phi^2}\left(\frac{1}{r}\right) + \frac{1}{r} = 0$ is

$$\frac{1}{r} = \frac{1}{r_0} \cos\phi, \tag{5.70}$$

a straight line (Fig. 5.4). Equation (5.65) is solved by substituting (5.70) into the right hand side:

$$\frac{d^2}{d\phi^2}\left(\frac{1}{r}\right) + \frac{1}{r} = \frac{3m}{{r_0}^2} \cos^2\phi. \tag{5.71}$$

This has the particular solution

$$\frac{1}{r} = \frac{m}{{r_0}^2}(1 + \sin^2\phi), \tag{5.72}$$

so (5.65) has the general solution

$$\frac{1}{r} = \frac{1}{r_0}\cos\phi + \frac{m}{r_0^2}(1 + \sin^2\phi). \tag{5.73}$$

Now consider the asymptotes:

$$(5.70): r \to \infty,\ \phi \to \pm\, \pi/2; \quad (5.73): r \to \infty,\ \phi \to \pm\,(\pi/2 + \delta)$$

[8] These figures are taken from Robertson & Noonan (1968), p. 239.

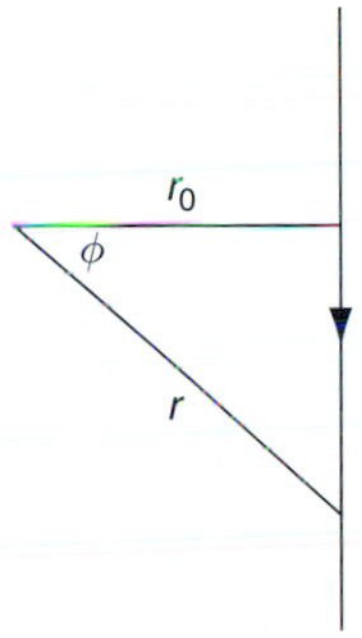

Fig. 5.4 A straight line – the path of a light ray in flat space-time.

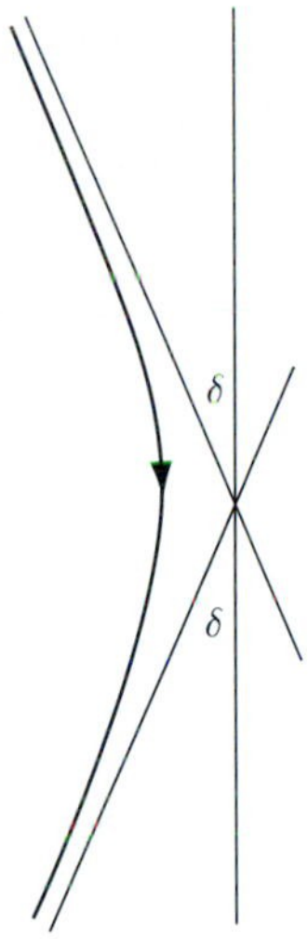

Fig. 5.5 The path of a light ray in a Schwarzschild field.

with

$$-\frac{1}{r_0}\sin\delta + \frac{m}{r_0^2}(1+\cos^2\delta) = 0 \Rightarrow \delta = \frac{2m}{r_0},$$

since δ is small. Hence the path of the light ray is as shown in Fig. 5.5 and the total deflection is, in the case of the Sun,

$$\Delta = 2\delta = \frac{4m}{r_0} = \frac{4MG}{r_0 c^2}. \tag{5.74}$$

For light just grazing the Sun the distance of nearest approach, r_0, is effectively the Sun's radius R, and

$$\Delta = \frac{4MG}{Rc^2} = 1.75''. \tag{5.75}$$

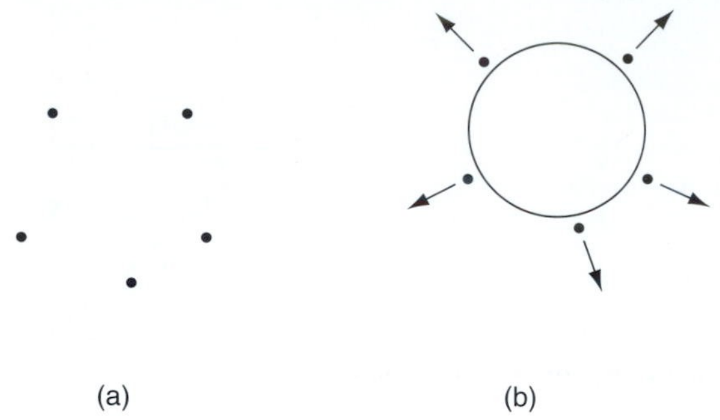

Fig. 5.6 (a) Stars in the night sky; (b) the same stars seen near the Sun's edge at a time of total solar eclipse – the Sun's light is blocked by the Moon.

Stars seen close to the Sun's edge are of course not visible by day, but become visible at the time of a total eclipse. Their position relative to the background of other stars then appears shifted relative to what it is in the usual night sky – they appear to move out from the Sun, as in Fig. 5.6.

This prediction of General Relativity was first verified in 1919. Two separate expeditions, to Brazil and Guinea, reported deflections of $1.98 \pm 0.16''$ and $1.61 \pm 0.40''$, in reasonable accord with Einstein's prediction. It was this 'bending of light' that really made Einstein famous. Many measurements of Δ were made in succeeding years, but the accuracy did not really increase until the advent of very long baseline (more than 800 km) radio interferometry in 1972, using quasar sources. A 1995 measurement on 3C273 and 3C279 gives[9]

$$\frac{\Delta_{\text{observed}}}{\Delta_{\text{predicted}}} = 0.9996 \pm 0.0017. \tag{5.76}$$

5.9 Note on PPN formalism

Since Einstein proposed the theory of General Relativity various other people have proposed modifications of it. They have in common with it that they are all *geometric* theories of gravity, i.e. the gravitational field is manifested in the geometry of space-time itself – in particular it is curved. These theories then give, in the case of a static, spherically symmetric mass, a metric similar to the Schwarzschild metric, but with some different numerical coefficients. A convenient way of discussing the results of these theories is to introduce two parameters β, γ into the Schwarzschild solution, so that it now has the form

[9] Lebach *et al.* (1995): see also Will (2001) for more information. The result quoted is cast in the language of the PPN formalism, which is explained in the next section.

$$ds^2 = -\left(1 - \frac{2m}{r} + 2(\beta - \gamma)\frac{m^2}{r^2}\right)c^2\,dt^2 + \frac{dr^2}{1 - \frac{2\gamma m}{r}} + r^2(d\theta^2 + \sin^2\theta\,d\phi^2). \tag{5.77}$$

In General Relativity

$$\text{GR: } \gamma = 1, \quad \beta = 1. \tag{5.78}$$

Einstein's theory and its rivals are all post-Newtonian theories, and these parameters γ, β are then post-Newtonian parameters.

The most commonly cited alternative to General Relativity is the Brans–Dicke theory, which includes a scalar field as a way of incorporating Mach's Principle. This theory has, however, together with all the other rivals to General Relativity, been to all intents and purposes ruled out on empirical grounds.[10] Many writers, nevertheless, pursue an interest in these theories, and the PPN (parametrised post-Newtonian) formalism does actually provide a convenient way of presenting the results of tests of General Relativity. For example the perihelion precession of a planetary orbit is

$$\delta\phi = \left(\frac{2 + 2\gamma - \beta}{3}\right)\frac{6\pi m^2 c^2}{a^2} \tag{5.79}$$

which reduces to (5.69) when $\beta = \gamma = 1$. In a similar way the deflection of light is given by

$$\Delta = \left(\frac{1 + \gamma}{2}\right)\frac{4m}{r_0} \tag{5.80}$$

which reduces to (5.74) when $\gamma = 1$.

5.10 Gravitational lenses

The power of large masses to bend light paths invites comparison with optical lenses – hence the topic of 'gravitational lenses'. The analogy, however, is not complete, as we shall see.

First, consider light coming from infinity and bent by a 'lens' L to reach an observer O. Call the distance of closest approach between the light path and the lens b – an 'impact parameter'. The bending of the light path may be simply represented as taking place at one point, discontinuously, as shown in Fig. 5.7. (Note that the angles involved here are extremely small; of the order of a few seconds of arc at most, as we saw above. The diagram is therefore greatly exaggerated.) Light is detected by the observer at a distance D from L:

$$D = \frac{b}{\theta}. \tag{5.81}$$

However from (5.74) above $\theta = \Delta = \dfrac{4m}{b}$ so

[10] Some years ago there was a problem connected with solar oblateness, and reason to hope that the Brans–Dicke theory might clear this up. It now appears, though, that the problem has probably gone away. See for example d'Inverno (1992), p. 206.

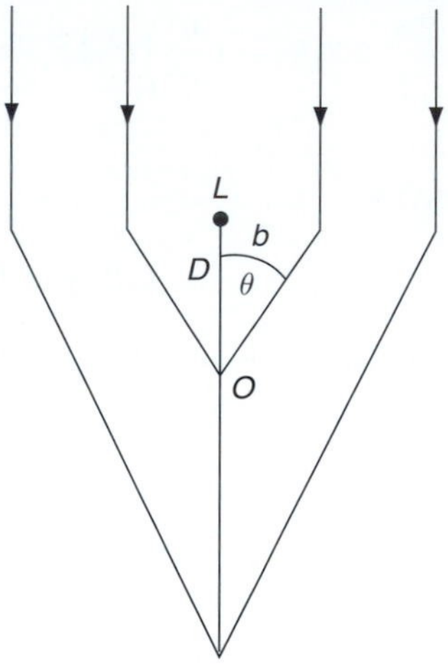

Fig. 5.7 **Light coming from infinity is bent by a lens *L* to reach an observer *O*.**

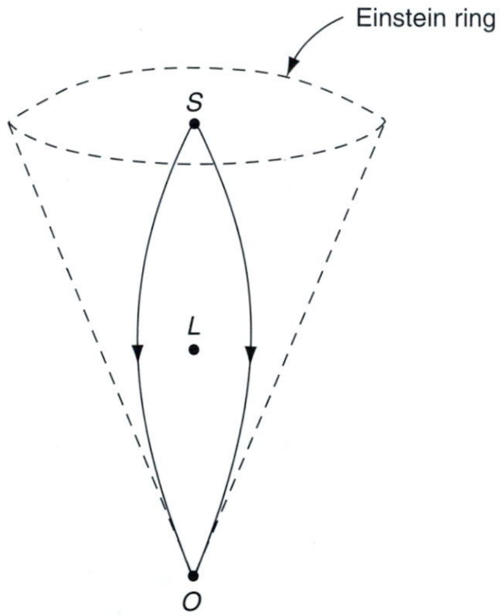

Fig. 5.8 **In a symmetrical configuration light from a source *S* is bent by a lens *L* to reach an observer *O*, who then sees the source as an 'Einstein ring'.**

$$D = \frac{b^2}{4m}, \tag{5.82}$$

where, as always, $m = MG/c^2$, M being the mass of the lens. D therefore increases with b: rays 'further out' from L reach the axis at a different point. The lens *does not focus* the rays to one point.

Consider the symmetric situation where the source S, lens L and observer O are in a straight line, as in Fig. 5.8. In a planar cross section the source S appears to the observer O as two images. There is cylindrical symmetry, however, so the image is actually a ring – the

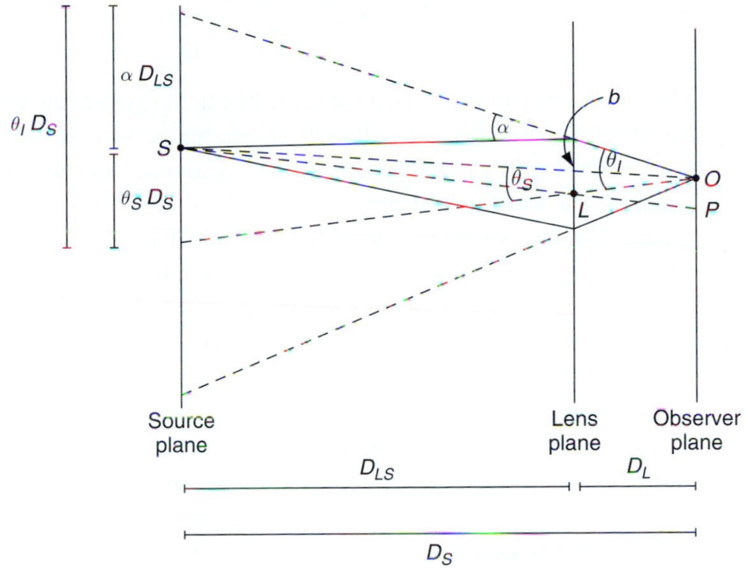

Fig. 5.9 **Observer *O* sees light from the source *S* bent by the lens *L*: *O*, *S* and *L* are not in a straight line.**

'Einstein ring'. What is its (angular) radius? The general situation is sketched in Fig. 5.9: here the source S, lens L and observer O are not in a straight line. The distances between the planes are as marked: D_{LS} is the distance from the lens to the source, etc. These distances depend on the angles, also as marked. Then

$$\theta_I D_S = \theta_S D_S + \alpha D_{LS} \tag{5.83}$$

but (see (5.74))

$$\alpha = \frac{4m}{b} = \frac{4m}{\theta_I D_L}$$

so

$$\theta_I = \theta_S + \frac{4}{\theta_I} \cdot \frac{m D_{LS}}{D_L D_S},$$

which on relabelling $\theta_I \to \theta$ becomes

$$\theta = \theta_S + \frac{{\theta_E}^2}{\theta}, \tag{5.84}$$

where

$${\theta_E}^2 = \frac{4m D_{LS}}{D_L D_S}, \tag{5.85}$$

and θ_E is called the 'Einstein angle'. The solutions to this equation give the angular positions of images in the sky – there are generally two solutions. In the situation where $D_{LS}, D_S \gg D_L$ (the source is at an 'infinite' distance) we have

$$\theta_E{}^2 \approx \frac{4m}{D_L}. \tag{5.86}$$

The significance of the Einstein angle is that in the symmetric case, when the source, lens and observer are collinear, $\theta_S = 0$ so $\theta = \theta_E$: the image is a *ring* with this angular radius. In the case of lensing in the galaxy (of one star by another, not the Sun), $\theta_E \approx 10^{-3''}$, an angle which is too small to be resolvable by current telescopes. This is an example of *microlensing*. On the cosmological scale, however, (the lensing of one galaxy by another) $\theta_E \approx 1''$, which *is* resolvable, and indeed there is observational evidence for lensing on this scale, known as *macrolensing*. (For the figures quoted above, see Problem 5.4.)

In a cosmological context lensing may be used – at least in principle, and practice seems now not far behind – to find the masses and distances away of galaxies acting as lenses. This is important information, for example in the study of dark matter; the determination of masses by this method is *non-dynamical*, in contrast with the usual method of detecting dark matter. It will be shown here, in a very simple (and somewhat unrealistic) model how this works.

In the approximation $D_{LS} \sim D_S \gg D_L = D$ Equation (5.84) holds with

$$\theta_E{}^2 = \frac{4m}{D} \tag{5.87}$$

where $m = MG/c^2$, M being the mass of the lens, and D being its distance away. Hence

$$\theta^2 - \theta_S\,\theta - \theta_E{}^2 = 0. \tag{5.88}$$

This equation has two solutions, θ_1 and θ_2 (which are of different signs). Their product is

$$|\theta_1\,\theta_2| = \theta_E{}^2 = \frac{4m}{D}. \tag{5.89}$$

We now find one more relation between θ_1 and θ_2. This depends on the differing *path lengths* of the two rays. Consider a point P in the observer plane on the line SL extrapolated, as in Fig. 5.10. From P an Einstein ring is seen: so in the planar diagram light fronts arriving

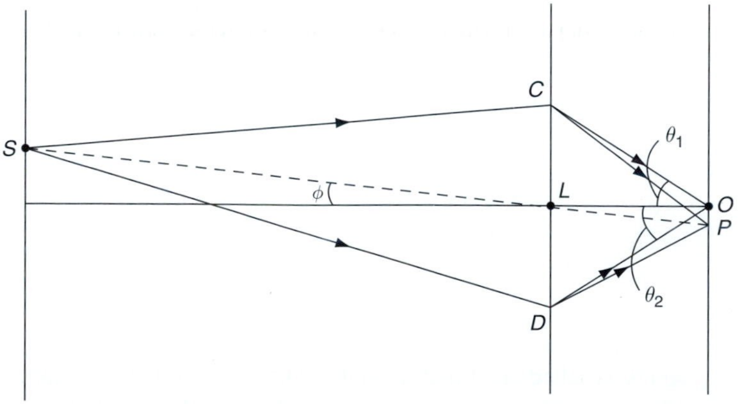

Fig. 5.10 **Light rays from the source S to observer O.**

along the two paths do so at the same time. In addition, to a good approximation, α, the angle between these two rays, is equal to $\theta_1+\theta_2$, the corresponding angle at O. Let $OP=d$, $CP=l_1$, $DP=l_2$. Also let $CL=LD=h$ (and $LO=D$, above). Then by Pythagoras

$$l_1=\sqrt{D^2+(h+d)^2}=D\left[1+\frac{(h+d)^2}{D^2}\right]^{1/2}\approx D+\frac{(h+d)^2}{2D}.$$

Similarly

$$l_2\approx D+\frac{(h-d)^2}{2D},$$

so

$$l_1-l_2=\frac{2hd}{D}=2h\phi.$$

But $2h=D(\theta_1+\theta_2)$, so $l_1-l_2=D(\theta_1+\theta_2)\phi$ and the time difference between the two signals arriving at P is

$$\Delta t=\frac{l_1-l_2}{c}=\frac{D}{c}(\theta_1+\theta_2)\phi. \tag{5.90}$$

On the other hand $\phi\approx\theta_S$ (compare Figs. 5.9 and 5.10) and from the theory of quadratic equations we have from (5.88)

$$|\theta_1-\theta_2|=\theta_S$$

(recall once more that θ_1 and θ_2 are of different signs), hence $\phi=|\theta_1-\theta_2|$ and (5.90) gives $\Delta t=\frac{D}{c}\left|\theta_1{}^2-\theta_2{}^2\right|$, hence

$$\left|\theta_1{}^2-\theta_2{}^2\right|=\frac{c\,\Delta t}{D}. \tag{5.91}$$

This is a second relation between θ_1 and θ_2; so by measuring θ_1, θ_2 and Δt, the Equations (5.89) and (5.91) allow us in principle to determine D and m, the distance away and mass of the lens. It was mentioned above that this model is too simplistic; two effects which have been ignored are the echo delay effect (the subject of the next section) and the expansion of the Universe. Nevertheless we have shown how gravitational lensing is a useful tool for cosmology.

5.11 Radar echoes from planets

A previous test for General Relativity, that of the bending of light paths, is a test of the *orbit* of a light ray. Modern techniques, however, allow the possibility of measuring directly the *time* for light to travel from a source to a reflector (planet or space probe) and back. In appropriate circumstances this differs from the Newtonian expression by an amount of the order of 10^{-4} seconds, an easily measurable time. The best reflector is a planet (which is less subject to non-gravitational forces like the solar wind than a space

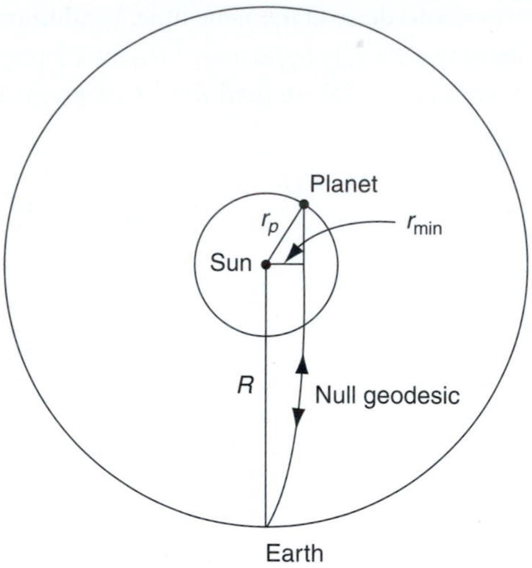

Fig. 5.11 Radar travelling from the Earth to a planet and back, with r_{min} the distance of closest approach to the Sun.

probe is), and the situation in which relativistic effects are greatest is when the planet is in 'superior conjunction', i.e. the Sun is almost on a straight line joining the Earth and the planet. This is the situation sketched in Fig. 5.11. General relativity predicts that the echo is *delayed* relative to the Newtonian prediction, that is, relative to the signal travelling out and back in a straight line.

Let $t(r_1, r_2)$ be the coordinate time between the emission of a light signal at $r = r_1$ and its reception at $r = r_2$. Then the coordinate time interval between the emission of a signal from the Earth and reception of the echo is

$$T = 2\left[t\left(R, r_{min}\right) + t\left(r_P,\ r_{min}\right)\right] \tag{5.92}$$

where r_{min} is the distance of closest approach of the path to the Sun – see Fig. 5.11. The corresponding *proper time* interval is

$$\tau = T\sqrt{1 - \frac{2GM}{rc^2}},$$

and since $\frac{GM}{rc^2}T$ is of the order of 10^{-8} s, we ignore this correction factor; we shall see below that the time delay is of the order of 100 μs, so the dominant effect is due to the fact that the path is not a straight line. If light travelled in straight lines we should have $t\ (r,\ r_{min}) = t_E\ (r,\ r_{min})$ (E stands for Euclidean), with

$$t_E(r, r_{min}) = \frac{1}{c}\sqrt{r^2 - r_{min}{}^2}, \tag{5.93}$$

by Pythagoras.

To obtain $t\ (r,\ r_{min})$ in the relativistic case we return to the Schwarzschild metric (5.36). Put $ds^2 = 0$ and $d\theta = 0$ (motion of light in a plane); then we have

$$0 = -\left(1 - \frac{2m}{r}\right) + \frac{1}{c^2}\left[\frac{\left(\frac{dr}{dt}\right)^2}{1 - \frac{2m}{r}} + r^2\left(\frac{d\phi}{dt}\right)^2\right]. \tag{5.94}$$

We can eliminate $\frac{d\phi}{dt}$ by using Equations (5.57) and (5.62), from which

$$\frac{r^2\frac{d\phi}{d\tau}}{\left(1 - \frac{2m}{r}\right)\frac{dt}{d\tau}} = \frac{r^2}{\left(1 - \frac{2m}{r}\right)}\frac{d\phi}{dt} = \frac{a}{b} = B = \text{const},$$

and hence

$$\frac{d\phi}{dt} = \frac{B}{r^2}\left(1 - \frac{2m}{r}\right).$$

Then (5.94) becomes

$$0 = -\left(1 - \frac{2m}{r}\right) + \frac{1}{c^2}\left[\frac{\left(\frac{dr}{dt}\right)^2}{1 - \frac{2m}{r}} + \frac{B^2}{r^2}\left(1 - \frac{2m}{r}\right)^2\right]. \tag{5.95}$$

We can find B by noting that $\frac{dr}{dt} = 0$ when $r = r_{\min}$, since then the velocity has no radial component (see the diagram). This gives

$$B^2 = \frac{c^2 {r_{\min}}^2}{\left(1 - \frac{2m}{r_{\min}}\right)}$$

and (5.95) gives, after a bit of rearrangement,

$$\frac{dr}{dt} = c\left(1 - \frac{2m}{r}\right)\left[1 - \frac{{r_{\min}}^2\left(1 - \frac{2m}{r}\right)^2}{r^2\left(1 - \frac{2m}{r_{\min}}\right)}\right]^{1/2}.$$

Then

$$t(r, r_{\min}) = \frac{1}{c}\int_{r_{\min}}^{r} \frac{dr'}{\left(1 - \frac{2m}{r'}\right)\left[1 - \frac{{r_{\min}}^2\left(1 - \frac{2m}{r'}\right)}{r'^2\left(1 - \frac{2m}{r_{\min}}\right)}\right]^{1/2}}. \tag{5.96}$$

This is the time taken for light to travel from $r_{\min}$ to r. To first order in the small quantities $\frac{m}{r}$ and $\frac{m}{r_{\min}}$ this gives

$$t(r, r_{min}) \approx \frac{1}{c}\int_{r_{min}}^{r} \frac{r'\,dr'}{\sqrt{r'^2 - r_{min}^2}}\left[1 + \frac{2m}{r'} + \frac{m r_{min}}{r'(r' + r_{min})}\right], \tag{5.97}$$

and on integration

$$t(r, r_{min}) = \frac{1}{c}\left[\sqrt{r^2 - r_{min}^2} + 2m\ln\left(\frac{r + \sqrt{r^2 - r_{min}^2}}{r_{min}}\right) + m\sqrt{\frac{r - r_{min}}{r + r_{min}}}\right]. \tag{5.98}$$

(Confirmation of the above expressions is left to Problems 5.5 and 5.6.) Note that the first term in (5.98) is the Euclidean expression (5.93); and note also that the extra terms are *positive*: t is *greater* than in the flat space result – light travels in a curve.

We can then state that the time for light to travel from the Earth (at $r = R$) to $r = r_{min}$ is, since $R \gg r_{min}$, r_{min} being taken to be of the order of the Sun's radius,

$$t(R, r_{min}) = t_1 = \frac{1}{c}\left[\sqrt{R^2 - r_{min}^2} + 2m\ln\left(\frac{2R}{r_{min}}\right) + m\right]. \tag{5.99}$$

Similarly the time from $r = r_{min}$ to the planet at $r = r_P$ is

$$t(r_P, r_{min}) = t_2 = \frac{1}{c}\left[\sqrt{r_P^2 - r_{min}^2} + 2m\ln\left(\frac{2r_P}{r_{min}}\right) + m\right], \tag{5.100}$$

and the total time of travel from the Earth back to the Earth is

$$\begin{aligned} T &= 2(t_1 + t_2) \\ &= \frac{2}{c}\left[\sqrt{R^2 - r_{min}^2} + \sqrt{r_P^2 - r_{min}^2} + 2m\ln\left(\frac{4Rr_P}{r_{min}^2}\right) + 2m\right] \\ &\approx \frac{2}{c}\left[R + r_P + 2m\left\{\ln\left(\frac{4Rr_P}{r_{min}^2}\right) + 1\right\}\right]. \end{aligned} \tag{5.101}$$

It is instructive to calculate this quantity for a specific case: let us take Mars, whose distance from the Sun is $r_P = 1.52$ AU $= 2.28 \times 10^{11}$ m. Then

$$\frac{2}{c}(R + r_P) = 2.52 \times 10^3\ \text{s} \approx 42\ \text{min}. \tag{5.102}$$

This is the 'Euclidean' time. The excess time, or delay, is clearly greatest when r_{min} takes its smallest value, which is the radius of the Sun R_S. In that case

$$\frac{4Rr_P}{R_S^2} = 2.82 \times 10^5, \quad \ln\left(\frac{4R\,r_P}{R_S^2}\right) = 12.55$$

and the 'echo delay' is

$$\begin{aligned} \Delta T &\equiv \frac{4m}{c}\left[\ln\left(\frac{4R\,r_P}{R_S^2}\right) + 1\right] \\ &= 2.66 \times 10^{-4}\ \text{s} = 266\ \mu\text{s}. \end{aligned} \tag{5.103}$$

It is clear that to detect this effect an accuracy of 1 part in 10^7 is necessary: to measure it to within 1%, measurement is required to 1 part in 10^9. Atomic clocks give an accuracy of 1 part in 10^{12}, so this is achievable, but it means on the other hand that $R + r_P$ must be known to an accuracy of the order of 1 km, which presents something of a challenge. Nevertheless the results are impressive; and are often cast in the PPN formalism. In that formalism the echo delay above is

$$\Delta T = \left(\frac{1+\gamma}{2}\right)\frac{4m}{c}\left[\ln\left(\frac{4R\,r_P}{{R_S}^2}\right)+1\right], \tag{5.104}$$

where, of course (see (5.78)), $\dfrac{(1+\gamma)}{2} = 1$ in General Relativity. Will (2001) displays the measured values of $\dfrac{(1+\gamma)}{2}$ and concludes that agreement with General Relativity holds to 0.1 per cent.

5.12 Radial motion in a Schwarzschild field: black holes – frozen stars

As observed already, the form of the Schwarzschild metric (5.36) suggests that there may be interesting phenomena associated with the surface $r = 2GM/c^2 = 2m$. In this section we consider the radial fall of objects in a Schwarzschild field and see the first indication of the bizarre nature of this surface.

Consider the motion of a particle falling radially in a Schwarzschild field – into a star, for example. Suppose it starts at $r = R$ with $\dfrac{\mathrm{d}r}{\mathrm{d}t} = 0$, as shown in Fig. 5.12, where the spherical surface $r = 2m$ is also marked. For radial motion we have

$$\mathrm{d}s^2 = -\left(1-\frac{2m}{r}\right)c^2\,\mathrm{d}t^2 + \left(1-\frac{2m}{r}\right)^{-1}\mathrm{d}r^2.$$

With $\mathrm{d}s^2 = -\,c^2\,\mathrm{d}\tau^2$, $\dot{t} = \dfrac{\mathrm{d}t}{\mathrm{d}\tau}$, $\dot{r} = \dfrac{\mathrm{d}r}{\mathrm{d}\tau}$ this gives

$$\left(1-\frac{2m}{r}\right)\dot{t}^2 - \frac{1}{c2}\left(1-\frac{2m}{r}\right)^{-1}\dot{r}^2 = 1, \tag{5.105}$$

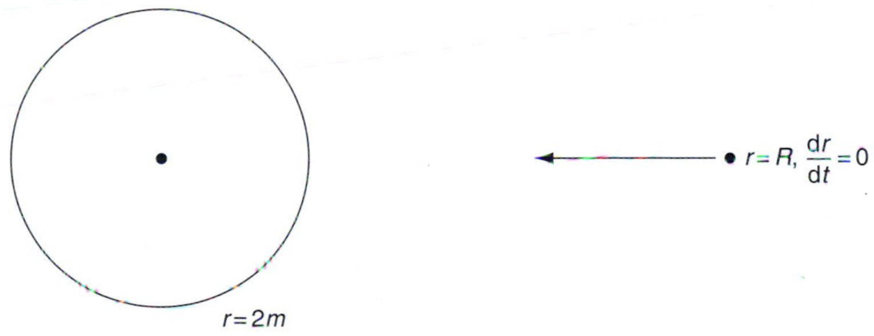

Fig. 5.12 A particle falling radially in a Schwarzschild field.

which, on putting $\dot{r} = \frac{\mathrm{d}r}{\mathrm{d}\tau}\dot{t}$ gives

$$\left[c^2\left(\frac{r-2m}{r}\right) - \left(\frac{r}{r-2m}\right)\left(\frac{\mathrm{d}r}{\mathrm{d}t}\right)^2\right]\dot{t}^2 = c^2. \tag{5.106}$$

The boundary condition $r = R$, $\frac{\mathrm{d}r}{\mathrm{d}t} = 0$ gives

$$\left(\frac{\mathrm{d}t}{\mathrm{d}\tau}\right)_{r=R} = \left(\frac{R}{R-2m}\right)^{1/2}. \tag{5.107}$$

From (5.57) above

$$\left(1 - \frac{2m}{r}\right)\dot{t} = b = \text{const},$$

hence

$$b = \left(1 - \frac{2m}{R}\right)\left(\frac{\mathrm{d}t}{\mathrm{d}\tau}\right)_R = \left(\frac{R-2m}{R}\right)^{1/2}$$

and therefore

$$\frac{\mathrm{d}t}{\mathrm{d}\tau} = \left(\frac{r}{r-2m}\right)b = \left(\frac{r}{r-2m}\right)\left(\frac{R-2m}{R}\right)^{1/2}. \tag{5.108}$$

Substituting this into (5.106) gives, on rearrangement

$$\frac{\mathrm{d}r}{\mathrm{d}t} = -c\frac{(r-2m)(2m)^{1/2}(R-r)^{1/2}}{r^{3/2}\,(R-2m)^{1/2}}. \tag{5.109}$$

Note the minus sign, selected on taking the square root: it reflects the fact that we are considering a 'fall' into the star, so r decreases as t increases. Finally we have

$$t = -\frac{1}{c}\left(\frac{R-2m}{2m}\right)^{1/2}\int_R^r \frac{\rho^{3/2}\,\mathrm{d}\rho}{(\rho-2m)(R-\rho)^{1/2}}. \tag{5.110}$$

This is the time taken for a particle to travel from $r = R$ to an arbitary r, as measured in Schwarzschild coordinates, in which t is the time parameter, which is the time measured on a clock 'at infinity' – by a distant observer. It is clear that the form of the integrand above will diverge when $\rho \to 2m$. To investigate this, put

$$\rho = 2m + \varepsilon$$

with ε small. Then (5.109) gives

$$\begin{aligned} ct &= -\left(\frac{R-2m}{2m}\right)^{1/2}\int_{R-2m}^{r-2m}\frac{(2m+\varepsilon)^{3/2}\,\mathrm{d}\varepsilon}{\varepsilon(R-2m)^{1/2}} = -2m\int_{R-2m}^{r-2m}\frac{\mathrm{d}\varepsilon}{\varepsilon} \\ &= -2m\ln\left(\frac{r-2m}{R-2m}\right), \end{aligned}$$

or

$$r - 2m = (R - 2m)\ \exp(-ct/2m). \tag{5.111}$$

So, as $t \to \infty$, $r \to 2m$: the radial infall of a particle to the Schwarzschild radius takes an *infinite* time as seen from a safe distance. This is always assuming, of course, that the surface $r = 2m$ is *outside* the star, in the vacuum. If we transfer this conclusion to the outer layers of a collapsing star itself, we then learn that the collapse of a star, as seen from the outside, takes an infinite time to reach a *finite* size. The star seems to collapse more and more slowly, and is *never* seen collapsing to a point. Zel'dovich and Novikov (1996) call this phenomenon a 'frozen star'.

This describes the fall of a particle (or of the star itself) as seen by a distant observer. What about the fall of the particle measured by its own clock? We need to calculate the *proper time* for the particle to reach $r = 2m$, and therefore want $\frac{\mathrm{d}r}{\mathrm{d}\tau}$. From (5.108) and (5.109)

$$\frac{\mathrm{d}r}{\mathrm{d}\tau} = \frac{\mathrm{d}r}{\mathrm{d}t} \cdot \frac{\mathrm{d}t}{\mathrm{d}\tau} = -c\left(\frac{2m(R-r)}{Rr}\right)^{1/2} = -c\left(\frac{2m}{R}\right)^{1/2}\left(\frac{R}{r} - 1\right)^{1/2}$$

so that

$$\tau = -\frac{1}{c}\sqrt{\frac{R}{2m}}\int_R^r \frac{\mathrm{d}r'}{\left(\frac{R}{r'} - 1\right)^{1/2}}.$$

On putting $\rho = \frac{r}{R}$ this gives

$$\tau = -\frac{1}{c}\sqrt{\frac{R^3}{2m}}\int_1^\rho \frac{\mathrm{d}\rho'}{\left(\frac{1}{\rho'} - 1\right)^{1/2}}.$$

The integral is straightforward to perform and we find

$$\tau = \frac{1}{c}\sqrt{\frac{R^3}{2m}}\left[\sqrt{\rho(1-\rho)} + \cos^{-1}\sqrt{\rho}\right]. \tag{5.112}$$

This is the proper time for collapse from $\frac{r}{R} = \rho = 1$ down to an arbitary value of ρ (<1). It is clearly *finite* for all r, even down to $r = 0$. In fact the proper time to reach $r = 0$ is

$$\tau_0 = \frac{\pi}{2c}\sqrt{\frac{R^3}{2m}}. \tag{5.113}$$

So in summary: an object falls into a collapsed star. Viewed from the outside, the object may still be seen, even after an infinite time, as it asymptotically approaches the Schwarzschild radius $r = 2m$. Measured on its own clock, however, the object only exists for a time given by (5.113), after which it is crushed to unlimited density at the centre of the star. It seems as if nothing strange happens at $r = 2m$.[11] In a scenario like this General Relativity provides a spectacular illustration of the relativity of time.

[11] It may be, however, that the scalar quantity constructed by contracting the covariant derivative of the Riemann tensor with itself changes sign at $r = 2m$, so that in principle a detector (accelerometer) with sufficient sensitivity should be able to register when the Schwarzschild surface is crossed. I am grateful to Brian Steadman for this remark.

5.13 A gravitational clock effect

After the rather lofty topics above, consider the following simpler, but equally interesting, experiment. Suppose there are two synchronised atomic clocks on the equator. One is placed on board an aircraft which flies once round the Earth (at a height h) while the other remains on the ground. After the flight, what is the time discrepancy between the clocks?[12]

We use the Schwarzschild metric on the equator ($\theta = \pi/2$, r = const), so that (5.36) gives

$$ds^2 = -\left(1 - \frac{2GM}{rc^2}\right)c^2\,dt^2 + r^2\,d\phi^2 = -c^2\,d\tau^2 \tag{5.114}$$

where τ is proper time. Referring to Fig. 5.13, observer A remains at rest on the Earth, rotating with angular velocity $\omega = \frac{d\phi}{dt}$; hence $d\phi = \omega\,dt$ and from above

$$d\tau_A{}^2 = \left[\left(1 - \frac{2GM}{Rc^2}\right) - \frac{R^2\omega^2}{c^2}\right]dt^2;$$

and since $\frac{2GM}{c^2} \ll 1$, $\frac{R^2\omega^2}{c^2} \ll 1$, then

$$d\tau_A \approx \left[1 - \frac{GM}{Rc^2} - \frac{R^2\omega^2}{2c^2}\right]dt. \tag{5.115}$$

(Note the logic implied by this derivation: A is moving in a static Schwarzschild field with velocity $R\omega = R\frac{d\phi}{dt}$.) Observer B, on the other hand, flies round the Earth at a height h: as shown in Fig. 5.12 she flies eastward. She moves at a speed v relative to the ground, and so with a speed $\approx v + (R+h)\omega$ with respect to the 'static' frame of the Schwarzschild field. Then

$$d\tau_B{}^2 = \left[1 - \frac{2GM}{(R+h)c^2} - \frac{\{(R+h)\omega + v\}^2}{c^2}\right]dt^2$$

and

$$d\tau_B \approx \left[1 - \frac{GM}{(R+h)c^2} - \frac{R^2\omega^2 + 2R\omega v + v^2}{2c^2}\right]dt. \tag{5.116}$$

Hence

$$d\tau_A - d\tau_B = \left[-\frac{GMh}{R^2c^2} + \frac{(2R\omega + v)v}{2c^2}\right]dt. \tag{5.117}$$

[12] The following treatment follows closely that of Berry (1976), Section 5.2.

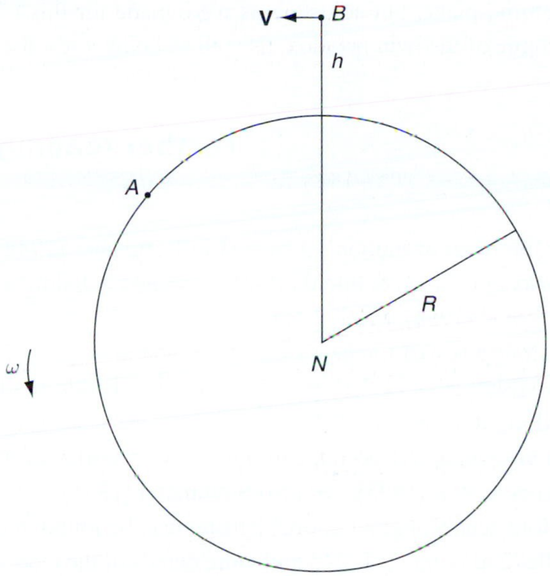

Fig. 5.13 The Earth, viewed from above: N is the North Pole. Observer A is at rest on the equator, while B travels with speed v at a height h above the Earth in the equatorial plane. They each carry a clock.

Then, defining the quantity

$$\Delta = \frac{\mathrm{d}\tau_A - \mathrm{d}\tau_B}{\mathrm{d}\tau_A}, \tag{5.118}$$

we have

$$\Delta \approx -\frac{GMh}{R^2c^2} + \frac{(2R\omega + v)v}{2c^2}. \tag{5.119}$$

Putting in numbers, with $h = 10^4\,\mathrm{m}$, $v = 300\,\mathrm{m\,s^{-1}}$, $\frac{GM}{R^2} = g = 9.81\ \mathrm{m\,s^{-2}}$, then $\frac{gh}{c^2} = 1.09 \times 10^{-12}$, $2R\omega = 931\,\mathrm{m\,s^{-1}}$ and $\frac{(2R\omega + v)v}{2c^2} = 2.1 \times 10^{-12}$ and

$$\Delta_{\mathrm{Eastward}} = 1.0 \times 10^{-12}; \tag{5.120}$$

'eastward' because this is what we are considering. With a *westward* journey $v \to -v$ and $\frac{(2R\omega - v)v}{2c^2} = 1.05 \times 10^{-12}$, giving

$$\Delta_{\mathrm{Westward}} = -2.1 \times 10^{-12}. \tag{5.121}$$

These quantities are measurable: the fractional accuracy of cesium clocks is about 1 part in 10^{13}. In 1971 Hafele and Keating[13] placed very accurate cesium clocks aboard commercial airliners and verified the above predictions to about 10%. (The orbits were not in the

[13] Hafele & Keating (1972).

equatorial plane, but adjustments were made for this.) This clock effect is a gravitational analogue of the twin paradox, as well as being a test for the Schwarzschild solution.

Further reading

The 'problem of motion' is treated in Bergmann (1942), Chapter 15. A later review of it appears in Einstein & Infeld (1949), reprinted in Kilmister (1973). See also Bażański (1962) and Synge (1964), Section 4.6.

A translation of Laplace's essay is given in Appendix A of Hawking & Ellis (1973); see also Misner *et al.* (1973), p. 623. A proof of Birkhoff's theorem appears in Appendix B of Hawking & Ellis (1973).

A very complete account of alternative theories of gravitation and the PPN formalism appears in Will (1993); see also Straumann (1991).

More details of gravitational lensing may be found in Chapter 4 of Peacock (1999) and in Hartle (2003), pp. 234–243 and more details of the radar echo data are given in Will (2001).

Problems

5.1 Laplace claimed that a star of the same density as the Earth with a diameter 250 times that of the Sun would be unable to shine. Verify this claim.

5.2 In the text it was claimed that the relative frequency of light from an atomic transition in the Sun (ν_1) and in a laboratory on Earth (ν_2) is approximately

$$\frac{\nu_1}{\nu_2} = 1 + \frac{GM_S}{c^2}\left(\frac{1}{R} - \frac{1}{R_S}\right).$$

Obtain a higher approximation for this ratio by taking into account the gravitational field of the Earth itself, and estimate the relative magnitude of this contribution.

5.3 Show that, according to Newtonian theory, for a planet of mass μ in a non-circular orbit around a Sun of mass M, the laws of conservation of energy (kinetic plus potential) and angular momentum $L = \mu a$ require that the equation for the orbit, $r(\phi)$, be (with $u = 1/r$)

$$\frac{d^2u}{d\phi^2} + u = \frac{GM}{a^2}.$$

5.4 Calculate the Einstein angle for (i) lensing within the galaxy of a star by an object of solar mass between us and the star, (ii) lensing of a source at a cosmological distance by a galaxy.

5.5 Prove Equation (5.97).

5.6 Prove Equation (5.98).

5.7 Find the radius at which light travels in a circular path round a body of mass M, (i) according the General Relativity, (ii) according to Newtonian theory.

5.8 Show that the lines r variable (θ, ϕ, t constant), are geodesics of the Schwarzschild metric.

5.9 Consider two identical clocks A and B, synchronised and placed on the equator of the Earth. Clock A is taken to the North Pole and stays there for a year after which it returns to clock B on the equator. Ignoring the difference between the equatorial and polar radii of the Earth, use the Schwarzschild metric to calculate the resulting time discrepancy between A and B.

5.10 Recalculate the time discrepancy in the previous problem, taking into account the difference between the polar and equatorial radii of the Earth, given by the 'flattening parameter'

$$\delta = \frac{R_E - R_P}{R_E} = 0.0034.$$

5.11 Using the symbols M, L and T for mass, length and time we may define *dimensions* of physical quantities, so that for example density ρ has dimensions $[\rho] = \mathrm{M\,L^{-3}}$ – mass divided by volume. Check that the right and left hand sides of the field equation (5.24) have the same dimension.

6 Gravitomagnetic effects: gyroscopes and clocks

We have already explored some features of the Schwarzschild solution, including the tests of General Relativity that it allows. In the Schwarzschild solution the Sun is taken to be static, that is, non-rotating. In fact, however, the Sun does rotate, and this suggests the question, is there another exact solution, a generalisation of the Schwarzschild solution, describing a rotating source? And, if there is, does it suggest any additional tests of General Relativity? It turns out that a generalisation of the Schwarzschild solution does exist – the Kerr solution. This is a rather complicated solution, however; it will be discussed further in the next chapter. In this chapter we shall find an *approximate* solution for a rotating source (which of course will also turn out to be an approximation of the Kerr solution). The tests for this solution include a prediction for the precession of gyroscopes in orbit round the Earth (which of course also rotates). This is a tiny effect, but in April 2004 a satellite was launched to look for this precession, which goes by the names of Lense and Thirring. We shall see that there is a parallel between the Lense–Thirring effect and magnetism, just as there is between 'ordinary' gravity (not involving rotations) and electricity – hence the name 'gravitomagnetism'. After a discussion of these matters the chapter finishes with a more theoretical look at the nature of the distinction between 'static' (Schwarzschild) and 'stationary' (Kerr) space-times. We begin with a description of the linear approximation; this is important in its own right, as well as allowing us to find the space-time metric round a rotating object.

6.1 Linear approximation

This approximation is appropriate for weak gravitational fields. The metric tensor is then very nearly the Minkowski metric tensor, so we put

$$g_{\mu\nu} = \eta_{\mu\nu} + h_{\mu\nu}, \qquad h_{\mu\nu} \ll 1. \tag{6.1}$$

This approximation was introduced in Section 5.1 to derive the Newtonian limit. Here we develop it more thoroughly; we shall find the form of the field equations in this approximation. We assume that at large distances from the source space-time becomes Minkowski, so

$$\lim_{r\to\infty} h_{\mu\nu} = 0. \tag{6.2}$$

Because of the smallness of $h_{\mu\nu}$ we neglect terms quadratic or of higher order in $h_{\mu\nu}$ – this is precisely the linear approximation. In this approximation we *define* the raising and lowering of indices to be performed with $\eta_{\mu\nu}$, not $g_{\mu\nu}$; for example

$$A^{\mu}{}_{\kappa} = \eta^{\mu\nu} A_{\nu\kappa}, \text{ etc.}$$

In this case we *cannot* assume that $g^{\mu\nu} = \eta^{\mu\nu} + h^{\mu\nu}$, so put

$$g^{\mu\nu} = \eta^{\mu\nu} + \chi^{\mu\nu}.$$

Then from $g^{\mu\nu} g_{\nu\kappa} = \delta^{\mu}{}_{\kappa}$ it follows that

$$(\eta^{\mu\nu} + \chi^{\mu\nu})(\eta_{\nu\kappa} + h_{\nu\kappa}) = \delta^{\mu}{}_{\kappa},$$

hence

$$\chi^{\mu}{}_{\kappa} + h^{\mu}{}_{\kappa} = 0$$

and

$$g^{\mu\nu} = \eta^{\mu\nu} - h^{\mu\nu}. \tag{6.3}$$

The connection coefficients then are, to first order in h,

$$\Gamma^{\kappa}{}_{\lambda\mu} = {}^1\!/\!_2\, \eta^{\kappa\rho}(h_{\rho\lambda,\mu} + h_{\rho\mu,\lambda} - h_{\lambda\mu,\rho})$$

and the Ricci tensor is

$$\begin{aligned} R_{\mu\nu} &= \Gamma^{\lambda}{}_{\mu\nu,\lambda} - \Gamma^{\lambda}{}_{\mu\lambda,\nu} + \mathrm{O}(h^2) \\ &= {}^1\!/\!_2(h^{\lambda}{}_{\mu,\nu\lambda} + h^{\lambda}{}_{\nu,\mu\lambda} - \Box h_{\mu\nu} - h^{\lambda}{}_{\lambda,\mu\nu}) + \mathrm{O}(h^2), \end{aligned} \tag{6.4}$$

where we are using the convention that all the indices following a comma are differentiating indices. The field equations are then, from (6.4) and (5.25)

$$h^{\lambda}{}_{\mu,\nu\lambda} + h^{\lambda}{}_{\nu,\mu\lambda} - \Box h_{\mu\nu} - h^{\lambda}{}_{\lambda,\mu\nu} = -\frac{16\pi G}{c^2} S_{\mu\nu} \tag{6.5}$$

where

$$S_{\mu\nu} = T_{\mu\nu} - {}^1\!/\!_2\, g_{\mu\nu} T^{\lambda}{}_{\lambda}.$$

To lowest (dominant) order in h, $T_{\mu\nu}$ and therefore $S_{\mu\nu}$ is *independent* of h (see (5.27)); and the conservation law (5.19) becomes

$$\partial_{\mu} S^{\mu}{}_{\nu} = 0 = \partial_{\mu} T^{\mu}{}_{\nu} \tag{6.6}$$

i.e. the conservation law of Special Relativity. So in the linearised theory the gravitational field has no influence on the motion of matter that produces the field. We can therefore specify $T_{\mu\nu}$ arbitrarily provided only that the conservation law (6.6) holds; and we can then calculate $h_{\mu\nu}$ from it using the field equations (6.5). (It is therefore possible in principle, as pointed out by Stephani,[1] that an *exact* solution, provided it could be found, could differ appreciably from the linearised solution. So we must beware, especially since the linear approximation may be used in cases where an exact solution is not known; and therefore the conclusions drawn may not be reliable.)

[1] Stephani (1982), p. 121; Stephani (2004), p. 217.

We now want to write the field equation (6.5) in a neater form. Instead of $h^{\mu\nu}$ we introduce the quantities $f^{\mu\nu}$:

$$\sqrt{-g}\,g^{\mu\nu} = \eta^{\mu\nu} - f^{\mu\nu}. \tag{6.7}$$

Now

$$g_{\mu\nu} = \eta_{\mu\nu} + h_{\mu\nu} = \begin{pmatrix} -1+h_{00} & h_{01} & h_{02} & h_{03} \\ h_{10} & 1+h_{11} & h_{12} & h_{13} \\ h_{20} & h_{21} & 1+h_{22} & h_{23} \\ h_{30} & h_{31} & h_{32} & 1+h_{33} \end{pmatrix}$$

so

$$\begin{aligned} g &= (-1+h_{00})(1+h_{11})(1+h_{22})(1+h_{33}) + \mathrm{O}(h^2) \\ &= -1 + h_{00} - h_{11} - h_{22} - h_{33} + \mathrm{O}(h^2) \\ &= -1 - h^0{}_0 - h^1{}_1 - h^2{}_2 - h^3{}_3 + \mathrm{O}(h^2) \\ &= -1 - h^\mu{}_\mu + \mathrm{O}(h^2) \end{aligned}$$

and

$$\sqrt{-g} = 1 + {}^1\!/\!_2\, h^\mu{}_\mu + \mathrm{O}(h^2),$$

giving

$$\begin{aligned} \sqrt{-g}\,g^{\mu\nu} &= (1 + {}^1\!/\!_2\, h^\lambda{}_\lambda)(\eta^{\mu\nu} - h^{\mu\nu}) + \mathrm{O}(h^2) \\ &= \eta^{\mu\nu} + {}^1\!/\!_2\, \eta^{\mu\nu} h^\lambda{}_\lambda - h^{\mu\nu} + \mathrm{O}(h^2). \end{aligned} \tag{6.8}$$

From (6.7) and (6.8) we find

$$f^{\mu\nu} = h^{\mu\nu} - {}^1\!/\!_2\, \eta^{\mu\nu} h^\lambda{}_\lambda, \tag{6.9}$$

hence

$$f^\mu{}_\mu = -h^\mu{}_\mu \tag{6.10}$$

and

$$h^{\mu\nu} = f^{\mu\nu} - {}^1\!/\!_2\, \eta^{\mu\nu} f^\lambda{}_\lambda. \tag{6.11}$$

Equations (6.10) and (6.11) substituted into (6.4) give

$$R_{\mu\nu} = {}^1\!/\!_2\left[\left(f^\lambda{}_\mu - {}^1\!/\!_2\eta^\lambda_\mu f^\rho{}_\rho\right)_{,\nu\lambda} + \left(f^\lambda{}_\nu - {}^1\!/\!_2\eta^\lambda{}_\nu f^\rho{}_\rho\right)_{,\mu\lambda} - \Box f_{\mu\nu} + {}^1\!/\!_2\eta_{\mu\nu}\Box f^\rho{}_\rho + f^\rho{}_{\rho,\mu\nu}\right]$$

which becomes, on noting that the second, fourth and final terms cancel,

$$R_{\mu\nu} = {}^1\!/\!_2\left(f^\lambda{}_{\mu,\nu\lambda} + f^\lambda{}_{\nu,\mu\lambda} - \Box f_{\mu\nu} + {}^1\!/\!_2\,\eta_{\mu\nu}\,\Box f^\rho{}_\rho\right).$$

We then have

$$\begin{aligned}\tfrac{1}{2}\eta_{\mu\nu}R &= \tfrac{1}{4}\eta_{\mu\nu}\eta^{\rho\sigma}[f^{\lambda}{}_{\rho,\sigma\lambda} + f^{\lambda}{}_{\sigma,\rho\lambda} - \Box f_{\rho\sigma} + \tfrac{1}{2}\eta_{\rho\sigma}\Box f^{\lambda}{}_{\lambda}] \\ &= \tfrac{1}{2}\eta_{\mu\nu}f^{\rho\sigma}{}_{,\rho\sigma} + \tfrac{1}{4}\eta_{\mu\nu}\Box f^{\lambda}{}_{\lambda}\end{aligned}$$

and the field equations $R_{\mu\nu} - \tfrac{1}{2}\eta_{\mu\nu}R = -\dfrac{8\pi G}{c^2}T_{\mu\nu}$ give

$$f^{\lambda}{}_{\mu,\nu\lambda} + f^{\lambda}{}_{\nu,\mu\lambda} - \Box f_{\mu\nu} - \eta_{\mu\nu}f^{\lambda\kappa}{}_{,\lambda\kappa} = -\frac{16\pi G}{c^2}T_{\mu\nu}. \tag{6.12}$$

These can now be simplified by making a coordinate transformation

$$x^{\mu} \to x'^{\mu} = x^{\mu} + b^{\mu}(x); \tag{6.13}$$

the function $b^{\mu}(x)$ will later be chosen to satisfy some condition resulting in the simplification of (6.12). The coordinate transformation (6.13), we shall see below, bears some resemblance to gauge transformations in electrodynamics. Under (6.13)

$$\frac{\partial x'^{\mu}}{\partial x^{\nu}} = \delta^{\mu}{}_{\nu} + b^{\mu}{}_{,\nu}$$

and

$$g^{\mu\nu} \to g'^{\mu\nu} = \frac{\partial x'^{\mu}}{\partial x^{\rho}}\frac{\partial x'^{\nu}}{\partial x^{\sigma}}g^{\rho\sigma} = g^{\mu\nu} + g^{\mu\sigma}b^{\nu}{}_{,\sigma} + g^{\rho\nu}b^{\mu}{}_{,\rho} + \mathrm{O}(b^2).$$

We shall choose b^{μ} to be $\ll 1$, so the $\mathrm{O}(b^2)$ terms may be ignored. Written out as a matrix the right hand side of the above equation has the appearance, to order b,

$$\begin{pmatrix} g^{00} + 2g^{0\sigma}b^{0}{}_{,\sigma} & g^{01} + g^{0\sigma}b^{1}{}_{,\sigma} + g^{1\sigma}b^{0}{}_{,\sigma} & \cdot\cdot & \cdot\cdot \\ \cdot\cdot & g^{11} + 2g^{1\sigma}b^{1}{}_{,\sigma} & \cdot\cdot & \cdot\cdot \\ \cdot\cdot & \cdot\cdot & g^{22} + 2g^{2\sigma}b^{2}{}_{,\sigma} & \cdot\cdot \\ \cdot\cdot & \cdot\cdot & \cdot\cdot & g^{33} + 2g^{3\sigma}b^{3}{}_{,\sigma} \end{pmatrix}.$$

The off-diagonal terms are of first order in h (see (5.3)), or b, so to leading order the determinant is $\left(\text{with } g = |g_{\mu\nu}| = |g^{\mu\nu}|^{-1}\right)$

$$\begin{aligned} g'^{-1} &= (g^{00} + 2g^{0\sigma}b^{0}{}_{,\sigma})(g^{11} + 2g^{1\sigma}b^{1}{}_{,\sigma})(g^{22} + 2g^{2\sigma}b^{2}{}_{,\sigma})(g^{33} + 2g^{3\sigma}b^{3}{}_{,\sigma}) \\ &= g^{-1} + 2(g^{11}g^{22}g^{33}g^{0\sigma}b^{0}{}_{,\sigma} + g^{00}g^{1\sigma}b^{1}{}_{,\sigma}g^{22}g^{33} \\ &\qquad + g^{00}g^{11}g^{2\sigma}b^{2}{}_{,\sigma}g^{33} + g^{00}g^{11}g^{22}g^{3\sigma}b^{3}{}_{,\sigma}) \\ &= g^{-1} + 2g^{-1}(b^{0}{}_{,0} + b^{1}{}_{,1} + b^{2}{}_{,2} + b^{3}{}_{,3}) \\ &= g^{-1}(1 + 2b^{\lambda}{}_{,\lambda}). \end{aligned}$$

Then $g' = g(1 + 2b^{\lambda}{}_{,\lambda})^{-1}$ and

$$\sqrt{-g'} = \sqrt{-g}\,(1 + 2b^{\lambda}{}_{,\lambda})^{-1/2} = \sqrt{-g}\,(1 - b^{\lambda}{}_{,\lambda}),$$

so that to leading order

$$\begin{aligned}\sqrt{-g'}\, g'^{\mu\nu} &= \sqrt{-g}\,(1 - b^{\lambda}{}_{,\lambda})(g^{\mu\nu} + g^{\mu\sigma} b^{\nu}{}_{,\sigma} + g^{\rho\nu} b^{\mu}{}_{,\rho}) \\ &= \sqrt{-g}\,(g^{\mu\nu} - g^{\mu\nu} b^{\lambda}{}_{,\lambda} + g^{\mu\sigma} b^{\nu}{}_{,\sigma} + g^{\rho\nu} b^{\mu}{}_{,\rho}).\end{aligned}$$

On the other hand, from (6.7) we may define $f'^{\mu\nu}$

$$\sqrt{-g'}\, g'^{\mu\nu} = \eta^{\mu\nu} - f'^{\mu\nu} \tag{6.14}$$

so that

$$\begin{aligned} f'^{\mu\nu} - f^{\mu\nu} &= -\sqrt{-g'}\, g'^{\mu\nu} + \sqrt{-g}\, g^{\mu\nu} \\ &= \sqrt{-g}\,(g^{\mu\nu} b^{\lambda}{}_{,\lambda} - g^{\mu\sigma} b^{\nu}{}_{,\sigma} - g^{\rho\nu} b^{\mu}{}_{,\rho}) \\ &= \eta^{\mu\nu} b^{\lambda}{}_{,\lambda} - b^{\nu,\mu} - b^{\mu,\nu}\end{aligned}$$

to leading order. So

$$f'^{\mu\nu} = f^{\mu\nu} - b^{\mu,\nu} - b^{\nu,\mu} + \eta^{\mu\nu} b^{\lambda}{}_{,\lambda} \tag{6.15}$$

and

$$f'^{\mu\nu}{}_{,\nu} = f^{\mu\nu}{}_{,\nu} - \Box b^{\mu} - (b^{\lambda}{}_{,\lambda})^{,\mu} + (b^{\lambda}{}_{,\lambda})^{,\mu} = f^{\mu\nu}{}_{,\nu} - \Box b^{\mu}. \tag{6.16}$$

Now choose b^{μ} to satisfy

$$\Box b^{\mu} = f^{\mu\nu}{}_{,\nu}$$

so that

$$f'^{\mu\nu}{}_{,\nu} = 0, \tag{6.17}$$

and, from (6.14)

$$(\sqrt{-g'}\, g'^{\mu\nu})_{,\nu} = 0. \tag{6.18}$$

This is the *harmonic condition* and the coordinates in which it holds are called *harmonic coordinates*. Substituting (6.15) into (6.12) gives

$$\Box f'_{\mu\nu} + \eta_{\mu\nu} f'^{\kappa\lambda}{}_{,\kappa\lambda} - f'^{\kappa}{}_{\mu,\nu\kappa} - f'^{\kappa}{}_{\nu,\mu\kappa} = +\frac{16\pi G}{c^2} T_{\mu\nu},$$

so that, with (6.17), and dropping the prime on f, the field equations become

$$\Box f_{\mu\nu} = \frac{16\pi G}{c^2} T_{\mu\nu}, \tag{6.19}$$

whose solutions obey the harmonic coordinate condition (see (6.14) and (6.18))

$$f^{\mu\nu}{}_{,\nu} = (\sqrt{-g}\, g^{\mu\nu})_{,\nu} = 0 \tag{6.20}$$

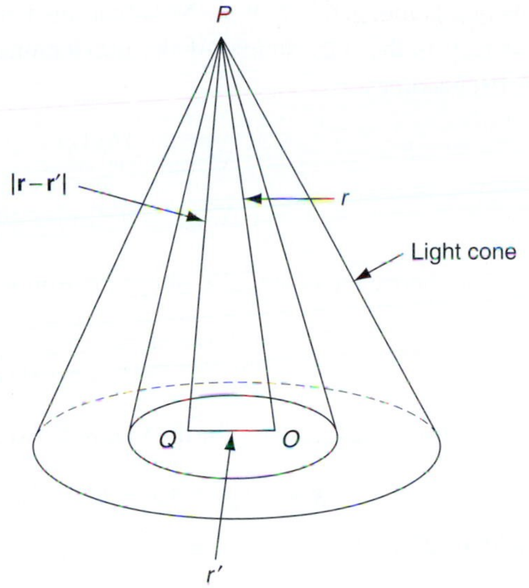

Fig. 6.1 **The retarded potential at *P* depends on the energy-momentum distribution at *Q*, which is within the past light-cone of *P*.**

where (see (6.9), (6.1) and (6.3))

$$
\begin{aligned}
f^{\mu\nu} &= h^{\mu\nu} - {}^1\!/_2\,\eta^{\mu\nu} h^{\lambda}{}_{\lambda}, \\
g_{\mu\nu} &= \eta_{\mu\nu} + h_{\mu\nu}, \quad g^{\mu\nu} = \eta^{\mu\nu} - h^{\mu\nu}.
\end{aligned}
\tag{6.21}
$$

We may note the analogy of the field equation (6.19) under the *coordinate condition* (6.20) with Maxwell's equations under the *Lorenz gauge condition*

$$\Box A_\mu = J_\mu, \quad A^\mu{}_{,\mu} = 0;$$

the coordinate condition in General Relativity is analogous to the gauge condition in electromagnetism. Guided by this analogy we may write the solution to (6.19) in terms of *retarded potentials*

$$f_{\mu\nu}(r, t) = \frac{1}{4\pi} \cdot \frac{16\pi G}{c^2} \int \frac{1}{|r - r'|} T_{\mu\nu}\left(r', t - \frac{|r - r'|}{c}\right) d^3x', \tag{6.22}$$

where, as shown in Fig. 6.1, Q, at coordinate distance r' from the origin, is within the past light-cone of P, at a distance r. We now consider two particular solutions; the first involving a static distribution of matter.

6.1.1 Static case: mass

Consider a static distribution of matter of density ρ. From Equation (4.15)

$$T^{00} = T_{00} = \rho, \quad \text{other } T_{\mu\nu} = 0.$$

The density ρ is independent of x^0. Note that this tensor obeys the conservation law (6.6). If this density is the only source of the gravitational field, the field must also be static. Then (6.19) becomes

$$\nabla^2 f_{00} = \frac{16\pi G\rho}{c^2},$$
$$\nabla^2 f_{\mu\nu} = 0\,((\mu\nu) \neq (00)). \tag{6.23}$$

In Newtonian theory the potential ϕ obeys Poisson's equation $\nabla^2\phi = -\,4\pi\rho G$, so we have

$$f_{00} = -\frac{4}{c^2}\phi, \quad \text{other } f_{\mu\nu} = 0. \tag{6.24}$$

We may now find an expression for the metric tensor. We have, from (6.1) and (6.11)

$$g_{\mu\nu} = \eta_{\mu\nu} + h_{\mu\nu} = \eta_{\mu\nu} + f_{\mu\nu} - {}^1\!/\!_2\,\eta_{\mu\nu} f^\lambda{}_\lambda,$$

hence, with (6.24),

$$g_{00} = -1 + f_{00} + {}^1\!/\!_2 f^0{}_0 = -1 + {}^1\!/\!_2 f_{00} = -\left(1 + \frac{2\phi}{c^2}\right), \tag{6.25}$$

$$g_{ik} = \delta_{ik} + h_{ik} = \delta_{ik} + f_{ik} - {}^1\!/\!_2\,\delta_{ik}(-f_{00}) = \delta_{ik}\left(1 - \frac{2\phi}{c^2}\right), \tag{6.26}$$

and

$$g_{i0} = 0, \tag{6.27}$$

so the space-time line element is

$$ds^2 = -\left(1 + \frac{2\phi}{c^2}\right)c^2\,dt^2 + \left(1 - \frac{2\phi}{c^2}\right)(dx^2 + dy^2 + dz^2). \tag{6.28}$$

This is the space-time metric corresponding to a static distribution of matter in the linear approximation. It is to be noted that it holds both inside and outside the matter distribution. In the case of a spherically symmetric distribution of matter, $\rho = \rho(r)$, with mass M, we have $\phi = -\frac{MG}{r}$, and $\left(\text{with } m = \frac{MG}{c^2}\right)$

$$ds^2 = -\left(1 - \frac{2m}{r}\right)c^2\,dt^2 + \left(1 + \frac{2m}{r}\right)(dx^2 + dy^2 + dz^2). \tag{6.29}$$

Note that this agrees with the Schwarschild solution (5.37): first replace $dx^2 + dy^2 + dz^2$ by $dr^2 + r^2(d\theta^2 + \sin^2\theta\, d\phi^2)$, then, concerning the coefficient of $(d\theta^2 + \sin^2\theta\, d\phi^2)$, redefine the parameter r, as explained below Equation (5.30), to make the coefficient simply $\hat{r}^2$, then remove the hat; and finally note that $\left(1 - \frac{2m}{r}\right)^{-1} = \left(1 + \frac{2m}{r}\right)$ in the linear approximation. The way we have proceeded is, it may be claimed, a more proper way to derive the Schwarzschild solution; in the derivation in Chapter 5, this solution was found to be a solution of the *vacuum* field equations, but we are searching for a solution corresponding to a static and spherically symmetric distribution of matter. Finally it should be remarked that if the distribution of matter is not spherically symmetric, we must put

$$\phi = -\frac{MG}{r} + \cdots .$$

6.1.2 Rotating body: angular momentum

Consider a body rotating with constant angular velocity $\omega = \frac{d\phi}{dt}$ about the $z = x^3$ axis and assume that $v \ll c$, where v is the resulting velocity of a typical part of the body. Retaining terms linear in v/c, the energy-momentum tensor is, from (5.15)

$$\begin{gathered} T^{00} = \rho, \\ T^{01} = -T_{01} = \frac{\rho v_x}{c} = -\frac{\rho v}{c}\sin\phi, \quad T^{02} = \frac{\rho v_y}{c} = \frac{\rho v}{c}\cos\phi, \\ \text{other } T^{\mu\nu} = 0. \end{gathered} \tag{6.30}$$

Denote the coordinates of a point Q inside the body by $(X, Y, Z) = (X^1, X^2, X^3)$; and those of a point P outside the body by $(x, y, z) = (x^1, x^2, x^3)$ – see Fig. 6.2. Put $r^2 = x^i x_i$ and $R^2 = X^i X_i$. With $R \ll r$ we have

$$\begin{aligned} |\mathbf{r} - \mathbf{R}| &= (r^2 - 2\,\mathbf{r}\cdot R + R^2)^{1/2} \approx r\left(1 - \frac{\mathbf{r}\cdot\mathbf{R}}{r^2}\right) + \cdots, \\ |\mathbf{r} - \mathbf{R}|^{-1} &= (r^2 - 2\,\mathbf{r}\cdot R + R^2)^{-1/2} \approx \frac{1}{r}\left(1 + \frac{\mathbf{r}\cdot\mathbf{R}}{r^2}\right) + \cdots. \end{aligned} \tag{6.31}$$

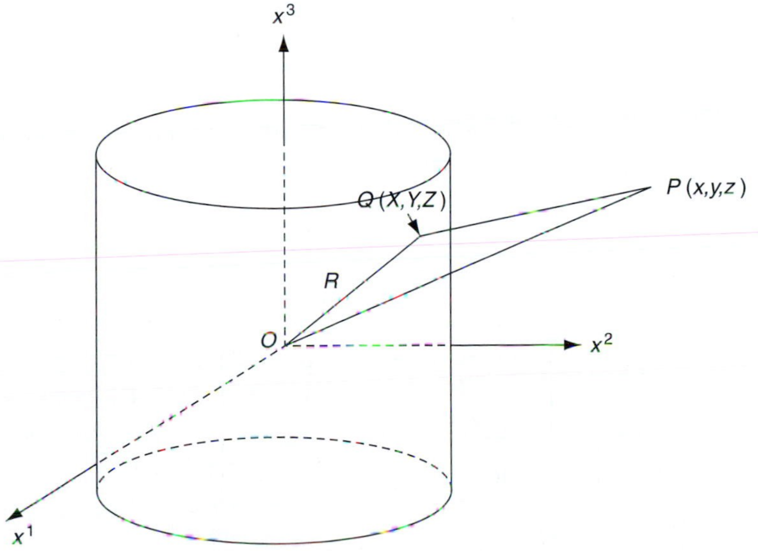

Fig. 6.2 **P is outside, and Q inside, the rotating body.**

The field equation (6.19) then gives, with (6.31)

$$\Box f_{00} = \frac{16\pi G}{c^2}\rho, \quad \Box f_{0i} = \frac{16\pi G}{c^2} T_{0i}, \quad \text{other } f_{\mu\nu} = 0, \tag{6.32}$$

with $i = 1, 2$. Then

$$\nabla^2 f_{00} = \frac{16\pi G}{c^2}\rho, \tag{6.33}$$

with the solutions, as above,

$$g_{00} = -\left(1 + \frac{2\phi}{c^2}\right), \quad \phi = -\frac{GM}{r} + \cdots \tag{6.34}$$

and, from (6.22) and (6.31),

$$\begin{aligned} f_{01} &= \frac{4G}{c^2}\int \frac{1}{r}\left(1 + \frac{r.R}{r^2}\right) T_{01}\, \mathrm{d}^3X \\ &= \frac{4G}{c^2}\left\{\frac{1}{r}\int T_{01}\mathrm{d}^3X + \frac{x^i}{r^3}\int X_i T_{01}\, \mathrm{d}^3X + \cdots\right\}. \end{aligned} \tag{6.35}$$

Now consider the first integral. We have, from (6.30),

$$T_{01} = -\frac{\rho v_x}{c} = -\frac{\rho}{c}\frac{\mathrm{d}X^1}{\mathrm{d}t}.$$

On performing the first integral above we are essentially integrating $\frac{\mathrm{d}X^1}{\mathrm{d}t}$ over the circular motion in the (X^1, X^2) plane. In the four quadrants the quantity $\frac{dX^1}{dt}$ itself has two negative and two positive signs, as shown in Fig. 6.3(a); it will therefore integrate to zero, and the first

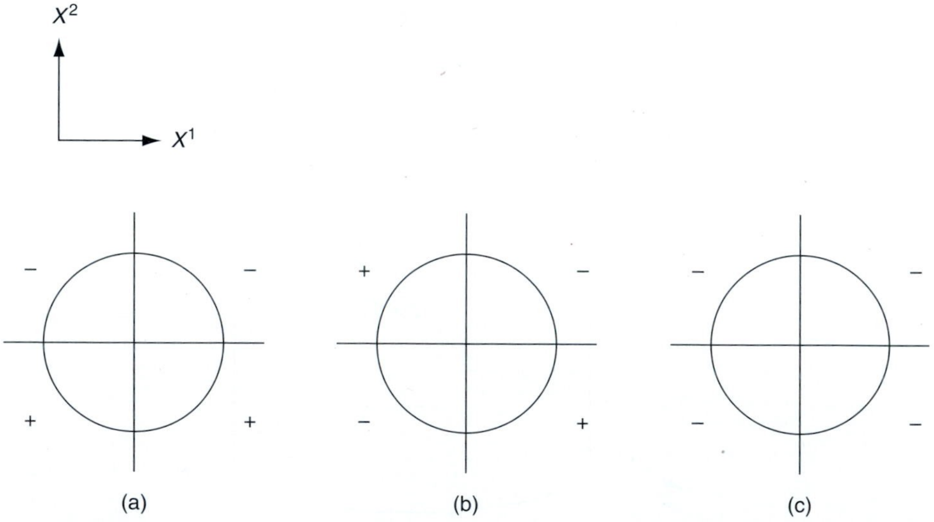

Fig. 6.3 The signs of the contributions of the integrals (6.35) in the four quadrants.

term in (6.35) gives no contribution. Now consider the second term, consisting of three integrals. The quantity $X^1\dfrac{\mathrm{d}X^1}{\mathrm{d}t}$ makes positive and negative contributions as shown in Fig. 6.3(b), and this clearly integrates to zero. The quantity $X^2\dfrac{\mathrm{d}X^1}{\mathrm{d}t}$, shown in Fig. 6.3(c), is negative in all four quadrants, so gives a non-zero contribution. Since X^3 always has the same sign, however, the quantity $X^3\dfrac{\mathrm{d}X^1}{\mathrm{d}t}$ integrates to zero, just as $\dfrac{\mathrm{d}X^1}{\mathrm{d}t}$ itself does. The only non-vanishing term is for $i=2$, hence

$$f_{01}=\frac{4G}{c^2}\frac{y}{r^3}\int Y\,T_{01}\,\mathrm{d}^3X=-\frac{4G}{c^2}\frac{y}{r^3}\int YT^{01}\,\mathrm{d}^3X. \qquad (6.36)$$

Similarly,

$$f_{02}=\frac{4G}{c^2}\frac{x}{r^3}\int XT_{02}\,\mathrm{d}^3X=-\frac{4G}{c^2}\frac{x}{r^3}\int XT^{02}\,\mathrm{d}^3X. \qquad (6.37)$$

These quantities may be related to the angular momentum of the source J^3, as follows. First recall that, from Special Relativity,[2]

$$J^3=\int(X^1P^2-X^2P^1)\,\mathrm{d}^3X=c\int(X^1T^{02}-X^2T^{01})\,\mathrm{d}^3X. \qquad (6.38)$$

(Note in passing that the dimensions of the above are correct: $T^{\mu\nu}$ has the dimension of mass density, $\mathrm{M\,L^{-3}}$, and the dimensions of angular momentum are $[J]=\mathrm{M\,L^2\,T^{-1}}$.) The law of conservation of angular momentum requires

$$T^{\mu\nu}{}_{,\nu}=0.$$

For a static distribution of matter $T^{\mu0}{}_{,0}=0$; and from cylindrical symmetry, as we have here, $T^{\mu3}=0$, hence with $\mu=0$ we have

$$T^{01}{}_{,1}+T^{02}{}_{,2}=0. \qquad (6.39)$$

Let us verify this in our example with cylindrical (axial) symmetry:

$$T^{\mu\nu}(R,x^3),\quad R=\sqrt{(x^1)^2+(x^2)^2}.$$

For the rotating system above (see (5.30)),

$$T^{01}=-\frac{\rho}{c}v(R,x^3)\sin\phi.$$

Then

$$T^{01}{}_{,1}=\frac{\partial T^{01}}{\partial x}=\frac{\partial T^{01}}{\partial R}\frac{\partial R}{\partial x}+\frac{\partial T^{01}}{\partial\phi}\frac{\partial\phi}{\partial x}=-\frac{\rho}{c}\sin\phi\cos\phi\left(\frac{\partial v}{\partial R}-\frac{v}{R}\right).$$

[2] See for example Landau & Lifshitz (1971) p. 79, or Weinberg (1972) p. 46.

Similarly

$$T^{02}{}_{,2} = +\frac{\rho}{c}\sin\phi\cos\phi\left(\frac{\partial v}{\partial R} - \frac{v}{R}\right),$$

so

$$T^{01}{}_{,1} + T^{02}{}_{,2} = 0,$$

or $T^{0i}{}_{,i} = 0$. It follows that

$$\int X^k X^m T^{0i}{}_{,i}\, d^3X = 0,$$

i.e.

$$\int \{\partial_i(X^k X^m T^{0i}) - (\partial_i X^k) X^m T^{0i} - X^k(\partial_i X^m) T^{0i}\}\, d^3X = 0.$$

The first term vanishes by Gauss's theorem, leaving

$$\int (X^m T^{0k} + X^k T^{0m})\, d^3X = 0,$$

and hence

$$\int YT^{01}\, d^3X = -\int XT^{02}\, d^3X = \tfrac{1}{2}\int (YT^{01} - XT^{02})\, d^3X = -\frac{J^3}{2c},$$

using (6.38). Then (6.36) becomes

$$f_{01} = \frac{2G}{c^3}\frac{y}{r^3}J^3. \tag{6.40}$$

Similarly

$$f_{02} = -\frac{2G}{c^3}\frac{x}{r^3}J^3. \tag{6.41}$$

These are the solutions to the field equations in our case of cylindrical symmetry of a rotating source. Practically speaking, however, we shall be concerned with the rotating Earth, and to a good approximation this exhibits *spherical symmetry*; and the two equations above can then be written as

$$f_{0i} = \varepsilon_{ikm} x^k J^m. \tag{6.42}$$

We then have, in the linear approximation

$$\begin{aligned} g_{00} &= -\left(1 - \frac{2m}{r}\right) = -\left(1 - \frac{2GM}{rc^2}\right), \\ g_{ik} &= \left(1 + \frac{2m}{r}\right)\delta_{ik} = \left(1 + \frac{2GM}{rc^2}\right)\delta_{ik}, \\ g_{0i} &= \frac{2G}{r^3c^3}\,\varepsilon_{ikm} x^k J^m. \end{aligned} \tag{6.43}$$

These are the components of the metric tensor outside a rotating body of mass M and angular momentum J in the linear approximation. It is not an exact solution to the Einstein field equations, though we shall see later that there *is* an exact solution – the Kerr solution – representing the space-time metric outside a rotating body, which yields the above components in the appropriate approximation.

The above metric gives rise to physical effects associated with g_{0i}; spin precession and the clock effect, to be described below. These of course are *gravitational* consequences of a rotating source, not simply kinematic ones. This raises the question of the Equivalence Principle (EP) again. For a static source EP is the statement that, to some approximation (i.e. ignoring curvature, or tidal effects) a gravitational field is equivalent to an accelerating frame of reference. Now an additional question is raised: are the gravitational effects of a rotating source equivalent to a rotating frame of reference? We shall return to this matter later. We now consider spin precession, measured in gyroscopes.

6.2 Precession of gyroscopes: the Lense–Thirring effect

We are concerned with precession of angular momentum, commonly stated as spin precession, in a gravitational field. How do we describe spin? Intuitively – and non-relativistically – we denote it by a 3-vector **S**, like **J**. Following a familiar and slightly simple-minded logic we could then suppose that relativistically we should define a 4-vector S_μ. This is the usual procedure adopted in this subject and indeed the one we shall follow here, but it is worth pausing a minute to describe why the logic is not actually so straightforward. From the point of view of quantum mechanics spin operators should obey the commutation relations of SU(2). There should therefore be three such operators; and non-relativistically, as the reader is doubtless already aware, the electron spin is described by $(\hbar/2)\boldsymbol{\sigma}$, where $\boldsymbol{\sigma}$ are the three Pauli matrices. To generalise this to the relativistic domain is not so easy, but the first step is to define the so-called Pauli–Lubanski 4-vector (actually pseudovector) W_μ[3]

$$W_\mu = -{}^1\!/\!_2\, \varepsilon_{\mu\nu\kappa\lambda} J^{\nu\kappa} P^\lambda. \tag{6.44}$$

Since there are four of these quantities, rather than three, they cannot obey the required SU(2) commutation relations, but some progress has been made to manufacture operators derived from W_μ which do obey the required relations.[4]

These considerations arise, of course, from quantum-mechanical reasoning, and so are perhaps not so important for our present purpose, which is to describe a spinning gyroscope – hardly a quantum object! Nevertheless we may observe that by virtue of the totally antisymmetric symbol $\varepsilon_{\mu\nu\kappa\lambda}$ above it follows that

$$W_\mu P^\mu = 0; \tag{6.45}$$

[3] See for example Pauli (1965).

[4] See Ryder (1999), Gürsey (1965).

W_μ is orthogonal to momentum, or velocity. This property is retained in the present context. We therefore define spin by a 4-vector S_μ which is orthogonal to momentum and therefore velocity $v^\mu = \dfrac{dx^\mu}{d\tau}$:

$$\frac{dx^\mu}{d\tau} S_\mu = 0,$$

hence $c\dfrac{dt}{d\tau} S_0 = -\dfrac{dx^i}{d\tau} S_i$, or

$$S_0 = -\frac{1}{c}\frac{dx^i}{dt} S_i. \tag{6.46}$$

We also suppose that the spin vector S_μ is covariantly constant (has vanishing absolute derivative) and is therefore parallel-transported along a geodesic

$$\frac{dS_\mu}{d\tau} = \Gamma^\lambda{}_{\mu\nu} S_\lambda \frac{dx^\mu}{d\tau}.$$

This means that

$$\begin{aligned}
\frac{dS_i}{dt} &= \frac{dS_i}{d\tau}\frac{d\tau}{dt} \\
&= (\Gamma^0{}_{i\nu} S_0 + \Gamma^k{}_{i\nu} S_k)\frac{dx^\nu}{dt} \\
&= \left(-\frac{1}{c}\Gamma^0{}_{i\nu}\frac{dx^k}{dt} S_k + \Gamma^k{}_{i\nu} S_k\right)\frac{dx^\nu}{dt} \\
&= \left(-\Gamma^0{}_{i0}\frac{dx^k}{dt} - \frac{1}{c}\Gamma^0{}_{im}\frac{dx^m}{dt}\frac{dx^k}{dt} + c\Gamma^k{}_{i0} + \Gamma^k{}_{im}\frac{dx^m}{dt}\right) S_k.
\end{aligned} \tag{6.47}$$

The connection coefficients $\Gamma^\lambda{}_{\mu\nu}$ are now calculated from the metric (6.43) above. We are already working in the linear approximation (weak gravitational field); and we shall apply our eventual result (of spin precession) to a gyroscope in orbit round the Earth. It is therefore a good idea to find numerical orders of magnitude for the terms in the metric. The metric tensor is

$$g_{\mu\nu} = \begin{pmatrix} -\left(1+\dfrac{2\phi}{c^2}\right) & \zeta_1 & \zeta_2 & \zeta_3 \\ \zeta_1 & 1-\dfrac{2\phi}{c^2} & 0 & 0 \\ \zeta_2 & 0 & 1-\dfrac{2\phi}{c_2} & 0 \\ \zeta_3 & 0 & 0 & 1-\dfrac{2\phi}{c^2} \end{pmatrix}, \tag{6.48}$$

with

$$\phi = -\frac{MG}{r}, \quad \zeta_i = \frac{2G}{r^3 c^3}\varepsilon_{ikm} x^k J^m. \tag{6.49}$$

The quantities $\frac{\phi}{c^2}$ and ζ_i are dimensionless and in the case of the Earth are of the order $\frac{\phi}{c^2} \sim 10^{-9}$ and $\zeta_i \sim 10^{-17}$. So in calculating $g^{\mu\nu}$ we may ignore terms in ϕ^2, $\phi\zeta_i$, $\zeta_i\,\zeta_k$. Then

$$g = \det g_{\mu\nu} \approx -\left(1 - \frac{4\phi}{c^2}\right)$$

and

$$g^{\mu\nu} = \begin{pmatrix} -\left(1-\frac{2\phi}{c^2}\right) & \zeta_1 & \zeta_2 & \zeta_3 \\ \zeta_1 & 1+\frac{2\phi}{c^2} & 0 & 0 \\ \zeta_2 & 0 & 1+\frac{2\phi}{c^2} & 0 \\ \zeta_3 & 0 & 0 & 1+\frac{2\phi}{c^2} \end{pmatrix}. \tag{6.50}$$

The quantities ϕ and ζ are time-independent and we get, to leading order

$$\begin{aligned} \Gamma^0{}_{i0} &= \frac{1}{c^2}\nabla_i\phi \\ \Gamma^k{}_{i0} &= {}^1\!/\!{}_2(\zeta_{k,i} - \zeta_{i,k}) \\ \Gamma^k{}_{im} &= \frac{1}{c^2}(\delta_{im}\nabla_k\phi - \delta_i{}^k\nabla_m\phi - \delta_m{}^k\nabla_i\phi) \\ \Gamma^0{}_{im} &= -{}^1\!/\!{}_2(\zeta_{i,m} + \zeta_{m,i}) \end{aligned} \tag{6.51}$$

(Note that these all have the dimension L^{-1} ($L = \text{length}$).) Substituting these expressions into (6.47) gives

$$\frac{dS_i}{dt} = \left\{-\frac{1}{c^2}(\nabla_i\phi)v^k + \frac{c}{2}(\zeta_{k,i} - \zeta_{i,k}) + \frac{1}{c^2}(\delta_{im}\nabla_k\phi - \delta_i^k\nabla_m\phi - \delta_m^k\nabla_i\phi)\,v^m\right\} S_k \tag{6.52}$$

or

$$\frac{d\mathbf{S}}{dt} = -\frac{2}{c^2}(\mathbf{v}\cdot\mathbf{S})\nabla\phi + \frac{1}{c^2}(\nabla\phi\cdot\mathbf{S})\mathbf{v} - \frac{1}{c^2}(\mathbf{v}\cdot\nabla\phi)\,\mathbf{S} + \frac{c}{2}\,[\mathbf{S}\times(\nabla\times\zeta)]. \tag{6.53}$$

This formula will lead to spin precession. Note that all the terms except the last one depend on $\mathbf{v}$, the velocity of the gyroscope in orbit. The final term depends on $\nabla \times \boldsymbol{\zeta}$, the angular momentum of the rotating source. This leads to the famous Lense–Thirring effect, as we shall see below.

Equation (6.52) must now be solved, and this is done by introducing a *new* spin operator. The reasoning is as follows. Parallel transport preserves the value of $S^\mu S_\mu$:

$$\frac{d}{dt}(g^{\mu\nu}S_\mu S_\nu) = 0, \quad g^{\mu\nu}S_\mu S_\nu = \text{const},$$

or, ignoring the terms in g^{0i},

$$g^{00}(S_0)^2 + g^{ik}S_iS_k = \text{const.}$$

Substituting for S_0 from (6.46) and for $g^{\mu\nu}$ from (6.49) and working to order $v^2\mathbf{S}^2$ or $\phi\mathbf{S}^2$ then gives

$$\mathbf{S}^2 + \frac{2\phi}{c^2}\mathbf{S}^2 - \frac{1}{c^2}(\mathbf{v}\cdot\mathbf{S})^2 = \text{const.} \tag{6.54}$$

We now introduce a new spin vector $\boldsymbol{\Sigma}$ through the equation[5]

$$\mathbf{S} = \left(1 - \frac{\phi}{c^2}\right)\boldsymbol{\Sigma} + \frac{1}{2c^2}\mathbf{v}(\mathbf{v}\cdot\boldsymbol{\Sigma}). \tag{6.55}$$

Then, to order $v^2\mathbf{S}^2$ or $\phi\mathbf{S}^2$ we have

$$\mathbf{S}^2 = \left(1 - \frac{2\phi}{c^2}\right)\boldsymbol{\Sigma}^2 + \frac{1}{c^2}(\mathbf{v}\cdot\boldsymbol{\Sigma})^2, \quad 2\phi\mathbf{S}^2 = 2\phi\boldsymbol{\Sigma}^2,$$

$$\mathbf{v}\cdot\mathbf{S} = \mathbf{v}.\boldsymbol{\Sigma}\left[1 - \frac{\phi}{c^2} + \frac{v^2}{2c^2}\right], \quad (\mathbf{v}\cdot\mathbf{S})^2 = (\mathbf{v}\cdot\boldsymbol{\Sigma})^2,$$

and (6.54) becomes

$$\boldsymbol{\Sigma}^2 = \text{const}: \tag{6.56}$$

that is, $\boldsymbol{\Sigma}$ is a vector whose *magnitude* is constant. It will therefore precess, changing only in direction. To the required order, inverting Equation (6.55) gives

$$\boldsymbol{\Sigma} = \left(1 + \frac{\phi}{c^2}\right)\mathbf{S} - \frac{1}{2c^2}\mathbf{v}(\mathbf{v}\cdot\mathbf{S}). \tag{6.57}$$

We must now calculate $\frac{\mathrm{d}\boldsymbol{\Sigma}}{\mathrm{d}t}$ to find the precession formula. Ignoring terms in $\frac{v^2}{c^2} \times \frac{\mathrm{d}\mathbf{S}}{\mathrm{d}t}$ we have

$$\frac{\mathrm{d}\boldsymbol{\Sigma}}{\mathrm{d}\tau} = \frac{\mathrm{d}\mathbf{S}}{\mathrm{d}t} + \frac{1}{c^2}\frac{\mathrm{d}\phi}{\mathrm{d}t}\mathbf{S} - \frac{1}{2c^2}\frac{\mathrm{d}\mathbf{v}}{\mathrm{d}t}(\mathbf{v}\cdot\mathbf{S}) - \frac{1}{2c^2}\mathbf{v}\left(\frac{\mathrm{d}\mathbf{v}}{\mathrm{d}t}\cdot\mathbf{S}\right). \tag{6.58}$$

We now put

$$\frac{\mathrm{d}\phi}{\mathrm{d}t} = \frac{\partial\phi}{\partial t} + \nabla\phi\cdot\mathbf{v} = \nabla\phi\cdot\mathbf{v}; \quad \frac{\mathrm{d}\mathbf{v}}{\mathrm{d}t} = -\nabla\phi\,(\text{acceleration})$$

so that (6.58) becomes

$$\frac{\mathrm{d}\boldsymbol{\Sigma}}{\mathrm{d}\tau} = \frac{\mathrm{d}\mathbf{S}}{\mathrm{d}t} + \frac{1}{c^2}(\nabla\phi\cdot\mathbf{v})\mathbf{S} + \frac{1}{2c^2}(\mathbf{v}\cdot\mathbf{S})\nabla\phi + \frac{1}{2c^2}(\nabla\phi\cdot\mathbf{S})\mathbf{v}.$$

[5] We follow here the procedure of Weinberg (1972), p. 234.

Substituting for $\frac{d\mathbf{S}}{dt}$ from (6.53) then gives

$$\frac{d\boldsymbol{\Sigma}}{dt} = \frac{c}{2}\mathbf{S} \times (\nabla \times \boldsymbol{\zeta}) - \frac{3}{2c^2}[\nabla\phi(\mathbf{v}\cdot\mathbf{S}) - \mathbf{v}(\nabla\phi\cdot\mathbf{S})].$$

The term in square brackets above is $\mathbf{S} \times (\nabla\phi \times \mathbf{v})$ so we have

$$\frac{d\boldsymbol{\Sigma}}{dt} = \mathbf{S} \times \left[\frac{c}{2}(\nabla \times \boldsymbol{\zeta}) + \frac{3}{2c^2}(\mathbf{v} \times \nabla\phi)\right].$$

To the required order of approximation $\mathbf{S}$ on the right hand side above may be replaced by $\boldsymbol{\Sigma}$, so finally

$$\frac{d\boldsymbol{\Sigma}}{dt} = \boldsymbol{\Omega} \times \boldsymbol{\Sigma}, \quad \boldsymbol{\Omega} = -\frac{c}{2}\nabla \times \boldsymbol{\zeta} - \frac{3}{2c^2}\mathbf{v} \times \nabla\phi. \tag{6.59}$$

The spin $\boldsymbol{\Sigma}$ precesses at a rate $|\boldsymbol{\Omega}|$ around the direction of $\boldsymbol{\Omega}$, with no change of magnitude. This is the solution to our spin precession problem.

Thinking now specifically of the motion of a gyroscope orbiting the Earth, with $\mathbf{r}$ and $\mathbf{v}$ the position and velocity of the gyroscope and $\mathbf{J}$ the angular momentum of the Earth, we may substitute (6.49) into (6.59), giving

$$\boldsymbol{\Omega} = -\frac{c}{2}\nabla \times \left\{\frac{2G}{r^3c^3}\mathbf{r} \times \mathbf{J}\right\} + \frac{3GM}{2c^2}\mathbf{v} \times \nabla\left(\frac{1}{r}\right)$$

or

$$\boldsymbol{\Omega} = \frac{G}{c^2r^3}\left\{\frac{3(\mathbf{J}\cdot\mathbf{r})\mathbf{r}}{r^2} - \mathbf{J}\right\} + \frac{3GM}{2c^2r^3}\mathbf{r} \times \mathbf{v}. \tag{6.60}$$

The first term above, dependent on $\mathbf{J}$, gives rise to the so-called Lense–Thirring effect. The second term, dependent on M but not on $\mathbf{J}$, is known as the de Sitter–Fokker effect, or simply geodetic precession; it is the precession caused by motion around the geodesic. Putting $\mathbf{J} = I\,\boldsymbol{\omega}$ where I is the moment of inertia, gives

$$\boldsymbol{\Omega} = \boldsymbol{\Omega}_{\text{LT}} + \boldsymbol{\Omega}_{\text{de Sitter}},$$

$$\boldsymbol{\Omega}_{\text{LT}} = \frac{GI}{c^2r^3}\left\{\frac{3(\boldsymbol{\omega}\cdot\mathbf{r})\mathbf{r}}{r^2} - \boldsymbol{\omega}\right\}, \tag{6.61}$$

$$\boldsymbol{\Omega}_{\text{de Sitter}} = \frac{3GM}{2c^2r^3}\mathbf{r} \times \mathbf{v}. \tag{6.62}$$

We recall from Chapter 2 that the spin (intrinsic angular momentum) of an electron in orbit round a proton nucleus also precesses. This is Thomas precession, and the precession rate is given by (2.82)

$$\boldsymbol{\Omega}_{\text{Thomas}} = \frac{1}{2c^2}\mathbf{v} \times \mathbf{a} = \frac{1}{2mc^2}\mathbf{v} \times \mathbf{F}, \tag{6.63}$$

where **a** is the acceleration and $\mathbf{F} = m\mathbf{a}$ the force exerted on the electron. So for the sake of comparison with (6.59) we may rewrite (2.83) as

$$\frac{D\mathbf{S}}{Dt} = \frac{d\mathbf{S}}{dt} + \mathbf{\Omega}_{\text{Thomas}} \times \mathbf{S}$$

(with a + instead of − sign), and $\mathbf{\Omega}_{\text{Thomas}}$ is given by (6.62). Putting these results together, the *total* precession rate of an object in orbit, subject to both gravitational and non-gravitational forces, is

$$\begin{aligned}\mathbf{\Omega} &= \mathbf{\Omega}_{\text{Thomas}} + \mathbf{\Omega}_{\text{de Sitter}} + \mathbf{\Omega}_{\text{Lense–Thirring}} \\ &= \frac{1}{2mc^2}\mathbf{F} \times \mathbf{v} + \frac{3GM}{2c^2r^3}\mathbf{r} \times \mathbf{v} + \frac{GI}{c^2r^3}\left\{\frac{3(\boldsymbol{\omega} \cdot \mathbf{r})\mathbf{r}}{r^2} - \boldsymbol{\omega}\right\}.\end{aligned} \tag{6.64}$$

Here **F** is the non-gravitational force, M and I are the mass and moment of inertia of the Earth (or other gravitating body) and m is the mass of the gyroscope. For *geodesic* motion $\mathbf{F} = 0$; there is no Thomas precession. On a *Newtonian* view the gravitational force is $\mathbf{F} = \frac{GMm}{r^3}\mathbf{r}$, so the de Sitter precession could be described as being like Thomas precession due to the gravitational force, but with an extra factor of 3. In General Relativity, however, a particle (satellite, gyroscope) in geodesic motion has no absolute acceleration, so no Thomas precession. On the other hand, there *is* a precession – the geodetic precession – given by $\mathbf{\Omega}_{\text{de Sitter}}$. And, of course, one of the main purposes of the present section is to show that we have, in addition to geodetic precession, the Lense–Thirring precession caused simply by the rotation of the Earth. This is something Newton would not have dreamed of!

Let us now calculate some orders of magnitude. For a body in circular motion the time for one revolution is $2\pi r/v$, so the de Sitter precession rate of $2GMv/2c^2r^2$ radians per second gives a precession of

$$\delta\phi_{\text{de Sitter}} = \frac{3\pi GM}{rc^2} \text{ radians per revolution.}$$

This gives, for the Earth orbiting the Sun, 0.019 arc seconds per year, which is too small to observe. On the other hand a satellite skimming the Earth has a period $\tau = 2\pi\sqrt{\frac{R^3}{MG}} = 84.5$ minutes, giving 6.2×10^3 orbits per year. The precession is then

$$\delta\phi_{\text{de Sitter}} = \frac{3(MG)^{3/2}}{2c^2}\frac{1}{R^{5/2}} \text{radians/second} \tag{6.65}$$

which works out at

$$\delta\phi_{\text{de Sitter}} = 8.4 \text{ arc seconds/year.} \tag{6.66}$$

For a satellite in an orbit of radius $r > R$ (R = radius of Earth) this clearly becomes amended to

$$\delta\phi_{\text{de Sitter}} = 8.4\left(\frac{R}{r}\right)^{5/2} \text{arc seconds/year,} \tag{6.67}$$

so that at a height of 650 km, $R = 6.38 \times 10^6$ m, $r = 7.03 \times 10^6$ m and

$$\delta\phi_{\text{de Sitter}} = 6.6 \text{ arc seconds/year.} \tag{6.68}$$

The frequency $\mathbf{\Omega}_{\text{de Sitter}}$ (and therefore $\delta\phi_{\text{de Sitter}}$) does not depend on $\boldsymbol{\omega}$: it is therefore the same for all orbits of the same radius – polar, equatorial or intermediate. But the orientation of $\mathbf{S}$ in the orbit is important. Let $\mathbf{r} \times \mathbf{v} = |\mathbf{r} \times \mathbf{v}|\mathbf{h}$; so $\mathbf{h}$ is a unit vector perpendicular to the orbital plane. Then $\mathbf{\Omega}_{\text{de Sitter}} \sim \mathbf{h}$ and $\delta\mathbf{S}_{\text{de Sitter}} \sim \mathbf{h} \times \mathbf{S}$, so to maximise the precession $\mathbf{S}$ must be in the orbital plane. We must now choose the orbit so that both $\delta\phi_{\text{de Sitter}}$ and $\delta\phi_{\text{LT}}$ (Lense–Thirring) are both measurable – and measurable *separately*. So we now consider $\delta\mathbf{S}_{\text{LT}}$.

First, consider an orbit in the equatorial plane. From above we have seen that $\mathbf{S}$ should be in the orbital plane, so it follows that $\mathbf{S} \perp \boldsymbol{\omega}$ (and of course $\mathbf{h} \parallel \boldsymbol{\omega}$). In an equatorial orbit $\boldsymbol{\omega} \cdot \mathbf{r} = 0$ hence (see (6.64)) $\mathbf{\Omega}_{\text{LT}} \sim \boldsymbol{\omega}$ and $\delta\mathbf{S}_{\text{LT}} \sim \boldsymbol{\omega} \times \mathbf{S}$. It then follows that $\delta\mathbf{S}_{\text{LT}} \parallel \delta\mathbf{S}_{\text{de Sitter}}$ – these vectors are parallel to each other, so the precessions are simply additive and are not separately measurable. We therefore turn to an orbit in a polar plane.

There are two separate contributions to the Lense–Thirring precession rate, coming from the terms in $\mathbf{r}$ and $\boldsymbol{\omega}$ above:

$$\delta\mathbf{S}_{\text{LT}}(\text{i}) \sim \mathbf{r} \times \mathbf{S}, \quad \delta\mathbf{S}_{\text{LT}}(\text{ii}) \sim \omega \times \mathbf{S};$$

and note further that $\boldsymbol{\omega} \cdot \mathbf{r}$ varies over the orbit. To maximise (ii) we want $\boldsymbol{\omega} \perp \mathbf{S}$, as shown in Fig. 6.4 – that is, the gyroscope spins in the orbital plane, in a direction which is tangential to the direction of motion when the satellite is over the poles (and perpendicular to the Earth's surface when the satellite passes over the equator). Then $\delta\mathbf{S}_{\text{LT}} \parallel \mathbf{h}$ – the spin vector is moved *out of* the orbital plane. On the other hand the de Sitter precession $\delta\mathbf{S}_{\text{de Sitter}} \sim \mathbf{S} \times \mathbf{h}$, which is *in* the orbital plane. The two precessions are therefore at right angles and are separately measurable. We conclude that to measure these predictions of General Relativity the satellite should be in a polar orbit and the gyroscope spinning in the orbital plane as described above.

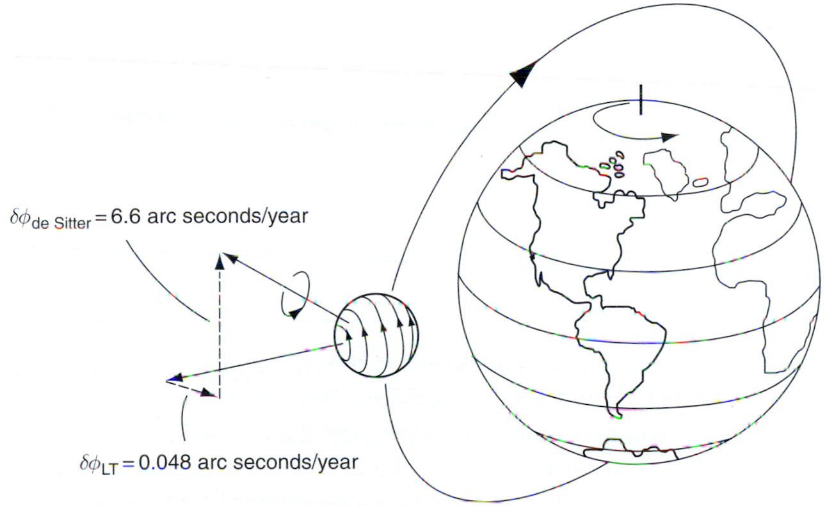

Fig. 6.4 The two types of precession of a gyroscope in polar orbit: motional, or Lense–Thirring, and geodetic, or de Sitter. Reproduced from Fairbank *et al.* (1988), with permission of Henry Holt and Company.

We now calculate the Lense–Thirring precession. It is not entirely straightforward since $\boldsymbol{\omega} \cdot \mathbf{r}$ varies over the orbit, as noted above. Taking an average value over one revolution we have, from (6.64)

$$\langle \Omega_{\mathrm{LT}} \rangle = \frac{GI}{c^2 r^3} \left\langle \frac{3(\boldsymbol{\omega} \cdot \mathrm{k})\mathrm{r}}{r^2} - \boldsymbol{\omega} \right\rangle .$$

Let us consider motion in the XZ plane: $\boldsymbol{\omega} = \omega \mathbf{k}$, $\mathbf{r} = r(\mathbf{i} \cos \omega t + \mathbf{k} \sin \omega t)$, hence $\boldsymbol{\omega} \cdot \mathbf{r} = \omega r \sin \omega t$. Then

$$\mathbf{r}(\boldsymbol{\omega} \cdot r) = \omega r^2 (\mathbf{i} \cos \omega t \sin \omega t + \mathbf{k} \sin^2 \omega t),$$

$$\langle \mathbf{r}(\boldsymbol{\omega} \cdot \mathbf{r}) \rangle = \frac{\omega r^2}{2} \mathbf{k}$$

and

$$\left\langle \frac{3(\boldsymbol{\omega}\mathbf{r})\mathbf{r}}{r^2} - \boldsymbol{\omega} \right\rangle = \frac{\omega}{2} \mathbf{k}.$$

Then for a sphere $I = \frac{2}{5} MR^2$ and

$$\langle \Omega_{\mathrm{LT}} \rangle = \frac{MGR^2 \omega}{5c^2 r^3} .$$

Hence for an orbit which skims the Earth $r = R$ and

$$\langle \Omega_{\mathrm{LT}} \rangle = \frac{MG\omega}{5c^2 R} = 0.065 \text{ arc seconds/year.}$$

Otherwise, with $r > R$,

$$\langle \Omega_{\mathrm{LT}} \rangle = 0.065 \left(\frac{R}{r} \right)^3 \text{ arc seconds/year.}$$

For example, with $R = 6.37 \times 10^6$ m and $r = 7.02 \times 10^6$ m,

$$\langle \Omega_{\mathrm{LT}} \rangle = 0.048 \text{ arc seconds /year.} \qquad (6.69)$$

6.2.1 Gravity Probe B

The precession rates (6.68) and (6.69) are very small but they are being looked for in the current Gravity Probe B experiment. This experiment, first conceived by Leonard Schiff in 1960, consists of four gyroscopes carried on a satellite. The gyros are spheres of diameter 3.8 cm coated with a 1.2 μm film of niobium, which at 6.5 K becomes superconducting. The spheres are held in place by a magnetic field, with a gap of 32 μm between them. The smoothness of the spheres is such that, if they were scaled up to the size of the Earth the maximum roughness would be that of a ploughed field (!). The precession is measured by locking a telescope onto a guide star, and the rotation of the gyros required to stay in this locked-on position is measured by the London moment of the superconducting material, as measured on a SQUID. A preliminary result gives, specifically for the Lense–Thirring precession, a measurement which is 99 ± 5 per cent of the predicted value.[6]

6.2.2 'Inertial drag'

Under the influence of Schiff[7] the Lense–Thirring effect is commonly described by the 'dragging' of inertial frames – almost as if inertial frames were a fluid in which the rotating body is immersed, and its rotation causes rotation of the space-time frames. Note that the *sense* of the rotation is of the opposite sign near the poles (where $\boldsymbol{\omega}\cdot\mathbf{r} = \omega r$) and near the equator (where $\boldsymbol{\omega}\cdot\mathbf{r} = 0$): from (6.61)

$$\begin{aligned} \text{above N pole:}\ (\boldsymbol{\Omega}_{\mathrm{LT}})_{\mathrm{pol}} &= \frac{2GI}{c^2 r^3}\boldsymbol{\omega} = \frac{4GM}{5c^2}\left(\frac{R^2}{r^3}\right)\boldsymbol{\omega}, \\ \text{in equatorial plane:}\ (\boldsymbol{\Omega}_{\mathrm{LT}})_{\mathrm{eq}} &= -\frac{GI}{c^2 r^3}\boldsymbol{\omega} = -\frac{2GM}{5c^2}\left(\frac{R^2}{r^3}\right)\boldsymbol{\omega}. \end{aligned} \tag{6.70}$$

Then, at $r \approx R$,

$$\begin{aligned} (\Omega_{\mathrm{LT}})_{\mathrm{pol}} &= \frac{4GM}{5c^2 R}\boldsymbol{\omega} = 5.52 \times 10^{-10}\boldsymbol{\omega}, \\ (\Omega_{\mathrm{LT}})_{\mathrm{eq}} &= -\frac{2GM}{5c^2 R}\boldsymbol{\omega} = -2.76 \times 10^{-10}\boldsymbol{\omega}. \end{aligned}$$

As remarked by Schiff, 'At the poles, this tends to drag the spin around in the same direction as the rotation of the earth. But at the equator, since the gravitational field falls off with increasing r, the side of the spinning particle nearest the earth is dragged more than the side away from the earth, so that the spin precesses in the opposite direction.' This view, and its implied analogy with the behaviour of fluids, has been taken up by many authors, in particular Misner, Thorne and Wheeler,[8] but more recently has come in for criticism.[9]

6.2.3 Lense–Thirring effect and Mach's Principle

Mach's Principle asserts that it is only the relative motion of *bodies* which can play a role in mechanics: the motion of a body relative to its surrounding *space* can have no effect, since space is not 'real' – space is only a way of describing the separation of bodies. With this in mind let us consider the motion of a gyroscope in the Earth's equatorial plane. From (6.61) and (6.70)

6 Ciufolini & Pavlis (2004).
7 Schiff (1960).
8 Misner, Thorne & Wheeler (1973).
9 By Rindler (1997).

$$\Omega = \frac{GM}{c^2 r^3}\left[\frac{3}{2}\,\mathbf{r}\times\mathbf{v} - \frac{2}{5}\,R^2\boldsymbol{\omega}\right].$$

If $\mathbf{h}$ is the unit vector perpendicular to the equatorial plane, $\mathbf{r}\times\mathbf{v} = rv\mathbf{h}$, $\boldsymbol{\omega} = \omega\mathbf{h}$. We also have $\omega = 2\pi/T$ ($T = 1$ day) and $v = 2\pi\, r/\tau_s$, (τ_s = period of satellite). Then $\boldsymbol{\Omega} = \Omega\mathbf{h}$ with

$$\Omega = \frac{\pi GM}{c^2}\left[\frac{3}{r\tau_s} - \frac{4R^2}{5r^3 T}\right].$$

For a *geostationary* satellite $\tau_s = T$ and

$$\Omega = \frac{\pi GM}{c^2 Tr}\left[3 - \frac{4}{5}\left(\frac{R}{r}\right)^2\right].$$

The height of a geostationary satellite is $r = \left(\frac{MG\tau^2}{4\pi^2}\right)^{1/3} \approx 4.23\times 10^4$ km, so $\frac{R}{r} \approx 0.15$ and

$$\Omega \approx 2.98\,\frac{\pi GM}{c^2 Tr} \neq 0.$$

This satellite and the Earth are not rotating relative to each other and yet the Lense–Thirring effect predicts that the satellite processes. This certainly seems to violate Mach's Principle, according to which $\Omega = 0$, exactly as if neither body were rotating; since of course they are 'only' rotating relative to space, and that doesn't count, according to Mach. One could argue that although $\Omega \neq 0$, Ω is actually *small*; and one would then try and argue that the satellite is rotating relative to the rest of the Universe (the 'background stars', in Mach's phrase), and this contributes a small but non-vanishing effect. The trouble with this argument is that the solution to the field equations which we are working with makes no reference at all to the 'rest of the Universe', so this type of reasoning would seem to be rather disingenuous. For more discussion of the rather vexed question of Mach's principle and the Lense–Thirring effect see 'Further reading'.

6.3 Gravitomagnetism

The reader will clearly have noticed that the Lense–Thirring effect differs from the other consequences of General Relativity, the so-called 'classic tests' discussed in Chapter 5. It involves rotations, and correspondingly the time-space components g_{0i} of the metric tensor, and gives rise to spin precession. These effects are reminiscent of magnetism; for example the reader will immediately recall the connections between magnetism and rotations. The present section is devoted to exploring this analogy.

We begin by reminding ourselves of the (very simple) correspondence between gravity and electricity in the static case. A charge Q gives rise to an electostatic potential ϕ_e a distance r away:

$$\phi_e = \frac{Q}{4\pi\varepsilon_0 r},$$

and an associated electric field

$$\mathbf{E} = -\nabla\phi_e = \frac{Q}{4\pi\varepsilon_0 r^2}\hat{\mathbf{r}}. \tag{6.71}$$

The force on a charge q in this field is

$$\mathbf{F} = q\mathbf{E} = \frac{Qq}{4\pi\varepsilon_0 r^2}\hat{\mathbf{r}}. \tag{6.72}$$

Like charges repel, unlike charges attract. A mass M gives rise to a gravitational potential ϕ_g:

$$\phi_g = -\frac{MG}{r} \tag{6.73}$$

and an associated gravitational field $\mathbf{g}$

$$\mathbf{g} = -\nabla\phi_g = -\frac{MG}{r^2}\hat{\mathbf{r}}. \tag{6.74}$$

The force on a mass m is

$$\mathbf{F} = m\,\mathbf{g} = -\frac{MmG}{r^2}\hat{\mathbf{r}}. \tag{6.75}$$

Masses (always positive) attract: gravity is always attractive, unlike electricity; it is this, of course, which is ultimately the source of instability in gravity.

In the *non-static* case, the Lorentz force law is

$$\mathbf{F} = q\mathbf{E} + q\mathbf{v}\times\mathbf{B}. \tag{6.76}$$

We define the 4-vector $A^\mu = (\phi_e, \mathbf{A})$ with

$$\mathbf{B} = \nabla\times\mathbf{A}, \quad \mathbf{E} = -\nabla\phi_e - \frac{\partial\mathbf{A}}{\partial t} \tag{6.77}$$

and

$$\phi_e = \int\frac{\rho}{r}\mathrm{d}V, \quad \mathbf{A} = \int\frac{\mathbf{j}}{r}\mathrm{d}V, \tag{6.78}$$

where ρ is the charge density and $\mathbf{j} = \rho\mathbf{u}$ the current density. The Lorentz force is then

$$\mathbf{F} = q\left(-\nabla\phi_e - \frac{\partial\mathbf{A}}{\partial t}\right) + q\mathbf{v}\times(\nabla\times\mathbf{A}), \tag{6.79}$$

giving an acceleration

$$\mathbf{a} = \frac{q}{m}\left[-\nabla\phi_e - \frac{\partial\mathbf{A}}{\partial t} + \mathbf{v}\times(\nabla\times\mathbf{A})\right]. \tag{6.80}$$

The gravitational analogue of this equation for acceleration is, of course, the geodesic equation, which we will write in the linear approximation. The geodesic equation is

$$\frac{\mathrm{d}^2x^\mu}{\mathrm{d}\tau^2} + \Gamma^\mu{}_{\nu\lambda}\frac{\mathrm{d}x^\nu}{\mathrm{d}\tau}\frac{\mathrm{d}x^\lambda}{\mathrm{d}\tau} = 0.$$

In the *Newtonian* approximation we have (see (5.2) and (5.3))

$$\frac{d^2x^i}{dt^2} \approx -c^2\,\Gamma^i_{\ 00} = \frac{c^2}{2}\,g_{00,i} = -\nabla_i\,\phi.$$

We need to work to a higher approximation. We have, then,

$$\begin{aligned}
a^i = \frac{d^2x^i}{dt^2} &= \left(\frac{dt}{d\tau}\right)^{-1}\frac{d}{d\tau}\left[\left(\frac{dt}{d\tau}\right)^{-1}\frac{dx^i}{d\tau}\right] \\
&= \left(\frac{dt}{d\tau}\right)^{-2}\frac{d^2x^i}{d\tau^2} - \left(\frac{dt}{d\tau}\right)^{-3}\frac{d^2t}{d\tau^2}\frac{dx^i}{d\tau} \\
&= -\Gamma^i_{\ \nu\lambda}\frac{dx^\nu}{dt}\frac{dx^\lambda}{dt} + \Gamma^0_{\ \nu\lambda}\frac{dx^\nu}{dt}\frac{dx^\lambda}{dt}\frac{dx^i}{d\tau} \\
&= -c^2\,\Gamma^i_{\ 00} - 2\,c\,\Gamma^i_{\ 0k}\,v^k - \Gamma^i_{\ km}\,v^k\,v^m \\
&\quad + [c^2\Gamma^0_{\ 00} + 2\,c\,\Gamma^0_{\ 0k}\,v^k + \Gamma^0_{\ km}\,v^k\,v^m]\,v^i,
\end{aligned} \tag{6.81}$$

where $v^i = \frac{dx^i}{dt}$ and in the third line we have used the geodesic equation. The equation above is exact. We now calculate the connection coefficients $\Gamma^\mu_{\ \nu\lambda}$ in the linear approximation. Many of these have already been found – Equation (6.51) above – but that referred to the static case. When time dependence is included the coefficients turn out to be

$$\begin{aligned}
\Gamma^0_{\ 00} &= -\frac{1}{c^3}\frac{\partial\phi}{\partial t}, \\
\Gamma^0_{\ i0} &= \frac{1}{c^2}\,\nabla_i\,\phi, \\
\Gamma^0_{\ im} &= -{}^1\!/\!_2\,(\zeta_{i,m} + \zeta_{m,i}) - \frac{1}{c^2}\,\delta_{km}\,\frac{\partial\phi}{\partial t}, \\
\Gamma^i_{\ 00} &= \frac{1}{c}\frac{\partial\zeta^i}{\partial t} + \frac{1}{c^2}\nabla_i\,\phi, \\
\Gamma^k_{\ i0} &= {}^1\!/\!_2\,(\zeta_{k,i} - \zeta_{i,k}) - \frac{1}{c^3}\frac{\partial\phi}{\partial t}\,\delta_{ik}, \\
\Gamma^k_{\ im} &= \frac{1}{c^2}\,(\delta_{im}\,\nabla_k\,\phi - \delta^k_i\,\nabla_m\,\phi - \delta^k_m\,\nabla_i\,\phi).
\end{aligned} \tag{6.82}$$

Substituting these into (6.81) gives, ignoring terms in v/c^2, $\partial\phi/\partial t$, $(v^2/c^2)\,\nabla_i\phi$,

$$a^i = -\nabla_i\phi - c\frac{\partial\zeta^i}{\partial t} - c(\zeta_{i,k} - \zeta_{k,i})v^k$$

or

$$\begin{aligned}
\mathbf{a} &= -\nabla\phi - c\,\frac{\partial\zeta}{\partial t} + c\,\mathbf{v}\times(\nabla\times\boldsymbol{\zeta}) \\
&= \mathbf{g} + c\,\mathbf{v}\times(\nabla\times\boldsymbol{\zeta})
\end{aligned} \tag{6.83}$$

where

$$\mathbf{g} = -\nabla\phi_{\mathrm{g}} - c\frac{\partial\boldsymbol{\zeta}}{\partial t}. \tag{6.84}$$

Comparing this with Equation (6.80) there is a clear analogy with the Lorentz force equation, with

$$\begin{aligned} \phi_e &\leftrightarrow \phi_g, \\ \mathbf{A} &\leftrightarrow \boldsymbol{\zeta}. \end{aligned} \tag{6.85}$$

(In passing, we may remark that the coefficient (q/m) in the Lorentz equation (6.80) becomes unity in the gravitational case, since the analogue of electric charge is gravitational mass, equal to inertial mass by the Equivalence Principle; and we must remember that this analogy has been derived, and only holds, in the *linear approximation* – it is not exact.) The quantity $\nabla \times \mathbf{A}$ describes a magnetic field. What does $\nabla \times \boldsymbol{\zeta}$ (a 'gravitomagnetic' field) describe? Consider a particle which, in an inertial frame, has velocity $\mathbf{v}'$ and acceleration $\mathbf{a}'$. In a frame rotating with angular velocity $\boldsymbol{\omega}$ its acceleration is, ignoring the centrifugal force[10]

$$\mathbf{a} = \mathbf{a}' - 2\mathbf{v}' \times \boldsymbol{\omega}; \tag{6.86}$$

the second term is the Coriolis force. This is of the same form as (6.83) with

$$\nabla \times \boldsymbol{\zeta} = -2\,\boldsymbol{\omega}.$$

We would therefore expect that this term in the linear approximation corresponds to the precession of inertial frames. This turns out to be correct; since from (6.49) and the above we have

$$\begin{aligned} \boldsymbol{\omega} &= -{}^1\!/\!_2 \nabla \times \boldsymbol{\zeta} = \frac{G}{c^3} \nabla \times \left\{ \frac{\mathbf{J} \times \mathbf{r}}{r^3} \right\} \\ &= \frac{G}{c^3 r^5} \left[3\mathbf{r}(\mathbf{J} \cdot \mathbf{r}) - r^2 \mathbf{J} \right] = \boldsymbol{\Omega}_{\mathrm{LT}} \end{aligned} \tag{6.87}$$

from (6.61); this is precisely Lense–Thirring precession.

The analogy, however, even in the linear approximation, is not exact. Let us write Equation (6.22) in the symbolic form

$$f_{\mu\nu} = \frac{4G}{c^2} \int \frac{T_{\mu\nu}}{r} \, \mathrm{dV}.$$

Now from (6.24) $\phi = -\dfrac{c^2}{4} f_{00}$, hence

$$\phi = -G \int \frac{T_{00}}{r} \, \mathrm{d}V = -G \int \frac{\rho}{r} \, \mathrm{d}V, \tag{6.88}$$

where ρ is the matter density. This equation is analogous to (6.78). On the other hand, with $g_{0i} = \zeta_i = f_{0i}$ so

$$\zeta_i = \frac{4G}{c^2} \int \frac{T_{0i}}{r} \, \mathrm{d}V = -\frac{4G}{c^2} \int \frac{T^{0i}}{r} \, \mathrm{d}V = -\frac{4G}{c^2} \int \frac{\rho u^i}{r} \, \mathrm{d}V, \tag{6.89}$$

and the factor of 4 spoils the similarity with the formula (6.78) for $\mathbf{A}$.

[10] See for example Kibble & Berkshire (1996), p. 91.

6.4 Gravitomagnetic clock effect

We now investigate the following problem. Two clocks, initially synchronised, are sent in circular orbits in the equatorial plane of the Earth, in opposite directions. When they return to the starting point, will they still tell the same time? Clearly, if the Earth is not rotating, they will, but what if the Earth is rotating?

Take the Earth to be spinning about the z axis: then $\mathbf{J} = J\mathbf{k}$ and (6.49) gives

$$\zeta_1 = \frac{2G}{r^3c^3}\, yJ, \quad \zeta_2 = -\frac{2G}{r^3c^3}\, xJ, \quad \zeta_3 = 0$$

and the metric (6.48) is

$$g_{\mu\nu} = \begin{pmatrix} -\left(1+\frac{2\phi}{c^2}\right) & \beta y & -\beta x & \zeta_3 \\ \beta y & 1-\frac{2\phi}{c^2} & 0 & 0 \\ -\beta x & 0 & 1-\frac{2\phi}{c^2} & 0 \\ \zeta_3 & 0 & 0 & 1-\frac{2\phi}{c^2} \end{pmatrix} \tag{6.90}$$

with

$$\beta = \frac{2GJ}{r^3c^3}; \tag{6.91}$$

or

$$\mathrm{d}s^2 = -\left(1-\frac{2GM}{rc^2}\right)c^2\,\mathrm{d}t^2 + 2\beta(y\,\mathrm{d}x - x\,\mathrm{d}y) + \left(1+\frac{2GM}{rc^2}\right)(\mathrm{d}x^2+\mathrm{d}y^2+\mathrm{d}z^2). \tag{6.92}$$

We now convert the line element into spherical polar coordinates (obviously convenient for our problem) and find (see Problem 6.2)

$$g_{rr} = 1+\frac{2GM}{rc^2}, \quad g_{\theta\theta} = r^2\left(1+\frac{2GM}{rc^2}\right), \quad g_{\phi\phi} = r^2\sin^2\theta\left(1+\frac{2GM}{rc^2}\right),$$
$$g_{tr} = g_{t\theta} = 0, \quad g_{t\phi} = -2r^2\beta\sin^2\theta,$$

so that

$$\mathrm{d}s^2 = -\left(1-\frac{2GM}{rc^2}\right)c^2\,\mathrm{d}t^2 + \left(1+\frac{2GM}{rc^2}\right)(\mathrm{d}r^2 + r^2\mathrm{d}\theta^2 + r^2\sin^2\theta\,\mathrm{d}\phi^2)$$
$$-\frac{4GJ}{rc^3}\sin^2\theta\,\mathrm{d}\phi(c\,\mathrm{d}t). \tag{6.93}$$

Then the covariant metric tensor

$$g_{\mu\nu} = \begin{pmatrix} -\left(1-\frac{2GM}{rc^2}\right) & 0 & 0 & -\frac{2GJ}{rc^3}\sin^2\theta \\ 0 & 1+\frac{2GM}{rc^2} & 0 & 0 \\ 0 & 0 & r^2\left(1+\frac{2GM}{rc^2}\right) & 0 \\ -\frac{2GJ}{rc^3}\sin^2\theta & 0 & 0 & r^2\sin^2\theta\left(1+\frac{2GM}{rc^2}\right) \end{pmatrix}$$

has determinant

$$g \approx -r^4\sin^2\theta\left(1+\frac{4GM}{rc^2}\right), \quad g^{-1} \approx -\frac{1}{r^4\sin^2\theta}\left(1-\frac{4GM}{rc^2}\right)$$

and contravariant components

$$g^{\mu\nu} = \begin{pmatrix} -\left(1+\frac{2GM}{rc^2}\right) & 0 & 0 & -\frac{1}{r^2\sin^2\theta}\frac{2GM}{rc^3} \\ 0 & 1-\frac{2GM}{rc^2} & 0 & 0 \\ 0 & 0 & \frac{1}{r^2}\left(1-\frac{2GM}{rc^2}\right) & 0 \\ -\frac{1}{r^2\sin^2\theta}\frac{2GM}{rc^3} & 0 & 0 & \frac{1}{r^2\sin^2\theta}\left(1-\frac{2GM}{rc^2}\right) \end{pmatrix}.$$

The geodesic equation for r is $[(ct, r, \theta, \phi) = (x^0, x^1, x^2, x^3)]$

$$\ddot{r} + \Gamma^1{}_{\mu\nu}\dot{x}^\mu\dot{x}^\nu = 0.$$

For a circular orbit in the equatorial plane $\dot{x}^1 = 0$, $\dot{x}^2 = 0$ so only $\mu, \nu = 0, 3$ contribute to the sum above, hence

$$\ddot{r} + \Gamma^1{}_{00}c^2\dot{t}^2 + 2\Gamma^1{}_{03}\,c\dot{t}\dot{\phi} + \Gamma^1{}_{33}\dot{\phi}^2 = 0,$$

or, with $\ddot{r} = 0$, $\dot{\phi} = \omega\dot{t}$,

$$\dot{t}^2[c^2\Gamma^1{}_{00} + 2\omega c\Gamma^1{}_{03} + \omega^2\Gamma^1{}_{33}] = 0.$$

It is straightforward to calculate the connection coefficients; to lowest order we have

$$\Gamma^1{}_{00} \approx \frac{GM}{r^2c^2}, \quad \Gamma^1{}_{03} \approx -\left(1-\frac{2GM}{rc^2}\right)\frac{GJ}{r^2c^3}, \quad \Gamma^1{}_{33} \approx -r\left(1-\frac{2GM}{rc^2}\right),$$

and thence, to lowest order

$$\omega^2 + \frac{2GJ}{r^3c^2}\omega - \frac{GM}{r^3} = 0. \tag{6.94}$$

(The reader might find it interesting to check that each term in the above equation has dimension T^{-2} (T = time).) This quadratic equation has the two solutions

$$\omega_1 = \sqrt{\left(\frac{GJ}{r^3c^2}\right)^2 + \frac{GM}{r^3}} + \frac{GJ}{r^3c^2}, \quad \omega_2 = \sqrt{\left(\frac{GJ}{r^3c^2}\right)^2 + \frac{GM}{r^3}} - \frac{GJ}{r^3c^2}, \tag{6.95}$$

which correspond to prograde and retrograde motion.

Now for a circular orbit (r = const) in the equatorial plane ($\theta = \pi/2$) we have, from (6.93)

$$\mathrm{d}s^2 = -\left(1 - \frac{2GM}{rc^2}\right)c^2\,\mathrm{d}t^2 + \left(1 + \frac{2GM}{rc^2}\right)r^2\,\mathrm{d}\phi^2 - \frac{4GJ}{rc^2}\,\mathrm{d}\phi\,\mathrm{d}t$$

and with $\mathrm{d}\phi = \omega\,\mathrm{d}t$, the proper time τ is given by, ignoring the term in $r^2\,\mathrm{d}\phi^2$ (and with $\mathrm{d}s^2 = -c^2\,\mathrm{d}\tau^2$)

$$\mathrm{d}\tau^2 \approx \frac{1}{\omega^2}\left(1 - \frac{2GM}{rc^2} - \frac{4GJ\omega}{rc^2}\right)\mathrm{d}\phi^2,$$

$$\mathrm{d}\tau \approx \frac{1}{\omega}\left(1 - \frac{GM}{rc^2} - \frac{2GJ\omega}{rc^2}\right)\mathrm{d}\phi.$$

For one revolution $\mathrm{d}\phi = 2\pi$ and the period is

$$T = \frac{2\pi}{\omega}\left(1 - \frac{GM}{rc^2}\right) + \frac{4\pi GJ}{rc^2}.$$

Hence, corresponding to the two values of ω,

$$T_1 = \frac{2\pi}{\omega_1}\left(1 - \frac{GM}{rc^2}\right) + \frac{4\pi GJ}{rc^2}, \quad T_2 = \frac{2\pi}{\omega_2}\left(1 - \frac{GM}{rc^2}\right) + \frac{4\pi GJ}{rc^2},$$

and the 'clock effect' is given by

$$T_1 - T_2 = \left(\frac{1}{\omega_1} - \frac{1}{\omega_2}\right)2\pi\left(1 - \frac{GM}{rc^2}\right).$$

We have, from (6.94), (6.95)

$$\frac{1}{\omega_1} - \frac{1}{\omega_2} = \frac{\omega_2 - \omega_1}{\omega_1\,\omega_2} = \frac{2J}{Mc^2}.$$

and finally

$$T_1 - T_2 \approx \frac{4\pi J}{Mc^2}. \tag{6.96}$$

This is the prediction: after one revolution the two clocks, originally synchronised, will differ in time by the above amount (to lowest order). A first remarkable feature of this prediction is that the result is independent of G; we have a consequence of a gravitational theory that does not involve Newton's constant of gravitation! For the Earth, $J = \frac{2}{5}MR^2\omega$ and with $R = 6.4 \times 10^6$ m, $\omega = 7.3 \times 10^{-5}$ s,

$$T_1 - T_2 = \frac{8\pi}{5}\frac{R^2}{c^2}\omega = 1.7 \times 10^{-7}\ \text{s}.$$

There has been a proposal to measure this, known as Gravity Probe C(lock) – see Gronwald *et al.* (1997).

The distinction between this clock effect and the one discussed in Section 5.13 should be understood. This ('gravitomagnetic') one is a time difference between two clocks on *geodesics*, travelling round a *rotating* Earth in opposite directions. The previous effect was a difference in time between two clocks being transported (not on geodesics) round the Earth. The rotation of the Earth is involved, but not as a 'dynamic' effect, since the metric is the Schwarzschild metric.

6.5 Fermi–Walker transport: tetrad formalism

Parallel displacement of a vector appears the most natural way of comparing vectors at different points in a space, or of transporting a vector from one point to another. But there are physically important cases in which another kind of transport law is more useful for the formulation of physical theories.

Consider an observer moving along an arbitrary timelike curve $x^\mu(\tau)$, under the action of forces (unless the curve happens to be a geodesic). He will regard as 'natural' a coordinate system in which he is *at rest* and his spatial axes *do not rotate*. The rest condition is

$$\frac{\mathrm{d}x^i}{\mathrm{d}\tau} = t^i = 0, \tag{6.97}$$

where t^μ is the tangent vector and Latin indices refer to the spatial components. The tangent vector then possesses only a timelike component

$$t^\mu = \frac{\mathrm{d}x^\mu}{\mathrm{d}\tau} = (t^0, \mathbf{0}). \tag{6.98}$$

The 'natural' coordinate system is then given by the tangent vector to the worldline; but the tangent vector to a curve at one point is *not* parallel-transported into the tangent vector at another point. If $V^\mu(x)$ is the tangent vector to a curve at one point x, $V^\mu(x) = \frac{\mathrm{d}x^\mu}{\mathrm{d}\sigma}$, where σ is some parameter along the curve C, then under parallel transport the absolute derivative of V^μ is

$$\frac{\mathrm{D}V^\mu}{\mathrm{d}\sigma} = \frac{\mathrm{d}V^\mu}{\mathrm{d}\sigma} + \Gamma^\mu{}_{\nu\kappa} V^\nu V^\kappa = \frac{\mathrm{d}^2x^\mu}{\mathrm{d}\sigma^2} + \Gamma^\mu{}_{\nu\kappa}\frac{\mathrm{d}x^\nu}{\mathrm{d}\sigma}\frac{\mathrm{d}x^\kappa}{\mathrm{d}\sigma} \neq 0, \tag{6.99}$$

assuming that C is not a geodesic – see Fig. 6.5. What is the transport law to convert a non-rotating tangent vector into a non-rotating tangent vector? (By which is meant: a vector whose space components are non-rotating and whose time component is tangent to the timelike worldline.) It is Fermi–Walker transport, which will now be defined.

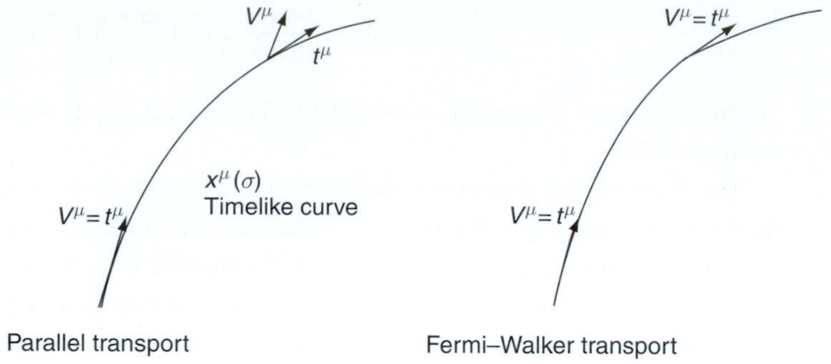

Fig. 6.5 **Parallel transport and Fermi–Walker transport along a timelike curve, not a geodesic.**

The Fermi derivative of a vector V^μ is

$$\frac{\mathrm{D_F}V^\mu}{\mathrm{d}\sigma} = \frac{\mathrm{D}V^\mu}{\mathrm{d}\sigma} - V_\nu\left(t^\mu\frac{\mathrm{D}t^\nu}{\mathrm{d}\sigma} - t^\nu\frac{\mathrm{D}t^\mu}{\mathrm{d}\sigma}\right). \tag{6.100}$$

A vector is Fermi–Walker (FW) transported if $\frac{\mathrm{D_F}V^\mu}{\mathrm{d}\sigma}=0$. $\left(\text{It is } \textit{Fermi transported} \text{ if } \frac{\mathrm{D}V^\mu}{\mathrm{d}\sigma} - V_\nu t^\mu\frac{\mathrm{D}t^\nu}{\mathrm{d}\sigma}=0.\right)$ We have to show that the tangent vector to a timelike curve is FW transported into a tangent vector. Hence if

$$t^\mu = \frac{\mathrm{d}V^\mu}{\mathrm{d}\sigma}: \quad t^\mu t_\mu = -1, \tag{6.101}$$

we must show that $\frac{\mathrm{D_F}t^\mu}{\mathrm{d}\sigma}=0$. On differentiation of (6.101) we have

$$\frac{\mathrm{d}}{\mathrm{d}\sigma}(t^\mu t_\mu) = \frac{\mathrm{D}}{\mathrm{d}\sigma}(t^\mu t_\mu) = 0,$$

or, in an obvious notation, $t^\mu \mathrm{D}t_\mu = 0$. Then, continuing with this notation

$$\mathrm{D_F}t^\mu = \mathrm{D}t^\mu - t_\nu(t^\mu\mathrm{D}t^\nu - t^\nu\mathrm{D}t^\mu) = \mathrm{D}t^\mu(1 + t_\nu t^\nu) = 0, \tag{6.102}$$

which proves our assertion. Writing (6.100) in the form

$$\frac{\mathrm{D}_F\mathbf{V}}{\mathrm{d}\sigma} = \frac{\mathrm{D}\mathbf{V}}{\mathrm{d}\sigma} - \mathbf{t}\left(\mathbf{V}\cdot\frac{\mathrm{D}\mathbf{t}}{d\sigma}\right) + \frac{\mathrm{D}\mathbf{t}}{\mathrm{d}\sigma}(\mathbf{V}\cdot\mathbf{t}), \tag{6.103}$$

the Fermi–Walker transport condition is

$$\frac{\mathrm{D}\mathbf{V}}{\mathrm{d}\sigma} = \mathbf{t}\left(\mathbf{V}\cdot\frac{\mathrm{D}\mathbf{t}}{\mathrm{d}\sigma}\right) - \frac{\mathrm{D}\mathbf{t}}{\mathrm{d}\sigma}(\mathbf{V}\cdot\mathbf{t}). \tag{6.104}$$

We shall now make use of FW transport to deal with the case of accelerated motion with no rotation. In Section 2.2 we considered a body subjected to constant acceleration. Such a body moves along a hyperbola in space-time given by (see (2.49))

$$x^\mu = \frac{c^2}{g}\left(\sinh\frac{g\tau}{c},\ \cosh\frac{g\tau}{c},\ 0,\ 0\right), \tag{6.105}$$

corresponding to acceleration in the x direction. The 4-velocity is

$$u^\mu = \frac{1}{c}\frac{\mathrm{d}x^\mu}{\mathrm{d}\tau} = \left(\cosh\frac{g\tau}{c},\ \sinh\frac{g\tau}{c},\ 0,\ 0\right) \tag{6.106}$$

and the acceleration is

$$a^\mu = c\frac{\mathrm{d}u^\mu}{\mathrm{d}\tau} = \left(g\ \sinh\frac{g\tau}{c},\ g\ \cosh\frac{g\tau}{c},\ 0,\ 0\right). \tag{6.107}$$

The 4-velocity u^μ is a tangent vector to the worldline and is timelike:

$$u^\mu u_\mu = -\cosh^2\frac{g\tau}{c} + \sinh^2\frac{g\tau}{c} = -1, \tag{6.108}$$

and a^μ is orthogonal to u^μ and spacelike:

$$a^\mu u_\mu = 0, \quad a^\mu a_\mu = g^2\left(-\sinh^2\frac{g\tau}{c} + \cosh^2\frac{g\tau}{c}\right) = g^2. \tag{6.109}$$

At any point in a 4-dimensional space-time (in this case Minkowski space-time) we may erect a set of four mutually orthogonal vectors, one timelike and three spacelike. This is a straightforward generalisation of the fact that in, say R^3, we may erect at each point the three mutually orthogonal basis vectors **i**, **j**, **k**:

$$\mathbf{i} = (1,\ 0,\ 0), \quad \mathbf{j} = (0,\ 1,\ 0), \quad \mathbf{k} = (0,\ 0,\ 1),$$
$$\mathbf{i}\cdot\mathbf{i} = \mathbf{j}\cdot\mathbf{j} = \mathbf{k}\cdot\mathbf{k} = 1, \quad \mathbf{i}\cdot\mathbf{j} = \mathbf{j}\cdot\mathbf{k} = \mathbf{k}\cdot\mathbf{i} = 0.$$

Let us now put

$$\mathbf{i} = \mathbf{e}_{(1)}, \quad \mathbf{j} = \mathbf{e}_{(2)}, \quad \mathbf{k} = \mathbf{e}_{(3)};$$

we now have a *triad* of vectors $\mathbf{e}_{(a)}$. A typical component is

$e_{(a)}{}^i$: a denotes which vector, i denotes which component of that vector.

For simplicity this may be denoted $e_a{}^i$: letters from the beginning of the (Latin) alphabet tell which vector, and from the middle of the alphabet, which component of that vector. It should be emphasised: $e_a{}^i$ is a triad of vectors, *not* a rank 2 tensor. We may simply mimic this exercise in Minkowski space-time. We may erect at each point a *tetrad* (or vierbein – four legs) of four orthonormal vectors $\mathbf{h}_{(\alpha)}$:

$h_{(\alpha)}{}^\mu$: α denotes which vector, μ denotes which component of that vector (6.110)

or, for simplicity $h_\alpha{}^\mu$ (with letters from the beginning of the (Greek) alphabet to tell which vector, and from the middle to tell which component). This tetrad of vectors comprises one timelike vector and three spacelike ones. In the present context – of a particle undergoing uniform acceleration – we may put

$$\begin{aligned} h_{(0)}{}^{\mu} &= u^{\mu} = \left(\cosh\frac{g\tau}{c}, \sinh\frac{g\tau}{c}, 0, 0\right), \\ h_{(1)}{}^{\mu} &= \frac{1}{g}a^{\mu} = \left(\sinh\frac{g\tau}{c}, \cosh\frac{g\tau}{c}, 0, 0\right), \\ h_{(2)}{}^{\mu} &= (0, 0, 1, 0), \\ h_{(3)}{}^{\mu} &= (0, 0, 0, 1), \end{aligned} \tag{6.111}$$

obeying

$$\begin{aligned} h_{(0)}{}^{\mu} h_{(0)\,\mu} &= -1 = \eta_{00}, \\ h_{(1)}{}^{\mu} h_{(1)\,\mu} &= +1 = \eta_{11}, \\ h_{(2)}{}^{\mu} h_{(2)\,\mu} &= +1 = \eta_{22}, \\ h_{(3)}{}^{\mu} h_{(3)\,\mu} &= +1 = \eta_{33}, \\ h_{(\alpha)}{}^{\mu} h_{(\beta)\,\mu} &= 0, \alpha \neq \beta, \end{aligned}$$

or

$$h_{(\alpha)}{}^{\mu} h_{(\beta)\mu} = \eta_{\alpha\beta}; \tag{6.112}$$

or, indeed

$$\eta_{\mu\nu} h_{(\alpha)}{}^{\mu} h_{(\beta)}{}^{\nu} = \eta_{\alpha\beta}. \tag{6.113}$$

This relation holds in Minkowski space but its significance is perhaps more clearly seen if we generalise to an arbitrary Riemannian space with metric $g_{\mu\nu}$. Then (6.113) is replaced by

$$g_{\mu\nu} h_{(\alpha)}{}^{\mu} h_{(\beta)}{}^{\nu} = \eta_{\alpha\beta}. \tag{6.114}$$

The indices μ and ν are *world indices*: at each point (x^{μ}) of the space we can erect a tetrad $h_{(\alpha)}{}^{\mu}$ $(\alpha=0, 1, 2, 3)$ obeying (6.114). The indices α and β are, on the other hand, *tangent space indices*. The erection of a local frame at any point in an arbitrary space-time is an expression of the Equivalence Principle – the locally inertial frame has a Minkowski metric tensor.

As long as $g_{\mu\nu}$ is non-singular we may define the 'inverse' of $h_{(\alpha)}{}^{\mu}$, denoted $h^{(\alpha)}{}_{\mu}$, by

$$g_{\mu\nu} = h^{(\alpha)}{}_{\mu} h^{(\beta)}{}_{\nu} \eta_{\alpha\beta},$$

or, more precisely,

$$g_{\mu\nu}(x) = h^{(\alpha)}{}_{\mu}(x)\, h^{(\beta)}{}_{\nu}(x)\, \eta_{\alpha\beta}. \tag{6.115}$$

It turns out that tetrads are an essential device to express the Dirac equation (for spin ½ particles) in a general Riemannian space-time – see Section 11.4 below.

Let us now return to the connection between FW transport and accelerating frames. We want to define a tetrad which an observer, subject to an acceleration, carries with her, and which defines a set of orthonormal basis vectors forming a rest-frame (so that $\mathbf{h}_{(0)} = \mathbf{u}$, a tangent to the world-line), and the tetrad is *non-rotating*. Now rotation in non-relativistic physics is given by

$$\frac{\mathrm{d}v^i}{\mathrm{d}t} = -\Omega^i{}_k v^k, \quad \Omega^{ik} = -\Omega^{ki}. \tag{6.116}$$

This is easily generalised to four dimensions:

$$\frac{\mathrm{d}v^\mu}{\mathrm{d}t} = -\Omega^\mu{}_\nu v^\nu, \quad \Omega^{\mu\nu} = -\Omega^{\nu\mu}. \tag{6.117}$$

It is clear that this leaves the length of the vector v^μ unchanged:

$$\frac{\mathrm{d}}{\mathrm{d}\tau}(v^\mu v_\mu) = 2\frac{\mathrm{d}v^\mu}{\mathrm{d}\tau}v_\mu = -2\,\Omega^{\mu\nu}\,v_\mu v_\nu = 0.$$

The tensor $\Omega^{\mu\nu}$ has six components, corresponding to the generators of the Lorentz group (including rotations). We now claim that the expression for $\Omega^{\mu\nu}$ giving (a) the correct Lorentz transformation appropriate to an acceleration, and (b) no (spatial) rotation, is

$$\Omega^{\mu\nu} = a^\mu u^\nu - a^\nu u^\mu. \tag{6.118}$$

To show that (a) holds, let us calculate $\frac{\mathrm{d}u^\mu}{\mathrm{d}\tau}$ from (6.117):

$$\frac{\mathrm{d}u^\mu}{\mathrm{d}\tau} = -(a^\mu u^\nu - a^\nu u^\mu)\,u_\nu = -a^\mu(u^\nu u_\nu) + u^\mu(a^\nu u_\nu) = a^\mu,$$

as desired, and where (6.108) and (6.109) have been used. To show (b) consider a spacelike vector W^μ, orthogonal to a^μ (and to u^μ), so that

$$\Omega^{\mu\nu}W_\nu = a^\mu(u^\nu W_\nu) - u^\mu(a^\nu W_\nu) = 0$$

and hence

$$\frac{\mathrm{d}W^\mu}{\mathrm{d}\tau} = 0; \tag{6.119}$$

there is no rotation of W^μ. For a general vector V^μ (6.117) and (6.118) give

$$\frac{\mathrm{d}V^\mu}{\mathrm{d}\tau} = -(a^\mu u^\nu - a^\nu u^\mu)V_\nu = u^\mu(\mathbf{a}\cdot\mathbf{V}) - a^\mu(\mathbf{u}\cdot\mathbf{V}) \tag{6.120}$$

which is the law of FW transport – see (6.104) with $\mathbf{t} = \mathbf{u}$, $\mathrm{D}\mathbf{t} = \mathbf{a}$. Hence Fermi–Walker transport of a vector describes the propagation of a frame that is accelerating but not rotating, and is therefore described by a triad of three orthogonal gyroscopes and a timelike vector (the 4-velocity) orthogonal to the triad.

6.6 Lie derivatives, Killing vectors and groups of motion

Consider the Schwarzschild solution. It has a very high degree of symmetry, being spherically symmetric and static. It was just these features that made it possible for us to find the form of the Schwarzschild metric, by working in a coordinate system designed for the symmetry we wanted – spherical polar coordinates and a simple time coordinate, with no terms in $dx^i\,dt$. We could, however, if we were perverse, re-express the Schwarzschild solution in a very different coordinate system – say, for example, confocal ellipsoidal coordinates, and a 'time' coordinate that mixed up our previous t with a space coordinate. The Schwarzschild metric would then look fairly unpleasant, and the symmetry that it still possessed would not be at all obvious. The whole spirit of General Relativity, of course, is that coordinate systems do not matter – they have no intrinsic significance – so the question then arises, is there a way of treating the Schwarzschild solution in a coordinate-independent manner; in other words, to express its symmetry in such a way? The answer is that there is such a method of approaching matters of symmetry in General Relativity and it is based on the notion of Lie derivatives. This section is devoted to this topic, which, as well as helping to understand the Schwarzschild solution, is of great use in cosmology, which is treated in Chapter 10.

If a space possesses a symmetry of some sort this means that it has some property (usually some property of the metric tensor) which is the 'same' at different points of the space. We therefore have to have a means of comparing two points in the space (for our purposes, a space-time, but I use the word 'space' for simplicity and generality). To make this comparison, let us introduce a *vector field* allowing us to pass from one point to another one. If a vector field $a^\mu(x)$ is defined on a manifold we can use it to define the coordinates of nearby points. For example, referring to Fig. 6.6, P has coordinates x^μ and $\bar{P}$ has coordinates $\bar{x}^\mu$ then

$$\bar{x}^\mu = x^\mu - \varepsilon\, a^\mu(x), \tag{6.121}$$

where ε is infinitesimal. This is a change of *point*, not of label. The vector field a^μ then maps P onto $\bar{P}$, i.e. maps the manifold onto itself. It also maps geometric objects defined on the manifold into similar geometric objects; for example vectors onto vectors. If $V^\mu(x)$ is a vector field, then

$$\bar{V}^\mu(\bar{x}) = \frac{\partial \bar{x}^\mu}{\partial x^\nu}\, V^\nu(x) = (\delta^\mu{}_\nu - \varepsilon\, a^\mu{}_{,\nu})\, V^\nu(x) = V^\mu(x) - \varepsilon\, a^\mu{}_{,\nu}\, V^\nu(x). \tag{6.122}$$

(This transformation law refers *either* to a change of label *or* to a change of point, but our present considerations are focussed on a change of point.) The quantity $\bar{V}^\mu(\bar{x}) - V^\mu(x)$ is, as the reader will appreciate, *not* a vector, since it is the difference between two vectors at *different* points, so we expand $\bar{V}^\mu(\bar{x})$:

$$\bar{V}^\mu(\bar{x}) = \bar{V}^\mu(\bar{x}) + (\bar{x} - x)^\nu\, \bar{V}^\mu{}_{,\nu}(x) + \cdots = \bar{V}^\mu(x) - \varepsilon\, a^\nu\, \bar{V}^\mu{}_{,\nu} + \mathrm{O}(\varepsilon^2),$$

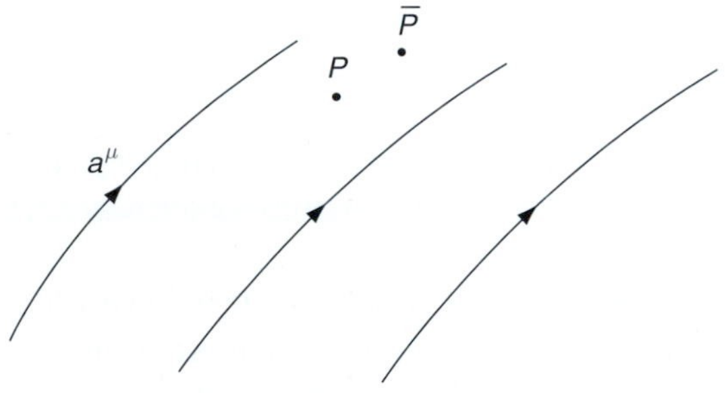

Fig. 6.6 The vector field a^μ maps the point P onto the point $\bar{P}$.

then together with (6.101),

$$V^{\mu}(x) = \bar{V}^{\mu}(x) - \varepsilon a^{\nu} \bar{V}^{\mu}{}_{,\nu} + \varepsilon a^{\mu}{}_{,\nu} V^{\nu} + \mathrm{O}(\varepsilon^2)$$

and

$$\lim_{\varepsilon \to 0} \left[\frac{V^{\mu}(x) - \bar{V}^{\mu}(x)}{\varepsilon} \right] = \lim_{\varepsilon \to 0} (-a^{\nu} \bar{V}^{\mu}{}_{,\nu} + a^{\mu}{}_{,\nu} V^{\nu}) = -a^{\nu} V^{\mu}{}_{,\nu} + a^{\mu}{}_{,\nu} V^{\nu}.$$

This is the *Lie derivative* of V with respect to a:

$$(L_{\mathbf{a}} \mathbf{V})^{\mu} = -\frac{\partial V^{\mu}}{\partial x^{\nu}} a^{\nu} + \frac{\partial a^{\mu}}{\partial x^{\nu}} V^{\nu}, \tag{6.123}$$

or

$$L_a V^{\mu} = -V^{\mu}{}_{,\nu} a^{\nu} + V^{\nu} a^{\mu}{}_{,\nu}. \tag{6.124}$$

Analogous formulae for covariant vectors and tensors follow, for example

$$L_a V_{\mu} = V_{\mu,\nu} a^{\nu} + V_{\nu} a^{\nu}{}_{,\mu}, \tag{6.125}$$

$$L_a g_{\mu\nu} = g_{\mu\lambda} a^{\lambda}{}_{,\nu} + g_{\lambda\nu} a^{\lambda}{}_{,\mu} + g_{\mu\nu,\lambda} a^{\lambda}. \tag{6.126}$$

It is clear that the Lie derivatives are defined without the use of affine connections (connection coefficients), so the formulae above do not involve Christoffel symbols. Nevertheless similar formulae to the above hold, which feature covariant, rather than ordinary, derivatives. We have, for example,

$$V^{\mu}{}_{;\nu} a^{\nu} - V^{\nu} a^{\mu}{}_{;\nu} = V^{\mu}{}_{,\nu} a^{\nu} - V^{\nu} a^{\mu}{}_{,\nu} + V^{\kappa} a^{\lambda} (\Gamma^{\mu}{}_{\kappa\lambda} - \Gamma^{\mu}{}_{\lambda\kappa})$$

and the last term vanishes in a space with no torsion (in a holonomic basis). Hence Equations (6.124)–(6.126) yield

$$L_a V^{\mu} = -V^{\mu}{}_{;\nu} a^{\nu} + V^{\nu} a^{\mu}{}_{;\nu}, \tag{6.127}$$

$$L_a V_{\mu} = V_{\mu;\nu} a^{\nu} + V_{\nu} a^{\nu}{}_{;\mu}, \tag{6.128}$$

$$L_a g_{\mu\nu} = g_{\mu\lambda} a^{\lambda}{}_{;\nu} + g_{\lambda\nu} a^{\lambda}{}_{;\mu}, \tag{6.129}$$

where in the last equation we have used the fact that the covariant derivative of the metric tensor vanishes. It is clear that the Lie derivatives of tensors are *tensors*.

Let us now expresss the above in coordinate-free notation. We put

$$\mathbf{V} = V^{\mu} \mathbf{e}_{\mu}, \quad \mathbf{A} = a^{\nu} \mathbf{e}_{\nu}. \tag{6.130}$$

In a coordinate basis $\mathbf{e}_{\mu} = \frac{\partial}{\partial x^{\mu}}$, $\mathbf{e}_{\nu} = \frac{\partial}{\partial x^{\nu}}$ and it is easy to see that

$$[\mathbf{A}, \mathbf{V}] = (a^{\nu} V^{\mu}{}_{,\nu} - V^{\nu} a^{\mu}{}_{,\nu}),$$

which is the right hand side of (6.124), which therefore becomes, in coordinate-free notation

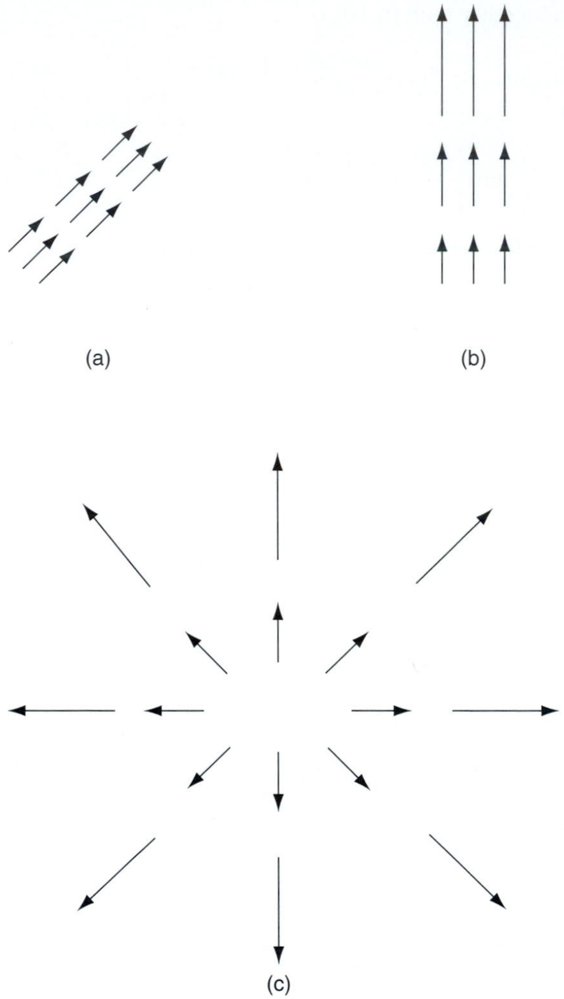

Fig. 6.7 **Three vector fields, invariant under particular symmetries – see text.**

$$L_{\mathbf{A}}\,\mathbf{V} = [\mathbf{A}\,,\,\mathbf{V}]; \tag{6.131}$$

a Lie bracket appears in the definition of a Lie derivative! (An equation involving Lie brackets has already been quoted above – see (3.10)). In a similar way Equation (6.128), when expressed in intrinsic notation, would involve the Lie derivative of a 1-form.

We now turn our attention to symmetries, and remark that particular types of vector fields are *invariant* under particular symmetry operations. For example, referring to the three diagrams in Fig. 6.7, (a) is invariant under any translation, (b) is invariant under translations in the x direction, and (c) is invariant under rotations. These are statements that particular Lie derivatives vanish, $L_{\mathbf{A}}\,\mathbf{V}=0$, where $\mathbf{V}$ is the vector field above and $\mathbf{A}$ is the generator of the relevant transformation (or motion, as it is sometimes called). To see this, consider the diagrams in turn:

(a) $\mathbf{V}=a\mathbf{i}+b\mathbf{j}=a\dfrac{\partial}{\partial x}+b\dfrac{\partial}{\partial y}$. An arbitrary translation is $\mathbf{A}=f\dfrac{\partial}{\partial x}+g\dfrac{\partial}{\partial y}$, where f and g are *constants*. Clearly

$$\left[a\frac{\partial}{\partial x}+b\frac{\partial}{\partial y},\ f\frac{\partial}{\partial x}+g\frac{\partial}{\partial y}\right]=0,$$

i.e.

$$L_{\mathbf{A}}\mathbf{V}=[\mathbf{A},\mathbf{V}]=0.$$

(b) $\mathbf{V}=\mathbf{y}=y\mathbf{j}=y\dfrac{\partial}{\partial y}$. A translation in the x direction is $a\dfrac{\partial}{\partial x}$:

$$\left[a\frac{\partial}{\partial x},\ y\frac{\partial}{\partial y}\right]=0.$$

(c) $\mathbf{V}=\mathbf{r}=r\mathbf{e}_r=r\left(\cos\phi\dfrac{\partial}{\partial x}+\sin\phi\dfrac{\partial}{\partial y}\right)=x\dfrac{\partial}{\partial x}+y\dfrac{\partial}{\partial y}$. The generator of rotations is $\dfrac{\partial}{\partial\phi}=-y\dfrac{\partial}{\partial x}+x\dfrac{\partial}{\partial y}$. It is easily proved that

$$\left[-y\frac{\partial}{\partial x}+x\frac{\partial}{\partial y},\ x\frac{\partial}{\partial x}+y\frac{\partial}{\partial y}\right]f(x,y)=0,$$

where f is any function of x and y.

Now consider not a vector field, as in (a), (b) or (c) above, but space or space-time, given by a metric. We want to know what symmetries it possesses, so we find the vector fields for which the Lie derivatives of the metric tensor *vanish*. Such symmetries are *isometries*. If the vector field is denoted ξ then the condition for an isometry is $L_\xi g_{\mu\nu}=0$, i.e.

$$\xi_{\mu;\nu}+\xi_{\nu;\mu}=0; \tag{6.132}$$

or equivalently

$$g_{\mu\nu,\lambda}\,\xi^\lambda+g_{\mu\lambda}\,\xi^\lambda{}_{,\nu}+g_{\lambda\nu}\,\xi^\lambda{}_{,\mu}=0. \tag{6.133}$$

Such vectors are called *Killing vectors*. In any given space there is a relation between its Killing vectors and its curvature tensor. For *any* vector ξ we have (see (4.59))

$$\xi_{\mu;\kappa;\lambda}-\xi_{\mu;\lambda;\kappa}=R^\rho{}_{\mu\kappa\lambda}\,\xi_\rho=R_{\rho\mu\kappa\lambda}\,\xi^\rho. \tag{6.134}$$

It follows that

$$\begin{aligned}(\xi_{\mu;\nu}-\xi_{\nu;\mu})_{;\kappa}&+(\xi_{\kappa;\mu}-\xi_{\mu;\kappa})_{;\nu}+(\xi_{\nu;\kappa}-\xi_{\kappa;\nu})_{;\mu}\\&=(R_{\rho\mu\nu\kappa}+R_{\rho\nu\kappa\mu}+R_{\rho\kappa\mu\nu})\,\xi^\rho\\&=0,\end{aligned} \tag{6.135}$$

where we have used (4.34iv). Then the Killing condition (6.132) implies that

$$\xi_{\mu;\nu;\kappa}+\xi_{\kappa;\mu;\nu}+\xi_{\nu;\kappa;\mu}=0,$$

i.e., using (6.132) again

$$\xi_{\mu;\nu;\kappa}-\xi_{\mu;\kappa;\nu}-\xi_{\kappa;\nu;\mu}=0,$$

so, appealing to (6.134)

$$\xi_{\kappa;\nu;\mu} = R^{\rho}_{\ \mu\nu k}\, \xi_{\rho}. \tag{6.136}$$

Spaces endowed with a 'symmetry' will possess corresponding Killing vectors – the more symmetry, the more vectors. Let us find the Killing vectors for a familiar space, R^3,

$$ds^2 = dx^2 + dy^2 + dz^2,$$

in Cartesian coordinates. Its Killing vectors, denoted ξ_i, obey (6.132) and (6.136). In this space $R^i_{\ jkl} = 0$ and in Cartesian coordinates the connection coefficients vanish, so (6.136) gives

$$\xi_{i,k,l} = 0,$$

hence

$$\xi_{i,k} = \beta_{ik} = \text{const} \quad (\beta_{ik} = -\beta_{ki})$$

and

$$\xi_i = \beta_{ik} x^k + \alpha_i.$$

There are three constants β_{ik} and three constants α_i so there are six Killing vectors, corresponding to different choices of α_i and β_{ik}.

(1) $\alpha=(1, 0, 0), \beta_{ik}=0, \xi^{(1)}=(1, 0, 0)$ $\left[\text{i.e. } \xi^{(1)}{}_1=1, \xi^{(1)}{}_2=\xi^{(1)}{}_3=0\right]$
(2) $\alpha=(0, 1, 0), \beta_{ik}=0, \xi^{(2)}=(0, 1, 0)$
(3) $\alpha=(0, 0, 1), \beta_{ik}=0, \xi^{(3)}=(0, 0, 1)$
(4) $\alpha=(0, 0, 0), \beta_{12}=\beta_{13}=0, \beta_{23}=-1=-\beta_{32}$,
$\xi^{(4)}{}_1=0, \xi^{(4)}{}_2=-z, \xi^{(4)}{}_3=y, \xi^{(4)}=(0, -z, y)$
(5) $\alpha=(0, 0, 0), \beta_{13}=1, \beta_{12}=\beta_{23}=0, \xi^{(5)}=(z, 0, -x)$
(6) $\alpha=(0, 0, 0), \beta_{12}=-1, \beta_{13}=\beta_{23}=0, \xi^{(6)}=(-y, x, 0)$.

6.6.1 Groups of motion

From these (Killing) vectors we may construct the operators

$$X_a = \xi^{(a)i} \frac{\partial}{\partial x^i}. \tag{6.137}$$

Continuing with the example of the space R^3 we then have

$$\begin{gathered} X_1 = \frac{\partial}{\partial x}, \quad X_2 = \frac{\partial}{\partial y}, \quad X_3 = \frac{\partial}{\partial z}, \\ X_4 = -z\frac{\partial}{\partial y} + y\frac{\partial}{\partial z}, \quad X_5 = z\frac{\partial}{\partial x} - x\frac{\partial}{\partial z}, \quad X_6 = -y\frac{\partial}{\partial x} + x\frac{\partial}{\partial y}, \end{gathered} \tag{6.138}$$

which obey the relations

$$[X_4, X_5] = -X_6 \ \text{and cyclic perms on} \ (4, 5, 6). \tag{6.139}$$

Relabelling these latter three operators,

$$\begin{aligned} J_1 &= -\mathrm{i}X_4 = -\mathrm{i}\left(y\frac{\partial}{\partial z} - z\frac{\partial}{\partial y}\right), \\ J_2 &= -\mathrm{i}X_5 = -\mathrm{i}\left(z\frac{\partial}{\partial x} - x\frac{\partial}{\partial z}\right), \\ J_3 &= -\mathrm{i}X_6 = -\mathrm{i}\left(x\frac{\partial}{\partial y} - y\frac{\partial}{\partial x}\right), \end{aligned} \tag{6.140}$$

gives the commutation relations

$$[J_1, J_2] = \mathrm{i}J_3 \text{ and cyclic,} \tag{6.141}$$

which are those of SU(2). In addition, relabelling the operators X_1 to X_3,

$$P_1 = \mathrm{i}X_1 = \mathrm{i}\frac{\partial}{\partial x}, \quad P_2 = \mathrm{i}X_2 = \mathrm{i}\frac{\partial}{\partial y}, \quad P_3 = \mathrm{i}X_3 = \mathrm{i}\frac{\partial}{\partial z},$$

gives (i, $k=1, 2, 3$)

$$[P_i,\ P_k] = 0; \tag{6.142}$$

as well as

$$[P_1,\ J_2] = \mathrm{i}P_3 \quad \text{and cyclic,} \quad [P_i,\ J_i] = 0. \tag{6.143}$$

These commutation relations (6.141)–(6.143) will be recognised as those of the groups of rotations and translations, so we find the result that the group of motions (isometry group) of R^3 is the group of rotations and translations in three dimensions. The rotation subgroup is non-abelian, the translation subgroup abelian, and rotations and translations do not commute with each other. Up to a factor $\hbar$ these of course are the generators of rotations and translations in quantum mechanics. This might at first seem a bizarre coincidence, since the subject of General Relativity would seem a million miles from quantum mechanics; but actually the finding is not completely surprising since the whole philosophy of quantum mechanics is that observables are represented by operators which operate directly on a linear vector space. In our present considerations we have the same apparatus of generators of symmetries of a space, but the space concerned is some space or space-time, not a Hilbert space of a quantum system.

As a second example of groups of motion, consider the sphere S^2. It has line element

$$\mathrm{d}s^2 = \mathrm{d}\theta^2 + \sin^2\theta\,\mathrm{d}\phi^2,$$

and with $x^1=\theta$, $x^2=\phi$ we have

$$g_{ik} = \begin{pmatrix} 1 & 0 \\ 0 & \sin^2\theta \end{pmatrix}.$$

Equation (6.133) gives for the Killing vectors ξ^i

$$g_{ik,m}\xi^m + g_{im}\xi^m{}_{,k} + g_{km}\xi^m{}_{,i} = 0.$$

Now take special values for i, k:

$$
\begin{aligned}
i = k = 1: &\quad \xi^1{}_{,1} = 0, && \text{(i)}\\
i = k = 2: &\quad 2\sin\theta\cos\theta\,\xi^1 + 2\sin^2\theta\xi^2{}_{,2} = 0 &&\\
&\quad \Rightarrow\ \cos\theta\,\xi^1 + \sin\theta\,\xi^2{}_{,2} = 0, && \text{(ii)}\\
i = 1,\ k = 2: &\quad \sin^2\theta\,\xi^2{}_{,1} + \xi^1{}_{,2} = 0. && \text{(iii)}
\end{aligned}
$$

These equations yield

$$
\begin{aligned}
\text{(i)} &\quad \to \quad \xi^1 = \mathrm{A}\sin(\phi + \phi_0),\\
\text{(ii)} &\quad \to \quad \xi^2 = \mathrm{A}\cot\theta\cos(\phi + \phi_0) + \mathrm{B}(\theta),\\
\text{(iii)} &\quad \to \quad \mathrm{B}'(\theta)\sin^2\theta = 0 \Rightarrow \mathrm{B} = \text{const.}
\end{aligned}
$$

So the Killing vectors are

$$
\begin{aligned}
\xi &= (\xi^1,\ \xi^2);\\
\xi^1 &= \mathrm{A}\sin(\phi + \phi_0), \quad \xi^2 = \mathrm{A}\cot\theta\cos(\phi + \phi_0) + \mathrm{B}.
\end{aligned}
$$

There are three constants of integration (A, B, ϕ_0), and so three Killing vectors, which we can now find.

$$
\begin{aligned}
\phi_0 = 0,\ \mathrm{A} = 1,\ \mathrm{B} = 0: &\quad \xi^{(1)} = (\sin\phi,\ \cot\theta\cos\phi)\\
\phi_0 = \pi/2,\ \mathrm{A} = 1,\ \mathrm{B} = 0: &\quad \xi^{(2)} = (\cos\phi,\ -\cot\theta\ \cos\phi)\\
\mathrm{A} = 0,\ \mathrm{B} = 1: &\quad \xi^{(3)} = (0,\ 1).
\end{aligned}
$$

We now construct the operators (6.137), to find

$$
\begin{aligned}
X_1 &= \sin\phi\frac{\partial}{\partial\theta} + \cot\theta\cos\phi\frac{\partial}{\partial\phi},\\
X_2 &= \cos\phi\frac{\partial}{\partial\theta} - \cot\phi\sin\phi\frac{\partial}{\partial\phi},\\
X_3 &= \frac{\partial}{\partial\phi},
\end{aligned}
$$

which obey

$$
[X_i,\ X_k] = \varepsilon_{ikm}X_m,
$$

and so generate SU(2): *the group of motions of the sphere S^2 is* SU(2). This 2-dimensional space has three Killing vectors. The plane R^2 also has three Killing vectors; by a simplified version of the calculation above they are

$$
X_1 = \frac{\partial}{\partial x}, \quad X_2 = \frac{\partial}{\partial y}, \quad X_3 = x\frac{\partial}{\partial y} - y\frac{\partial}{\partial x};
$$

two translations and one rotation. Note that

$$\begin{aligned}\text{rotation symmetry} &\rightarrow \text{isotropy}\\ \text{translation symmetry} &\rightarrow \text{homogeneity}\end{aligned}$$

of a space or space-time.

It can be shown that the maximum number of Killing vectors of a space of dimensionality n is $\frac{n(n+1)}{2}$.[11] A space with this maximal number is called *maximally symmetric*. R^2, S^2 and R^3 are maximally symmetric spaces. So is Minkowski space-time, with 10 Killing vectors (see Problem 6.1).

The apparatus of Killing vectors and groups of motion provides a suitable language in which to discuss the symmetry properties of a space. This will be taken up again in the study of cosmological space-times, in Chapter 10. The present chapter ends, however, with a consideration of static and stationary space-times – symmetries with respect to time translation.

6.7 Static and stationary space-times

Consider the Schwarzschild metric (5.37)

$$ds^2 = -\left(1 - \frac{2m}{r}\right) c^2\, dt^2 + \left(1 - \frac{2m}{r}\right)^{-1} dr^2 + r^2(d\theta^2 + \sin^2\theta\, d\phi^2) \tag{6.144}$$

in which $m = \frac{MG}{c^2}$. With $(x^0, x^1, x^2, x^3) = (ct, r, \theta, \phi)$ the metric tensor components are

$$\begin{aligned} &g_{00} = -\left(1 - \frac{2m}{r}\right), \quad g_{11} = \left(1 - \frac{2m}{r}\right)^{-1}, \quad g_{22} = r^2, \quad g_{33} = r^2 \sin^2\theta \\ &g_{\mu\nu} = 0, \quad \mu \neq \nu. \end{aligned} \tag{6.145}$$

In contrast consider what we may call the Lense–Thirring metric (6.48)

$$ds^2 = -\left(1 - \frac{2m}{r}\right) c^2\, dt^2 + \left(1 - \frac{2m}{r}\right)(dx^2 + dy^2 + dz^2) + 2\zeta_i (c\, dt) dx^i \tag{6.146}$$

or, with $(x^0, x^1, x^2, x^3) = (ct, x, y, z)$,

$$\begin{aligned} &g_{00} = -\left(1 - \frac{2m}{r}\right), \quad g_{11} = g_{22} = g_{33} = \left(1 - \frac{2m}{r}\right), \\ &g_{0i} = \zeta_i = \frac{2G}{r^3 c^3}\, \varepsilon_{ikm}\, x^k J^m . \end{aligned} \tag{6.147}$$

The first metric is *static*, the second one *stationary*. Imagine photographing the source of the field. For a star rotating with constant angular velocity the motion of the particles within

[11] See for example Weinberg (1972), Chapter 13.

the source is the same at all times, and the metric shares this property. A naive definition of a *stationary* space-time is that $g_{\mu\nu}$ is independent of t, or

$$\frac{\partial}{\partial t} g_{\mu\nu} = g_{\mu\nu,0} = 0. \qquad (6.148)$$

A *static* space-time has the *further* property

$$g_{0i} = 0. \qquad (6.149)$$

Both of these conditions, however, refer to a particular coordinate system, so we must reformulate them in an invariant, or geometric language. We begin with stationary space-times, and use the language of Killing vectors.

Suppose a space-time possesses a Killing vector $\mathbf{K} = K^\mu \, \mathbf{e}_\mu$ which is *timelike*

$$K^\mu K_\mu < 0. \qquad (6.150)$$

Then there is a coordinate system in which $K^\mu = (1, \mathbf{0})$, i.e. $K^\mu = \delta^\mu{}_0$. Since $\mathbf{K}$ is a Killing vector $L_\mathbf{K}\, g_{\mu\nu} = 0$ or, from (6.126)

$$L_\mathbf{K}\, g_{\mu\nu} = g_{\mu\lambda} K^\lambda{}_{,\nu} + g_{\lambda\nu} K^\lambda{}_{,\mu} + g_{\mu\nu,\lambda} K^\lambda = 0. \qquad (6.151)$$

In the coordinate system $K^\mu = \delta^\mu{}_0$ this becomes

$$g_{\mu\nu,\lambda} K^\lambda = g_{\mu\nu,0} = 0,$$

as in (6.148) above. Hence

A stationary space-time is one which posseses a timelike Killing vector. (6.152)

Moreoever, in the frame $K^\mu = \delta^\mu{}_0$, in a coordinate basis, we have

$$\mathbf{K} = \frac{1}{c}\frac{\partial}{\partial t}. \qquad (6.153)$$

A *static* space-time has a further restriction: $g_{0i} = 0$. We must find an invariant characterisation of this condition. Space-time is a 4-dimensional manifold and we may consider 3-dimensional 'hypersurfaces' in it characterised by t = const, where t is, in some coordinate system, time. These hypersurfaces will be *spacelike*: the invariant separation between any two points ('events') (t, x_1, y_1, z_1) and (t, x_2, y_2, z_2) is spacelike. Let the equation of the hypersurface be $f(x^\mu) = a$ = const (which can be recast as t = const). Then consider two points, at x^μ and $x^\mu + \mathrm{d}x^\mu$, in the hypersurface (see Fig. 6.8). We have

$$a = f(x^\mu) = f(x^\mu + \mathrm{d}x^\mu) = f(x^\mu) + \frac{\partial f}{\partial x^\mu}\mathrm{d}x^\mu,$$

hence $\frac{\partial f}{\partial x^\mu}\mathrm{d}x^\mu = 0$, so the gradient vector

$$n_\mu = \frac{\partial f}{\partial x^\mu} = \partial_\mu f \qquad (6.154)$$

is orthogonal to $\mathrm{d}x^\mu$ (which is here a vector, not a 1-form!)

$$n_\mu \, \mathrm{d}x^\mu = g_{\mu\nu}\, n^\mu \, \mathrm{d}x^\nu = 0, \qquad (6.155)$$

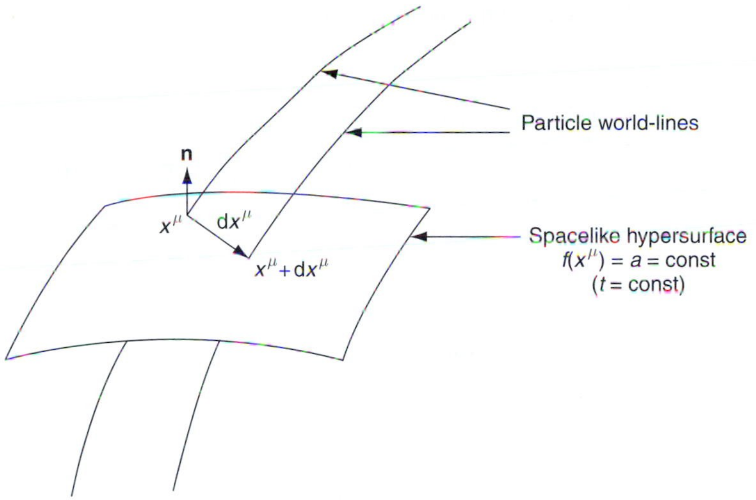

Fig. 6.8 Particle world-lines and a spacelike hypersurface. The vector **n** is normal to the hypersurface.

and hence to the surface, as shown in Fig. 6.8. n^μ is a *normal vector field*. Now our claim is that a space-time is *static* if the Killing vector **K** is proportional to **n** at each point – that the Killing vector is not only timelike but also hypersurface-orthogonal. Such a vector would be of the form

$$\mathbf{X} = \lambda(x)\ \mathbf{n}; \quad X^\mu = \lambda(x)\ n^\mu, \tag{6.156}$$

or, in view of (6.154),

$$X_\mu = \lambda(x)\ \partial_\mu f = \lambda(x) f_{,\mu}. \tag{6.157}$$

To prove this is slightly long. First observe that from this last equation it follows that

$$X_\rho X_{\mu,\,\nu} = \lambda\lambda_{,\nu}\, f_{,\mu}\, f_{,\rho} + \lambda^2 f_{,\mu\nu}\, f_{,\rho} \tag{6.158}$$

and then, taking the completely antisymmetric part of this (see (3.51)) it follows straightforwardly that

$$X_{[\rho}\, X_{\mu,\,\nu]} = 0 \tag{6.159}$$

(which can easily be seen by noting that the first term is symmetric under $\mu \leftrightarrow \rho$ and the second symmetric under $\mu \leftrightarrow \nu$). The reader may also verify that the same condition holds, by virtue of the symmetry under $\mu \leftrightarrow \nu$ of $\Gamma^\sigma{}_{\mu\nu}$, when ordinary derivatives are replaced by covariant ones, i.e.

$$X_{[\rho}\, X_{\mu;\,\nu]} = 0. \tag{6.160}$$

We have now shown that a hypersurface-orthogonal vector satisfies (6.160); in other words, that (6.156) implies (6.160). But the converse does not hold; (6.160) does not imply (6.156). We shall, however, now show that (6.160) together with (6.151) does imply (6.156): a Killing vector field obeying (6.160) is hypersurface orthogonal. Equation (6.160) written out in full is

$$X_\rho X_{\mu;\,\nu} + X_\mu\, X_{\nu;\,\rho} + X_\nu\, X_{\rho;\,\mu} - X_\rho X_{\nu;\,\mu} - X_\nu\, X_{\mu;\,\rho} - X_\mu\, X_{\rho;\,\nu} = 0. \tag{6.161}$$

Now, if $\mathbf{X}$ is a Killing vector

$$X_{\mu;\nu} + X_{\nu;\mu} = 0, \tag{6.162}$$

which reduces (6.161) to

$$X_\rho X_{\mu;\nu} + X_\mu X_{\nu;\rho} + X_\nu X_{\rho;\mu} = 0.$$

Multiplying this by X^ν gives

$$X_\rho X^\nu X_{\mu;\nu} + X_\mu X^\nu X_{\nu;\rho} + (X^2) X_{\rho;\mu} = 0,$$

or, in virtue of (6.162),

$$X_\mu X^\nu X_{\nu;\rho} - X_\rho X^\nu X_{\nu;\mu} + (X^2) X_{\rho;\mu} = 0.$$

Exchanging upper and lower indices and rewriting the last term gives

$$X_\mu X_\nu X^\nu_{\ ;\rho} - X_\rho X_\nu X^\nu_{\ ;\mu} - (X^2) X_{\mu;\rho} = 0.$$

Adding the last two equations gives

$$X_\mu (X^2)_{;\rho} - X_\rho (X^2)_{;\mu} + X^2 (X_{\rho;\mu} - X_{\mu;\rho}) = 0.$$

Replacing covariant derivatives by ordinary ones (a valid operation here) gives

$$X_\mu \partial_\rho (X^2) - X_\rho \partial_\mu (X^2) + X^2 (\partial_\mu X_\rho - \partial_\rho X_\mu) = 0,$$

or

$$X_\mu \partial_\rho (X^2) - X^2 \partial_\rho X_\mu = X_\rho \partial_\mu (X^2) - X^2 \partial_\mu X_\rho. \tag{6.163}$$

Now

$$\partial_\rho \left(\frac{X_\mu}{X^2} \right) = \frac{1}{X^2} \partial_\rho X_\mu - \frac{1}{X^4} X_\mu \partial_\rho (X^2),$$

so dividing (6.163) by X^4 gives

$$\partial_\rho \left(\frac{X_\mu}{X^2} \right) = \partial_\mu \left(\frac{X_\rho}{X^2} \right). \tag{6.164}$$

(Note that $X^2 \neq 0$; $\mathbf{X}$ is a timelike Killing vector.) Hence $\frac{X_\mu}{X^2}$ is the gradient of some scalar field f

$$\frac{X_\mu}{X^2} = \partial_\mu f; \quad X_\mu = X^2 \partial_\mu f,$$

which is (6.157) with $\lambda = X^2$: the hypersurface orthogonality condition. We have shown that a Killing vector field obeying (6.160) is hypersurface orthogonal. We have, finally, to show that this condition of hypersurface orthogonality implies that the space-time is *static*.

We have seen that in a stationary space-time the Killing vector is, in some coordinate system

$$K^\mu = (1, \mathbf{0}),$$

hence $K_\mu = g_{\mu\nu} K^\nu = g_{\mu 0}$ and

$$K^2 = K^\mu K_\mu = K^0 K_0 = g_{00}.$$

We also have the hypersurface orthogonality condition $K_\mu = K^2 \partial_\mu f(x)$. With $\mu = 0$ this gives

$$g_{00} = g_{00}\, \partial_0 f,$$

hence $\partial_0 f = 1$ and

$$f(x) = x^0 + h(x^i).$$

With $\mu = i$, on the other hand, we find

$$g_{i0} = g_{00}\, \partial_i f = g_{00}\, \partial_i h. \tag{6.165}$$

Now perform a coordinate transformation

$$x^0 \to x'^0 = x^0 + h(x^i), \quad x^i \to x'^i = x^i$$

with inverse

$$x^0 = x'^0 - h(x^i), \quad x^i = x'^i.$$

Under this transformation the Killing vector is unchanged:

$$K'^\mu = \frac{\partial x'^\mu}{\partial x^\nu} K^\nu = \frac{\partial x'^\mu}{\partial x^0} \Rightarrow K'^0 = 1, \ K'^i = 0, \ \mathbf{K}' = (1, \mathbf{0}).$$

The metric tensor transforms as follows:

$$g'_{00} = \frac{\partial x^\rho}{\partial x'^0} \frac{\partial x^\sigma}{\partial x'^0} g_{\rho\sigma} = g_{00}, \tag{6.166}$$

and

$$\begin{aligned} g'_{0i} &= \frac{\partial x^\rho}{\partial x'^0} \frac{\partial x^\sigma}{\partial x'^i} g_{\rho\sigma} = \frac{\partial x^\sigma}{\partial x'^i} g_{0\sigma} = \frac{\partial x^0}{\partial x'^i} g_{00} + \frac{\partial x^m}{\partial x'^i} g_{0m} \\ &= (-\partial_i h)\, g_{00} + g_{0i} \\ &= 0, \end{aligned} \tag{6.167}$$

where in the last step we have used (6.165). Equation (6.167) is the condition for a static space-time, as noted in (6.149). We conclude that a space-time is static if it admits a hypersurface-orthogonal (timelike) Killing vector field; and in a static space-time there exists a coordinate system in which $g_{0i} = 0$. Hence there are no mixed terms $\mathrm{d}x^i\, \mathrm{d}t$ in the expression for $\mathrm{d}s^2$.

6.8 Killing vectors and conservation laws

We conclude this chapter by noting a general result, which will be of use in later chapters, that connects symmetries of the metric tensor with conservation laws, for particles moving along a geodesic. The result is straightforward to derive. The Killing vector ξ_μ associated with a metric isometry obeys Equation (6.132)

$$\xi_{\mu;\nu} + \xi_{\nu;\mu} = 0,$$

which we write in the form

$$\nabla_\nu \xi_\mu + \nabla_\mu \xi_\nu = 0, \tag{6.168}$$

using the symbol ∇ for absolute, or covariant, derivative, as introduced in Chapter 3. If, now, a particle moves along a geodesic γ, with tangent vector u^μ, then since the geodesic is an autoparallel curve, we have from (4.6) $\nabla_\nu u^\mu = 0$, hence

$$u^\nu \nabla_\nu u^\mu = 0. \tag{6.169}$$

Then

$$u^\lambda \nabla_\lambda(\boldsymbol{\xi}\cdot\mathbf{u}) = u^\lambda \nabla_\lambda(\xi_\mu u^\mu) = u^\lambda u^\mu \nabla_\lambda \xi_\mu + \xi_\mu u^\lambda \nabla_\lambda u^\mu.$$

These terms vanish separately; the first by virtue of (6.168) and the second from (6.169). We therefore have

$$\nabla_\lambda(\boldsymbol{\xi}\cdot\mathbf{u}) = 0 \Rightarrow \boldsymbol{\xi}\cdot\mathbf{u} = \text{const.} \tag{6.170}$$

This is the conservation law. It is instructive to see how it operates in particular situations. Consider first the case of a particle moving along a straight line (geodesic) in 3-dimensional Euclidean space. The metric is, in Cartesian coordinates,

$$g_{ij} = \begin{pmatrix} 1 & 0 & 0 \\ 0 & 1 & 0 \\ 0 & 0 & 1 \end{pmatrix}.$$

Choosing the direction of motion to be the x axis, the tangent vector is $\frac{\mathrm{d}x}{\mathrm{d}t} = u_x$ and the velocity is $\mathbf{u} = (u_x, 0, 0)$. The metric above is clearly invariant under displacements along the x axis, so the Killing vector is

$$\boldsymbol{\zeta} = \frac{\partial}{\partial x} = (1, 0, 0)$$

and the conservation law (6.170) is

$$\boldsymbol{\zeta}\cdot\mathbf{u} = u_x = \text{const},$$

or equivalently $mu_x = p_x = \text{const}$, conservation of momentum – Newton's first law of motion. As a second example, let us stay in 3-dimensional Euclidean space, but now consider a particle moving in a central field of force. As we know, such particles move in a plane, so we may take that to be the equatorial plane $\theta = \pi/2$ and the metric tensor is, in spherical polars,

$$g_{ij} = \begin{pmatrix} 1 & 0 & 0 \\ 0 & r^2 & 0 \\ 0 & 0 & r^2 \sin^2\theta \end{pmatrix}.$$

This is independent of ϕ; $\dfrac{\partial g_{ij}}{\partial \phi} = 0$, and the Killing vector is

$$\boldsymbol{\xi} = \frac{\partial}{\partial \phi} = (0,\ 0,\ 1)$$

in coordinates $(x^1, x^2, x^3) = (r, \theta, \phi)$. The tangent vector to the geodesic is $u^i = \dfrac{dx^i}{dt}$ and the conserved quantity is (recalling that $\theta = \pi/2$)

$$\boldsymbol{\xi} \cdot \mathbf{u} = g_{ik}\, \xi^i u^k = g_{33}\, u^3 = r^2 \dot{\phi} = l,$$

angular momentum – or rather, angular momentum per unit mass. Then conservation of angular momentum follows from isometry of the metric tensor under change of azimuthal angle ('rotation'), just as, in the previous example, conservation of linear momentum followed from isometry of the metric under translations in space.

It is interesting to note the difference between these arguments, connecting invariance under space translations and rotations with conservation of linear and angular momentum, with what for many physics students is the more standard argument, cast in the language of quantum theory, which connects these fundamental ideas. The above argument is couched in purely geometric language, connecting a symmetry of the metric with a conservation law for a particle moving along a geodesic. In the quantum theory approach the connection is made between a symmetry in coordinate space and the generator of this symmetry as an operator in Hilbert space; for example the momentum operator in Hilbert space is the generator of translations in coordinate space.

Further reading

The Lense–Thirring effect originates in Thirring & Lense (1918), of which an English translation appears in Mashhoon, Hehl & Theiss (1984). A good account of the precession of gyroscopes appears (in French) in Tonnelat (1964), and good accounts of Lense–Thirring precession are to be found in Ciufolini & Wheeler (1995) and Lämmerzahl & Neugebauer (2001). Schiff's suggestion of searching for the Lense-Thirring effect in a gyroscope experiment appears in Schiff (1960); see also Schiff (1939).

Detailed descriptions of the Gravity Probe B experiment appear in Everitt *et al.* (2001), in Ciufolini & Wheeler (1995) and in Fairbank *et al.* (1988). This last reference comprises seven separate articles on The Stanford Relativity Gyroscope Experiment by: C. W. F. Everitt (History and overview), J. A. Lipa & G. M. Keiser (Gyroscope development), J. T. Anderson (London moment readout of the gyroscope), J. P. Turneaure, E. A. Cornell, P. D. Levine & J. A. Lipa (Ultrahigh vacuum techniques for the experiment), R. A. Van Patten (Flight gyro suspension system), J. V. Breakwell (Correction to the predicted geodetic precession of the gyroscope resulting from the Earth's oblateness) and D. B. DeBra (Translation and orientation control). For up-to-date information on the Gravity Probe B experiment consult http://www.gravityprobeb.com, or http://einstein.stanford.edu. Further arguments for and against Mach's principle in the context of the Lense–Thirring effect can be found in Rindler (1994), Ciufolini (1995) and Bondi & Samuel (1997).

A thorough discussion of gravitomagnetism appears in Ciufolini & Wheeler (1995), Chapter 6, and a more recent review is Ruggiero & Tartaglia (2002). See also Mashhoon (1993). The gravitomagnetic clock effect is discussed by Bonnor & Steadman (1999), Mashhoon & Santos (2000), Mashhoon *et al.* (2001), Ruggiero & Tartaglia (2002) and Tartaglia (2002).

Good expositions of Lie derivatives and Killing vectors may be found in Robertson & Noonan (1968), Chapter 13 and Schutz (1980), Chapter 3. More mathematical accounts may be found in Lichnérowicz (1958), Wald (1984), Appendix C and Martin (1991), Section 7.7.

An elegant account of static and stationary space-times may be found in Rindler (2001), Chapter 9.

Problems

6.1 Show that

$$\nabla \times \left\{ \frac{\mathbf{r}}{r^3} \times \mathbf{J} \right\} = \frac{\mathbf{J}}{r^3} - \frac{3(\mathbf{J} \cdot \mathbf{r})\mathbf{r}}{r^5}$$

as in (6.60) above, and

$$\mathbf{v} \times \nabla \left(\frac{1}{r} \right) = \frac{1}{r^3} \mathbf{r} \times \mathbf{v}.$$

6.2 Prove Equation (6.93).

6.3 Show that the group of motions of Minkowski space-time is the inhomogeneous Lorentz group (= Poincaré group).

7 Gravitational collapse and black holes

Black is black. It's mysterious.

Ellsworth Kelly
American artist
(Television interview with Nicholas Glass, 20 March 2006)

A striking feature of the vacuum Schwarzschild metric, as seen in Chapter 5, is the surface $r = 2m$. In the case of the Sun, $2m = 2.95$ km and this surface is *inside* the Sun, where in any case the vacuum field equations do not apply. This might encourage the belief that any problems posed by the Schwarzschild surface were unreal and unphysical. We also noted in Chapter 5 that, if there were any circumstances in which the Schwarzschild surface were *outside* the material of a star, the time taken for an object falling radially into the star to reach this surface would be infinite, so the region $r < 2m$ is 'out of bounds', at least in this coordinate system. These considerations suggest two questions: first, are there any circumstances in which 'stars' might actually be confined to the region $r < 2m$; and second, if so, how do we investigate this region mathematically?

The twentieth century has seen spectacular progress in astrophysics. From the work of Bethe[1] onwards we know that star light is produced by nuclear fusion reactions, the simplest of which is

$$\mathrm{p} + \mathrm{n} \rightarrow \mathrm{d} + \gamma.$$

The deuteron is a proton–neutron bound state and consequently has a smaller mass than the combined masses of these particles, the 'missing' mass appearing as the energy of the photon γ. The pressure exerted by these photons, both directly, as radiation pressure, and indirectly, as thermal pressure, is what keeps the star in equilibrium, balancing the inward gravitational pull. The reaction above is only the first of many; after deuterium, heavier elements in the periodic table are produced – lithium, beryllium, and so on, through carbon to iron. At each stage the nucleus on the right hand side has a greater binding energy per nucleon than that on the left, so a mass defect is converted into energy, emitted as a photon of light. When, however, the centre of a star becomes dominated by iron, the fusion reactions producing heavier elements are no longer exothermic. Iron has the greatest binding energy per nucleon of all the chemical elements so the subsequent fusion reactions do not emit light. The star begins to cool, the thermal pressure to drop and, as first realised by Oppenheimer, Snyder and Volkoff,[2] this will result in gravitational collapse at the end-point of thermonuclear evolution. How far does the star collapse? There are two states of possible equilibrium, in which the gravitational inward

[1] Bethe (1939).
[2] Oppenheimer & Snyder (1939), Oppenheimer & Volkoff (1939).

pressure is balanced by the *Fermi pressure*, of either electrons or neutrons; these configurations are known respectively as white dwarfs and neutron stars. In retrospect, this crucial role played by quantum theory in the physics of stars is one of the most surprising aspects of the whole subject. Detailed consideration of these equilibrium states enabled Chandrasekhar to show that for these configurations there is a limiting mass – the so-called Chandrasekhar mass – above which equilibrium is no longer possible.[3] It was in the 1960s that these considerations began to be taken seriously: the discovery of quasars, and especially of pulsars, played a key role in persuading physicists that a proper consideration of gravitational collapse was crucial in understanding these objects, and from that time General Relativity could be said to have entered physics proper, having previously been largely a pursuit of applied mathematicians. A consequence of the Chandrasekhar limit is that for heavier collapsing stars (for example the collapse of a star with a mass 10 times that of the Sun) there is *no equilibrium configuration*. There is simply continued collapse, which will inevitably result in the stellar material – or more properly some fraction of it – being confined eventually to a region $r<2m$. This state was christened by Wheeler a 'black hole'.

This chapter begins with a consideration of the *internal* Schwarzschild solution – the space-time metric inside a spherically symmetric and static star – and goes on to treat white dwarfs and neutron stars. We continue on to the investigation of rotating black holes (the Kerr solution) and finish with brief accounts of the thermodynamics of black holes, Hawking radiation and some associated matters.[4]

7.1 The interior Schwarzschild solution and the Tolman–Oppenheimer–Volkoff equation

We want to set up the general relativistic equations for computing the pressure and density of matter in a spherically symmetric, static star. In Chapter 5 we found the Schwarzschild solution that describes the space-time *outside* such a star. We are now concerned with a solution for the stellar *interior*.[5] Since we are assuming spherical symmetry the metric will be of the form

$$\begin{gathered} \mathrm{d}s^2 = -\mathrm{e}^{2\nu(r)}c^2\,\mathrm{d}t^2 + \mathrm{e}^{2\lambda(r)}\,\mathrm{d}r^2 + r^2(\mathrm{d}\theta^2 + \sin^2\theta\,\mathrm{d}\phi^2), \\ g_{\mu\nu} = 0\,(\mu \neq \nu). \end{gathered} \tag{7.1}$$

The two unknown functions $\nu(r)$ and $\lambda(r)$ will not turn out to be the same as in Chapter 5, since they must satisfy the field equations inside the star. For this we need the energy-momentum tensor for the stellar material, which is taken to be a *perfect fluid*. Let us recall that the energy-momentum tensor for *dust* is given by (5.10):

$$T^{\mu\nu}(x) = \rho_0(x)\,u^\mu(x)\,u^\nu(x),$$

[3] Chandrasekhar (1931a, b).

[4] I should like to record here my indebtedness to the late Dan Martin, whose lecture notes (1978, Glasgow University, unpublished) were a great help in preparing this chapter.

[5] Schwarzschild (1916b).

where u^{μ} is the 4-velocity of the dust. For a *perfect fluid*, in which both a density ρ and a pressure p are defined, the energy-momentum tensor for the fluid at rest is

$$\tilde{T}^{\mu\nu} = \begin{pmatrix} \rho & 0 & 0 & 0 \\ 0 & \frac{p}{c^2} & 0 & 0 \\ 0 & 0 & \frac{p}{c^2} & 0 \\ 0 & 0 & 0 & \frac{p}{c^2} \end{pmatrix};$$

or

$$\tilde{T}^{00} = \rho, \quad \tilde{T}^{ij} = \frac{p}{c^2}\delta^{ij}, \quad \tilde{T}^{\mu\nu} = 0, \quad \mu \neq \nu. \tag{7.2}$$

We need a covariant expression for this tensor, for use in the field equations. We start by going to a moving frame, by using the Lorentz boost transformation (2.26) (but with $v \to -v$)

$$x'^{i} = x^{i} + (\gamma - 1)\frac{x^{k} v_k}{v^2} v^{i} + \gamma \frac{v^{i}}{c} ct = \left(\delta^{i}{}_{k} + \frac{\gamma - 1}{v^2} v^{i} v_k\right) x^{k} + \gamma \frac{v^{i}}{c} x^{0},$$

$$ct' = \gamma c t + \gamma \frac{v_k}{c} x^{k},$$

from which we may read off the elements of the Lorentz matrix

$$\Lambda^{i}{}_{k} = \delta^{i}{}_{k} + \frac{\gamma - 1}{v^2} v^{i} v_k, \quad \Lambda^{i}{}_{0} = \gamma \frac{v^{i}}{c}, \quad \Lambda^{0}{}_{0} = \gamma, \tag{7.3}$$

and thereby calculate the elements of $T^{\mu\nu}$ in the moving frame:

$$\begin{aligned} T^{ij} &= \Lambda^{i}{}_{\mu} \Lambda^{j}{}_{\nu} \tilde{T}^{\mu\nu} \\ &= \Lambda^{i}{}_{k} \Lambda^{j}{}_{m} \tilde{T}^{km} + \Lambda^{i}{}_{0} \Lambda^{j}{}_{0} \tilde{T}^{00} \\ &= \left(\delta^{i}{}_{k} + \frac{\gamma - 1}{v^2} v^{i} v_k\right)\left(\delta^{j}{}_{m} + \frac{\gamma - 1}{v^2} v^{j} v_m\right) \frac{p}{c^2} \delta^{jm} + \gamma^2 \frac{v^{i} v^{j}}{c^2} \rho \\ &= \frac{p}{c^2}\delta^{ij} + \frac{v^{i} v^{j} p}{c^2 v^2}\left[2(\gamma - 1) + (\gamma - 1)^2\right] + \gamma^2 \frac{v^{i} v^{j}}{c^2} \rho \\ &= \frac{p}{c^2}\delta^{ij} + \frac{v^{i} v^{j}}{c^2}\gamma^2 \left[\frac{p}{c^2} + \rho\right]; \end{aligned} \tag{7.4}$$

$$\begin{aligned} T^{i0} &= \Lambda^{i}{}_{\mu} \Lambda^{0}{}_{\nu} \tilde{T}^{\mu\nu} \\ &= \Lambda^{i}{}_{k} \Lambda^{0}{}_{m} \tilde{T}^{km} + \Lambda^{i}{}_{0} \Lambda^{0}{}_{0} \tilde{T}^{00} \\ &= \left(\delta^{i}{}_{j} + \frac{\gamma - 1}{v^2} v^{i} v_j\right) \gamma \frac{v^{m}}{c} \frac{p}{c^2} \delta^{jm} + \gamma^2 \frac{v^{i}}{c} \rho \\ &= \gamma^2 \frac{v^{i}}{c}\left[\frac{p}{c^2} + \rho\right]; \end{aligned} \tag{7.5}$$

$$\begin{aligned}T^{00} &= \Lambda^0{}_\mu \Lambda^0{}_\nu \tilde{T}^{\mu\nu} \\ &= \Lambda^0{}_i \Lambda^0{}_j \tilde{T}^{ij} + (\Lambda^0{}_0)^2 \tilde{T}^{00} \\ &= \gamma^2 \frac{v^i v^j}{c^2} \frac{p}{c^2} \delta^{ij} + \gamma^2 \rho \\ &= (\gamma^2 - 1) \frac{p}{c^2} + \gamma^2 \rho.\end{aligned} \tag{7.6}$$

Equations (7.4) to (7.6) may be expressed in the form

$$T^{\mu\nu} = \frac{p}{c^2} \eta^{\mu\nu} + \left(\rho + \frac{p}{c^2}\right) u^\mu u^\nu, \tag{7.7}$$

with (cf. (5.11))

$$u^\mu = \frac{1}{c} \frac{\mathrm{d}x^\mu}{\mathrm{d}\tau}; \quad u^0 = \frac{\mathrm{d}t}{\mathrm{d}\tau} = \gamma, \quad u^i = \gamma \frac{v^i}{c},$$

as may easily be checked. Having found an expression for the energy-momentum tensor in a moving frame, it is now straightforward to make this generally covariant; then (7.7) becomes

$$T^{\mu\nu} = \frac{p}{c^2} g^{\mu\nu} + \left(\rho + \frac{p}{c^2}\right) u^\mu u^\nu, \tag{7.8}$$

with

$$g^{\mu\nu} u_\mu u_\nu = -1 = g_{\mu\nu} u^\mu u^\nu, \tag{7.9}$$

and from (7.1)

$$g_{\mu\nu} = \begin{pmatrix} -\mathrm{e}^{2\nu} & 0 & 0 & 0 \\ 0 & \mathrm{e}^{2\lambda} & 0 & 0 \\ 0 & 0 & r^2 & 0 \\ 0 & 0 & 0 & r^2 \sin^2\theta \end{pmatrix}, \quad g^{\mu\nu} = \begin{pmatrix} -\mathrm{e}^{-2\nu} & 0 & 0 & 0 \\ 0 & \mathrm{e}^{-2\lambda} & 0 & 0 \\ 0 & 0 & \frac{1}{r^2} & 0 \\ 0 & 0 & 0 & \dfrac{1}{r^2 \sin^2\theta} \end{pmatrix}. \tag{7.10}$$

In our star, which is a fluid at rest, we have

$$\begin{aligned} &u_r = u_1 = 0, \quad u_\theta = u_2 = 0, \quad u_\phi = u_3 = 0; \\ &u_0 = -(-g^{00})^{-1/2} = -\mathrm{e}^{\nu(r)}. \end{aligned} \tag{7.11}$$

Then

$$\begin{aligned} T^{00} &= \frac{p}{c^2} g^{00} + \left(\rho + \frac{p}{c^2}\right)(u^0)^2 \\ &= \frac{p}{c^2} g^{00} + \left(\rho + \frac{p}{c^2}\right)(-g^{00}) \\ &= -\rho g^{00} = \rho e^{-2\nu}, \\ T^{11} &= \frac{p}{c^2} g^{11} = \frac{p}{c^2} e^{-2\lambda}, \text{ etc.;} \end{aligned}$$

so

$$T^{\mu\nu} = \begin{pmatrix} \rho e^{-2\nu} & 0 & 0 & 0 \\ 0 & \frac{p}{c^2} e^{-2\lambda} & 0 & 0 \\ 0 & 0 & \frac{p}{c^2}\frac{1}{r^2} & 0 \\ 0 & 0 & 0 & \frac{p}{c^2}\frac{1}{r^2 \sin^2\theta} \end{pmatrix}. \tag{7.12}$$

We write the field equations (5.24) in the form

$$R^\mu{}_\nu - {}^1\!/_2\, \delta^\mu{}_\nu R = \frac{8\pi G}{c^2} T^\mu{}_\nu. \tag{7.13}$$

The components of the mixed tensor $T^\mu{}_\nu$ are easily found;

$$T^0{}_0 = g_{00} T^{00} = -\rho; \quad T^1{}_1 = g_{11} T^{11} = \frac{p}{c^2}, \text{ etc.},$$

so

$$T^\mu{}_\nu = \begin{pmatrix} -\rho & 0 & 0 & 0 \\ 0 & \frac{p}{c^2} & 0 & 0 \\ 0 & 0 & \frac{p}{c^2} & 0 \\ 0 & 0 & 0 & \frac{p}{c^2} \end{pmatrix}. \tag{7.14}$$

The components of the Ricci tensor $R_{\mu\nu}$ are given by (5.35). Then

$$R^0{}_0 = g^{00} R_{00} = -e^{-2\nu} R_{00} = e^{-2\lambda}\left(-\nu'' - \nu'^2 + \nu'\lambda' - \frac{2\nu'}{r}\right), \tag{7.15}$$

$$R^1{}_1 = g^{11} R_{11} = e^{-2\lambda}\left(-\nu'' - \nu'^2 + \nu'\lambda' + \frac{2\lambda'}{r}\right), \tag{7.16}$$

$$R^2{}_2 = e^{-2\lambda}\left(-\frac{1}{r^2} - \frac{\nu'}{r} + \frac{\lambda'}{r}\right) + \frac{1}{r^2}. \tag{7.17}$$

It turns out that $R^3{}_3 = R^2{}_2$: also $T^3{}_3 = T^2{}_2$, so the field equations for these indices are the same. The curvature scalar is

$$\begin{aligned} R = R^\mu{}_\mu &= R^0{}_0 + R^1{}_1 + R^2{}_2 + R^3{}_3 \\ &= e^{-2\lambda}\left(-2\nu'' - 2\nu'^2 + 2\nu'\lambda' - \frac{4}{r}(\nu' - \lambda') - \frac{2}{r^2}\right) + \frac{2}{r^2}. \end{aligned} \tag{7.18}$$

Then

$$R^0{}_0 - {}^1\!/_2 R = -e^{-2\lambda}\left(\frac{2\lambda'}{r} - \frac{1}{r^2}\right) - \frac{1}{r^2}$$

and the field equation (7.13) with $\mu = \nu = 0$ gives

$$e^{-2\lambda}\left(\frac{2\lambda'}{r}-\frac{1}{r^2}\right)+\frac{1}{r^2}=\frac{8\pi G}{c^2\rho}. \tag{7.19}$$

Similarly

$$R^1{}_1-{}^1\!/_2 R=-e^{-2\lambda}\left(\frac{2\nu'}{r}-\frac{1}{r^2}\right)-\frac{1}{r^2}$$

and the field equation with $\mu=\nu=1$ gives

$$e^{-2\lambda}\left(\frac{2\nu'}{r}+\frac{1}{r^2}\right)-\frac{1}{r^2}=\frac{8\pi G}{c^2}\frac{p}{c^2}. \tag{7.20}$$

Likewise the $\mu=\nu=2$ and $\mu=\nu=3$ field equations both give

$$e^{-2\lambda}\left(\nu''+\nu'^2-\nu'\lambda'+\frac{\nu'}{r}-\frac{\lambda'}{r}\right)=\frac{8\pi G}{c^2}\frac{p}{c^2}. \tag{7.21}$$

The three equations (7.19)–(7.21) may now be used to find the functions $\nu(r)$ and $\lambda(r)$, as long as some additional information about the pressure p and matter density ρ is available. To tackle the problem in full generality a relation between these quantities must be assumed; that is, an *equation of state*, normally taken to be of the form

$$\frac{p}{c^2}=K\rho^\gamma. \tag{7.22}$$

This is called a *polytropic* equation of state, with γ the polytropic exponent (and K a constant). We shall here adopt a simpler procedure and assume that the fluid of the star is *incompressible*: $\rho=$ const. This is not such a bad assumption for light stars such as the Sun, and it has the additional virtue of simplicity – as well as being the assumption made by Schwarzschild himself. Since

$$\frac{d}{dr}\left(r\,e^{-2\lambda(r)}\right)=e^{-2\lambda}-2r\lambda'\,e^{-2\lambda},$$

Equation (7.19) may be written in the form

$$\frac{d}{dr}\left(r\,e^{-2\lambda}\right)=1-\frac{8\pi G\rho}{c^2}r^2, \tag{7.23}$$

which under the assumption $\rho=$ const may be integrated to give

$$e^{-2\lambda}=1-\frac{8\pi G\rho}{3c^2}r^2+\frac{C}{r},$$

where C is a constant. To eliminate the singularity at $r=0$ put $C=0$, then

$$e^{-2\lambda}=1-\frac{8\pi G\rho}{3c^2}r^2\equiv 1-Ar^2, \tag{7.24}$$

with

$$A=\frac{8\pi G\rho}{3c^2}. \tag{7.25}$$

We now have a solution for $g_{11} = e^{2\lambda}$. To proceed further, differentiate Equation (7.20) and substitute for ν'' from (7.21), yielding, after a bit of algebra

$$\frac{8\pi G}{c^2}\frac{p'}{c^2} = -\frac{2e^{-2\lambda}}{r}\nu'(\nu' + \lambda'), \tag{7.26}$$

where $p' = \frac{dp}{dr}$: note that p is not constant, even though ρ is. Adding Equations (7.19) and (7.20) gives

$$\frac{8\pi G}{c^2}\left(\rho + \frac{p}{c^2}\right) = \frac{2e^{-2\lambda}}{r}(\nu' + \lambda'). \tag{7.27}$$

These last two equations give

$$\frac{p'}{c^2} = -\nu'\left(\rho + \frac{p}{c^2}\right),$$

which may be integrated to give

$$\rho + \frac{p(r)}{c^2} = D\,e^{-\nu(r)}, \tag{7.28}$$

where D is a constant. This may be rearranged, using (7.24), to give

$$\frac{8\pi G}{c^2}Dr = 2\nu' e^{\nu}(1 - Ar^2) + 2Ar\,e^{\nu}. \tag{7.29}$$

To solve (7.29) put

$$e^{\nu(r)} = \gamma(r),$$

then $\gamma' = \nu'\,e^{\nu}$ and (7.29) becomes

$$2\gamma'(1 - Ar^2) + 2Ar\,\gamma(r) = \frac{8\pi G}{c^2}Dr. \tag{7.30}$$

This is an inhomogeneous differential equation, which has a particular solution

$$\gamma = \frac{4\pi G}{c^2}\frac{D}{A}.$$

The corresponding homogeneous equation

$$\frac{du}{dr}(1 - Ar^2) + Ar\,u(r) = 0$$

has the solution

$$u(r) = B(1 - Ar^2)^{1/2},$$

(B = const), hence (7.29) has the solution

$$e^{\nu(r)} = \gamma(r) = \frac{4\pi G}{c^2}\frac{D}{A} - B(1 - Ar^2)^{1/2} \equiv C - B(1 - Ar^2)^{1/2}, \tag{7.31}$$

where C is a further constant, or

$$-g_{00} = e^{2\nu(r)} = \left[C - B\left(1 - A\,r^2\right)^{1/2}\right]^2.$$

This equation, together with (7.24), gives the interior Schwarzschild solution in the form

$$ds^2 = -\left[C - B(1 - Ar^2)^{1/2}\right]^2 c^2\,dt^2 + \frac{dr^2}{1 - Ar^2} + r^2\,d\Omega^2. \tag{7.32}$$

This metric contains two constants B and C – the quantity A is essentially the density – see (7.25). These constants may now be found from two requirements; firstly, that $p = 0$ at $r = R$, the radius of the star (this is surely a reasonable definition of radius!), and secondly that (7.32) matches the Schwarzschild vacuum solution (5.37) at $r = R$.

For the first requirement, put $p = 0$ at $r = R$ in (7.28) to give

$$\rho = De^{-\nu(R)} \Rightarrow \rho\,e^{\nu(R)} = D$$

and hence, from (7.31),

$$\rho\left[\frac{4\pi G}{c^2}\frac{D}{A} - B(1 - AR^2)^{1/2}\right] = D.$$

Substitute for A in the first term above, from (7.25), to give

$$D = 2\rho B(1 - AR^2)^{1/2}$$

and then (7.31) gives

$$e^{\nu(r)} = B\left[3(1 - AR^2)^{1/2} - (1 - Ar^2)^{1/2}\right],$$

so the line element becomes

$$ds^2 = -B^2\left[3(1 - AR^2)^{1/2} - (1 - Ar^2)^{1/2}\right]^2 c^2\,dt^2 + \frac{dr^2}{1 - Ar^2} + r^2\,d\Omega^2, \tag{7.33}$$

with A given by (7.25). This line element now features the constant B, which is found from the requirement that (7.33) match the line element (5.37) corresponding to the exterior solution, at $r = R$. This vacuum line element is

$$ds^2 = -\left(1 - \frac{2m}{r}\right)c^2\,dt^2 + \left(1 - \frac{2m}{r}\right)^{-1} dr^2 + r^2\,d\Omega^2 \tag{7.34}$$

with

$$m = \frac{MG}{c^2}.$$

Matching g_{11} at $r = R$ gives

$$\frac{2m}{R} = AR^2 \Rightarrow R = \left(\frac{3M}{4\pi\rho}\right)^{1/3}, \tag{7.35}$$

on substituting from (7.25), which is of course the expected expression for the radius of a star with mass M and constant density. Then matching g_{00} gives

$$B^2\left[3(1-AR^2)^{1/2}-(1-AR^2)^{1/2}\right]^2=1-\frac{2m}{R}=1-AR^2,$$

using (7.35); and this gives

$$B=\frac{1}{2}.$$

Finally, then, the internal Schwarzschild solution for the space-time metric inside a spherically symmetric star of radius R and density ρ is

$$\mathrm{d}s^2=-\left[\frac{3}{2}(1-AR^2)^{1/2}-\frac{1}{2}(1-Ar^2)^{1/2}\right]^2c^2\,\mathrm{d}t^2+\frac{\mathrm{d}r^2}{1-Ar^2}+r^2\,\mathrm{d}\Omega^2, \tag{7.36}$$

with

$$A=\frac{8\pi G\rho}{3c^2}.$$

The constant A evidently has the dimensions of $(\text{length})^{-2}$. For the Sun

$$A=\frac{8\pi G\rho}{3c^2}=\frac{c^2R_S^3}{2GM_S}=(3.38\times10^{11}\ \mathrm{m})^{-2}.$$

Since the actual radius of the Sun is 6.96×10^8 m,

$$AR^2\ll1\quad(\text{Sun}) \tag{7.37}$$

and g_{00} and g_{11} differ little from unity, so the Schwarzschild metric inside the Sun is, like that outside it, close to the Minkowski metric.

In understanding the nature of gravitational collapse in General Relativity, and to perform explicit calculations, the Tolman–Oppenheimer–Volkoff equation has played a crucial role.[6] We close this section by giving a simplified derivation of it, restricting ourselves to the case of constant density, as assumed above. We start from the condition for energy-momentum conservation

$$T^{\mu\nu}{}_{;\nu}=0. \tag{7.38}$$

From (7.8) we have

$$T^{\mu\nu}{}_{;\nu}=\frac{1}{c^2}\frac{\partial p}{\partial x^\nu}g^{\mu\nu}+\left[\left(\rho+\frac{p}{c^2}\right)u^\mu u^\nu\right]_{;\nu} \tag{7.39}$$

since $g^{\mu\nu}{}_{;\nu}=0$. For a general (2, 0) tensor $S^{\mu\nu}$ we have

$$\begin{aligned}S^{\mu\nu}{}_{;\nu}&=S^{\mu\nu}{}_{,\nu}+\Gamma^\mu{}_{\lambda\nu}S^{\lambda\nu}+\Gamma^\nu_{\lambda\nu}S^{\mu\lambda}\\&=S^{\mu\nu}{}_{,\nu}+\frac{1}{\sqrt{-g}}\partial_\lambda\left(\sqrt{-g}\right)S^{\mu\lambda}+\Gamma^\mu{}_{\lambda\nu}S^{\lambda\nu}\\&=\frac{1}{\sqrt{-g}}\partial_\lambda\left(\sqrt{-g}\,S^{\mu\nu}\right)+\Gamma^\mu{}_{\lambda\nu}S^{\lambda\nu},\end{aligned} \tag{7.40}$$

[6] Tolman (1939), Oppenheimer & Volkoff (1939).

where in the second line we have used Equation (3.212). Applying this to the second term in (7.39) gives

$$T^{\mu\nu}{}_{;\nu} = \frac{1}{c^2}\frac{\partial p}{\partial x^\nu} g^{\mu\nu} + \frac{1}{\sqrt{-g}}\partial_\nu\left[\sqrt{-g}\left(\rho + \frac{p}{c^2}\right)u^\mu u^\nu\right] + \Gamma^\mu{}_{\lambda\nu}\left(\rho + \frac{p}{c^2}\right)u^\nu u^\lambda.$$

Referring to the second term on the right hand side above, $u^\nu = 0$ unless $\nu = 0$ (see (7.11)), but the condition of equilibrium requires that $\partial_0[\ldots] = 0$, so this second term vanishes. Multiplying by $g_{\mu\lambda}$ then gives, in view of (7.38)

$$\frac{1}{c^2}\frac{\partial p}{\partial x^\nu}\delta^\nu{}_\lambda = -\Gamma^\mu{}_{00} g_{\mu\lambda}\left(\rho + \frac{p}{c^2}\right)(u^0)^2.$$

From (7.11), however, we see that $(u^0)^2 = -(g_{00})^{-1}$; and since

$$\Gamma^\mu{}_{00} = -{}^1/{}_2 g^{\mu\sigma}(\partial_\sigma g_{00})$$

we obtain

$$\frac{1}{c^2}\frac{\partial p}{\partial x^\lambda} = -{}^1/{}_2(\partial_\lambda g_{00})(g_{00})^{-1}\left(\rho + \frac{p}{c^2}\right). \tag{7.41}$$

On the other hand,

$$\partial_\lambda \ln\sqrt{-g_{00}} = \frac{1}{2g_{00}}\partial_\lambda g_{00},$$

so (7.41) gives

$$\frac{1}{c^2}\frac{\partial p}{\partial x^\lambda} = -\left(\rho + \frac{p}{c^2}\right)\partial_\lambda \ln\sqrt{-g_{00}}. \tag{7.42}$$

Putting $g_{00} = -\mathrm{e}^{2\nu(r)}$, the $\lambda = 1$ component of this equation gives

$$\frac{\mathrm{d}p}{\mathrm{d}r} = -(\rho c^2 + p)\nu'. \tag{7.43}$$

We shall now find an expression for $\frac{\mathrm{d}p}{\mathrm{d}r}$ in the case of a star of constant density. From Equations (7.19) and (7.20) we have

$$\mathrm{e}^{-2\lambda}\left(\frac{2\lambda'}{r} - \frac{2\nu'}{r} - \frac{2}{r^2}\right) + \frac{2}{r^2} = \frac{8\pi G}{c^2}\left(\rho - \frac{p}{c^2}\right),$$

or, multiplying by ½r^2,

$$1 - \mathrm{e}^{-2\lambda}(1 + r\nu') + r\lambda'\mathrm{e}^{-2\lambda} = \frac{4\pi G}{c^2}r^2\left(\rho - \frac{p}{c^2}\right). \tag{7.44}$$

Now write (7.19) as

$$\mathrm{e}^{-2\lambda}.\frac{\lambda'}{r} = \frac{1}{2}\left[\frac{\mathrm{e}^{-2\lambda}}{r^2} - \frac{1}{r^2} + \frac{8\pi G\rho}{c^2}\right],$$

hence

$$r\lambda' e^{-2\lambda} = \frac{1}{2}\left[e^{-2\lambda} - 1 + \frac{8\pi G\rho r^2}{c^2}\right] = \frac{4\pi G\rho r^2}{c^2} - \frac{m}{r}, \tag{7.45}$$

where (7.34) has been used in the last step. Now substitute (7.45) into the appropriate term in (7.44), again making use of (7.34), and also (7.43):

$$1 - \left(1 - \frac{2m}{r}\right)\left(1 - \frac{r p'}{p + \rho c^2}\right) + \frac{4\pi G\rho r^2}{c^2} - \frac{m}{r} = \frac{4\pi G}{c^2} r^2 \left(\rho - \frac{p}{c^2}\right).$$

This equation gives, on rearrangement, and on substituting $m = \frac{MG}{c^2} = \frac{4\pi G}{3c^2} r^3 \rho$,

$$\frac{\mathrm{d}p}{\mathrm{d}r} = -\frac{4\pi G\left(\rho + \frac{p}{c^2}\right)\left(\frac{\rho}{3} + \frac{p}{c^2}\right) r^2}{(r - 2m)}. \tag{7.46}$$

This is the Tolman–Oppenheimer–Volkoff (TOV) equation. It has here been derived under the assumption of constant density ρ, but actually holds in the general case where ρ depends on r. It is clear from this equation that a higher density ρ gives rise to a larger pressure gradient, which then aids the collapse of the star. This is, of course, completely consistent with intuition. But the interesting feature of the TOV equation is that the same is true of pressure. Like mass density, *pressure also aids collapse*: it gives rise to a *gravitational* force which contributes to collapse. This is a characteristic of General Relativity not shared by Newtonian gravity. In Newtonian theory pressure has no gravitational effect; and of course in *any* theory, pressure gives rise to *non-gravitational* forces which oppose collapse (successfully or not): for example through the gas laws, or Fermi pressure.

To solve the TOV equation in the general case, an equation of state relating ρ and p is needed, and this is taken to be a polytropic equation like (7.22) above. It is then clear that the TOV equation is non-linear; it may be solved numerically, giving the so-called Lane–Emden solutions. For more details of this the reader is referred to the literature. It is worth noting, finally, that in the non-relativistic limit $r \gg 2m$, $\rho \gg p/c^2$, Equation (7.46) becomes

$$\frac{\mathrm{d}p}{\mathrm{d}r} = -\frac{4\pi G}{3}\rho^2 r, \tag{7.47}$$

as expected – see (7.80) below.

7.2 Energy density and binding energy

> On a visit to Einstein in Princeton … [George Gamow] casually mentioned, when they were out walking, that … Pascual Jordan had realised that a star might be made out of nothing, since … its negative gravitational energy is numerically equal to its positive rest-mass energy. Einstein stopped in his tracks and, since we were crossing a street, several cars had to stop to avoid running us down.[7]

[7] Gamow (1970).

A star is of course a bound system, and its mass, being the quantity that determines the motion of a planet moving in its gravitational field, includes a contribution from the gravitational binding energy. This is in complete analogy with, for example, nuclear mass: the mass of an atomic nucleus is the number of protons multiplied by the proton mass plus the same for neutrons, minus the nuclear binding energy. This binding energy, by Special Relativity, contributes to the inertial mass of a nucleus, and in the same way, in General Relativity, the gravitational binding energy will contribute to the gravitational mass of a star.

With this sort of consideration in mind, let us consider once more the exterior and interior Schwarzschild solutions (7.34) and (7.36) above. The 'mass' M of the star, as measured by a planetary orbit (the planet following a geodesic in the Schwarzschild vacuum space-time), is, from the above equations, simply $(4\pi/3)\rho R^3$, or, in the general case where the density is not constant,

$$M = \int_0^R 4\pi r^2 \rho(r)\, \mathrm{d}r. \tag{7.48}$$

There is at first sight something odd about this expression, however. Would it not be more proper, more 'covariant', to include a factor $\sqrt{g_{11}}$, with g_{11} given by (7.36), which would give a proper 3-dimensional (t = const) volume element? We would then write

$$M_1 = \int_0^R 4\pi r^2 \left(1 - \frac{8\pi G\rho r^2}{3c^2}\right)^{-1/2} \rho(r)\, \mathrm{d}r. \tag{7.49}$$

What we want to claim is that *something like* this expression is, in fact, the 'bare mass' of the star – the mass arrived at by *neglecting* the gravitational binding energy. Because of the extra factor in the integrand in (7.49) it is clear that $M_1 > M$, so it is indeed plausible that M_1 could represent the bare mass, with M being the actual, physical, mass, as measured by orbiting planets, including, as noted above, the (negative) gravitational binding energy. This idea is almost satisfactory, but not quite: it is in fact better to replace $\rho(r)$ with $n(r)$, the *baryon number density*, and then to define

$$B = \frac{1}{m_\mathrm{N}} \int_0^R 4\pi r^2 \left(1 - \frac{8\pi G\rho r^2}{3\,c^2}\right)^{-1/2} n\,(r)\, \mathrm{d}r, \tag{7.50}$$

(where m_N is the nucleon mass) as being the *baryon number* of a star. (It is, of course, understood that the vast bulk of the mass of stars comes from protons and neutrons.) Then

$$m_\mathrm{N} B = M_0 = \int_0^R 4\pi r^2 \left(1 - \frac{8\pi G\rho r^2}{3c^2}\right)^{-1/2} n(r)\, \mathrm{d}r \tag{7.51}$$

is the 'preassembly' or 'bare' mass of the star. The virtue of proceeding in this way is that baryon number is a conserved quantity, with the consequence that $n(r)$ obeys a simpler conservation law than does $\rho(r)$, as we shall see below.

We now consider the simplified case $n(r)$ = const, analogous to $\rho(r)$ = const considered above; then (7.50) becomes

$$B = \frac{4\pi n}{m_{\rm N}} \int_0^R r^2 (1 - Ar^2)^{-1/2}\, \mathrm{d}r, \tag{7.52}$$

with A given by (7.25). This integral may be evaluated by standard trigonometrical substitutions to give

$$B = \frac{4\pi n R^3}{3m_{\rm N}} \cdot \frac{3}{2} \frac{\sin^{-1} X - X\sqrt{1 - X^2}}{X^3} \equiv \frac{4\pi n R^3}{3m_{\rm N}} f(X), \tag{7.53}$$

where

$$X = R\sqrt{A} \Rightarrow X^2 = \frac{2GM}{Rc^2} = \frac{2m}{R}. \tag{7.54}$$

and the function $f(X)$ is defined by (7.53). We see from (7.37) that for normal stars such as the Sun

$$X^2 \ll 1. \tag{7.55}$$

This allows us to find a good approximation for $f(X)$. Putting $X = \sin\theta$ it is straightforward to check that up to the relevant order

$$\theta = \sin^{-1} X = X + \frac{X^3}{6} + \frac{3X^5}{40}, \qquad X \ll 1$$

and

$$f(X) = 1 + \frac{3}{10} X^2 + \cdots = 1 + \frac{3}{5}\frac{m}{R} + \cdots. \tag{7.56}$$

We can now check our claim that the difference between M, defined in (7.48), and the preassembly mass (or bare mass) M_0 is, in the Newtonian limit, equal to the gravitational binding energy. Working in the regime of constant ρ and n the preassembly mass becomes, from (7.52), with $n = \rho$,

$$M_0 = m_{\rm N} B = \frac{4\pi\rho R^3}{3} f(X) = M f(X)$$

where M is given by (7.48), so

$$M_0 - M = M[f(x) - 1] = \frac{3}{5}\frac{GM^2}{Rc^2}, \tag{7.57}$$

which is the gravitational binding energy in the Newtonian approximation (see Problem 7.2).

We now need to justify the manoeuvre above, of defining the preassembly mass in terms of $n(r)$ rather than $\rho(r)$. We turn to the more general case of a *compressible* material and define

$$M(r) = \int_0^r 4\,\pi r'^2 \rho(r')\,\mathrm{d}r', \tag{7.58}$$

the mass contained within a sphere of radius r. It follows from this equation that

$$\mathrm{d}M(r) = 4\pi r^2 \rho(r)\,\mathrm{d}r. \tag{7.59}$$

We may then write (7.48) as

$$\begin{aligned} M &= \int_0^R 4\pi r^2\,\rho(r)\,\mathrm{d}r \\ &= \int_0^R 4\pi r^2 \left[1 - \left(1 - \frac{2GM(r)}{rc^2}\right)^{-1/2}\right] \rho(r)\,\mathrm{d}r \\ &\quad + \int_0^R 4\pi r^2 \left(1 - \frac{2GM(r)}{rc^2}\right)^{-1/2} \rho(r)\,\mathrm{d}r. \end{aligned} \tag{7.60}$$

Let us write the relation between $\rho(r)$, the mass density, and $n(r)$, the baryon number density, as

$$\rho(r) = n(r)\left(m_\mathrm{N} + \frac{\varepsilon}{c^2}\right), \tag{7.61}$$

where ε is the *specific internal energy* (thermal, compressional, etc.) per particle. Using a binomial approximation on the first term in (7.60) and substituting (7.61) into the second term gives

$$\begin{aligned} M &\approx \int 4\pi r^2 \frac{GM(r)}{rc^2} \rho(r)\,\mathrm{d}r + \int 4\pi r^2 \left(1 - \frac{2GM(r)}{rc^2}\right)^{-1/2} m_\mathrm{N}\, n(r)\,\mathrm{d}r \\ &\quad + \frac{1}{c^2} \int 4\pi r^2 \left(1 - \frac{2GM(r)}{rc^2}\right)^{-1/2} \varepsilon\, n(r)\,\mathrm{d}r \\ &= \int \frac{GM(r)}{rc^2}\,\mathrm{d}M + M_0 + \frac{U}{c^2}, \end{aligned} \tag{7.62}$$

where (7.59) and (7.51) have been used in the last step and

$$U = \int 4\pi r^2 \left(1 - \frac{2GM(r)}{r\,c^2}\right)^{-1/2} \varepsilon\, n(r)\,\mathrm{d}r \tag{7.63}$$

is the *internal energy* of the stellar material. Defining

$$E_\mathrm{G} = \int \frac{GM(r)}{r}\,\mathrm{d}M \tag{7.64}$$

as the *gravitational binding energy* we may write (7.62) in the form

$$M_{\mathrm{PN}} = M_0 + \frac{E_{\mathrm{G}}}{c^2} + \frac{U}{c^2}. \tag{7.65}$$

This is the Post-Newtonian (PN) approximation to the expression for the mass of a star, since the gravitational binding energy (7.64) becomes, in the Newtonian approximation of constant density, equal to (7.57) – and (7.65) represents an improvement on this.

We shall show finally that the relations above imply that *entropy* is conserved in this hydrodynamic flow. We begin with the energy-momentum tensor (7.8), which obeys the covariant conservation law $T^{\mu\nu}{}_{;\nu}=0$, hence

$$T_{\mu;\nu}^{\ \nu} = \left[\left(\rho + \frac{p}{c^2}\right)u^\nu\right]_{;\nu} u_\mu + \left(\rho + \frac{p}{c^2}\right)u^\nu\, u_{\mu;\nu} + p_{,\nu}\, g_\mu{}^\nu = 0,$$

or

$$\left[\left(\rho + \frac{p}{c^2}\right)u^\nu\right]_{;\nu} u_\mu + \left[\left(\rho + \frac{p}{c^2}\right)u_\nu\, u_{\mu;}{}^\nu\right] + \frac{1}{c^2}p_{,\mu} = 0. \tag{7.66}$$

Now *define*, for any tensor $T^{\cdots}{}_{\cdots}$

$$\dot{T}^{\cdots}{}_{\cdots} = T^{\cdots}{}_{\cdots;\mu}\, u^\mu. \tag{7.67}$$

This is a 'generalised' time derivative, which reduces to $\frac{\mathrm{d}}{\mathrm{d}\tau}$ in the comoving frame. Then $u_{\mu;\nu}u^\nu = \dot{u}_\mu$, hence

$$\left[\left(\rho + \frac{p}{c^2}\right)u^\nu\right]_{;\nu} u_\mu + \left(\rho + \frac{p}{c^2}\right)\dot{u}_\mu + \frac{1}{c^2}p_{,\mu} = 0,$$

or

$$\dot{u}_\mu = -\frac{p_{,\mu}}{\rho c^2 + p} - \frac{\left[\left(\rho + \frac{p}{c^2}\right)u^\nu\right]_{;\nu}u_\mu}{\rho + \frac{p}{c^2}}. \tag{7.68}$$

Multiply this equation by u^μ, and recall that since $u^\mu\, u_\mu = -1$, $\dot{u}_\mu\, u^\mu = 0$; then

$$\begin{aligned}\frac{1}{c^2}\dot{p} &= \left[\left(\rho + \frac{p}{c^2}\right)u^\nu\right]_{;\nu}\\ &= \left(\rho + \frac{p}{c^2}\right)_{;\nu} u^\nu + \left(\rho + \frac{p}{c^2}\right)u^\nu{}_{;\nu}\\ &= \frac{1}{c^2}\dot{p} + \dot{\rho} + \left(\rho + \frac{p}{c^2}\right)u^\nu{}_{;\nu}\end{aligned} \tag{7.69}$$

or

$$\dot{\rho} = -\left(\rho + \frac{p}{c^2}\right)u^\nu{}_{;\nu}. \tag{7.70}$$

We may express this in a different way: remove the brackets above and put, as in (7.67),

$$\dot{\rho} = \rho_{;\mu}u^\mu,$$

then

$$\rho u^{\nu}{}_{;\nu} + \rho_{;\nu} u^{\nu} = -\frac{1}{c^2} p\, u^{\nu}{}_{;\nu},$$

$$(\rho u^{\mu})_{;\mu} + \frac{1}{c^2} p u^{\mu}{}_{;\mu} = 0. \tag{7.71}$$

This equation, derived from the covariant conservation of the energy-momentum tensor, is, however, *not* a continuity equation. A continuity equation is of the form

$$(n u^{\mu})_{;\mu} = 0, \tag{7.72}$$

which indeed we may write if n is identified with baryon number density. Consider, then, a gas of N particles, occupying a volume V, so

$$n = \frac{N}{V}.$$

The connection between the (mass) density ρ, baryon number density n and pressure p follows from the equation of state

$$p = p(\rho)$$

and the definition of pressure

$$p = -\frac{\mathrm{d(energy/baryon)}}{\mathrm{d(volume/baryon)}} = -c^2 \frac{\mathrm{d}(\rho/n)}{\mathrm{d}(1/n)} = c^2 n \frac{\partial \rho}{\partial n} - c^2 \rho$$

or

$$\rho + \frac{p}{c^2} = n \frac{\partial \rho}{\partial n}. \tag{7.73}$$

The first law of thermodynamics gives the entropy change dS of a gas as

$$T\,\mathrm{d}S = \mathrm{d}U + p\,\mathrm{d}V. \tag{7.74}$$

If the gas contains N particles then the specific entropy σ and specific internal energy ε are given by

$$S = N\sigma, \quad U = N\varepsilon$$

and (7.74) gives

$$T\,\mathrm{d}\sigma = \mathrm{d}\varepsilon + p\,\mathrm{d}\left(\frac{1}{n}\right). \tag{7.75}$$

Now from (7.69) and (7.61)

$$\begin{aligned}
\frac{1}{c^2}\dot{p} &= \left[\left(n(r)\, m_{\mathrm{N}} + \frac{1}{c^2} n(r)\varepsilon + \frac{P}{c^2}\right) u^{\mu}\right]_{;\mu} \\
&= m_{\mathrm{N}}\dot{\varepsilon} + \frac{\varepsilon}{c^2}(n u^{\mu})_{;\mu} + \frac{1}{c^2} n\varepsilon_{,\mu} u^{\mu} + \frac{1}{c^2} P_{,\mu} u^{\mu} + \frac{p}{c^2} u^{\mu}{}_{;\mu} \\
&= \frac{1}{c^2} n\dot{\varepsilon} + \frac{1}{c^2}\dot{p} + \frac{p}{c^2} u^{\mu}{}_{;\mu},
\end{aligned}$$

where (7.72) has been used. Hence

$$pu^{\mu}{}_{;\mu} + n\dot{\varepsilon} = 0. \tag{7.76}$$

Equation (7.72), however, implies

$$(nu^{\mu})_{;\mu} = n_{,\mu}u^{\mu} + nu^{\mu}{}_{;\mu} = \dot{n} + nu^{\mu}{}_{;\mu} = 0,$$

hence

$$u^{\mu}{}_{;\mu} = -\frac{\dot{n}}{n}. \tag{7.77}$$

Substituting this in (7.76) gives

$$-p\frac{\dot{n}}{n} + n\dot{\varepsilon} = 0,$$

hence

$$p\left(\frac{1}{n}\right)^{\cdot} + \dot{\varepsilon} = 0, \tag{7.78}$$

which is equivalent to (7.75), and shows that the entropy is constant.

7.3 Degenerate stars: white dwarfs and neutron stars

As mentioned above, stars are configurations of matter in equilibrium under the action of two forces; gravity pulling inwards and some other force in counterbalance. In 'normal' stars the counterbalancing force is provided by the thermal pressure of the stellar gas, but in degenerate stars it is provided by the Fermi pressure of either electrons or neutrons. It is this second case which is of principal interest in this section, but it is instructive to begin with some remarks about 'normal' stars, in a simple Newtonian approximation.

Consider a spherical, non-rotating star, of mass density $\rho(r)$, and consider within it a spherical shell a distance r from the centre, and thickness dr. An area A of this shell has mass $\rho A\,\mathrm{d}r$ and experiences a gravitational force

$$\mathrm{d}F = \frac{GM(r)}{r^2}\rho(r)\,A\,\mathrm{d}r,$$

where $M(r)$ is given by (7.58) – the mass of material inside the spherical shell. This force must be balanced by a pressure $p(r)$ such that dF = A dp, and hence

$$\frac{\mathrm{d}p}{\mathrm{d}r} = -\frac{GM(r)}{r^2}\rho(r). \tag{7.79}$$

In the case of constant density ρ = const, $M(r) = (4\pi/3)\rho r^3$ and the pressure gradient is

$$\frac{\mathrm{d}p}{\mathrm{d}r} = -\frac{4\pi}{3}G\rho^2 r, \tag{7.80}$$

whose solution is

$$p = p_0 - \frac{2\pi}{3} G\rho^2 r^2,$$

where $p_0 = p(0)$, the pressure at the centre. The stellar radius R may be defined as the (minimum) value of r for which $p(r) = 0$, giving

$$p_0 = \frac{2\pi}{3} G\rho^2 R^2. \tag{7.81}$$

Then

$$\frac{p_0}{\rho c^2} = \frac{2\pi}{3c^2} G\rho R^2 = \frac{GM}{2Rc^2} \equiv \frac{m}{2R}, \tag{7.82}$$

where $m = \frac{GM}{c^2}$ is the Schwarzschild radius of the star.

A ‘normal’ star may be considered to be a gas of protons and neutrons, each of mass m_N. A volume V of N such particles obeys the ideal gas law $pV = NkT$ (where k is Boltzmann's constant), so since $\rho = \frac{Nm_N}{V}$ we have

$$\frac{p}{\rho c^2} = \frac{kT}{m_N c^2}.$$

At the centre of the Sun $T = 1.6 \times 10^7$ K, so, expressing both numerator and denominator in MeV,

$$\frac{kT}{m_N c^2} = \frac{1.6 \times 10^7 \times 8.6 \times 10^{-11}}{939} = 1.5 \times 10^{-6}.$$

On the other hand, for the Sun

$$\frac{m_S}{2R_S} = \frac{1.47 \times 10^3}{2 \times 6.96 \times 10^8} = 1.1 \times 10^{-6},$$

so (7.82), which is only an order-of-magnitude estimate, holds up well.

We now turn to the case of degenerate stars. As mentioned in the introductory section of this chapter, stars get their energy from nuclear fusion reactions, but when the stellar interior becomes dominated by iron the fusion reactions no longer produce photons, the temperature drops and the star begins to collapse. The question then arises: is there any other equilibrium configuration possible? The answer is yes: as the star collapses the electrons are squeezed together. In one dimension a particle confined to a region d has a momentum $p \sim \hbar/d$. If the particles are bosons any number of them may occupy the same state, but if they are fermions with spin ½, a maximum of two may occupy a given state, so if N electrons are confined to a volume V the momentum states become filled up to the Fermi level p_F. As an order of magnitude estimate, we have $n = N/V$ particles per unit volume, so one particle occupies a volume $1/n = d^3$ and the Fermi momentum is

$$p_F \sim \frac{h}{d} = \hbar n^{1/3} \Rightarrow n \sim {p_F}^3. \tag{7.83}$$

The electron becomes a degenerate gas, whose pressure may – or may not – be able to withstand the gravitational inward pressure of the star. Thus quantum considerations, in the shape of degenerate matter, offer the possibility of new types of objects in the sky.

Let us first make a more precise reckoning than the order-of-magnitude estimate (7.83). It is well known from elementary quantum theory that for a particle confined to a cubic box with sides of length L, periodic boundary conditions mean that the wave numbers k_x, k_y and k_z are integral multiples of $(2\pi/L)$:

$$k_x = \left(\frac{2\pi}{L}\right)l, \quad k_y = \left(\frac{2\pi}{L}\right)m, \quad k_z = \left(\frac{2\pi}{L}\right)n,$$

with l, m and n integers. Hence the possible states in k-space are given by the corners of cubes with $l, m, n = 1, 2, 3, 4$ and so on, and to each state there correspond two electrons (with spin up and spin down along some axis). It might help to visualise this by displacing the cubes by (π/L) in each of the x, y and z directions so that the points corresponding to possible quantum states are at the *centres* of cubes. There are then two electrons per cube, each cube being of volume $(2\pi/L)^3$. The electrons then occupy all states up to k_F; these states are contained within a sphere of radius k_F and therefore volume $(4\pi/3){k_F}^3$. The number of electrons (which must be an *integer*!) is, nevertheless, to an excellent approximation

$$N = 2 \times \frac{4\pi}{3} \times {k_F}^3 \times \left(\frac{L}{2\pi}\right)^3 = \frac{{k_F}^3}{3\pi^2}L^3 = \frac{{k_F}^3}{3\pi^2}V$$

and the number per unit volume (at the rate of two per cube of volume $V = L^3$) is

$$\frac{N}{V} = n = \frac{{k_F}^3}{3\pi^2}. \tag{7.84}$$

With $p_F = \hbar k_F$ this gives

$$n = \frac{1}{3\pi^2\hbar^3}{p_F}^3, \tag{7.85}$$

or

$$p_F = (3\pi^2)^{1/3}\hbar n^{1/3}, \tag{7.86}$$

as expected in (7.83). We may write (7.85) as

$$n = \frac{1}{\pi^2\hbar^3}\int_0^{p_F} p^2\,\mathrm{d}p = \frac{1}{3\pi^2\hbar^3}{p_F}^3. \tag{7.87}$$

This is the number of electrons per unit volume. To find the energy density of the electron gas we must include in this integral the (kinetic) energy of each electron, which is

$$T = (m^2c^4 + p^2c^2)^{1/2} - mc^2. \tag{7.88}$$

In the *non-relativistic limit* $p \ll mc$ this becomes

$$T = mc^2\left(1+\frac{p^2}{m^2c^2}\right)^{1/2} - mc^2 \approx \frac{p^2}{2m},$$

giving an energy density

$$e = \frac{1}{2m\pi^2\hbar^3}\int_0^{p_F} p^4\,dp = \frac{1}{10m\pi^2\hbar^3}p_F{}^5. \tag{7.89}$$

On substituting for p_F from (7.87) this gives

$$e = \frac{3}{10}(3\pi^2)^{2/3}\left(\frac{\hbar^2}{m_e}\right)n^{5/3} \quad \text{(non-rel)}. \tag{7.90}$$

Here the mass is explicitly shown as the electron mass m_e; since it is much smaller than the nucleon mass the electron will make by far the bigger contribution to the energy density of a degenerate gas.

In the opposite case, of the ultra-relativistic limit $p \gg mc$ the kinetic energy (7.88) becomes

$$T = pc\left(1+\frac{m^2c^2}{p^2}\right)^{1/2} - mc^2 \approx pc,$$

giving an energy density

$$e = \frac{1}{\pi^2\hbar^3}\int_0^{p_F} p^2\cdot pc\,dp = \frac{c}{4\pi^2\hbar^3}p_F^4,$$

which, on substituting for p_F from (7.87), gives

$$e = \frac{3}{4}(3\,\pi^2)^{1/3}(c\hbar)\,n^{4/3} \quad \text{(ultra-rel)}. \tag{7.91}$$

The *pressure* of the gas is given by (7.73), with $\rho c^2 \to e$:

$$p = n\frac{\partial e}{\partial n} - e, \tag{7.92}$$

from which it is straightforward to find the following expressions in the non-relativistic and ultra-relativistic limits:

$$p = \frac{2}{3}e = \frac{1}{5}(3\pi^2)^{2/3}\left(\frac{\hbar^2}{m_e}\right)n^{5/3} \quad \text{(non-rel)}, \tag{7.93}$$

$$p = \frac{1}{3}e = \frac{1}{4}(3\,\pi^2)^{1/3}(c\hbar)\,n^{4/3} \quad \text{(ultra-rel)}. \tag{7.94}$$

There is another way of characterising these (non-relativistic and ultra-relativistic) limits. The mass density of a star is

$$\rho = n\, m_{\rm N} \mu, \tag{7.95}$$

where n is, as above, the number density of electrons, $m_{\rm N}$ is the nucleon mass and μ is the number of nucleons per electron – an average taken throughout the star. For stars that have used up their hydrogen $\mu \approx 2$. As the star collapses ρ increases and the electrons are forced closer together. When the distance between electrons approaches their Compton wavelength $\hbar/m_{\rm e}c$ we may define a *critical density* $\rho_{\rm c}$:

$$\rho_{\rm c} \sim \frac{\mu m_{\rm N}}{d^3} = \frac{\mu m_{\rm N}}{\hbar^3} m_{\rm e}^3 c^3. \tag{7.96}$$

For a precise definition let us incorporate a factor $3\pi^2$ and define

$$\rho_{\rm c} = \frac{\mu m_{\rm N}}{3\pi^2 \hbar^3} m_{\rm e}^3 c^3. \tag{7.97}$$

From (7.85) and (7.95) this definition corresponds to

$$\rho = \rho_{\rm c} \Leftrightarrow p_{\rm F} = m_{\rm e} c. \tag{7.98}$$

The non-relativistic limit $p_{\rm F} \ll m_{\rm e}c$ is therefore $\rho \ll \rho_{\rm c}$, and the ultra-relativistic limit is $\rho \gg \rho_{\rm c}$. What is the value of $\rho_{\rm c}$? Working in SI units,

$$\begin{aligned}\rho_{\rm c} &= \frac{1.7 \times 10^{-27} \times 2 \times (9.11)^3 \times 10^{-93} \times 27 \times 10^{24}}{3\pi^2 \times (1.05)^3 \times 10^{-102}} \\ &= 2.03 \times 10^9\ {\rm kg\,m^{-3}} \\ &\approx 2 \times 10^6 \times \text{density of water}.\end{aligned} \tag{7.99}$$

Now let us calculate the ratio of central pressure to energy density, in this case of a star composed of a degenerate electron gas. From (7.95) and (7.97) we have

$$n = \left(\frac{\rho}{\rho_{\rm c}}\right) \cdot \frac{m_{\rm e}^3 c^3}{3\pi^2 \hbar^3}, \tag{7.100}$$

and hence, with a bit of algebra, it follows from (7.93) that in the non-relativistic limit

$$\rho \ll \rho_{\rm c}: \quad \frac{p}{\rho c^2} = \frac{m_{\rm e}}{5 m_{\rm N} \mu} \left(\frac{\rho}{\rho_{\rm c}}\right)^{2/3} \quad \text{(non-rel)} \tag{7.101}$$

and that in the ultra-relativistic limit, using (7.94),

$$\rho \gg \rho_{\rm c}: \quad \frac{p}{\rho c^2} = \frac{m_{\rm e}}{4 m_{\rm N} \mu} \left(\frac{\rho}{\rho_{\rm c}}\right)^{1/3} \quad \text{(ultra-rel)}. \tag{7.102}$$

We may convert these expressions (for the ratio of electron Fermi pressure to mass density) of the star into ones involving the stellar mass M. From (7.82),

$$\frac{p}{\rho c^2} = \frac{GM}{2Rc^2} = \frac{m}{2R}, \tag{7.103}$$

where m is the Schwarzschild radius of the star (not a mass!). For a star of constant density

$$R = \left(\frac{3M}{4\pi\rho}\right)^{1/3},$$

hence

$$\frac{m}{R} = \frac{GM^{2/3}}{c^2}\left(\frac{4\pi}{3}\right)^{1/3}\rho^{1/3}$$

and

$$M = \left(\frac{3}{4\pi}\right)^{1/2}\frac{c^3}{G^{3/2}}\rho^{-1/2}\left(\frac{m}{R}\right)^{3/2}. \tag{7.104}$$

Now insert (7.103) into (7.104) and find, in the two cases corresponding to (7.101) and (7.102),

$$\rho \ll \rho_c: \quad M = \left(\frac{18\pi}{125}\right)^{1/2}\left(\frac{c\hbar}{G}\right)^{3/2}\frac{1}{m_N{}^2\mu^2}\left(\frac{\rho}{\rho_c}\right)^{1/2}, \tag{7.105}$$

$$\rho \gg \rho_c: \quad M = \left(\frac{9\pi}{32}\right)^{1/2}\left(\frac{c\hbar}{G}\right)^{3/2}\frac{1}{m_N^2\mu^2}. \tag{7.106}$$

It is noteworthy that this latter expression is a *unique mass*, dependent only on constants of nature and μ. In particular it depends on the nucleon mass m_N but *not* on the electron mass, *even though* it is the electron Fermi pressure which is withstanding the gravitational inward pressure.

From dimensional considerations it is clear from (7.106) that the quantity $(c\hbar/G)^{1/2}$ is a *mass*; it generally goes by the name of the *Planck mass*:

$$m_{Pl} = \left(\frac{c\hbar}{G}\right)^{1/2} = 2.17 \times 10^{-8}\,\text{kg}. \tag{7.107}$$

The crucial ingredient in the formulae above is, in numerical terms,

$$\left(\frac{c\hbar}{G}\right)^{3/2}\frac{1}{m_N{}^2} = \frac{m_{Pl}{}^3}{m_N{}^2} = 3.66 \times 10^{30}\,\text{kg} = 1.84\,M_S. \tag{7.108}$$

It is then clear that the mass of an object which is held in equilibrium by a balancing of the gravitational force with the Fermi pressure of degenerate electrons is of the order of the *solar mass*.

As well as deriving formulae for the mass it is straightforward, though slightly messy, also to derive expressions for the radius of degenerate stars. In the case $\rho \ll \rho_c$ the balancing of gravitational and degeneracy pressures gives, from (7.81) and (7.93)

$$\frac{2\pi}{3}G\rho^2R^2 = \frac{1}{5}(3\pi^2)^{2/3}\left(\frac{\hbar^2}{m_e}\right)n^{5/3}. \tag{7.109}$$

Since $M = \dfrac{4\pi}{3}R^3\rho$, it follows that

$$R = \left(\frac{3}{4\pi}\right)^{1/3} M^{1/3} \rho^{-1/3} \tag{7.110}$$

and hence

$$\rho R = \left(\frac{3}{4\pi}\right)^{1/3} M^{1/3} \rho^{2/3}.$$

Substituting this, together with (7.95), into the left hand side of (7.109) gives

$$\left(\frac{\pi}{6}\right)^{1/3} G M^{2/3} (\mu m_{\rm N})^{4/3} n^{4/3} = \frac{(3\pi^2)^{2/3}}{5} \left(\frac{\hbar^2}{m_{\rm e}}\right) n^{5/3},$$

from which

$$n = \frac{125}{54\pi^3} G^3 M^2 (\mu m_{\rm N})^4 \frac{{m_{\rm e}}^3}{\hbar^6}$$

and hence

$$\rho = \mu m_{\rm N} n = \frac{125}{54\pi^3} G^3 M^2 (\mu m_{\rm N})^5 \frac{{m_{\rm e}}^3}{\hbar^6}.$$

Substituting this into (7.110) gives

$$R = \left(\frac{81\pi^2}{250}\right)^{1/3} \frac{1}{G} M^{-1/3} (\mu m_{\rm N})^{-5/3} \frac{\hbar^2}{m_{\rm e}} \tag{7.111}$$

and, finally, substituting for M from (7.105) gives

$$R = \left(\frac{9\pi}{10}\right)^{1/2} \left(\frac{\hbar^3}{cG}\right)^{1/2} \frac{1}{\mu m_{\rm N} m_{\rm e}} \left(\frac{\rho}{\rho_{\rm c}}\right)^{-1/6}. \tag{7.112}$$

In numerical terms this gives, taking $\mu = 2$,

$$\rho \ll \rho_{\rm c}: \quad R = 4.2 \times 10^6 \left(\frac{\rho}{\rho_{\rm c}}\right)^{-1/6} \text{ m}. \tag{7.113}$$

This is of the same order of magnitude as the radius of the Earth, which is 6.4×10^6 m. (Note the only slight dependence on ρ: if $\rho/\rho_{\rm c} = 0.1$, $(\rho/\rho_{\rm c})^{-1/6} \approx 1.47$.)

On the other hand, in the relativistic case $\rho \gg \rho_{\rm c}$, Equations (7.111) and (7.106) yield

$$R = \frac{6}{5} \left(\frac{\pi}{2}\right)^{1/2} \left(\frac{\hbar^3}{cG}\right)^{1/2} \frac{1}{\mu m_{\rm N} m_{\rm e}}. \tag{7.114}$$

In numerical terms, again with $\mu = 2$, we find

$$\rho \gg \rho_{\rm c}: \quad R = 3.8 \times 10^6 \text{ m}, \tag{7.115}$$

a *unique* radius, which is actually very close in value to the non-relativistic case (7.113).

Numerical estimates for the mass are easily obtained from (7.105) and (7.106) which, with (7.108) give (with $\mu=2$)

$$\rho \ll \rho_c: \quad M = 0.31\left(\frac{\rho}{\rho_c}\right)^{1/2} M_S, \tag{7.116}$$

$$\rho \gg \rho_c: \quad M = 0.43 M_S. \tag{7.117}$$

This reasoning indicates that there is a *maximum mass*, above which an object held in equilibrium by the balancing of gravity with the electron Fermi pressure becomes *unstable*. The actual values found above assume a constant density throughout the star. A more precise estimate yields a value of $1.44M_S$, called the *Chandrasekhar limit*.[8] Objects with these properties have been known for some time, and are *white dwarfs* – they typically have a mass similar to the Sun's and a radius similar to the Earth's. No white dwarf has been discovered with a mass exceeding the Chandrasekhar limit. Sirius B, for example, is quoted as having a mass of $1.05M_S$ and a radius of 5.5×10^6 m.[9]

Collapsing stars with a larger mass than the Chandrasekhar limit will therefore possess no white dwarf equilibrium state, so collapse will continue. As it does so, however, the electron energy increases and eventually the electrons in the star will react with protons

$$e^- + p \to n + \nu_e$$

to produce neutrons and electron neutrinos. The neutrinos, having an extremely weak interaction with matter, will escape from the star, which then simply becomes a collection of neutrons – a *neutron star*. Collapse then proceeds unimpeded by the electrons, as layer upon layer of the stellar material hurtles towards the centre. Neutrons, however, are also fermions and there will come a point at which neutron Fermi pressure will (successfully or not) resist the inward gravitational pull. If a neutron star is formed its radius is easy to calculate: since neutrons, and not electrons, are exerting a Fermi pressure, simply replace m_e by m_N in (7.112) and (7.114) (and put $\mu=1$), giving, in place of (7.115), a radius of about 4 km (!). The material of a collapsing star, raining down on this core, will be thrown back on impact, resulting in a *supernova explosion*. The earliest such explosion known is the famous Chinese observation in AD 1054 of an event in the Crab Nebula; it was so bright that it was visible in the day sky for about three weeks. Because of their small size neutron stars can rotate very fast, and it is now widely believed that *pulsars* are rotating neutron stars. A pulsar has indeed been found at the centre of the Crab Nebula and in fact many pulsars are now known. The calculation of the mass of a neutron star is more complicated than the method shown above. As well as non-linear effects in the TOV equation, there are also problems arising because the nuclear physics required is not well understood, but a recent estimate of the maximum mass quotes the value $6.7M_S$.[10] In this context a most interesting object is PSR 1913+16, the Hulse–Taylor *binary pulsar*. It consists of two neutron stars in orbit round each other, and details of this system are now known with amazing accuracy: for example the

[8] Chandrasekhar (1931a, b).
[9] By Carroll & Ostlie (1996), p. 578.
[10] Hartle (1978); see also Hartle (2003), p. 536.

masses of the neutron stars are $M_1 = (1.4410 \pm 0.0005)M_S$ and $M_2 = (1.3874 \pm 0.0005)M_S$. But perhaps the most interesting feature of this binary system is the *slow-down* in its rotation rate, since this is attributed to a loss of energy through gravitational radiation. This topic will be treated in Chapter 9.

Very heavy collapsing stars will, following the logic above, have *no* equilibrium. There will then be a state of *continued collapse*, in particular into a size smaller than the Schwarzschild radius $2m$. This is a black hole, on whose study we now embark.

7.4 Schwarzschild orbits: Eddington–Finkelstein coordinates

The extraordinary physics of black holes has its origin in the highly unexpected properties of the surface $r = 2m$ in the Schwarzschild solution. Let us first note that this really is a (2-dimensional) surface. In general one constraint ($r = 2m$) in the parameters of a 4-dimensional space (space-time) will leave a 3-dimensional manifold, or 'hypersurface'. But putting $r = 2m$ in the line element (5.36) causes g_{00} to vanish, so the time dimension collapses and we are actually left with a 2-dimensional 'surface'. It is in fact a null surface, as we shall see below.

We saw in Section 5.12 that the radial fall of a particle from $r > 2m$ to decreasing values of r takes an *infinite* time, measured in the Schwarzschild time coordinate t, to reach $r = 2m$; though it only takes a *finite* proper time. This seems to indicate that to explore the situation further, and in particular to understand the region $r < 2m$ we must consider alternative coordinate systems. Let us never forget that the whole spirit of General *Relativity* is that coordinate systems themselves have no direct relevance: the physics we are trying to understand is independent of any coordinate system. The coordinate transformation introduced by Eddington and Finkelstein is a significant advance in understanding the peculiarities of the surface $r = 2m$.

The most straightforward way to approach this transformation is to consider radial null geodesics in Schwarzschild space-time given by the metric (5.60)

$$ds^2 = -\left(1 - \frac{2m}{r}\right)c^2\,dt^2 + \left(1 - \frac{2m}{r}\right)^{-1} dr^2 + r^2(d\theta^2 + \sin^2\theta\, d\phi^2). \qquad (7.118)$$

For these geodesics $ds^2 = 0$ and θ and ϕ are both constant so we have ($\dot{t} = \frac{dt}{ds}$ etc.)

$$0 = -\left(1 - \frac{2m}{r}\right)c^2\,\dot{t}^2 + \left(1 - \frac{2m}{r}\right)^{-1}\dot{r}^2. \qquad (7.119)$$

We have in addition, for radial motion, Equation (5.57),

$$\left(1 - \frac{2m}{r}\right)\dot{t} = b = \text{const.} \qquad (7.120)$$

Substituting (7.120) into (7.119) gives $\dot{r}^2 = b^2$, so

$$\dot{r} = \pm cb. \qquad (7.121)$$

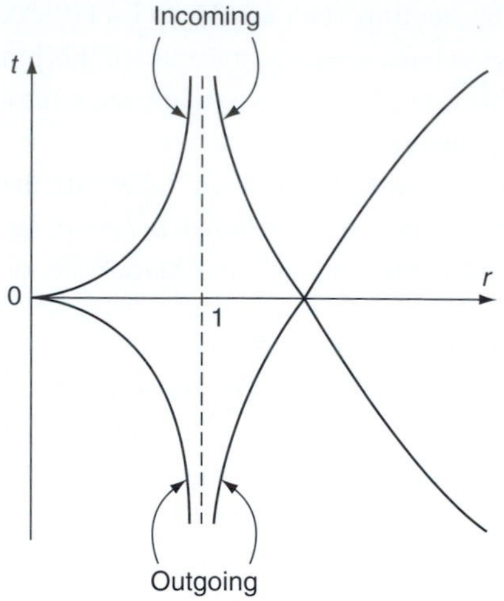

Fig. 7.1 **Incoming and outgoing light rays in a Schwarzschild field.**

The last two equations imply that

$$\frac{\mathrm{d}r}{\mathrm{d}t} = \frac{\dot{r}}{\dot{t}} = \pm c\frac{r-2m}{r}, \tag{7.122}$$

which can be integrated to give

$$ct = \pm(r + 2m \ln |r - 2m|) + \text{const.} \tag{7.123}$$

The + sign refers to outgoing, and the − sign to incoming, rays, and these rays are sketched in Fig. 7.1. Note that for outgoing rays, if $r>2m$, r increases as t increases, but when $r<2m$ (that is, *inside* the surface $r=2m$), r increases while t *decreases*. Similar and opposite remarks apply to incoming rays: for them, when $r>2m$, r decreases while t increases, but in the region $r<2m$, r and t decrease (or increase) together. In both cases the singular behaviour of the rays at $r=2m$ is evident. In fact from the metric (7.118) we see that the signature inside the Schwarzschild surface is

$$r<2m: + - + +,$$

in contrast with that outside

$$r>2m: - + + + .$$

As r crosses from $>2m$ to $<2m$, it changes from being a spacelike parameter to a timelike one; and the time t makes the opposite change, from timelike to spacelike. Hence, as in the world outside the Schwarzschild surface time *inevitably* increases (it cannot be made to run backwards), so *inside* the Schwarzschild surface it is r which will inevitably decrease.

Now let us introduce a new time coordinate

$$\bar{t} = t \pm \frac{2m}{c}\ln\left|\frac{r}{2m} - 1\right|, \quad \bar{r} = r, \tag{7.124}$$

so that

$$\mathrm{d}\bar{t} = \mathrm{d}t \pm \frac{1}{c}\frac{2m}{r-2m}\mathrm{d}r.$$

Taking the upper (+) sign above, the line element (7.118) becomes

$$\mathrm{d}s^2 = -\left(1-\frac{2m}{r}\right)c^2\,\mathrm{d}\bar{t}^2 + \left(1+\frac{2m}{r}\right)\mathrm{d}r^2 + \frac{4m}{r}c\,\mathrm{d}\bar{t}\,\mathrm{d}r + r^2\,\mathrm{d}\Omega^2. \qquad (7.125)$$

This is the *Eddington–Finkelstein* form of the metric.[11] Note that g_{11}, the coefficient of $\mathrm{d}r^2$, is regular at $r=2m$, so the introduction of $\bar{t}$ in effect extends the range of the radial coordinate in the Schwarzschild solution from $2m<r<\infty$ in (7.118) to $0<r<\infty$ in (7.125). Now let us consider radial null geodesics in advanced Eddington–Finkelstein coordinates. From (7.125) we have

$$0 = -\left(1-\frac{2m}{r}\right)c^2\,\mathrm{d}\bar{t}^2 + \left(1+\frac{2m}{r}\right)\mathrm{d}r^2 + \frac{4m}{r}c\,\mathrm{d}\bar{t}\,\mathrm{d}r,$$

from which

$$\frac{1}{c}\frac{\mathrm{d}r}{\mathrm{d}\bar{t}} = -1 \text{ or } \frac{r-2m}{r+2m}. \qquad (7.126)$$

These equations, giving the gradients of the light cones, are sketched in Fig. 7.2. It is seen that light rays coming in from $r>2m$ will cross the surface $r=2m$ smoothly and reach $r=0$.

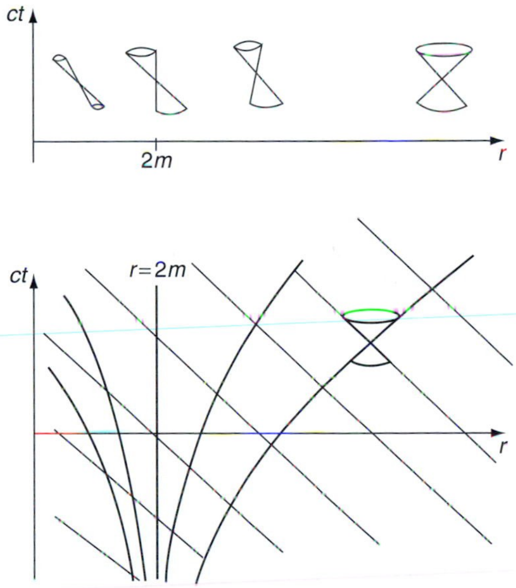

Fig. 7.2 **Light cones and light rays obeying (7.126) – a black hole.**

[11] Eddington (1924), Finkelstein (1958).

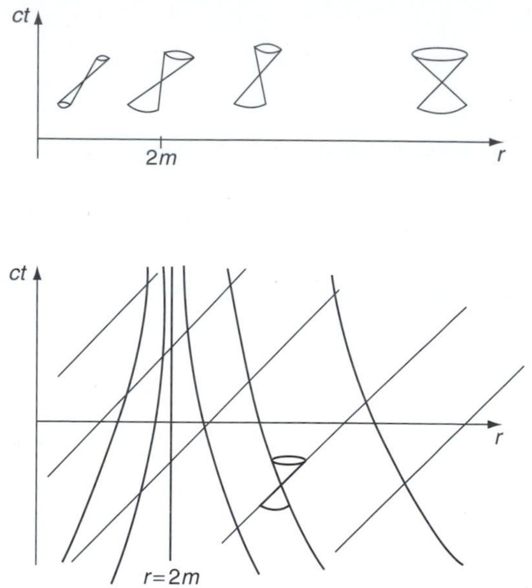

Fig. 7.3 **Light cones and light rays obeying (7.127) – a white hole.**

Outgoing rays, originating from $r > 2m$, will eventually reach $r = \infty$, the light cones tilting progressively towards their Minkowski value as $r \to \infty$. Rays originating from $r < 2m$, however, will never escape to the region $r > 2m$: they all eventually reach $r = 0$. Thus the surface $r = 2m$ acts as a *one-way membrane*: light may cross it going in but not coming out. A similar conclusion holds also for particles: since they travel along geodesics inside the light cone, they may cross the Schwarzschild surface from $r > 2m$ to $r < 2m$ but once inside they may not leave. Consequently an observer outside this surface will never be able to detect light signals or particles originating inside it. To an outside observer the Schwarzschild surface is in effect a *boundary* of space-time. It is an *event horizon*, such that timelike or null geodesics from events inside it will never reach the outside. An object with such an event horizon is a *black hole*.

These conclusions were reached taking the + sign in (7.124). Now consider what happens when we take the − sign. The sign of the $d\bar{t}\, dr$ term in (7.125) is changed so null geodesics (7.126) become

$$\frac{1}{c}\frac{dr}{d\bar{t}} = 1 \text{ or } -\left(\frac{r-2m}{r+2m}\right). \tag{7.127}$$

These rays are sketched in Fig. 7.3, and we see that light can pass through the surface $r = 2m$ from inside to outside but not vice versa; and, as before, a similar conclusion holds for particles. Particles and light rays approaching from $r = \infty$ are prevented from entering the region, so to an observer *inside* the Schwarzschild surface, this surface is an event horizon, which it is *not* to an observer outside. An object of this type is called a *white hole*. This is, in an obvious sense, a time-reversed black hole (Fig. 7.3 upside down looks like Fig. 7.2.) It seems likely that black holes exist in nature, but unlikely that white holes do. This might be thought strange, in view of the fact that Einstein's field equations are invariant under time reversal. On the other hand, the wave equations of Maxwell's electrodynamics are invariant under time reversal, but nature seems only to make use of the retarded potential solutions, not the advanced potential ones.

We may conclude that the Eddington–Finkelstein coordinates give us a fuller understanding of the Schwarzschild solution. The advanced and retarded time parameters give two differing versions of the space-time geometry of this solution. A further development, however, took place when the coordinates introduced by Kruskal and Szekeres were able to display *both* of these geometries – and two more also – in one diagram. This is the subject of the next section.

7.5 Kruskal–Szekeres coordinates

Instead of the coordinate $\bar{t}$ introduced in (7.124) above, let us now introduce the (closely related) *advanced time coordinate* v by

$$cv = ct + r + 2m \ln\left(\frac{r}{2m} - 1\right), \tag{7.128}$$

then

$$\mathrm{d}v = \mathrm{d}t + \frac{1}{c}\frac{r}{r-2m}\mathrm{d}r \equiv \mathrm{d}t - \mathrm{d}t^*. \tag{7.129}$$

From the Schwarzschild metric (7.118) we see that for radial photons

$$\mathrm{d}t = \frac{1}{c}\frac{r}{r-2m}\mathrm{d}r = -\mathrm{d}t^*,$$

so $\mathrm{d}t^*$ is the (Schwarzschild coordinate) time for a radial (incoming) photon to travel a distance $\mathrm{d}r$. Then, from (7.118)

$$\begin{aligned}\mathrm{d}s^2 &= -\left(1-\frac{2m}{r}\right)c^2\,\mathrm{d}t^2 + \left(1-\frac{2m}{r}\right)^{-1}\mathrm{d}r^2 + r^2\,\mathrm{d}\Omega^2\\ &= -\left(1-\frac{2m}{r}\right)\left[c\,\mathrm{d}v - \frac{r}{r-2m}\mathrm{d}r\right]^2 + \left(1-\frac{2m}{r}\right)^{-1}\mathrm{d}r^2 + r^2\,\mathrm{d}\Omega^2,\end{aligned}$$

or

$$\mathrm{d}s^2 = -\left(1-\frac{2m}{r}\right)c^2\,\mathrm{d}v^2 + 2c\,\mathrm{d}v\,\mathrm{d}r + r^2\,\mathrm{d}\Omega^2. \tag{7.130}$$

This line element is not invariant under $v \to -v$, which corresponds to $t \to -t$ and the substitution of inward-travelling rays by outward-travelling ones. Corresponding to this time reversal we find another section of the Schwarzschild solution. Introduce the *retarded time coordinate* u by

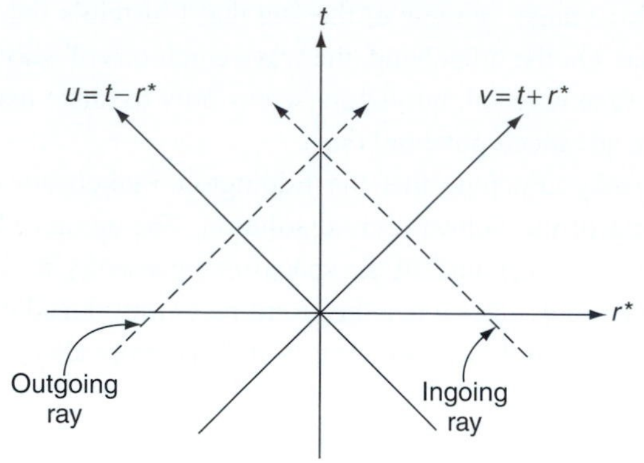

Fig. 7.4 Advanced (v) and retarded (u) null coordinates with ingoing (v = constant) and outgoing (u = constant) light rays.

$$cu = ct - r - 2m\ln\left(\frac{r}{2m} - 1\right), \tag{7.131}$$

so that

$$c(v - u) = 2r + 4m\ln\left(\frac{r}{2m} - 1\right). \tag{7.132}$$

Replacing (7.130) we have

$$ds^2 = -\left(1 - \frac{2m}{r}\right)c^2\,du^2 - 2c\,du\,dr + r^2\,d\Omega^2. \tag{7.133}$$

Writing the relations (7.128) and (7.131) as

$$u = t - r^*, \quad v = t + r^*, \tag{7.134}$$

the ingoing and outgoing light rays have the equations v = const and u = const respectively, as shown in Fig. 7.4, and u and v are therefore *null coordinates*. It is straightforward to see that

$$\begin{aligned} c^2\,du\,dv &= c^2\left(dt + \frac{1}{c}\frac{r}{r-2m}dr\right)\left(dt - \frac{1}{c}\frac{r}{r-2m}dr\right) \\ &= c^2\,dt^2 - \left(1 - \frac{2m}{r}\right)^{-2} dr^2 \end{aligned}$$

so that

$$c^2\left(1 - \frac{2m}{r}\right)du\,dv = \left(1 - \frac{2m}{r}\right)c^2\,dt^2 - \left(1 - \frac{2m}{r}\right)^{-1} dr^2$$

and the Schwarzschild metric is, in null coordinates

$$ds^2 = -c^2\left(1 - \frac{2m}{r}\right)du\,dv + r^2\,d\Omega^2. \tag{7.135}$$

Now instead of u and v let us introduce the coordinates

$$z = {}^1/_2(e^{cv/4m} + e^{-cu/4m}), \quad w = {}^1/_2(e^{cv/4m} - e^{-cu/4m}). \tag{7.136}$$

On substituting (7.128) and (7.131) we find

$$z = \left(\frac{r}{2m} - 1\right)^{1/2} e^{r/4m} \cosh(ct/4m), \tag{7.137}$$

$$w = \left(\frac{r}{2m} - 1\right)^{1/2} e^{r/4m} \sinh(ct/4m). \tag{7.138}$$

Clearly,

$$z^2 - w^2 = \left(\frac{r}{2m} - 1\right) e^{r/2m}, \tag{7.139}$$

$$\frac{w}{z} = \tanh(ct/4m) = \frac{1 - e^{-ct/2m}}{1 + e^{-ct/2m}}, \tag{7.140}$$

and after some algebra it follows that

$$dz = \alpha \cosh\left(\frac{ct}{4m}\right) dr + \beta \sinh\left(\frac{ct}{4m}\right) c\, dt, \tag{7.141}$$

$$dw = \alpha \sinh\left(\frac{ct}{4m}\right) dr + \beta \cosh\left(\frac{ct}{4m}\right) c\, dt, \tag{7.142}$$

where

$$\alpha^2 = \frac{r^2}{32m^3(r - 2m)} e^{r/2m}, \quad \beta^2 = \frac{(r - 2m)}{32m^3} e^{r/2m}. \tag{7.143}$$

Then

$$dz^2 - dw^2 = \alpha^2\, dr^2 - \beta^2 c^2\, dt^2;$$

hence

$$\frac{32m^3}{r} e^{-r/2m}(dz^2 - dw^2) = \frac{r}{r - 2m} dr^2 - \frac{r - 2m}{r} c^2\, dt^2$$

and the Schwarzschild line element is

$$ds^2 = \frac{32m^3}{r} e^{-r/2m}(dz^2 - dw^2) + r^2(z, w)\, d\Omega^2. \tag{7.144}$$

This is the Schwarzschild metric in Kruskal–Szekeres (or simply Kruskal) coordinates.[12] Restricted to radial motion (θ, ϕ constant) the line element is

[12] Kruskal (1960), Szekeres (1960).

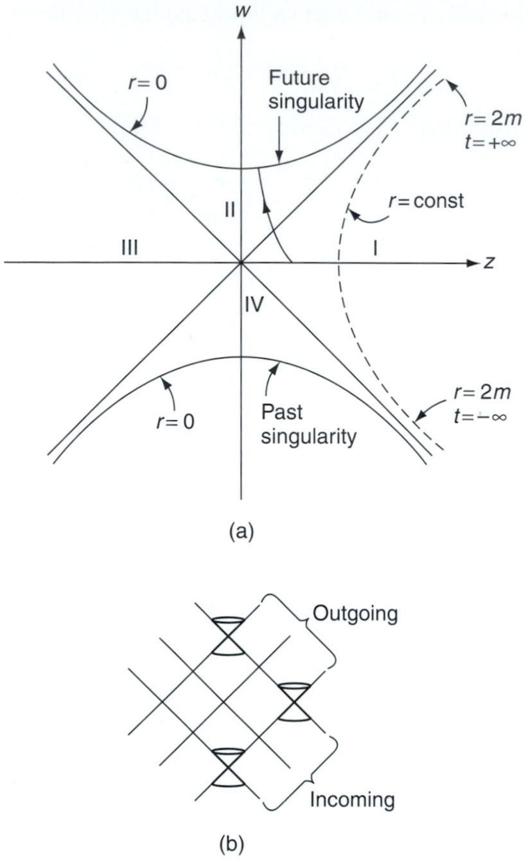

Fig. 7.5 **(a) Kruskal diagram with regions I, II, III and IV marked; (b) incoming (*r* increases, *t* increases) and outgoing (*r* decreases, *t* increases) light rays in the Kruskal diagram.**

$$ds^2 = \frac{32m^3}{r} e^{-r/2m}(dz^2 - dw^2), \tag{7.145}$$

which is conformally flat in the zw plane (i.e. an overall factor multiplied by the pseudo-Euclidean line element). Null radial geodesics $ds^2 = 0$ have the equations $dz = \pm\, dw$ – straight lines in the zw plane. This plane is shown in Fig. 7.5. The connection between z, w on the one hand and r, t on the other is given by Equations (7.139) and (7.140).

The Kruskal diagram has four sections, marked I, II, III, IV in Fig. 7.5. Region I represents the exterior $r > 2m$ of the Schwarzschild metric. It is bounded by $z = \pm w$, $z > 0$, i.e. by $r = 2m$, $t = \pm\infty$. The z axis is $t = 0$. The light cones are displayed, and one can see for example that a timelike geodesic could start on the z axis and cross the line $r = 2m$ from region I into region II. The Eddington–Finkelstein coordinates cover both of these regions. The geodesic ends on $r = 0$, a genuine singularity.

From the structure of the light cones one can see that starting from region I one can enter region II, but not regions III or IV. Once $r < 2m$, in region II, all null and timelike geodesics eventually reach $r = 0$; that is, r *inevitably* decreases – it plays a role like time in region

I (except that time inevitably *increases*!). Region II is a black hole: everything can go in, nothing can come out.

Our world (region I) cannot influence region III but it can be influenced by it: outgoing light rays can travel from III to I. In fact, all light rays and material objects (following timelike geodesics) eventually leave region III but nothing can enter it – it is a white hole. Region III is remote from us: nothing happening there can be influenced by or have an influence on anything in region I.

The Kruskal diagram makes evident the time inversion symmetry of the Schwarzschild solution (deriving, of course, from that of the Einstein field equations). In a white hole, time goes 'backwards' and gravity is repulsive rather than attractive. Nature does not make use of this; so perhaps the space-time represented by the whole Kruskal diagram does not correspond to anything in nature. One example of such a possibility, suggested by the diagram, is the Einstein–Rosen bridge and the phenomenon of 'wormholes'.

7.6 Einstein–Rosen bridge and wormholes

In Kruskal–Szekeres coordinates we are considering a 4-dimensional space-time with coordinates z, w, θ and ϕ, and line element (7.144). Let us take the time-slice (spacelike hypersurface) corresponding to $w=0$ (which, from (7.138) is also $t=0$) and at the same time put $\theta=\pi/2$ (the equator of a 2-sphere). Then

$$\mathrm{d}s^2 = \frac{32m^3}{r}\mathrm{e}^{-r/2m}\,\mathrm{d}z^2 + r^2\,\mathrm{d}\phi^2 = \left(\frac{r}{r-2m}\right)\mathrm{d}r^2 + r^2\,\mathrm{d}\phi^2. \qquad (7.146)$$

This is the metric on a (2-dimensional) surface which is a paraboloid of revolution, got by rotating the parabola $y=\dfrac{z^2}{8m}+2m$ about the z axis. This is shown in Fig. 7.6. Putting $y\to r$ we have

$$r = \frac{z^2}{8m} + 2m, \quad \mathrm{d}r = \frac{z}{4m}\mathrm{d}z, \quad \mathrm{d}r^2 = \frac{1}{2m}(r-2\,m)\,\mathrm{d}z^2,$$

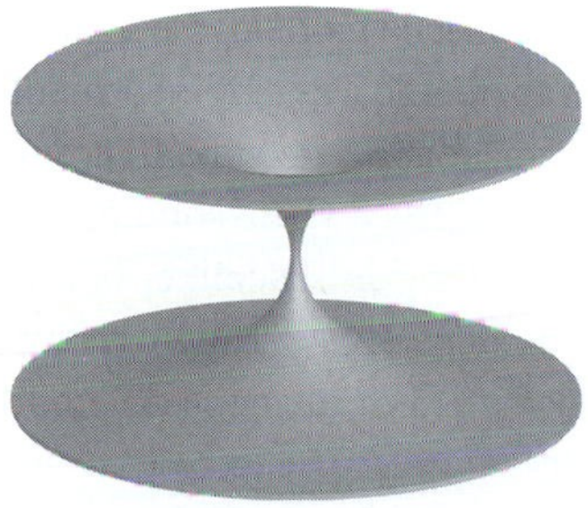

Fig. 7.6 A paraboloid of revolution about the vertical axis, representing a Schwarzschild throat, or wormhole, or Einstein–Rosen bridge.

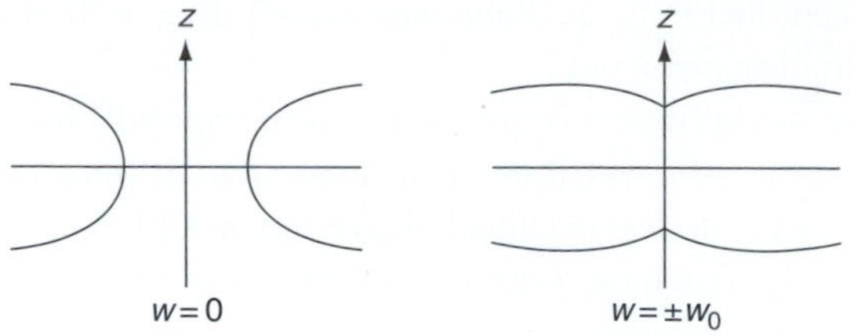

Fig. 7.7 A paraboloid of revolution at $w = 0$ has evolved from, and becomes, at $w = \pm w_0$, two separated space-times.

hence on the paraboloid surface

$$dz^2 = \left(\frac{2m}{r - 2m}\right) dr^2.$$

Taking a flat 3-dimensional space with cylindrical coordinates r, ϕ, z and metric

$$ds^2 = dr^2 + r^2\, d\phi^2 + dz^2,$$

then on the paraboloid surface in this space

$$ds^2 = dr^2\left(1 + \frac{2m}{r - 2m}\right) + r^2\, d\phi^2 = \left(\frac{r}{r - 2m}\right) dr^2 + r^2\, d\phi^2,$$

as in (7.146) above. Note that the flat 3-dimensional space into which this surface is embedded is *not* physical 3-dimensional space. Note also that one space dimension is suppressed in (7.146) so the diagram in Fig. 7.6 is not a 'picture' of the construction in any obvious sense, but only a representation of it.

This construction is called the 'Einstein–Rosen bridge', or Schwarzschild throat, or wormhole.[13] It is obtained by taking the spacelike hypersurface $w = 0$, i.e. passing from $z > 0$ to $z < 0$, from region I to region III in the Kruskal diagram. On taking a different hypersurface $w = \pm\, w_0$, such that the passage from I to III passes through II or IV and encounters the singularity $r = 0$, the paraboloids of revolution above are replaced by surfaces of revolution which are qualitatively different, and are shown in Fig. 7.7. In the second diagram here the two space-times are not connected. Hence the complete Kruskal diagram corresponds to an evolutionary, time-dependent, picture: as w (a timelike coordinate) evolves from $-\infty$ to 0, and then to $+\infty$, the two space-times, originally separate, become joined and then separate again. Hence the wormhole only lasts for a finite time – actually, a rather short time. It is initially surprising that the Kruskal extension of the Schwarzschild solution actually converts a static solution into a non-static one. This is because in regions I and III the parameter z is spacelike, but in II and IV it becomes timelike so a solution containing regions I and II is bound to feature time evolution. The germ of these observations was already contained in a remark in Finkelstein's paper (1958), about the lack of time reversal symmetry in a supposedly 'static' solution.

[13] Einstein & Rosen (1933).

7.7 Conformal treatment of infinity: Penrose diagrams

The technique of performing conformal transformations, or rescalings, of the metric gives a way of adjoining the 'points at infinity' in a space into the finite region. This means that the whole of the (infinite) manifold of space-time may be represented on a finite diagram. The resulting diagrams, known as Penrose diagrams, afford an elegant way of studying particular solutions, for example, the Kruskal diagram, as well as giving an insight into the causal structure of space-time. Let us begin by considering Minkowski space-time.

The metric for Minkowski space-time in spherical polar coordinates

$$d\tilde{s}^2 = -c^2\,dt^2 + dr^2 + r^2(d\theta^2 + \sin^2\theta\,d\phi^2)$$

may be re-expressed in null coordinates (advanced and retarded time coordinates)

$$u = ct - r, \qquad v = ct + r. \tag{7.147}$$

Since $du\,dv = c^2\,dt^2 - dr^2$ we have

$$d\tilde{s}^2 = -du\,dv + \frac{1}{4}(u - v)^2(d\theta^2 + \sin^2\theta\,d\phi^2). \tag{7.148}$$

Now introduce the *conformal factor*

$$\Omega^2 = \frac{1}{(1 + u^2)(1 + v^2)} \tag{7.149}$$

to give a rescaled metric

$$ds^2 = \Omega^2\,d\tilde{s}^2 = -\frac{du\,dv}{(1 + u^2)(1 + v^2)} + \frac{1}{4}\frac{(u - v)^2}{(1 + u^2)(1 + v^2)}(d\theta^2 + \sin^2\theta\,d\phi^2). \tag{7.150}$$

Introduce the new coordinates

$$u = \tan p, \qquad v = \tan q \tag{7.151}$$

so that

$$\frac{(u - v)^2}{(1 + u^2)(1 + v^2)} = \sin^2(p - q)$$

and (7.150) becomes

$$ds^2 = -dp\,dq + \frac{1}{4}\sin^2(p - q)[d\theta^2 + \sin^2\theta\,d\phi^2], \tag{7.152}$$

with

$$-\pi \le p, \quad q \le \pi. \tag{7.153}$$

This is the *conformally rescaled metric* of Minkowski space in null coordinates.[14] The first term ($\mathrm{d}p\,\mathrm{d}q$) has the same form as the term $\mathrm{d}u\,\mathrm{d}v$ in the usual Minkowski metric (7.148) but it covers only a compact region. Let us put (cf. (7.147))

$$p = cT - R, \quad q = cT + R. \tag{7.154}$$

In terms of the coordinates r, t we may represent the different types of infinity in Minkowski space as in Fig. 7.8(a). The past and future light cones extend to past and future null infinity, and in turn separate past and future timelike infinity on the one hand, and spacelike infinity on the other. In terms of the coordinates R and T, however, these infinities are now in the finite domain, as shown in Fig. 7.8(b). Null future infinity and null past infinity are represented by the lines $\mathscr{I}^+$ and $\mathscr{I}^-$ (pronounced scri plus and scri minus). Timelike future and timelike past infinity are represented by the points $\boldsymbol{i}^+$ and $\boldsymbol{i}^-$ and the two points $\boldsymbol{i}^0$ represent spacelike infinity. In terms of the coordinates these lines and points are

$$\begin{aligned}
&\boldsymbol{i}^+\colon cT = \pi,\\
&\boldsymbol{i}^-\colon cT = -\pi,\\
&\boldsymbol{i}^0\colon q = -p = \pi, \quad R = \pi; \; p = -q = \pi, \quad R = -\pi,\\
&\mathscr{I}^+\colon cT \pm R = \pi,\\
&\mathscr{I}^-\colon cT \pm R = -\pi.
\end{aligned} \tag{7.155}$$

In the compactification of the full 4-dimensional Minkowski space-time each point in the diagram of Fig. 7.8 becomes a 2-sphere. Putting $\theta = \pi/2$ in (7.152) (so with only one dimension suppressed), the compactified diagram becomes as shown in Fig. 7.9.

Let us now consider the Schwarzschild solution in Kruskal coordinates. We begin with the metric (7.135)

$$\mathrm{d}s^2 = -c^2\left(1 - \frac{2m}{r}\right)\mathrm{d}u\,\mathrm{d}v + r^2\,\mathrm{d}\Omega^2, \tag{7.156}$$

where u and v are the retarded and advanced time coordinates (7.128) and (7.131). Now introduce new coordinates U, V:

$$U = -4m\mathrm{e}^{-u/4m}, \quad V = 4m\mathrm{e}^{v/4m}, \tag{7.157}$$

so that

$$\frac{2m}{r}\mathrm{e}^{-r/2m}\,\mathrm{d}U\,\mathrm{d}V = \left(1 - \frac{2m}{r}\right)\mathrm{d}u\,\mathrm{d}v,$$

and the above metric becomes

$$\mathrm{d}s^2 = -\frac{2mc^2}{r}\mathrm{e}^{-r/2m}\,\mathrm{d}U\,\mathrm{d}V + r^2\,\mathrm{d}\Omega^2. \tag{7.158}$$

[14] It should be noted that conformal rescalings of the type discussed here are to be *distinguished* from conformal transformations on coordinates. These latter transformations may be combined with those of the Poincaré group to give an enlarged symmetry group under which some field theories (for example Maxwell's electrodynamics) are invariant. See for example Fulton *et al.* (1962), Wess (1960).

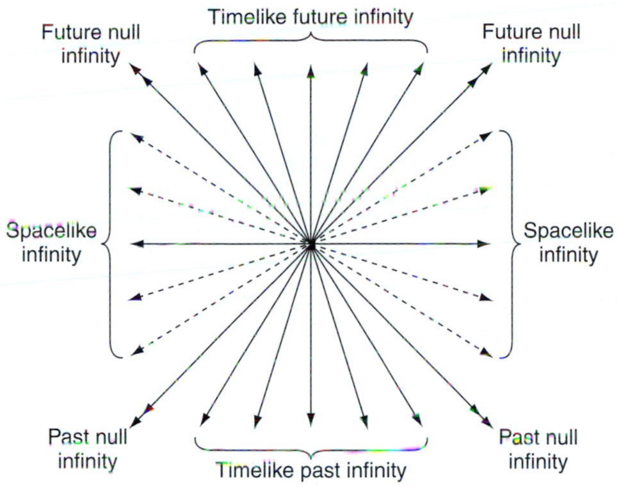

(a)

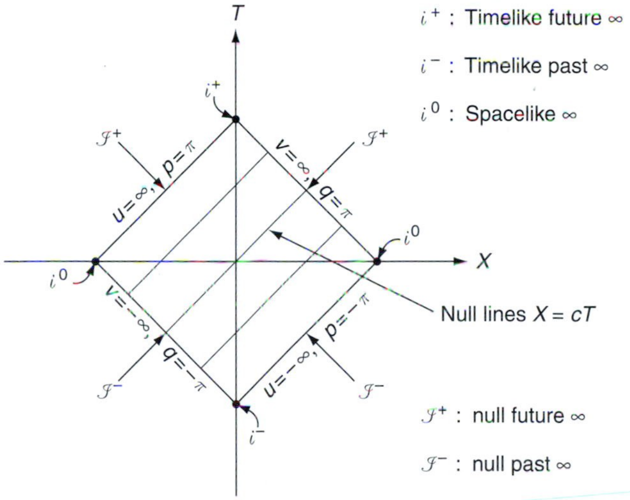

(b)

Fig. 7.8 **(a) Infinities in Minkowski space-time: timelike past and future, null past and future, and spacelike; (b) these infinities in the finite domain, in a compactified diagram.**

The parameters u, v have the same range of values as in the Penrose diagram for Minkowski space, Fig. 7.10, but this range is not shared by U and V:

$$\begin{array}{llcll} u=-\infty, & U=-\infty & : & v=-\infty, & V=0 \\ u=0, & U=-4m & : & v=0, & V=4m \\ u=\infty, & U=0 & : & v=\infty, & V=\infty \end{array}$$

We see that $-\infty \leq U \leq 0$, whereas $0 \leq V \leq \infty$, so the Penrose diagram becomes the one shown in Fig. 7.10, which is not maximal. It may be extended by adding another diagram

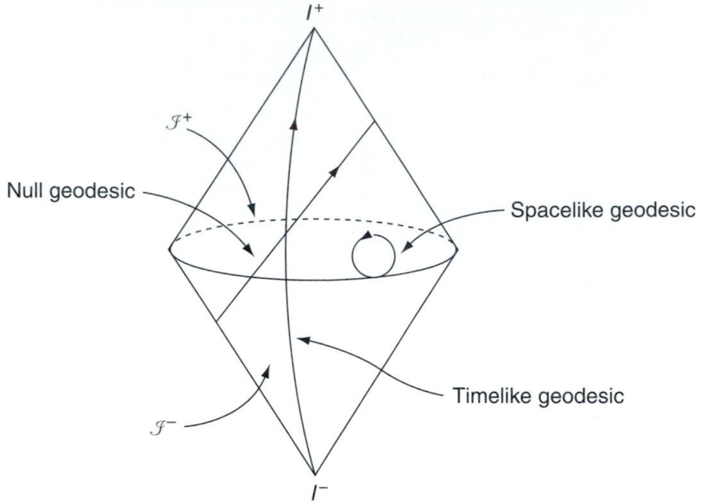

Fig. 7.9 Compactified Minkowski space-time with one dimension suppressed.

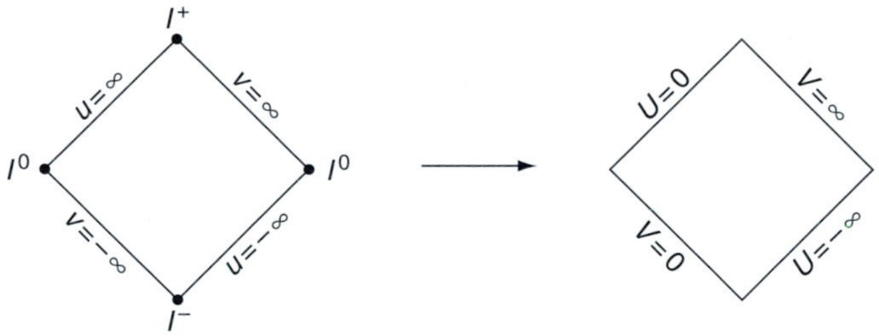

Fig. 7.10 The Penrose diagram for the parameters u, v translated into one for U, V.

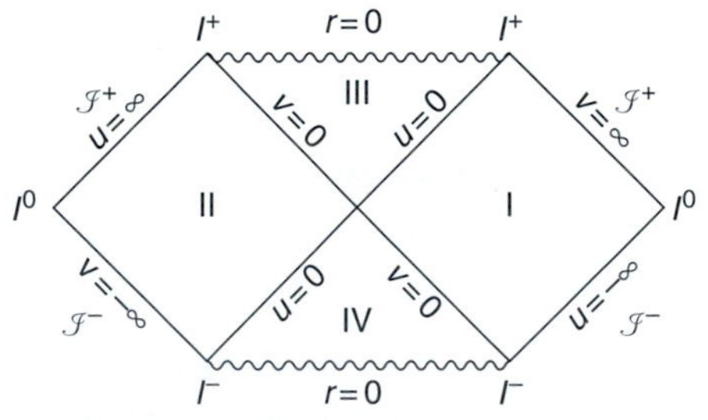

Fig. 7.11 Penrose diagram for the Kruskal solution.

with $U \geq 0$, $V \leq 0$, as in Fig. 7.11. The left half of this diagram differs from the right half by time reversal, $t \to -t$. The diagram then represents the Schwarzschild solution (in Kruskal coordinates) over the whole range $-\infty \leq U, V \leq \infty$. Region II is the time-reversed mirror of region I. The horizontal zig-zag lines represent the singularity $r=0$ in the past and the future.

For an observer in region I all null rays $U > 0$ reach null future infinity $\mathcal{J}^+$: the null ray $U = 0$ ($r = 2m$, $t = +\infty$) is the last (retarded) null ray to reach $\mathcal{J}^+$, hence the ray $U = 0$ is an event horizon for observers in region I. Similarly $V = 0$ is an event horizon for observers in region II. The regions III and IV are therefore black holes for these observers.

Let us revert to some more physical considerations regarding black holes. The discussion above was related to the (interior) Schwarzschild solution and therefore describes *static* black holes (modulo the observations made about wormholes). Since real stars, however, *rotate*, we must ask if there is a solution of the field equations which corresponds to a rotating source. And indeed there is; it is the Kerr solution and will be discussed in the next section. At the close of this section, however, it is perhaps appropriate to mention that there is also a solution of the field equations describing a star with *electric charge*. This is actually a solution of the field equations (5.24), with the energy-momentum tensor that of the electromagnetic field: it therefore describes a (non-rotating) black hole carrying electric charge and is known as the *Reissner–Nordstrøm* solution.[15] It has the form

$$ds^2 = -c^2\left(1 - \frac{2m}{r} + \frac{q^2}{r^2}\right)dt^2 + \left(1 - \frac{2m}{r} + \frac{q^2}{r^2}\right)^{-1} dr^2 + r^2\, d\Omega^2, \tag{7.159}$$

with

$$q^2 = \frac{GQ^2}{4\pi\varepsilon_0 c^4}, \quad Q = \text{electric charge}.$$

This solution describes the *gravitational effect* of electric charge; an electrically charged object (star) gives rise to an electromagnetic field, whose energy-momentum tensor contributes, via Einstein's field equations, to the curvature of space. The geometry of space-time is therefore modified by electric charge, as expressed in the Reissner–Nordstrøm solution. As far as the real world is concerned, however, I think it is reasonable to take the view that, so far as we know, this is a physically *unrealistic* situation. Stars are electrically neutral, so while the Reissner–Nordstrøm solution might be interesting as a mathematical exercise, it probably does not have physical relevance. For the interested reader there are many good accounts of this solution in the literature.

7.8 Rotating black holes: Kerr solution

We begin by introducing the Newman–Penrose *null tetrad*,[16] which is of great use in modern relativity theory and turns out to offer an approach to the Kerr solution. In R^3 the three basis vectors are, in Cartesian coordinates,

$$\mathbf{e}_1 = \mathbf{i} = (1, 0, 0), \quad \mathbf{e}_2 = \mathbf{j} = (0, 1, 0), \quad \mathbf{e}_3 = \mathbf{k} = (0, 0, 1),$$

[15] Reissner (1916), Nordstrøm (1918).
[16] Newman & Penrose (1962).

respectively perpendicular to the planes x = const, y = const and z = const. Then the components are $i_1=1, j_3=0$, etc. The metric tensor is δ_{mn} and we have the relation

$$\delta_{mn} = i_m\, i_n + j_m\, j_n + k_m\, k_n,$$

as may easily be checked. To progress to Minkowski space-time the 3-vectors above are upgraded to spacelike 4-vectors

$$i_\mu = (0,1,0,0), \quad j_\mu = (0,0,1,0), \quad k_\mu = (0,0,0,1) \tag{7.160}$$

and to them we add the timelike 4-vector

$$u_\mu = (-1,0,0,0) \tag{7.161}$$

so the Minkowski metric tensor may be written

$$\eta_{\mu\nu} = -u_\mu\, u_\nu + i_\mu\, i_\nu + j_\mu\, j_\nu + k_\mu\, k_\nu = \text{diag}(-1,1,1,1), \tag{7.162}$$

consistent with our metric convention. We also have $u^\mu = (1, 0, 0, 0)$, $i^\mu = (0, 1, 0, 0)$, etc., giving

$$u^\mu\, u_\mu = -1, \quad i^\mu\, i_\mu = j^\mu\, j_\mu = k^\mu\, k_\mu = +1, \tag{7.163}$$

$$u^\mu\, i_\mu = u^\mu\, j_\mu = u^\mu\, k_\mu = 0. \tag{7.164}$$

Then, due to the non-positive definite metric, we may construct two *null vectors*

$$l_\mu = \frac{1}{\sqrt{2}}(u_\mu + i_\mu), \quad n_\mu = \frac{1}{\sqrt{2}}(u_\mu - i_\mu), \tag{7.165}$$

with inverses

$$u_\mu = \frac{1}{\sqrt{2}}(l_\mu + n_\mu), \quad i_\mu = \frac{1}{\sqrt{2}}(l_\mu - n_\mu). \tag{7.166}$$

The following relations are easily checked:

$$\begin{aligned} l^\mu\, n_\mu &= \frac{1}{2}(u^\mu\, u_\mu - i^\mu\, i_\mu) = -1, \\ l^\mu\, l_\mu &= \frac{1}{2}(u^\mu\, u_\mu + i^\mu\, i_\mu) = 0, \quad n^\mu\, n_\mu = 0, \end{aligned} \tag{7.167}$$

showing that we have indeed constructed two null vectors from two vectors (u_μ and i_μ) of the basis set (7.160), (7.161) – one of which is timelike and one spacelike. We now want to construct two more null vectors, using combinations of j^μ and k^μ, which are both spacelike. To do this take the *complex* combinations

$$m^\mu = \frac{1}{\sqrt{2}}(j^\mu + i\,k^\mu), \quad \bar{m}^\mu = \frac{1}{\sqrt{2}}(j^\mu - i\,k^\mu), \tag{7.168a}$$

(with $i = \sqrt{-1}$), with their inverses

$$j^\mu = \frac{1}{\sqrt{2}}(m^\mu + \bar{m}^\mu), \quad k^\mu = \frac{-i}{\sqrt{2}}(m^\mu - \bar{m}^\mu). \tag{7.168b}$$

Then

$$m^\mu m_\mu = \bar{m}^\mu \bar{m}_\mu = 0, m^\mu \bar{m}^\mu = 1, \tag{7.169}$$

showing that m^μ and $\bar{m}^\mu$ are indeed null (but not orthogonal). We also have the relations

$$m^\mu l_\mu = m^\mu n_\mu = \bar{m}^\mu l_\mu = \bar{m}^\mu n_\mu = 0. \tag{7.170}$$

Then choosing for the tetrad components (see Section 6.5)

$$h^\mu{}_{(0)} = l^\mu, \quad h^\mu{}_{(1)} = n^\mu, \quad h^\mu{}_{(2)} = m^\mu, \quad h^\mu{}_{(3)} = \bar{m}^\mu, \tag{7.171}$$

we have the *tangent space metric* (or *frame metric*)

$$g_{(\alpha)(\beta)} = h^\mu{}_{(\alpha)} h_{\mu(\beta)}, \tag{7.172}$$

with components, for example,

$$g_{(0)(0)} = l^\mu l_\mu = 0, \;\; g_{(0)(1)} = g_{(1)(0)} = l^\mu n_\mu = -1, \;\; g_{(2)(3)} = g_{(3)(2)} = m^\mu \bar{m}_\mu = 1,$$

or

$$g_{(\alpha)(\beta)} = \begin{pmatrix} 0 & -1 & 0 & 0 \\ -1 & 0 & 0 & 0 \\ 0 & 0 & 0 & 1 \\ 0 & 0 & 1 & 0 \end{pmatrix}. \tag{7.173}$$

The *world space metric* is, from (7.162), (7.166), (7.168b),

$$\begin{aligned} g_{\mu\nu} = \eta_{\mu\nu} &= -\frac{1}{2}(l_\mu + n_\mu)(l_\nu + n_\nu) + \frac{1}{2}(l_\mu - n_\mu)(l_\nu - n_\nu) \\ &\quad + \frac{1}{2}(m_\mu + \bar{m}_\mu)(m_\nu + \bar{m}_\nu) - \frac{1}{2}(m_\mu - \bar{m}_\mu)(m_\nu - \bar{m}_\nu) \\ &= -l_\mu n_\nu - n_\mu l_\nu + \bar{m}_\mu m_\nu + m_\mu \bar{m}_\nu, \end{aligned} \tag{7.174}$$

with the equivalent contravariant form

$$g^{\mu\nu} = -l^\mu n^\nu - n^\mu l^\nu + m^\mu \bar{m}^\nu + \bar{m}^\mu m^\nu. \tag{7.175}$$

The above equations apply to *Minkowski* space-time. Let us now, as the next step towards the Kerr solution, find expressions for the tetrads l^μ, n^μ, m^μ and $\bar{m}^\mu$ relevant to the Schwarzschild metric in advanced Kruskal coordinates. This is, from (7.130),

$$g_{\mu\nu} = \begin{pmatrix} -c^2\left(1 - \frac{2m}{r}\right) & c & 0 & 0 \\ c & 0 & 0 & 0 \\ 0 & 0 & r^2 & 0 \\ 0 & 0 & 0 & r^2 \sin^2\theta \end{pmatrix}, \tag{7.176}$$

with the corresponding contravariant components

$$g^{\mu\nu} = \begin{pmatrix} 0 & \dfrac{1}{c} & 0 & 0 \\ \dfrac{1}{c} & \left(1 - \dfrac{2m}{r}\right) & 0 & 0 \\ 0 & 0 & \dfrac{1}{r^2} & 0 \\ 0 & 0 & 0 & \dfrac{1}{r^2 \sin^2\theta} \end{pmatrix}. \tag{7.177}$$

This is equivalent to the following null tetrad

$$\begin{aligned} l^\mu &= (0, 1, 0, 0) = \partial_r, \\ n^\mu &= \left(-\frac{1}{c}, -\frac{1}{2}\left(1 - \frac{2m}{r}\right), 0, 0\right) = -\frac{1}{c}\partial_v - \frac{1}{2}\left(1 - \frac{2m}{r}\right)\partial_r, \\ m^\mu &= \frac{1}{r\sqrt{2}}\left(0, 0, 1, \frac{\mathrm{i}}{\sin\theta}\right) = \frac{1}{r\sqrt{2}}(\partial_\theta + \mathrm{i}\,\mathrm{cosec}\,\theta\,\partial_\phi), \\ \bar{m}^\mu &= \frac{1}{r\sqrt{2}}\left(0, 0, 1, -\frac{\mathrm{i}}{\sin\theta}\right) = \frac{1}{r\sqrt{2}}(\partial_\theta - \mathrm{i}\,\mathrm{cosec}\,\theta\,\partial_\phi), \end{aligned} \tag{7.178}$$

as may easily be checked: for example

$$\begin{aligned} g^{00} &= -2l^0\, n^0 + 2m^0\, \bar{m}^0 = 0, \\ g^{01} &= -l^0\, n^1 - n^0\, n^1 + m^0\, \bar{m}^1 + \bar{m}^0\, m^1 = \frac{1}{c}, \\ g^{11} &= -2l^1\, n^1 + 2m^1\, \bar{m}^1 = 1 - \frac{2m}{r}, \end{aligned}$$

and so on, as in (7.177).

The key to finding the Kerr solution is, following Newman and Janis,[17] to make r complex (with $\bar{r}$ its complex conjugate) and replace the tetrad above by

$$\begin{aligned} l^\mu &= (0, 1, 0, 0) = \partial_r, \\ n^\mu &= \left(-\frac{1}{c}, -\frac{1}{2} + \frac{m}{2}\left(\frac{1}{r} + \frac{1}{\bar{r}}\right), 0, 0\right) = -\frac{1}{c}\partial_v - \frac{1}{2}M\,\partial_r, \\ m^\mu &= \frac{1}{\bar{r}\sqrt{2}}\left(0, 0, 1, \frac{\mathrm{i}}{\sin\theta}\right) = \frac{1}{\bar{r}\sqrt{2}}(\partial_\theta + \mathrm{i}\,\mathrm{cosec}\,\theta\,\partial_\phi), \\ \bar{m}^\mu &= \frac{1}{r\sqrt{2}}\left(0, 0, 1, -\frac{\mathrm{i}}{\sin\theta}\right) = \frac{1}{r\sqrt{2}}(\partial_\theta - \mathrm{i}\,\mathrm{cosec}\,\theta\,\partial_\phi), \end{aligned} \tag{7.179}$$

with

$$M = 1 - \frac{m}{r} - \frac{m}{\bar{r}}. \tag{7.180}$$

(Note that l^μ and n^μ are still real, and m^μ and $\bar{m}^\mu$ the complex conjugates of each other.) Now perform the transformations

[17] Newman & Janis (1965).

$$r = r' - \frac{ia}{c}\cos\theta, \qquad v = v' - \frac{ia}{c^2}\cos\theta, \text{(with } r',\ v' \text{ real)}; \tag{7.181}$$

then the tetrad becomes

$$\begin{aligned} l'^{\mu} &= (0,1,0,0), \\ n'^{\mu} &= \left(-\frac{1}{c}, -\frac{1}{2} + \frac{mr'}{r'^2 + \frac{a^2}{c^2}\cos^2\theta}, 0, 0\right), \\ m'^{\mu} &= \frac{1}{(r' + \frac{ia}{c}\cos\theta)\sqrt{2}}\left[-\frac{ia}{c}\sin\theta, -\frac{ia}{c}\sin\theta, 1, \frac{i}{\sin\theta}\right], \\ \bar{m}'^{\mu} &= \frac{1}{(r' - \frac{ia}{c}\cos\theta)\sqrt{2}}\left[\frac{ia}{c}\sin\theta, \frac{ia}{c}\sin\theta, 1, -\frac{i}{\sin\theta}\right]. \end{aligned} \tag{7.182}$$

(These equations follow quite straightforwardly from the usual formula

$$V'^{\mu} = \frac{\partial x'^{\mu}}{\partial x^{\nu}} V^{\nu};$$

so for example

$$m'^{0} = \frac{\partial(cv')}{\partial(cv)} m^0 + \frac{\partial(cv')}{\partial\theta} m^2 = -\frac{ia}{c}\sin\theta . \frac{1}{\bar{r}\sqrt{2}} .\Bigg)$$

The contravariant components of the metric tensor then follow from (7.175), with (7.182) (and dropping the primes); for example

$$g^{00} = -2l^0 n^0 + 2m^0\,\bar{m}^0 = \frac{a^2\sin^2\theta}{\rho^2},$$

where we have defined

$$\rho^2 = r^2 + \frac{a^2}{c^2}\cos^2\theta. \tag{7.183}$$

We find

$$g^{\mu\nu} = \begin{pmatrix} \dfrac{(a/c)^2\sin^2\theta}{\rho^2} & \dfrac{r^2 + a^2/c^2}{\rho^2} & 0 & -\dfrac{a}{c\rho^2} \\ \dfrac{r^2 + a^2/c^2}{\rho^2} & \dfrac{r^2 + a^2/c^2 - 2mr}{\rho^2} & 0 & -\dfrac{a}{c\rho^2} \\ 0 & 0 & \dfrac{1}{\rho^2} & 0 \\ -\dfrac{a}{c\rho^2} & -\dfrac{a}{c\rho^2} & 0 & \dfrac{1}{\rho^2\sin^2\theta} \end{pmatrix}. \tag{7.184}$$

The parameter a, introduced in the transformation (7.181), is a key feature of this metric. Below we shall justify its interpretation as the *angular momentum* of the source. It is straightforward to calculate the covariant components $g_{\mu\nu}$ by simply finding the inverse of (7.184) (which has determinant $-(\rho^4\sin^2\theta)^{-1}$). The resulting line element is

$$\mathrm{d}s^2 = -\left(1-\frac{2mr}{\rho^2}\right)c^2\,\mathrm{d}v^2 + 2c\,\mathrm{d}v\,\mathrm{d}r - \frac{4mra}{c\rho^2}\sin^2\theta\, c\,\mathrm{d}v\,\mathrm{d}\phi$$
$$-\frac{2a}{c}\sin^2\theta\,\mathrm{d}r\,\mathrm{d}\phi + \rho^2\,\mathrm{d}\theta^2 + \left\{\left(r^2+\frac{a^2}{c^2}\right)\sin^2\theta + \frac{2mra^2}{c^2\rho^2}\sin^4\theta\right\}\mathrm{d}\phi^2. \tag{7.185}$$

Now perform the coordinate transformations

$$c\,\mathrm{d}v = c\,\mathrm{d}t + \frac{r^2+a^2/c^2}{\Delta}\,\mathrm{d}r, \quad \mathrm{d}\phi = \mathrm{d}\psi + \frac{a}{c\Delta}\,\mathrm{d}r, \tag{7.186}$$

with

$$\Delta = r^2 + \frac{a^2}{c^2} - 2mr. \tag{7.187}$$

After some algebra (and relabelling $\psi \to \phi$) we find

$$\mathrm{d}s^2 = -\left(1-\frac{2mr}{\rho^2}\right)c^2\,\mathrm{d}t^2 - \frac{4mra\sin^2\theta}{\rho^2}\,\mathrm{d}\phi\,\mathrm{d}t + \frac{\rho^2}{\Delta}\mathrm{d}r^2 + \rho^2\,\mathrm{d}\theta^2$$
$$+\left(r^2+\frac{a^2}{c^2}+\frac{2mra^2\sin^2\theta}{c^2\rho^2}\right)\sin^2\theta\,\mathrm{d}\phi^2. \tag{7.188}$$

This is the Kerr solution in Boyer–Linquist coordinates.[18] Equivalent forms are

$$\mathrm{d}s^2 = -\frac{\Delta}{\rho^2}\left[c\,\mathrm{d}t - \frac{a}{c}\sin^2\theta\,\mathrm{d}\phi\right]^2 + \frac{\rho^2}{\Delta}\mathrm{d}r^2 + \rho^2\,\mathrm{d}\theta^2$$
$$+\frac{\sin^2\theta}{\rho^2}\left[\left(r^2+\frac{a^2}{c^2}\right)\mathrm{d}\phi - a\,\mathrm{d}t\right]^2 \tag{7.189}$$

and

$$\mathrm{d}s^2 = -c^2\,\mathrm{d}t^2 + \frac{2mr}{\rho^2}\left[c\,\mathrm{d}t - \frac{a}{c}\sin^2\theta\,\mathrm{d}\phi\right]^2 + \frac{\rho^2}{\Delta}\,\mathrm{d}r^2 + \rho^2\,\mathrm{d}\theta^2$$
$$+\left(r^2+\frac{a^2}{c^2}\right)\sin^2\theta\,\mathrm{d}\phi^2. \tag{7.190}$$

(Note that when $a=0$ these reduce, as expected, to the Schwarzschild form (5.37).) The metric tensor (7.188) is

$$g_{\mu\nu} = \begin{pmatrix} -\left(1-\frac{2mr}{\rho^2}\right) & 0 & 0 & -\frac{2mra\sin^2\theta}{c\rho^2} \\ 0 & \frac{\rho^2}{\Delta} & 0 & 0 \\ 0 & 0 & \rho^2 & 0 \\ -\frac{2mra\sin^2\theta}{c\rho^2} & 0 & 0 & \left(r^2+\frac{a^2}{c^2}+\frac{2mra^2\sin^2\theta}{c^2\rho^2}\right)\sin^2\theta \end{pmatrix}, \tag{7.191}$$

[18] Kerr (1963), Boyer & Lindquist (1967).

which may be compared with the metric (6.93) for the gravitomagnetic field. In fact, taking an approximation for large r (7.191) becomes

$$g_{\mu\nu} = \begin{pmatrix} -\left(1-\dfrac{2m}{r}\right) & 0 & 0 & -\dfrac{2ma}{cr}\sin^2\theta \\ 0 & 1+\dfrac{2m}{r} & 0 & 0 \\ 0 & 0 & r^2 & 0 \\ -\dfrac{2ma}{cr}\sin^2\theta & 0 & 0 & r^2\sin^2\theta \end{pmatrix}, \tag{7.192}$$

and comparing the g_{03} terms in (6.93) and above, recalling that $m = \dfrac{GM}{c^2}$, we see that $a = \dfrac{J}{Mc}$, essentially the angular momentum per unit mass of the source. This is the interpretation of the parameter a occurring in the Kerr solution, and introduced in Equation (7.181). It is clear that in the limit $r \to \infty$ (7.192) becomes the Minkowski metric. It is also clear, in general terms, that as $a \to 0$ the Kerr metric (7.191) becomes the Schwarzschild metric. This solution was found by Kerr in 1963, though his reasoning was different from that given above, being based on the fact that the solution is of an 'algebraically special' type of solution to Einstein's vacuum field equations. In any case we may remark that the Kerr solution is an exact two-parameter solution to the field equations, the Schwarzschild solution being a one-parameter one.

In closing this section it may be noted that the generalisation of the Kerr solution to that of a rotating source *with electric charge* is the Kerr–Newman solution

$$\begin{aligned} ds^2 = &-\frac{\Delta}{\rho^2}\left[c\,dt - \frac{a}{c}\sin^2\theta\,d\phi\right]^2 + \frac{\rho^2}{\Delta}\,dr^2 + \rho^2\,d\theta^2 \\ &+ \frac{\sin^2\theta}{\rho^2}\left[\left(r^2+\frac{a^2}{c^2}\right)d\phi - a\,dt\right]^2 \end{aligned}$$

where ρ^2 is given by (7.183), as before, but Δ is now given by

$$\Delta = r^2 + \frac{a^2}{c^2} - 2\,mr + q^2,$$

a generalisation of (7.187).

7.9 The ergosphere and energy extraction from a black hole

In the Schwarzschild solution the surface $r = 2m$ is an event horizon (EH) (see below Equation (7.126)). It is *also* a surface of infinite red-shift (SIR), as we shall now show. In the Kerr solution these surfaces are different. The area between them is called the *ergosphere* and it is a region in which there is some interesting physics.

We begin by investigating the event horizon more fully and then finding what it is in the Kerr solution. A hypersurface has the equation

$$f(x^0, x^1, x^2, x^3) = 0$$

and the normal to it is the (covariant) vector $n_\mu = \dfrac{\partial f}{\partial x^\mu}$. In the case of the Schwarzschild solution the EH is given by $r - 2m = 0$, so the function f may be written

$$f = r - 2m = x^1 - 2m,$$

and

$$n_\mu = (0, 1, 0, 0).$$

The hypersurface is *null* if $n^\mu n_\mu = 0$. To find n^μ we need $g^{\mu\nu}$. Working in Eddington–Finkelstein coordinates the covariant metric tensor is, from (7.125),

$$g_{\mu\nu} = \begin{pmatrix} -\left(1 - \dfrac{2m}{r}\right) & \dfrac{2m}{r} & 0 & 0 \\ \dfrac{2m}{r} & 1 + \dfrac{2m}{r} & 0 & 0 \\ 0 & 0 & r^2 & 0 \\ 0 & 0 & 0 & r^2 \sin^2\theta \end{pmatrix}, \tag{7.193}$$

from which $g = -r^4 \sin^2\theta$ and

$$g^{\mu\nu} = \begin{pmatrix} -\left(1 + \dfrac{2m}{r}\right) & \dfrac{2m}{r} & 0 & 0 \\ \dfrac{2m}{r} & 1 - \dfrac{2m}{r} & 0 & 0 \\ 0 & 0 & \dfrac{1}{r^2} & 0 \\ 0 & 0 & 0 & \dfrac{1}{r^2 \sin^2\theta} \end{pmatrix}; \tag{7.194}$$

so

$$n^\mu n_\mu = g^{\mu\nu} n_\mu n_\nu = g^{11} = 1 - \frac{2m}{r},$$

that is, $n^\mu n_\mu = 0$ on the hypersurface $r = 2m$, as we aimed to show.

Now let us turn to the Kerr solution, for which $g_{\mu\nu}$ is (7.191) above. The contravariant $g^{\mu\nu}$ is

$$g^{\mu\nu} = \begin{pmatrix} -\dfrac{\left(r^2 + \frac{a^2}{c^2}\right)^2 - \frac{a^2}{c^2}\Delta \sin^2\theta}{c^2\rho^2\Delta} & 0 & 0 & -\dfrac{2mra}{c\rho^2\Delta} \\ 0 & \dfrac{\Delta}{\rho^2} & 0 & 0 \\ 0 & 0 & \dfrac{1}{\rho^2} & 0 \\ -\dfrac{2mra}{c\rho^2\Delta} & 0 & 0 & -\dfrac{\frac{a^2}{c^2}\sin^2\theta - \Delta}{\rho^2\Delta \sin^2\theta} \end{pmatrix}. \tag{7.195}$$

We claim that the null hypersurface is $\Delta = 0$, whose solution is (see (7.187))

$$r = r_{\pm} = m \pm \sqrt{m^2 - \frac{a^2}{c^2}}; \tag{7.196}$$

equivalently

$$r_+{}^2 + \frac{a^2}{c^2} = 2mr_+, \tag{7.197}$$

where we have singled out the solution $r = r_+$, which is the one of greatest interest (the solution $r = r_-$ being further inside the source). Consider, then, the hypersurface

$$f = r - r_+ = r - m - \sqrt{m^2 - \frac{a^2}{c^2}}.$$

The normal is $n_\mu = f_{,\mu} = (0, 1, 0, 0)$, so

$$n^\mu n_\mu = g^{\mu\nu} n_\mu n_\nu = g^{11} = \frac{\Delta}{\rho^2}\ .$$

But $\Delta = 0$ at $r = r_+$, hence

$$n^\mu n_\mu = 0 \quad \text{on} \quad r = r_+\,,$$

justifying our claim that $r = r_+$ is a *null hypersurface*. Light will propagate within this surface, without leaving it – it will not escape from the Kerr black hole. We have therefore shown that the event horizon of the Kerr black hole is:

$$\text{Kerr event horizon:} \quad r = r_+ = m + \sqrt{m^2 - \frac{a^2}{c^2}}. \tag{7.198}$$

(As expected, when $a = 0$, this reduces to the Schwarzschild case $r = 2m$.)

We now turn to the other property of the surface $r = 2m$ in the Schwarzschild case, that it is a surface of infinite red-shift (SIR). The red-shift factor is

$$z = \frac{\nu_{\text{em}}}{\nu_{\text{rec}}} - 1,$$

the symbols referring to the frequencies of emitted and received light. If light is emitted at r_2 and received at r_1 we have, from (5.46),

$$z = \frac{\nu^2}{\nu^1} - 1 = \sqrt{\frac{g_{00}(r_1)}{g_{00}(r_2)}} - 1,$$

with $g_{00}(r) = 1 - \dfrac{2m}{r}$. So for light emitted at r and received at infinity, $g_{00}(r_1) = 1$ and

$$z = \frac{1}{\sqrt{g_{00}(r)}} - 1 = \sqrt{\frac{r}{r - 2m}} - 1,$$

and as $r \to 2m$, $z \to \infty$. The surface $r = 2m$ is therefore an SIR, defined by $g_{00} = 0$.

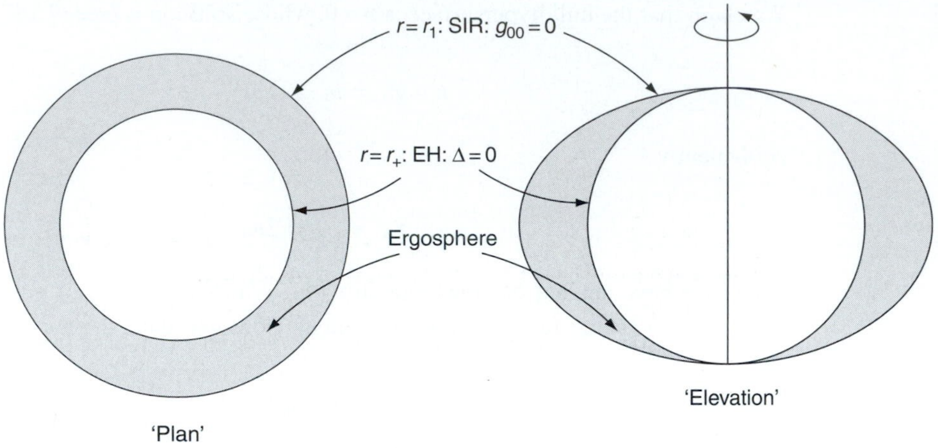

Fig. 7.12 **The ergosphere in plan and in elevation.**

In the case of the Kerr solution the condition $g_{00}=0$ gives, from (7.191), with (7.183),

$$\rho^2 = 2mr \Rightarrow r^2 - 2mr + \frac{a^2}{c^2}\cos^2\theta = 0$$

with the solutions

$$r_{1,2} = m \pm \sqrt{m^2 - \frac{a^2}{c^2}\cos^2\theta},$$

the interesting case of which is $r=r_1$: so

$$\text{Kerr SIR}: \quad r = r_1 = m + \sqrt{m^2 - \frac{a^2}{c^2}\cos^2\theta} \tag{7.199}$$

(which again, in the case $a=0$, reduces to the Schwarzschild case).

The surfaces EH and SIR are sketched in Fig. 7.12, in 'plan' and 'elevation' views. The *ergosphere* is the region between them, and therefore defined by

$$m + \sqrt{m^2 - \frac{a^2}{c^2}} = r_+ < r < r_1 = m + \sqrt{m^2 - \frac{a^2}{c^2}\cos^2\theta}\ . \tag{7.200}$$

Now consider a particle in the ergosphere. What constraints are there on its movement? A 'real' particle must move along a timelike geodesic; we shall first show that this is impossible if the particle is at rest in the ergosphere. For, from (7.188), if r = const, θ = const and ϕ = const, then

$$\mathrm{d}s^2 = -\left(1 - \frac{2mr}{\rho^2}\right)c^2\,\mathrm{d}t^2. \tag{7.201}$$

On the surface of infinite red-shift $\rho^2 = 2mr$ and $r = r_1$, so in the ergosphere, where $r < r_1$, (7.199) gives

$$r - m < \sqrt{m^2 - \frac{a^2}{c^2}\cos^2\theta}$$

which is equivalent to

$$r^2 + \frac{a^2}{c^2}\cos^2\theta < 2mr,$$

or, from (7.183), $\rho^2 < 2mr$, so (7.201) gives $\mathrm{d}s^2 > 0$ inside SIR. However, as just noted, for a real particle we must have $\mathrm{d}s^2 < 0$, so our assumption that r, θ and ϕ all be constant cannot be met; other terms in the line element must contribute and the particle must *change position*.

Consider then a particle with fixed r, θ but moving in the direction of increasing ϕ. Then the 4-velocity is

$$\frac{\mathrm{d}x^\mu}{\mathrm{d}\tau} = u^\mu = (u^t, u^r, u^\theta, u^\phi) = (u^0, 0, 0, u^3) = u^0\left(1, 0, 0, \frac{u^3}{u^0}\right) = u^0\left(1, 0, 0, \frac{\Omega}{c}\right),$$

since

$$\Omega = \frac{\mathrm{d}\phi}{\mathrm{d}t} = \frac{\mathrm{d}\phi/\mathrm{d}\tau}{\mathrm{d}t/\mathrm{d}\tau} = c\frac{u^3}{u^0}. \tag{7.202}$$

The condition that u^μ is *timelike* is

$$g_{\mu\nu}\, u^\mu\, u^\nu < 0,$$

i.e.

$$g_{00}(u^0)^2 + 2g_{03}\, u^0 u^3 + g_{33}(u^3)^2 < 0,$$

or, with (7.202),

$$\frac{\Omega^2}{c^2} + 2\frac{g_{03}}{g_{33}}\frac{\Omega}{c} + \frac{g_{00}}{g_{33}} < 0. \tag{7.203}$$

The solution to $\frac{\Omega^2}{c^2} + 2\frac{g_{03}}{g_{33}}\frac{\Omega}{c} + \frac{g_{00}}{g_{33}} = 0$ is

$$\frac{\Omega}{c} = -\frac{g_{03}}{g_{33}} \pm \sqrt{\left(\frac{g_{03}}{g_{33}}\right)^2 - \frac{g_{00}}{g_{33}}}. \tag{7.204}$$

The metric components g_{00}, g_{03} and g_{33} may be read off from (7.191). Putting (noting that $g_{03} < 0$)

$$-\frac{g_{03}}{g_{33}} = \frac{\omega}{c}, \tag{7.205}$$

the two solutions (7.204) may be written

$$\Omega_{\mathrm{min}} = \omega - \sqrt{\omega^2 - c^2\frac{g_{00}}{g_{33}}}, \qquad \Omega_{\mathrm{max}} = \omega + \sqrt{\omega^2 - c^2\frac{g_{00}}{g_{33}}}. \tag{7.206}$$

We conclude that for (7.203) to hold Ω must lie in the range

$$\Omega_{\min} < \Omega < \Omega_{\max}. \tag{7.207}$$

Ω is the angular velocity of a particle in Kerr space-time, as seen by a distant observer. It depends on r, since the metric coefficients depend on r. In general, since $g_{00} < 0$, $\Omega_{\min} < 0$, and a distant observer would be able to see a particle *counter-rotating*. As r approaches the surface of infinite red-shift, however, $g_{00} \to 0$, so $\Omega_{\min} = 0$ (and $\Omega_{\max} = 2\omega$), and a particle would not have the option of counter-rotating, but could still remain at rest. For this reason the surface of infinite red-shift is also called the *static limit*. It is the nearest a particle can get to the centre of the black hole and still not rotate (that is, not be seen as rotating by a distant observer). Inside it, and therefore in the ergosphere, $\Omega > 0$; a particle is forced to rotate, in the direction of the angular momentum of the source.

To explore more fully the nature of angular momentum and angular velocity in the Kerr solution, let us consider the part played by conservation laws and Killing vectors. In Section 6.8 we saw that the fact that the metric in 3-dimensional flat space

$$g_{ij} = \begin{pmatrix} 1 & 0 & 0 \\ 0 & r^2 & 0 \\ 0 & 0 & r^2 \sin^2\theta \end{pmatrix}$$

is independent of ϕ, $\dfrac{\partial g_{ij}}{\partial \phi} = 0$, leads to the conservation of angular momentum $\boldsymbol{\xi} \cdot \mathbf{u}$, where $\boldsymbol{\xi} = \dfrac{\partial}{\partial \phi}$ and $\mathbf{u}$ is the velocity of a particle moving along a geodesic. In the case of the Kerr metric, $g_{\mu\nu}$ in (7.191) above is also independent of ϕ, so there is a Killing vector $\boldsymbol{\xi} = \dfrac{\partial}{\partial \phi}$, i.e. $\xi^\mu = (0, 0, 0, 1)$. There is also a Killing vector $\dfrac{\partial}{\partial t} = \boldsymbol{\eta}$, corresponding to the fact that it is a stationary metric, with coefficients independent of t. Now consider a particle moving in a planar orbit with $\theta = \pi/2$:

$$u^\mu = (c\dot{t}, 0, 0, \dot{\phi}). \tag{7.208}$$

The conserved angular momentum is

$$L = \boldsymbol{\xi} \cdot \mathbf{p} = m\boldsymbol{\xi} \cdot \mathbf{u} = m(g_{\mu\nu}\xi^\mu u^\nu) = m\, g_{3\nu} u^\nu = m\,(g_{30}\, c\dot{t} + g_{33}\,\dot{\phi}) \tag{7.209}$$

and a particle with *zero* angular momentum has

$$g_{33}\,\dot{\phi} + g_{30}\, c\dot{t} = 0$$

or, from (7.191),

$$\left(r^2 + \frac{a^2}{c^2} + \frac{2mra^2 \sin^2\theta}{c^2\rho^2}\right)\dot{\phi} - \left(\frac{2mar \sin^2\theta}{c\rho^2}\right)c\dot{t} = 0.$$

Since $\sin\theta = 1$ here, this gives

$$\frac{\mathrm{d}\phi}{\mathrm{d}t} = \frac{\dot{\phi}}{\dot{t}} = \frac{2mra}{(r^2 + \frac{a^2}{c^2})\rho^2 + 2mr\frac{a^2}{c^2}} = -c\,\frac{g_{03}}{g_{33}} = \omega, \tag{7.210}$$

from (7.204) and (7.191), and we see that *particles with zero angular angular momentum possess non-zero angular velocity*. This is an example of the phenomenon that the space-time frames are being 'dragged' round by the source – the Lense–Thirring effect.

We have seen that, as observed by a distant observer, a particle in a circular orbit in a Kerr field may, so long as it is *outside* the surface of infinite red-shift, counter-rotate, since when $r<r_1$, $\Omega_{min}<0$. On the SIR, however, $\Omega_{min}=0$, and inside it, in the ergosphere, both Ω_{min} and Ω_{max} are positive, so the particle (even one with no angular momentum) is seen to rotate in the same direction as the black hole – the phenomenon of inertial drag. What happens as r decreases further and the particle's orbit approaches the event horizon? It turns out that in this case Ω_{min} and Ω_{max} tend to the same value, so any particle on the event horizon rotates with a *fixed* angular velocity – which may, conveniently, be taken as a *definition* of the angular velocity of the black hole. To see this, recall that the EH is a null hypersurface, that is, the normal vector n^μ is null, $n^\mu n_\mu=0$. A particle on a geodesic in the hypersurface follows a tangent vector orthogonal to n^μ. Any vector orthogonal to a null (lightlike) vector is either null or spacelike and the spacelike option cannot apply here, so the tangent vector $\mathbf{k}$ obeys $k^2(r_+)=\mathbf{k}\cdot\mathbf{k}(r_+)=0$: recall that $r=r_+$ on the EH – see (7.196). The vector $\mathbf{k}$ has components only in the t, θ and ϕ directions so let us put

$$k^\mu = (1,0,\alpha,\beta), \tag{7.211}$$

or, equivalently

$$u^\mu = u^0(1,0,\alpha,\beta) = u^0 k^\mu. \tag{7.212}$$

Then

$$\begin{aligned} k^2 &= g_{\mu\nu}k^\mu k^\nu = g_{00} + \alpha^2 g_{22} + \beta^2 g_{33} + 2\alpha\beta\, g_{03} \\ &\equiv \alpha^2 g_{22} + \frac{(g_{03}+\beta g_{33})^2}{g_{33}} + \frac{g_{00}\,g_{33} - g_{03}^2}{g_{33}}. \end{aligned}$$

On the event horizon, however, where $\Delta=0$, we also have $g_{00}\,g_{33}-g_{03}{}^2=0$ (see Problem 7.4), so the last term vanishes, and $k^2=0$ requires that $\alpha=0$ and

$$\begin{aligned} \beta &= -\frac{g_{03}(r_+)}{g_{33}(r_+)} \\ &= \frac{\dfrac{2mr_+a\sin^2\theta}{c\rho_+^2}}{\left(r_+^2+\dfrac{a^2}{c^2}+\dfrac{2mr+a^2\sin^2\theta}{c^2\rho_+^2}\right)\sin^2\theta} \\ &= \frac{2mr_+a}{c\rho_+{}^2(2mr_+)+2mr_+\frac{a^2}{c}\sin^2\theta} \\ &= \frac{a/c}{\rho_+^2+\frac{a^2}{c^2}\sin^2\theta} \\ &= \frac{a/c}{2mr_+}, \end{aligned}$$

where we have made use of Equations (7.183) and (7.197). We therefore have, from (7.212),

$$u^\mu = u^0\left(1, 0, 0, \frac{a/c}{2mr_+}\right), \tag{7.213}$$

and hence an angular velocity Ω_H which may be taken as a *definition of the angular velocity* of a Kerr black hole:

$$\Omega_H = \frac{a}{2mr_+}. \tag{7.214}$$

(It might be useful to remind ourselves that the units are correct: a/c has the dimension of length, as have m and r, so Ω_H has the dimension of inverse time.)

Our next task is to find an expression for the *area* of the event horizon. On EH $r = r_+$ so $dr = 0$, and since $\Delta = 0$ on EH we have, from (7.189)

$$ds^2 = \rho_+^2\, d\theta^2 + \frac{\sin^2\theta}{\rho_+{}^2}\left[\left(r_+^2 + \frac{a^2}{c^2}\right)d\phi - a\, dt\right]^2.$$

From Equation (7.197), however, $r_+{}^2 + \frac{a^2}{c^2} = 2mr_+$, the above equation becomes

$$ds^2 = \rho_+^2\, d\theta^2 + \left(\frac{2mr_+}{\rho_+}\right)^2 (\sin^2\theta)\left[d\phi - \frac{a}{2mr_+}dt\right]^2. \tag{7.215}$$

This is the separation between two points on the surface at differing values of θ, ϕ and t. For a *static* configuration, however, the term in dt disappears and the *area 2-form* of the event horizon is

$$dA = \rho_+\, d\theta \wedge \left(\frac{2mr_+}{\rho_+}\right)\sin\theta\, d\phi = 2mr_+ \sin\theta\, d\theta \wedge d\phi,$$

giving an area of

$$A = 2mr_+\int \sin\theta\, d\theta \wedge d\phi = 8\pi m r_+ = 8\pi\left(m^2 + \sqrt{m^4 - \frac{m^2a^2}{c^2}}\right). \tag{7.216}$$

With $m = \frac{MG}{c^2}$, $a = \frac{J}{Mc}$ this gives

$$A = \frac{8\pi G^2}{c^4}\left(M^2 + \sqrt{M^4 - \frac{J^2}{G^2}}\right), \tag{7.217}$$

or, in geometrical units $G = c = 1$,

$$A = 8\pi\left(M^2 + \sqrt{M^4 - J^2}\right). \tag{7.218}$$

This is the area of the event horizon of a Kerr black hole.

We come now to the highly interesting result, first pointed out by Penrose,[19] that it is possible to *extract energy* from a Kerr black hole. This is initially surprising, since after all a black hole is

[19] Penrose (1969).

a region from which no light can escape. The result follows, however, from the fact that the event horizon and the surface of infinite red-shift are distinct; the energy is extracted from the ergosphere. In parallel with the observation that the (conserved) angular momentum of a body is $\boldsymbol{\xi}\cdot\mathbf{p}$ (see (7.208)), we also note that its conserved energy is $E=-\boldsymbol{\eta}\cdot\mathbf{p}$. For a particle outside the static limit (SIR) $\boldsymbol{\eta}$ is timelike, so in the frame in which $\eta^\mu=(1, 0, 0, 0)$,

$$E = m\boldsymbol{\eta}\cdot\mathbf{u}\,.$$

This is *conserved*, so for a particle coming in from infinity, entering the ergosphere and leaving it again, the energy is $E>0$. But now consider a particle originating in the ergosphere. Here $\boldsymbol{\eta}$ is *spacelike*: this follows immediately by noting that (see (7.191))

$$\eta^\mu\eta_\mu = g_{\mu\nu}\eta^\mu\eta^\nu = g_{00} = -\left(\frac{r^2+\dfrac{a^2}{c^2}\cos^2\theta - 2mr}{\rho^2}\right). \tag{7.219}$$

Outside the SIR $r>r_1$ so $\eta^\mu\eta_\mu<0$ and $\boldsymbol{\eta}$ is timelike. On SIR $\boldsymbol{\eta}$ becomes null and inside it becomes spacelike, $\eta^\mu\eta_\mu>0$. In the ergosphere, then, there is a local frame in which $\eta^\mu=(0, 1, 0, 0)$ and the 'energy' is $m\boldsymbol{\eta}\cdot\mathbf{u}=m\dot{x}$ in Cartesian coordinates; it becomes in effect a *momentum*, which may be either positive or negative. To be clear, we are here *not* talking about a particle which can *leave* the ergosphere through the SIR and reach infinity, since such a particle must always have $E>0$. To say of a particle that it has $E<0$ means that this *would* be the energy measured at infinity *if* the particle could be brought there.

But now consider a particle entering the ergosphere from infinity (with $E>0$, of course) and then breaking up into two particles in the ergosphere; and suppose that one of these particles escapes back through the SIR, but the other one does not escape. This is sketched in Fig. 7.13, in which particle 1 enters the ergosphere, splitting up into particles 2 and 3. Particle 3 then leaves the ergosphere, escaping to infinity while particle 2, while also originating in the ergosphere, does not escape to infinity; it enters the black hole through the event horizon. It may then have negative energy, $E_2<0$. On the other hand E_1 and E_3 are

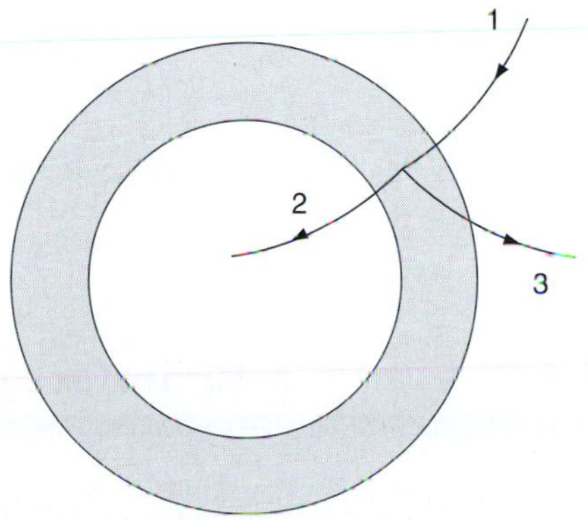

Fig. 7.13 The Penrose process: particle 3 may leave the black hole (ergosphere) with more energy than particle 1 had on entering.

both positive. Energy conservation at the point of disintegration, however, implies that $E_1 = E_2 + E_3$ and hence

$$E_3 = E_1 - E_2 > E_1:$$

the particle leaving the ergosphere may have a greater energy than the particle which entered: energy has been *extracted* from the black hole. This is commonly known as the Penrose process. As a result of it, the mass of the black hole must decrease. Note the surprising nature of this result!

This, however, is not all; the black hole *also* loses angular momentum through the Penrose process. The 4-velocity u^μ of a particle in the ergosphere is timelike, and from (7.210) may be written $\mathbf{u} = \boldsymbol{\eta} + \Omega_H \boldsymbol{\xi}$; hence the Killing vector

$$\boldsymbol{\zeta} = \boldsymbol{\eta} + \Omega_H \boldsymbol{\xi} \tag{7.220}$$

is timelike, from which it follows that $-\mathbf{p} \cdot \boldsymbol{\zeta} > 0$, i.e.

$$-\mathbf{p} \cdot \boldsymbol{\eta} - \Omega_H \boldsymbol{p} \cdot \boldsymbol{\xi} = E - \Omega_H L > 0,$$

where (7.209) has been used; hence

$$E > \Omega_H L. \tag{7.221}$$

Now in the Penrose process a black hole absorbs an amount of energy $E_2 < 0$, so it also absorbs the angular momentum

$$L_2 < \frac{E^2}{\Omega_H} < 0:$$

the angular momentum of the black hole also decreases. Eventually the angular momentum will become zero, at which point the surface of infinite red-shift and the event horizon will coincide and the ergosphere will disappear (will shrink to zero), and further extraction of energy (mass) from the black hole will become impossible.

Note that in the process in which a black hole absorbs (a particle of) energy E and angular momentum L its parameters change by $\delta M = E/c^2$ and $\delta J = L$, so by virtue of (7.221),

$$\delta M - \frac{\Omega_H}{c} \delta J > 0. \tag{7.222}$$

This inequality is actually a consequence of the first and second laws of black hole mechanics, as we see below.

7.10 Surface gravity

Surface gravity turns out to play an important role in the thermodynamics of black holes. It is not a difficult concept; the surface gravity on the Earth is $\frac{GM}{R^2}$, and this expression, suitable for the non-relativistic and weak field approximations, simply has to be generalised. So first consider a static, spherically symmetric space-time

$$ds^2 = -f(r)\,c^2\,dt^2 + f(r)^{-1}\,dr^2 + r^2\,d\Omega^2; \tag{7.223}$$

for example, in the Schwarzschild case, $f(r) = 1 - \frac{2m}{r}$. Hold a particle with mass m_0 stationary at a distance r from the centre, with coordinates and 4-velocity

$$x^\mu = (ct, r, \theta, \phi), \quad u^\mu = (c\dot{t}, 0, 0, 0)$$

and $ds^2 = -f(r)\,c^2\,dt^2 = -c^2\,d\tau^2$, so $\dot{t} = \frac{dt}{d\tau} = f^{-1/2}$ and

$$u^0 = cf^{-1/2}, \quad u^i = 0. \tag{7.224}$$

In Minkowski space acceleration is defined as (see Section 2.2) $a^\mu = \dot{u}^\mu$. We generalise this to $\mathbf{a} = \nabla_\mathbf{u}\mathbf{u}$, which from Equation (4.1a) may be written

$$\mathbf{a} = \nabla_\mathbf{u}\,\mathbf{u} = u^\kappa{}_{;\lambda}\,u^\lambda\,\mathbf{e}_\kappa.$$

Then

$$a^\mu = \langle \mathbf{a}, \boldsymbol{\theta}^\mu \rangle = u^\mu{}_{;\lambda}u^\lambda = u^\mu{}_{;0}\,u^0 = cf^{-1/2}\,u^\mu{}_{;0}.$$

In turn, $u^\mu{}_{;0} = u^\mu{}_{,0} + \Gamma^\mu{}_{\nu 0}\,u^\nu = \Gamma^\mu{}_{00}$. It is easily established that the only non-zero component of the connection coefficient is $\Gamma^1{}_{00} = \frac{1}{2}ff'\ \left(f' = \frac{df}{dr}\right)$, so the only non-zero component of acceleration is

$$a^1 = cf^{-1/2}\,u^1{}_{;0} = \frac{c^2}{2}f'.$$

The magnitude of the acceleration vector is

$$a = \sqrt{g_{\mu\nu}\,a^\mu\,a^\nu} = \sqrt{g_{11}}\,a^1 = \frac{c^2}{2}f^{-1/2}f'.$$

In the Schwarzschild case $f = 1 - \frac{2m}{r}$ and

$$a(r) = \frac{mc^2}{r^2}\frac{1}{\sqrt{1 - \frac{2m}{r}}}.$$

This is the force (when multiplied by m_0) required to hold the particle at its local position, and it diverges as $r \to 2m$. (Note in passing that when $r \gg 2m$, a becomes GM/r^2, as expected.) But consider instead that the particle is held in position by observer at infinity, holding the particle on a long massless string. If the observer moves the string by a small proper distance δs, he does an amount of work $m_0\,a(\infty)\,\delta s = \delta W_\infty$. At the position of the particle, however, the work done is $m_0\,a(r)\,\delta s = \delta W(r)$. These amounts of work are different, but one may imagine a machine converting work into radiation; the work at position r is converted into radiation which is then beamed to the observer at infinity – and in the process *red-shifted* by the factor $\sqrt{g_{00}}=f^{1/2}$, so that

$$\delta E_\infty = f^{1/2} m_0 a \ \delta s.$$

But $\delta E_\infty = \delta W_\infty$, by energy conservation, so

$$a_\infty(r) = f^{1/2}\, a(r) = \frac{c^2}{2} f'(r).$$

The quantity $m_0\, a_\infty(r)$ is the force applied by the observer at infinity to keep the particle in place. It is well behaved as $r \to 2m$ in the Schwarzschild case and is called the *surface gravity* κ of the black hole:

$$\kappa = a_\infty(r_{\mathrm{EH}}) = \frac{c^2}{2} f'(r_{\mathrm{EH}}) \tag{7.225}$$

where r_{EH} refers to the event horizon.[20] In the Schwarzschild case $f' = \dfrac{2m}{r^2}$ and $r_{\mathrm{EH}} = 2m$ so

$$\kappa = \frac{c^2}{4m} = \frac{c^4}{4MG} \quad \text{(Schwarschild)}. \tag{7.226}$$

(Note that the dimensions are those of acceleration, as expected.)

We now show that surface gravity may also be defined in terms of Killing vectors – in the static case we are considering at present, of the vector $\boldsymbol{\eta} = \dfrac{1}{c}\dfrac{\partial}{\partial t}$. For the moment we confine ourselves to the Schwarzschild black hole, so (cf. (7.219))

$$\eta^\mu \eta_\mu = -\left(1 - \frac{2m}{r}\right), \tag{7.227}$$

which is *null* on the event horizon. We then define the scalar quantity

$$\Phi = -\eta^\mu \eta_\mu, \quad \Phi = 1 - \frac{2m}{r}. \tag{7.228}$$

Clearly $\Phi = 0$ on the event horizon. The normal to this surface is $\Phi_{,\mu}$; but since the surface is null its normal is proportional to η_μ itself, so there will exist a quantity κ such that

$$c^2 \Phi_{,\mu} = 2\kappa \eta_\mu. \tag{7.229}$$

Let us work in advanced time coordinates

$$(cv, r, \theta, \phi) = (x^0, x^1, x^2, x^3), \tag{7.230}$$

so from (7.130) the line element is

$$\mathrm{d}s^2 = -\left(1 - \frac{2m}{r}\right) c^2\, \mathrm{d}v^2 + 2c\, \mathrm{d}v\, \mathrm{d}r + r^2\, \mathrm{d}\Omega^2,$$

or

[20] This argument is taken from Poisson (2004), p. 185.

$$g_{\mu\nu} = \begin{pmatrix} -\left(1-\frac{2m}{r}\right) & 1 & 0 & 0 \\ 1 & 0 & 0 & 0 \\ 0 & 0 & r^2 & 0 \\ 0 & 0 & 0 & r^2\sin^2\theta \end{pmatrix}, \tag{7.231}$$

with corresponding contravariant components

$$g^{\mu\nu} = \begin{pmatrix} 0 & 1 & 0 & 0 \\ 1 & 1-\frac{2m}{r} & 0 & 0 \\ 0 & 0 & \frac{1}{r^2} & 0 \\ 0 & 0 & 0 & \frac{1}{r^2\sin^2\theta} \end{pmatrix}. \tag{7.232}$$

The metric has a Killing vector $\boldsymbol{\eta} = \frac{1}{c}\frac{\partial}{\partial v}$, i.e. $\eta^\mu = (1, 0, 0, 0)$ in the coordinate system (7.230), hence

$$\eta_0 = g_{0\mu}\eta^\mu = g_{00} = -\left(1-\frac{2m}{r}\right), \quad \eta_1 = g_{1\mu}\eta^\mu = g_{10} = 1$$

and

$$\Phi = -\eta^0\eta_0 - \eta^1\eta_1 = 1 - \frac{2m}{r}, \quad \Phi_{,1} = \frac{2m}{r^2}$$

and (7.229) then gives, at $r = 2m$,

$$\kappa = \frac{c^2}{4m},$$

as in (7.226) above.

We now turn to the consideration of surface gravity in Kerr black holes. We want to work in a coordinate system which is a generalisation of the advanced time coordinate system (7.230). This is given by the line element (7.185), which may be written

$$g_{\mu\nu} = \begin{pmatrix} -\left(1-\frac{2mr}{\rho^2}\right) & 1 & 0 & -\frac{2mra}{c\rho^2}\sin^2\theta \\ 1 & 0 & 0 & -\frac{a}{c}\sin^2\theta \\ 0 & 0 & \rho^2 & 0 \\ -\frac{2mra}{c\rho^2}\sin^2\theta & -\frac{a}{c}\sin^2\theta & 0 & \frac{\sin^2\theta}{\rho^2}\left[\left(r^2+\frac{a^2}{c^2}\right)^2 - \frac{a^2}{c^2}\Delta\sin^2\theta\right] \end{pmatrix}. \tag{7.233}$$

The Killing vector (7.220) is $\zeta^\mu = (1, 0, 0, \frac{\Omega_H}{c})$, then

$$\zeta_0 = g_{00} + g_{03}, \quad \zeta_3 = g_{30} + g_{33}, \quad \zeta_1 = g_{10} + \frac{\Omega_H}{c}g_{13},$$

and the field Φ is, in this case,

$$\Phi = -\zeta^\mu \zeta_\mu = -\zeta^0 \zeta_0 - \zeta^3 \zeta_3 = -(g_{00} + 2g_{03} + g_{33}). \tag{7.234}$$

To find the surface gravity κ we must (see (7.229)) calculate $\Phi_{,r}$ and compare with $\zeta_r = \zeta_1$. We have, from above

$$\zeta_1 = 1 - \frac{a}{c^2}\Omega_H \sin^2\theta. \tag{7.235}$$

Now $\boldsymbol{\zeta} \cdot \boldsymbol{\zeta}$ is, from (7.220),

$$\begin{aligned}
\boldsymbol{\zeta}\cdot\boldsymbol{\zeta} &= \boldsymbol{\eta}\cdot\boldsymbol{\eta} + 2\,\Omega_H\,\boldsymbol{\eta}\cdot\boldsymbol{\xi} + \Omega_H^2\,\boldsymbol{\xi}\cdot\boldsymbol{\xi} \\
&= g_{00} + 2\,\Omega_H\, g_{03} + \Omega_H^2\, g_{33} \\
&= g_{00} - \frac{{g_{03}}^2}{g_{33}} + g_{33}\left[{\Omega_H}^2 + 2\Omega_H \frac{g_{03}}{g_{33}} + \frac{{g_{03}}^2}{g_{33}^2}\right] \\
&= \frac{1}{g_{33}}(-\Delta \sin^2\theta) + g_{33}(\Omega_H - \omega)^2,
\end{aligned}$$

where the result of Problem 7.4 and (7.205) have been used. Now define

$$\Sigma = \left(r^2 + \frac{a^2}{c^2}\right)^2 - \frac{a^2}{c^2}\Delta \sin^2\theta, \tag{7.236}$$

so that, from (7.233),

$$g_{33} = \frac{\Sigma}{\rho^2}\sin^2\theta,$$

and

$$\boldsymbol{\zeta}\cdot\boldsymbol{\zeta} = \frac{\Sigma \sin^2\theta}{\rho^2}\,(\Omega_H - \omega)^2 - \frac{\rho^2\Delta}{\Sigma}. \tag{7.237}$$

We must now differentiate this with respect to r, regarding ω as a parameter (not a function of r), so

$$(\boldsymbol{\zeta}\cdot\boldsymbol{\zeta})_{,1} = \left(\frac{\Sigma \sin^2\theta}{\rho^2}\right)_{,1}(\Omega_H - \omega^2) - \left(\frac{\rho^2}{\Sigma}\right)_{,1}\Delta - \frac{\rho^2}{\Sigma}\Delta_{,1}. \tag{7.238}$$

On the event horizon $\omega = \Omega_H$ and $\Delta = 0$, so

$$(\boldsymbol{\zeta}\cdot\boldsymbol{\zeta})_{,1} = \left[\frac{\rho^2}{\Sigma}\Delta_{,1}\right]_{r=r_+} = \left[\frac{2\rho^2(r-m)}{\Sigma}\right]_{r=r_+},$$

hence the formula (7.229) gives

$$2\kappa = c^2\left[\frac{2\rho^2(r-m)}{\Sigma}\right]_{r=r_+}\cdot\frac{1}{1 - \frac{a}{c^2}\Omega_H \sin^2\theta},$$

which, after a bit of algebra, yields

$$\kappa = \frac{c^2\sqrt{m^2 - \frac{a^2}{c^2}}}{r_+^2 + \frac{a^2}{c^2}}. \tag{7.239}$$

In the case $a = 0$, $r_+ \to 2m$, $\kappa \to \frac{c^2}{4m}$, as in (7.226) above, and as expected.

Let us now collect together the formulae for the angular momentum, (7.214), surface gravity (7.239) and area (7.216) of a Kerr black hole:

$$\Omega_{\rm H} = \frac{a}{2mr+}, \tag{7.214}$$

$$\kappa = \frac{c^2\sqrt{m^2 - \frac{a^2}{c^2}}}{r_+^2 + \frac{a^2}{c^2}}, \tag{7.239}$$

$$A = 8\pi\left(m^2 + \sqrt{m^4 - \frac{m^2a^2}{c^2}}\right), \tag{7.216}$$

where, from (7.196), $r_+ = m + \sqrt{m^2 - \frac{a^2}{c^2}}$. Putting $a = \frac{J}{Mc}$ and using geometric units $G = c = 1$ these become

$$\Omega_{\rm H} = \frac{J}{2M\left(M^2 + \sqrt{M^4 - J^2}\right)}, \tag{7.240}$$

$$\kappa = \frac{\sqrt{M^4 - J^2}}{2M\left(M^2 + \sqrt{M^4 - J^2}\right)}, \tag{7.241}$$

$$A = 8\pi\left[M^2 + \sqrt{M^4 - J^2}\right]. \tag{7.242}$$

In the Schwarzschild case $\kappa = \frac{1}{4M}$, $A = 16\pi M^2$ and $\frac{\kappa A}{4\pi} = M$. As J increases from zero, κ decreases until, when $J = M^2$, it becomes zero – like centrifugal force. Then the mass is given by $\Omega_{\rm H} = \frac{J}{2M^3}$, or

$$M = 2\,\Omega_{\rm H}\,J. \tag{7.243}$$

In the general case the formulae above give

$$\frac{\kappa A}{4\pi} = \frac{\sqrt{M^4 - J^2}}{M}, \qquad 2\,\Omega_{\rm H}\,J = \frac{J^2}{M\left(M^2 + \sqrt{M^4 - J^2}\right)},$$

and then

$$2\,\Omega_{\rm H}J + \frac{\kappa A}{4\pi} = M, \tag{7.244}$$

which is Smarr's formula for the mass of a black hole.[21]

It is clear that the area of a black hole is a function of its mass and angular momentum, $A = A(M, J)$. Now suppose that these quantities change, for example by the absorption of a particle, as considered in the Penrose process: $M \to M + \delta M, J \to J + \delta J$. We can then find a relation between δA, δM and δJ. Working in geometric units, so that $m = M$ and so on, from (7.216) we have

$$\frac{\delta A}{8\pi} = M\,\delta r_+ + r_+\,\delta M,$$

then, using (7.239) and (7.197) (with $c = 1$),

$$\frac{\kappa}{8\pi}\,\delta A = \frac{r_+ - M}{2Mr_+}(\,M\,\delta r_+ + r_+\,\delta M\,).$$

In addition, from $a = \dfrac{J}{M}$ we have $\delta J = a\,\delta M + M\,\delta a$ so that

$$\Omega_{\rm H}\,\delta J = \frac{a}{2mr_+}(\,a\,\delta M + M\,\delta a\,),$$

and it then follows from simple algebra that

$$\frac{\kappa}{8\pi}\delta A + \Omega_{\rm H}\,\delta J = \delta M. \tag{7.245}$$

This is the relation we were anticipating. The Penrose process is an example of a process in which M and J change. By virtue of the relation (7.222), found there, and the above relation we see that δA in this process – the area of the black hole – *increases*. This is actually an example of a more general result proved by Hawking in 1971 that in *any* process the surface area of a black hole can never decrease:[22]

$$\delta A \geq 0. \tag{7.246}$$

There is one further concept which it is useful to introduce, that of *irreducible mass*. Let us define[23]

$$M_{\rm ir}{}^2 = \frac{1}{2}\left(M^2 + \sqrt{M^4 - J^2}\right), \tag{7.247}$$

so that, from (7.242),

[21] Smarr (1973).
[22] Hawking (1971).
[23] Christodoulou (1970).

$$A = 16\pi M_{\text{ir}}^2. \tag{7.248}$$

From simple algebra it then follows that

$$M^2 = M_{\text{ir}}^2 + \frac{J^2}{4M_{\text{ir}}^2}. \tag{7.249}$$

The second term above represents the rotational contribution to the mass of a black hole. The maximum amount of energy to be extracted from a black hole by slowing its rotation down is $M - M_{\text{ir}}$. The irreducible mass $M_{\text{ir}} = \sqrt{\frac{A}{16\pi}}$ is the energy which cannot be extracted by the Penrose process – and by virtue of Hawking's result (7.246) the irreducible mass of a black hole will never decrease.

7.11 Thermodynamics of black holes and further observations

Results like (7.246) and (7.245), that the area of a black hole never decreases, and the relation between the changes in mass, area and angular momentum of a black hole, are reminiscent of the laws of thermodynamics and lead to the formulation of analogous laws of black hole dynamics.[24] The energy E, temperature T and entropy S of a gas are respectively analogous to the mass M, surface gravity κ and area A of a black hole, and the laws are as follows:

Zeroth law The surface gravity of a stationary black hole is constant over the horizon.

First law is given by (7.245)

$$\mathrm{d}M = \frac{\kappa}{8\pi}\mathrm{d}A + \Omega_{\text{H}}\,\mathrm{d}J,$$

and corresponds to the first law of thermodynamics

$$\mathrm{d}E = T\,\mathrm{d}S - P\,\mathrm{d}V,$$

which expresses conservation of energy.

Second law is $\delta A \geq 0$, as in (7.246) above. This law is actually slightly stronger than the thermodynamic analogy $\delta S \geq 0$. In thermodynamics entropy may be transferred from one system to another, and it is only required that the *total* entropy does not decrease. Since individual black holes cannot bifurcate,[25] however, area cannot be transferred from one to the other. The second law of black hole mechanics states that the area of each black hole separately cannot decrease.

Third law The Planck–Nernst form of the third law of thermodynamics states that $S \to 0$ as $T \to 0$. Israel has proposed an analogous law of black hole dynamics.[26]

[24] Bardeen *et al.* (1973).
[25] Hawking & Ellis (1973), pp. 315–316.
[26] Israel (1986).

At this stage we would consider the laws of black hole dynamics and thermodynamics to be merely *analogous*, but Bekenstein[27] pointed out that in this case there is a problem with the second law of thermodynamics in the presence of black holes. A system whose overall entropy is increasing, as it always will, may nevertheless be made up of individual parts, in some of which S is decreasing and in others of which S is increasing (and S still increases overall). In the presence of a black hole one could then throw the parts of the system in which S is *increasing* into the black hole. We must now, however, consider a fundamentally important feature of black holes, considered as solutions (Schwarzschild or Kerr) to the field equations. This is that these solutions are exact and are described *completely* by their mass m and angular momentum a. They are respectively one and two parameter solutions. This implies that, for example, two Schwarzschild black holes with the same mass must be *identical*; and similarly two Kerr black holes with the same mass and angular momentum must be identical. This result is the content of the theorem that a black hole 'has no hair' – the only labels it possesses are mass and angular momentum (and electric charge in the case of Kerr–Newman black holes). Anything else is 'hair'. A striking implication of this theorem, in the context of elementary particle physics, for example, concerns baryon number. As far as is known baryon number B is a conserved quantity. No reactions have yet been observed in which B is violated, though it should be noted that the SU(5) Grand Unified Theory predicts proton decay $\mathrm{p} \rightarrow \mathrm{e}^+ + \pi^0$, which would indeed violate this conservation law. Since black holes, however, defined as (Schwarzschild or Kerr) solutions to Einstein's field equations, do not possess a baryon number, then the process of throwing a collection of baryons into a black hole will, in effect, be one in which B is violated.

Returning to our example concerning the second law of thermodynamics, the implication of throwing the part of the system in which S is increasing into the black hole is, since the black hole itself has no 'hair', that the overall entropy has *decreased*, thus *violating* the second law of thermodynamics! This drastic and unthinkable conclusion may be avoided, however, if we suppose that the black hole *does actually possess entropy*, proportional to its area, so that as its area increases, so does its entropy. That is to say that the entropy of a black hole is *not* hair; it is simply a function of its mass and angular momentum. In *that* case the second law of thermodynamics will be saved. On the other hand, to say this is to say that what we supposed above was an analogy is actually more than that: a black hole does actually possess entropy, proportional to its area A. But in that case it must also possess a temperature T, proportional to its surface gravity κ (by comparing (7.245) and (7.250)). And in *that* case it must shine – it must emit radiation. But this conclusion goes against the whole understanding of black holes so far developed: in classical General Relativity black holes do *not* shine, since light cannot escape through the event horizon. How is it possible that black holes might shine?

A clue is to be found in looking for formulae relating the entropy and area, and temperature and surface gravity of black holes. Entropy has the dimensions of Bolzmann's constant, energy/temperature, so if we suppose that S and A are proportional, the proportionality constant must involve k. Apart from that it must presumably involve only fundamental constants. It is easy to see, however, that using G and c alone it is not possible to

[27] Bekenstein (1980).

obtain a quantity with the dimensions of area. We need another constant, and the only one available is Planck's. We may then write the following formulae – known as the Bekenstein–Hawking formulae – for the entropy and temperature of a black hole:

$$S = \frac{kc^3}{4G\hbar}A, \quad T = \frac{\hbar}{2\pi ck}\kappa. \tag{7.252}$$

In the case of a Schwarzschild black hole, putting $a=0$ in (7.239) gives for the second formula

$$T = \frac{\hbar c^3}{8\,\pi GkM} \text{ (Schwarzschild)}, \tag{7.253}$$

showing that the temperature of a black hole is inversely proportional to its mass – the lighter it is, the hotter it is. The key observation about these formulae is, of course, that they involve Planck's constant, which indicates that the mechanism by which black holes shine is a *quantum* mechanism. This mechanism was first derived by Hawking and relies, at bottom, on an observation in quantum field theory in curved spaces.[28] In flat (Minkowski) space we may express, for example, a massless Hermitian scalar field ϕ as

$$\phi = \sum_i [f_i a_i + f_i^* a_i^\dagger], \tag{7.254}$$

where the $\{f_i\}$ are a complete orthonormal family of complex valued solutions to the wave equation $f_{i;\mu\nu}\,\eta^{\mu\nu}=0$, which contain only positive frequencies with respect to the usual Minkowski time coordinate. The operators a_i and $a_i^\dagger$ are interpreted as the annihilation and creation operators, respectively, for the particles in the ith state. The vacuum $|0\rangle$ is defined as the state from which one cannot annihilate particles

$$a_i|0\rangle = 0.$$

In a curved space-time, however, one cannot unambiguously decompose ϕ into its positive and negative frequency components, since these terms cease to have an invariant meaning in a curved space-time. Hawking shows that in effect this means that particles may be created in the vacuum. To arrive at a simplified pictorial representation of this consider the Feynman 'bubble' diagram shown in Fig. 7.14. Diagrams like this characterise the vacuum in quantum field theory (QFT). They correspond to the production, at each point of space-time, of a particle–antiparticle pair, which shortly afterwards again annihilate each other. This process is allowed by QFT, and physics being what it is, what is allowed by its laws is actually compulsory, so the vacuum is full of these processes of creation and annihilation of particle–antiparticle pairs. Processes of this type break the degeneracy of the $2S_{1/2}$ and $2P_{1/2}$ energy levels in hydrogen, and explain the Lamb shift in atomic physics. Now consider such a process taking place just outside the event horizon of a black hole. A particle–antiparticle pair is created, for example e^+e^-. In flat space these particles will stay within each other's wavefunction and after a short time, governed by the uncertainty principle, will annihilate. Just outside an event horizon, however, not only is the gravitational field very strong, but so

[28] Hawking (1975).

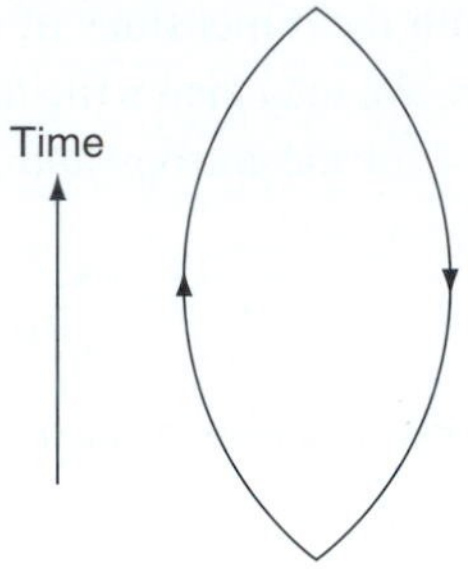

Fig. 7.14 A Feynman bubble diagram: creation and annihilation of a particle–antiparticle pair from the vacuum.

is its *gradient*, so *one* of the particles e^+ and e^-, which are created at slightly different spatial points, may find itself in a considerably stronger gravitational field than the other particle, so the two particles quickly become separated and at a later time, when in flat space they would have come together to annihilate each other, they are actually 'outside each other's wave function', so this is impossible. One of the particles may cross the event horizon into the black hole and acquire negative energy, as happens in the Penrose process, and the other particle, with equal and opposite (i.e. positive) energy, will escape from the hole altogether: it will be *emitted*. Hawking showed that this indeed happens, and that the spectrum of emitted particles is *thermal*, beautifully consistent with the notion that the black hole has a temperature. The conclusion, then, is that while in classical General Relativity black holes emit no radiation, by virtue of quantum theory they actually *do* emit radiation: they posses a temperature and shine, in the process losing mass (carried by the emitted particles), and from Equations (7.252) and (7.253), becoming *hotter*, and therefore emitting more thermal radiation. The emission process accelerates and the black hole eventually ends its life in an explosion. Black holes are therefore not completely stable entities, though their lifetimes are extremely long.

This connection between gravity, thermodynamics and quantum theory must be seen as indicative of something very deep in nature. Black holes are thermal objects (they have a temperature), but to understand *why* they are thermal or *how* they are thermal is impossible without a quantum theory of gravity, which at present we do not possess. Nevertheless it might be interesting to mention a couple of areas of research at present being pursued. The first is concerned with entropy. If black holes possess an entropy, what is the *statistical* origin of it? Entropy, after all, is related via Boltzmann's constant to degrees of freedom. Classically, however, black holes possess very few degrees of freedom; a Kerr black hole possesses two! So where does the entropy come from? This is seen as a question about *information theory*, and as one which inevitably involves quantum theory. The entropy of a black hole is proportional to its area, so its area must be the key to the information contained, counted in q-bits. The so-called 'holographic principle' is an area of current research addressed to this question.

Another interesting topic is in some sense orthogonal to this: the simplicity and structureless quality of black holes has lead 't Hooft to speculate on the role played by black holes in elementary particle physics. After all, if black holes evaporate there will be a point at which they are extremely small (just before they disappear): a mini black hole might be the smallest thing there is, so if we really want to understand the world of the very small – perhaps even to go beyond the Standard Model of elementary particle physics – might it not be a good idea to take this thought seriously?

7.12 Global matters: singularities, trapped surfaces and Cosmic Censorship

Our investigation of black holes above has been based entirely on two *exact solutions* of the Einstein field equations. These solutions possess a high degree of symmetry: spherical symmetry in the Schwarzschild solution, cylindrical in the case of Kerr. But real stars cannot be expected to exhibit any exact symmetry, though they may possess an approximate one. What, then, can we say about the collapse of real stars? Will they produce black holes? Of Schwarzschild or Kerr type? (Are there any other types?) These are obviously important questions and it is helpful to make them a bit more specific. In the case of an *exactly* spherically symmetric collapsing star, all the infalling matter will come together at $r=0$, producing a 'singularity', and this singularity is, as we saw, surrounded by a surface $r=2m$, an 'event horizon'. So we may ask: (1) what exactly is a singularity? (2) in the case of a star with only *approximate* spherical symmetry, is this singularity still produced? (or, equivalently, is the Schwarzschild solution stable under perturbations?) (3) in this same case, is an event horizon produced, which 'shields' the singularity? Similar questions may be asked about the Kerr solution, and we may go on to ask, are there any *general results*, independent of particular solutions of the field equations, relating to gravitational collapse?

There *are* general results, though unfortunately not everything that one might reasonably assume to be true has actually been proved to be true. First a word about singularities. The metric on the 2-sphere

$$g_{ik} = \begin{pmatrix} a^2 & 0 \\ 0 & a^2 \sin^2\theta \end{pmatrix}, \quad g^{ik} = \begin{pmatrix} \frac{1}{a^2} & 0 \\ 0 & \frac{1}{a^2 \sin^2\theta} \end{pmatrix},$$

is singular at $\theta=0, \pi$, where $\sin^2\theta=0$. But this is simply a singularity in the *coordinate system*: the surface S^2 itself is completely smooth at the north and south poles, where $\theta=0, \pi$. Similarly the metric $ds^2=\frac{1}{x}dx^2 + dy^2$ has a coordinate singularity at $x=0$, but this is removable by putting $x=u^2$, $y=v$, then $ds^2=4du^2 + dv^2$. In Riemannian spaces a singularity can often be detected by considering invariants that can be formed from the curvature scalar. There are

$$\frac{1}{12}n(n-1)(n-2)(n+3)$$

of these (independent) invariants in general, so 14 of them in the case of space-time. If any one of them becomes infinite – a condition independent of any coordinate system – then we can be said to have a true singularity. In Schwarzschild space-time,

$$R_{\kappa\lambda\mu\nu} R^{\kappa\lambda\mu\nu} \propto \frac{1}{r^6},$$

suggesting that $r=0$ is a true singularity but $r=2m$ is not. A precise definition of a singularity, however, is not so easy. Geroch (1960) outlines some of the difficulties, and Schmidt (1971) proposes a technical definition which in looser language amounts to *geodesic incompleteness*: that is, a geodesic cannot be continued to arbitarily large values of its parameter, but comes to an 'end', either in the future or in the past. The termination point is then a singularity. We then will expect that black holes contain a singularity at $r=0$, since the incoming matter and light 'cease to exist' in some sense at $r=0$. Perhaps we should say that *classical* General Relativity – which of course is the theory we are dealing with – exhibits this singularity in the (exact) Schwarzschild and Kerr solutions, though the expectation of physicists is surely that ultimately General Relativity will be replaced by a more refined theory, in which quantum effects are taken into account, so that what happens at $r=0$ loses its catastrophic and singular nature. But this is a question for the future! The question we are now asking is, does this singularity exist in more general situations (without any particular symmetry)? Penrose (1965) and Hawking and Penrose (1969) have shown that under rather weak conditions, not including any condition of symmetry, a space-time M is not timelike and null geodesically complete, i.e. there is a singularity present. These conditions are:

(i) $R_{\mu\nu} u^\mu u^\nu \geq 0$, where $R_{\mu\nu}$ is the Ricci tensor and u^μ is a tangent to the geodesic, or 4-velocity. By virtue of the field equations this condition may be written

$$(T_{\mu\nu} - {}^1/_2\, g_{\mu\nu} T)\, u^\mu u^\nu \geq 0,$$

which is a reasonable condition on the energy;

(ii) every timelike and null geodesic contains a point at which

$$u_{[\mu} R_{\nu]\kappa\lambda[\rho} u_{\sigma]} u^\kappa u^\lambda \neq 0;$$

(iii) the space-time M contains no closed timelike curves – a reasonable causality condition;[29]

(iv) M contains a trapped surface.

A trapped surface is the general term for the event horizon in the Schwarzschild and Kerr cases. Consider light being emitted from all points on a spherical surface in flat space-time, orthogonal to the surface. There is an outgoing wavefront, which diverges as it travels away, and an incoming wavefront, which converges to the point $r=0$. But now consider a surface *inside* the event horizon $r=2m$ (or $r=r_+$). From what we have seen above, it is clear that *both* the outgoing *and* the incoming wavefronts actually *converge*, and no light escapes through the event horizon. A surface for which this is true is defined by Penrose (1969) to be

[29] If a space-time contains closed timelike world-lines, then it will be possible to start a journey from a particular point in space and time and eventually return to the same point in space *and in time*. This must involve travelling into the past (as well as the future), so that one could arrange for one's parents never to meet, which clearly violates causality.

a *trapped surface*. So the Hawking–Penrose result is that if a trapped surface is present in a space-time M, with the other conditions listed above, then a singularity will be present. But we may ask: is it possible that a singularity will occur *without* a trapped surface? Penrose (1969) hypothesises not. This is the *Cosmic Censorship hypothesis*, that all singularities in nature are surrounded by event horizons, and are therefore not visible from the outside world – there are no *naked singularities*.

We have attempted, at least in brief outline, to answer questions (1) and (3) above. As far as (2) is concerned, it has been shown that an *equilibrium* state of a vacuum black hole (i.e. on an asymptotically flat space-time) must correspond to the Kerr solution, so the higher multipole moments present in any realistic star will be radiated away as the star collapses and settles down.

Further reading

An excellent account, both scholarly and very readable, of stellar collapse and black holes is Israel (1987). An early, and still arresting, set of lectures on the subject is to be found in Wheeler (1964). Detailed accounts of stellar equilibrium and the physics of white dwarfs and neutron stars may be found in Weinberg (1972) and Zel'dovich & Novikov (1996). Rather briefer, though very readable and authoritative, reviews may be found in Chandrasekhar (1969, 1972).

The Kruskal extension of the Schwarzschild solution receives a nice treatment in Rindler (2001), Section 12.5 and an interesting and technically undemanding account, including its implications for wormholes and time travel may be found in Thorne (1994).

Penrose diagrams are introduced in Penrose (1964) and other good accounts of them may be found in Wald (1984), Chapter 11, and Hawking & Ellis (1973), where also their contribution to understanding the causal structure of space-time is treated.

Accounts of the Reissner–Nordstrøm solution may be found in Misner *et al.* (1973), Papapetrou (1974), d'Inverno (1992) and Stephani *et al.* (2003).

Good accounts, in varying amount of detail, of the Kerr solution and the Penrose process are to be found in Wald (1984), Novikov & Frolov (1989), Poisson (2004), Hobson, Efstathiou & Lasenby (2006) and Plebański & Krasiński (2006).

An early and very readable account of black hole thermodynamics is Bekenstein (1980). Authoritative and more recent accounts of this subject, as well as Hawking radiation are Wald (1994, 2001). A more general but very stimulating book, with references to the scientific literature, including that on black holes, thermodynamics and the holographic principle, is Smolin (2001). For more on the holographic principle see Bousso (2002) and Susskind & Lindesay (2005). For interesting speculation on the relation between black holes and elementary particles see 't Hooft (1997).

Authoritative accounts of singularities and other global matters are Hawking & Ellis (1973), Ryan & Shepley (1975), Carter (1979), Wald (1984) and Clarke (1993). Less complete versions may be found in Misner *et al.* (1973), Chapter 34, Ludvigsen (1999), Chapter 14, and Poisson (2004). A good reference on Cosmic Censorship is Penrose (1998).

Problems

7.1 Show that the gravitational binding energy of a sphere of constant density, mass M and radius R is

$$E_B = \frac{3}{5}\frac{GM^2}{R}.$$

7.2 Suppose that Jupiter is held in equilibrium by the balancing of gravity and the (non-relativistic) Fermi pressure of its electrons. Given that its mass is about 10^{-3} times the solar mass, estimate its density.

7.3 By integrating the TOV equation for a star of incompressible fluid, show that such a star cannot be smaller than (9/8) times the Schwarzschild radius.

7.4 Show that in Kerr space-time, when $\Delta = 0$ (on the event horizon), then also

$$g_{00}\, g_{33} - g_{03}{}^2 = 0.$$

7.5 Consider the process in which two black holes, of masses M_1 and M_2, coalesce into a single hole of mass M_3. Show that the area theorem indicates that less than half of the initial energy can be emitted as (gravitational) radiation, and that in the Schwarzschild case the fraction emitted is

$$f < 1 - \frac{1}{\sqrt{2}} \approx 0.29.$$

8 Action principle, conservation laws and the Cauchy problem

If we take seriously – as of course we must – the notion that Einstein's theory of gravity has a status equal in validity to that of other major theories of physics, for example Maxwell's electrodynamics or the more modern gauge field theories of particle physics, then we shall want to ask how General Relativity may be formulated at a fundamental level. In Chapter 5 the field equations were introduced on a more or less ad hoc basis, arguing that what was wanted were equations relating space-time curvature to the energy and momentum of the source, that they should therefore involve second rank tensors; and that the equations also reduced to Newton's law in the non-relativistic limit. This approach is fine as far as it goes, but recall that Maxwell's equations, for example, may be derived from a principle of least action; a Lagrangian formulation. May Einstein's equations also be derived from a Lagrangian formulation? Indeed they may, and this is the subject of the first part of this chapter. We go on to investigate the tricky topic of conservation laws in General Relativity, a subject which has complications resulting from the fact that the (matter) energy-momentum tensor is *covariantly* conserved, whereas 'true' conservation laws involve simply partial, rather than covariant, differentiation. The chapter finishes with a consideration of the Cauchy, or initial value, problem. Einstein's field equations are second order differential equations, whose solutions will therefore involve specifying 'initial data' on a spacelike hypersurface. It turns out that a distinction can then be made between 'dynamical' and non-dynamical components of the metric tensor. This is important in the Hamiltonian formulation of General Relativity, and therefore in some approaches to quantum gravity – though these topics are not treated in this book.

8.1 Gravitational action and field equations

The casting of fundamental theories of physics into the language of an action principle has a long and distinguished history. Let us recall the excellent lecture by Feynman, 'The Principle of Least Action', reproduced in his *Lectures on Physics*,[1] in which he starts by recounting how his physics teacher, Mr Bader, said to him one day after a physics class, 'You look bored; I want to tell you something interesting.' Then says Feynman, 'he told me something which I found absolutely fascinating. Every time the subject comes up, I work on it.' Mr Bader talked about the parabolic path described by a particle in free motion in a gravitational field – which of course can be described by Newton's law $F = ma$. But it can

[1] Feynman *et al.* (1964), Chapter 19.

also be described, Mr Bader said, in terms of an action – call it S – which is the integral over the whole of the particle's path (equivalently over time) of the difference between the kinetic and potential energies at any point

$$S = \int (T - V)\mathrm{d}t. \tag{8.1}$$

One may draw any number of paths between the starting and finishing points, and calculate S for each path. The *actual* path taken is the one for which S is a *minimum*. This is an extraordinary formulation, but one whose spirit goes back to Maupertuis' principle – and indeed further back than that. The principle of least action now occupies such an honourable place in physics that almost all fundamental theories have been formulated in terms of it: we need only recall Hamilton's principle, variational principles in classical mechanics and Feynman's path-integral formulation of quantum mechanics to see examples of the principle at work.

In the area of (special) relativistic *field* theories the action is an integral of a Lagrange density $\mathcal{L}$ over space-time

$$S = \int \mathcal{L}\,\mathrm{d}^4x \tag{8.2}$$

and the principle of least action results in the Euler–Lagrange equations. In almost all cases $\mathcal{L}$ depends on the field in question and its first derivatives only, so that in the case of a scalar field $\phi(x^\mu)$ for example we have

$$\mathcal{L}(\phi, \partial_\mu\phi) = {}^1\!/_2(\partial^\mu\phi)(\partial_\mu\phi) - {}^1\!/_2 m^2\phi^2 \tag{8.3}$$

and the Euler–Lagrange equation

$$\frac{\partial\mathcal{L}}{\partial\phi} - \frac{\partial}{\partial x^\mu}\left(\frac{\partial}{\partial(\partial_\mu\phi)}\right) = 0 \tag{8.4}$$

leads to the Klein–Gordon equation of motion

$$(\partial^\mu\partial_\mu - m^2)\phi(x) = 0. \tag{8.5}$$

Electrodynamics may be formulated in a similar way.

The question we now wish to address is whether, and how, the Einstein field equations of General Relativity

$$R_{\mu\nu} - {}^1\!/_2\, g_{\mu\nu}R = \frac{8\pi G}{c^2}T_{\mu\nu} \tag{8.6}$$

may also be derived from a variational principle. In the example above the field ϕ is defined 'on' space-time, and similarly the electromagnetic field is conveniently represented by $A^\mu(x)$, the 4-vector potential, also a quantity defined as an *additional* structure existing in space-time. But in General Relativity the 'gravitational field' is space-time *itself*; so the first question is, what do we take as the field variable? The usual answer is the metric tensor $g_{\mu\nu}$, and we then have to construct a Lagrangian $\mathcal{L}$ which is a function of $g_{\mu\nu}$ and its derivatives, such that variation of this Lagrangian will yield the field equations. Because of the unusual nature of General Relativity, we expect to – and indeed shall – encounter some subtleties, compared, say, with electrodynamics.

In the field equations (8.6) $T_{\mu\nu}$ represents the contribution of matter to the gravitational field, so in the absence of matter

$$R_{\mu\nu} - {}^1\!/_2\, g_{\mu\nu} R = 0 \tag{8.7}$$

and there remains only the gravitational field itself. It would seem reasonable to express the total action S as a sum of contributions from the matter and from the gravitational field:

$$S = S_{\mathrm{m}} + S_{\mathrm{g}}. \tag{8.8}$$

Let us first show that if S_{g} is given by

$$S_{\mathrm{g}} = \frac{c^3}{16\pi G} \int R \sqrt{-g}\, \mathrm{d}^4 x, \tag{8.9}$$

where $R = R_{\mu\nu}\, g^{\mu\nu}$, the curvature scalar, then the action principle yields Equation (8.7). (The overall multiplicative constant is of course irrelevant at this stage.)

We then consider a variation in the gravitational field

$$g_{\mu\nu} \to g_{\mu\nu} + \delta g_{\mu\nu}. \tag{8.10}$$

Here it is to be understood that this is an *actual* change in the field at each point, not simply a variation resulting from a change in coordinate system. From Equation (3.194)

$$\Gamma^{\lambda}{}_{\mu\nu} = {}^1\!/_2\, g^{\lambda\sigma}(g_{\mu\sigma,\nu} + g_{\nu\sigma,\mu} - g_{\mu\nu,\sigma})$$

it follows that

$$\begin{aligned} \delta\Gamma^{\lambda}_{\mu\nu} = {}& {}^1\!/_2 (\delta g^{\lambda\sigma})(g_{\mu\sigma,\nu} + g_{\nu\sigma,\mu} - g_{\mu\nu,\sigma}) \\ & + {}^1\!/_2\, g^{\lambda\sigma}\left[(\delta g_{\mu\sigma})_{,\nu} + (\delta g_{\nu\sigma})_{,\mu} - (\delta g_{\mu\nu})_{,\sigma}\right]. \end{aligned} \tag{8.11}$$

Now

$$g^{\lambda\sigma} g_{\sigma\rho} = \delta^{\lambda}{}_{\rho} \Rightarrow \delta g^{\lambda\sigma} = -g^{\lambda\kappa} g^{\sigma\rho} (\delta g_{\kappa\rho}) \tag{8.12}$$

and when substituted into (8.11) this gives

$$\delta\Gamma^{\lambda}{}_{\mu\nu} = -g^{\lambda\kappa}\, \delta g_{\kappa\rho} \Gamma^{\rho}{}_{\mu\nu} + {}^1\!/_2\, g^{\lambda\sigma}[(\delta g_{\mu\sigma})_{,\nu} + (\delta g_{\nu\sigma})_{,\mu} - (\delta g_{\mu\nu})_{,\sigma}]. \tag{8.13}$$

From the usual formula for the covariant derivative of a rank 2 tensor

$$(\delta g_{\mu\sigma})_{;\nu} = (\delta g_{\mu\sigma})_{,\nu} - \Gamma^{\rho}{}_{\mu\nu}\, \delta g_{\rho\sigma} - \Gamma^{\rho}{}_{\sigma\nu}\, \delta g_{\rho\mu}, \tag{8.14}$$

hence

$$(\delta g_{\mu\sigma})_{,\nu} + (\delta g_{\nu\sigma})_{,\mu} - (\delta g_{\mu\nu})_{,\sigma} = (\delta g_{\mu\sigma})_{;\nu} + (\delta g_{\nu\sigma})_{;\mu} - (\delta g_{\mu\nu})_{;\sigma} + 2\Gamma^{\rho}{}_{\mu\nu}\, \delta g_{\rho\sigma}, \tag{8.15}$$

and from (8.13)

$$\delta\Gamma^{\lambda}{}_{\mu\nu} = {}^1\!/\!_2\, g^{\lambda\sigma}[(\delta g_{\mu\sigma})_{;\nu} + (\delta g_{\nu\sigma})_{;\mu} - (\delta g_{\mu\nu})_{;\sigma}], \tag{8.16}$$

where the covariant derivatives are calculated using the *unperturbed* connection coefficients $\Gamma^{\lambda}{}_{\mu\nu}$. This equation shows that $\delta\Gamma^{\lambda}{}_{\mu\nu}$ is a *tensor* – which $\Gamma^{\lambda}{}_{\mu\nu}$ is *not*. By virtue of this we may write for the covariant derivative

$$(\delta\Gamma^{\lambda}{}_{\mu\nu})_{;\kappa} = (\delta\Gamma^{\lambda}{}_{\mu\nu})_{,\kappa} - \Gamma^{\rho}{}_{\mu\kappa}\,\delta\Gamma^{\lambda}{}_{\rho\nu} - \Gamma^{\rho}{}_{\nu\kappa}\,\delta\Gamma^{\lambda}{}_{\rho\mu} + \Gamma^{\lambda}{}_{\rho\kappa}\,\delta\Gamma^{\rho}{}_{\mu\nu}$$

and hence, contracting the indices ν and λ (leaving μ and κ as the only free indices)

$$\begin{aligned}(\delta\Gamma^{\lambda}{}_{\mu\lambda})_{;\kappa} - (\delta\Gamma^{\lambda}{}_{\mu\kappa})_{;\lambda} = {} & (\delta\Gamma^{\lambda}{}_{\mu\lambda})_{,\kappa} - (\delta\Gamma^{\lambda}{}_{\mu\kappa})_{,\lambda} - \Gamma^{\rho}{}_{\mu\kappa}\,\delta\Gamma^{\lambda}{}_{\rho\lambda} + \Gamma^{\lambda}{}_{\rho\kappa}\,\delta\Gamma^{\rho}{}_{\mu\lambda} \\ & + \Gamma^{\rho}{}_{\mu\lambda}\,\delta\Gamma^{\lambda}{}_{\rho\kappa} - \Gamma^{\lambda}{}_{\rho\lambda}\,\delta\Gamma^{\rho}{}_{\mu\kappa}.\end{aligned} \tag{8.17}$$

On the other hand, since

$$R_{\mu\kappa} = R^{\lambda}{}_{\mu\lambda\kappa} = \Gamma^{\lambda}{}_{\mu\kappa,\lambda} - \Gamma^{\lambda}{}_{\mu\lambda,\kappa} + \Gamma^{\rho}{}_{\mu\kappa}\,\Gamma^{\lambda}{}_{\lambda\rho} - \Gamma^{\rho}{}_{\mu\lambda}\,\Gamma^{\lambda}{}_{\kappa\rho},$$

then

$$\begin{aligned}\delta R_{\mu\kappa} = {} & (\delta\Gamma^{\lambda}{}_{\mu\kappa})_{,\lambda} - (\delta\Gamma^{\lambda}{}_{\mu\lambda})_{,\kappa} + (\delta\Gamma^{\rho}{}_{\mu\kappa})\,\Gamma^{\lambda}{}_{\rho\lambda} + \Gamma^{\rho}{}_{\mu\kappa}(\delta\Gamma^{\lambda}{}_{\rho\lambda}) \\ & - (\delta\Gamma^{\rho}{}_{\mu\lambda})\,\Gamma^{\lambda}{}_{\rho\kappa} - \Gamma^{\rho}{}_{\mu\lambda}(\delta\Gamma^{\lambda}{}_{\rho\kappa}).\end{aligned} \tag{8.18}$$

Labelling the six terms on the right hand side of (8.17) by 1, 2, 3, 4, 5 and 6, the six terms on the rhs of (8.18) are respectively 2, 1, 6, 3, 4 and 5; so we have finally

$$\delta R_{\mu\kappa} = (\delta\Gamma^{\lambda}{}_{\mu\kappa})_{;\lambda} - (\delta\Gamma^{\lambda}{}_{\mu\lambda})_{;\kappa}, \tag{8.19}$$

known as the *Palatini identity*. Note that here the covariant derivatives are defined as if $\delta\Gamma$ were a tensor – which it is. Note also, in passing, that $\delta R_{\mu\kappa}$ may be written in the form (Problem 8.1)

$$\delta R_{\mu\kappa} = {}^1\!/\!_2\, g^{\lambda\sigma}[(\delta g_{\sigma\mu})_{;\kappa;\lambda} + (\delta g_{\sigma\kappa})_{;\mu;\lambda} - (\delta g_{\mu\kappa})_{;\sigma;\lambda} - (\delta g_{\lambda\sigma})_{;\mu;\kappa}]. \tag{8.20}$$

We are concerned to derive the field equations, so wish to vary the action (8.9). Defining

$$\mathcal{L}_g = \kappa\sqrt{-g}\,R, \quad \kappa = \frac{c^3}{16\pi G}, \tag{8.21}$$

we have

$$\begin{aligned}\delta\mathcal{L}_g &= \kappa\,\delta(\sqrt{-g}\,R) = \kappa\,\delta(\sqrt{-g}\,g^{\mu\nu}R_{\mu\nu}) \\ &= \kappa\{\sqrt{-g}\,R_{\mu\nu}\,\delta g^{\mu\nu} + R(\delta\sqrt{-g}) + \sqrt{-g}\,g^{\mu\nu}\,\delta R_{\mu\nu}\}.\end{aligned} \tag{8.22}$$

We first show that, by virtue of the Palatini identity, the last term in (8.22) is a total derivative. Indeed, from (8.19)

$$\begin{aligned}\sqrt{-g}g^{\mu\kappa}\,\delta R_{\mu\kappa} &= \sqrt{-g}g^{\mu\kappa}[(\delta\Gamma^{\lambda}{}_{\mu\kappa})_{;\lambda} - (\delta\Gamma^{\lambda}{}_{\mu\lambda})_{;\kappa}] \\ &= \sqrt{-g}\,[(g^{\mu\kappa}\,\delta\Gamma^{\lambda}{}_{\mu\kappa})_{;\lambda} - (g^{\mu\kappa}\,\delta\Gamma^{\lambda}{}_{\mu\lambda})_{;\kappa}]\end{aligned}$$

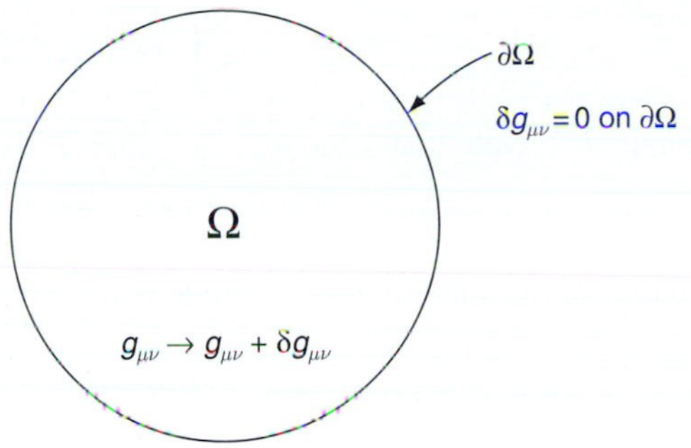

Fig. 8.1 A space-time Ω with boundary $\partial\Omega$; the variation of the metric tensor vanishes on the boundary.

since $g^{\mu\kappa}{}_{;\lambda}=0$. Now it follows from (3.202b) and (3.212) that

$$V^{\mu}{}_{;\mu} = \frac{1}{\sqrt{-g}}(\sqrt{-g}\,V^{\mu})_{,\mu}$$

hence

$$\begin{aligned}\sqrt{-g}\,g^{\mu\kappa}\,\delta R_{\mu\kappa} &= (\sqrt{-g}\,g^{\mu\kappa}\,\delta\Gamma^{\lambda}{}_{\mu\kappa})_{,\lambda} - (\sqrt{-g}\,g^{\mu\kappa}\,\delta\Gamma^{\lambda}{}_{\mu\lambda})_{,\kappa}\\ &= \partial_{\lambda}[\sqrt{-g}\,g^{\mu\kappa}\,\delta\Gamma^{\lambda}{}_{\mu\kappa} - \sqrt{-g}\,g^{\mu\lambda}\,\delta\Gamma^{\kappa}{}_{\mu\kappa}]\\ &\equiv \partial_{\lambda}\,W^{\lambda},\end{aligned} \tag{8.23}$$

a total divergence. Then, if we are evaluating $\int_{\Omega}\delta(\sqrt{-g}R)$ over the region Ω with $\delta g_{\mu\nu}=0$ on the boundary $\partial\Omega$ (see Fig. 8.1), by Gauss's theorem the total divergence does not contribute. Furthermore, since

$$\delta(\sqrt{-g}) = -{}^1\!/_2\,\sqrt{-g}\,g_{\mu\nu}\,\delta g^{\mu\nu},$$

then (8.22) and (8.23) give

$$\delta(\sqrt{-g}\,R) = \sqrt{-g}\,\delta g^{\mu\nu}(R_{\mu\nu} - {}^1\!/_2\,g_{\mu\nu}R) + \partial_{\lambda}W^{\lambda} \tag{8.24}$$

and, from (8.9)

$$\delta S_g = \frac{c^3}{16\pi G}\int_{\Omega}\delta g^{\mu\nu}(R_{\mu\nu} - {}^1\!/_2\,g_{\mu\nu}\,R)\;\mathrm{d}^4x. \tag{8.25}$$

The principle of least action dictates that $\delta S_g=0$ for arbitrary $\delta g^{\mu\nu}$, giving finally

$$R_{\mu\nu} - {}^1\!/_2\,g_{\mu\nu}R = 0, \tag{8.26}$$

the vacuum field equations. We conclude that (8.9) is a suitable expression for the action of a gravitational field.

Turning now to the inclusion of matter and radiation, if we accept that they interact with gravity through the *energy-momentum tensor*, this furnishes a *definition* of this tensor; in other words a change $\delta g_{\mu\nu}$ in the gravitational field will result in a change in the matter action S_{m} (Equation (8.8)) of

$$\delta S_{\rm m} = -\frac{c}{2}\int T^{\mu\nu}(x)\sqrt{-g}\,\delta g_{\mu\nu}\,{\rm d}^4x. \tag{8.27}$$

This *defines* $T^{\mu\nu}$. Note that since $g_{\mu\nu} = g_{\nu\mu}$, then $T^{\mu\nu}$ is also symmetric

$$T^{\mu\nu} = T^{\nu\mu}. \tag{8.28}$$

The equation above then implies a matter action equal to

$$S_{\rm m} = -\frac{c}{2}\int T^{\mu\nu}(x)\sqrt{-g}\,g_{\mu\nu}\,{\rm d}^4x, \tag{8.29}$$

giving a total action

$$S = S_{\rm g} + S_{\rm m} = \int\left\{\frac{c^3}{16\pi G}R - \frac{c}{2}T_{\mu\nu}g^{\mu\nu}\right\}\sqrt{-g}\,{\rm d}^4x. \tag{8.30}$$

On variation of $g_{\mu\nu}$

$$\delta S = \int\left\{\frac{c^3}{16\pi G}\left(R_{\mu\nu} - \frac{1}{2}g_{\mu\nu}R\right) - \frac{c}{2}T_{\mu\nu}\right\}\delta g^{\mu\nu}\sqrt{-g}\,{\rm d}^4x, \tag{8.31}$$

and $\delta S = 0$ gives

$$R_{\mu\nu} - {}^1\!/\!{}_2\, g_{\mu\nu}R = \frac{8\pi G}{c^2}\,T_{\mu\nu}, \tag{8.32}$$

the matter field equations. We conclude that (8.29) is a valid expression for the contribution of matter (and radiation) to the action.

8.2 Energy-momentum pseudotensor

The left hand side of the field equation (8.6) above obeys

$$(R^{\mu\nu} - {}^1\!/\!{}_2\, g^{\mu\nu}R)_{;\,\nu} = 0,$$

and it therefore follows that

$$T^{\mu\nu}{}_{;\,\nu} = 0, \tag{8.33}$$

which is a statement of energy-momentum conservation, but in *covariant* form. This is not ideal, however; if energy and momentum are truly conserved we should be looking for a conservation law of the form, involving simply a partial divergence,

$$W^{\mu\nu}{}_{,\nu} = \frac{\partial}{\partial x^\nu}W^{\mu\nu} = 0. \tag{8.34}$$

It is our task in this section to find a form for $W^{\mu\nu}$. It turns out that that is possible, and the extra contribution to $W^{\mu\nu}$, apart from $T^{\mu\nu}$, is interpreted as coming from the gravitational field itself.

To begin, let us consider again a variation in the metric tensor, given by Equation (8.10). The requirement that this variation is not simply a consequence of a coordinate change, but represents a 'real' change in the gravitational field, is, translated into mathematical language, the statement that it is described by a *Lie derivative*

$$\bar{x}^{\mu} = x^{\mu} - \varepsilon\, a^{\mu}(x) \equiv x^{\mu} - \varepsilon^{\mu}, \tag{8.35}$$

as in Equation (6.121) above. Then

$$\delta g_{\mu\nu} = \varepsilon(g_{\mu\lambda}\, a^{\lambda}{}_{,\nu} + g_{\lambda\nu}\, a^{\lambda}{}_{,\mu} + g_{\mu\nu,\lambda}\, a^{\lambda}). \tag{8.36}$$

Write the action (8.9) (or (8.21)) as

$$S_{\mathrm{g}} = \kappa \int \tilde{R}\, \mathrm{d}^4x, \tag{8.37}$$

$$\tilde{R} = \tilde{R}^{\mu\nu}\, g_{\mu\nu}, \quad \tilde{R}^{\mu\nu} = \sqrt{-g}\, R^{\mu\nu}. \tag{8.38}$$

Then under (8.36)

$$\delta S_{\mathrm{g}} = \varepsilon\, \kappa \int (\tilde{R}^{\mu\nu}\, g_{\mu\lambda}\; a^{\lambda}{}_{,\nu} + \tilde{R}^{\mu\nu}\, g_{\lambda\nu}\, a^{\lambda}{}_{,\mu} + \tilde{R}^{\mu\nu}\, g_{\mu\nu,\lambda}\, a^{\lambda})\, \mathrm{d}^4x.$$

The first term in the integrand may be written as

$$\partial_{\nu}(\tilde{R}^{\mu\nu}\, g_{\mu\lambda}\, \varepsilon^{\lambda}) - (\tilde{R}^{\mu\nu}\, g_{\mu\lambda})_{,\nu}\, \varepsilon^{\lambda}.$$

Re-expressing the total divergence as a surface term, which will vanish if $a^{\mu}(x)$ vanishes on the surface, and performing the same operation on the second term, gives

$$\delta S_{\mathrm{g}} = \varepsilon\kappa \int [-(\tilde{R}^{\mu\nu}\, g_{\mu\lambda})_{,\nu} - (\tilde{R}^{\mu\nu}\, g_{\nu\lambda})_{,\mu} + \tilde{R}^{\mu\nu} g_{\mu\nu,\lambda}]\, a^{\lambda}\, \mathrm{d}^4x, \tag{8.39}$$

and noting that a^{λ} is arbitrary, this implies that

$$2(\tilde{R}^{\mu\nu} g_{\mu\lambda})_{,\nu} - \tilde{R}^{\mu\nu}\, g_{\mu\nu,\lambda} = 0$$

or

$$(\tilde{R}^{\nu}{}_{\lambda})_{,\nu} - {}^{1}\!/\!_{2}\tilde{R}^{\mu\nu}\, g_{\mu\nu,\lambda} = 0. \tag{8.40}$$

This equation is actually

$$\tilde{R}^{\nu}{}_{\lambda;\nu} = 0 \tag{8.41}$$

(see Problem 8.2), or equivalently

$$\tilde{R}^{\mu\nu}{}_{;\nu} = 0.$$

We have already seen that because of the contracted Bianchi identities and the field equations, we have

$$(R^{\mu\nu} - {}^{1}\!/\!_{2} g^{\mu\nu} R)_{;\nu} = 0; \quad T^{\mu\nu}{}_{;\nu} = 0.$$

It follows immediately that

$$[\sqrt{-g}(R^{\mu\nu} - {}^1\!/\!_2\, g^{\mu\nu}R)]_{;\nu} = 0; \quad \tilde{T}^{\mu\nu}{}_{;\nu} = 0,$$

where $\tilde{T}^{\mu\nu} = \sqrt{-g}T^{\mu\nu}$; or, in virtue of the equivalence of (8.40) and (8.41),

$$\tilde{T}^{\nu}{}_{\mu\,,\nu} - {}^1\!/\!_2\,\tilde{T}^{\kappa\lambda}g_{\kappa\lambda,\mu} = 0. \tag{8.42}$$

This equation, which is a consequence of the covariant conservation law (8.33), is approaching, but is still not actually in, the most desirable form for a conservation law. We need to show that the second term in (8.42) can be expressed as a divergence. To do this, note that (see (8.29))

$$\frac{c}{2}\tilde{T}^{\mu\nu} = -\frac{\delta\mathcal{L}}{\delta g_{\mu\nu}} = -\frac{\partial\mathcal{L}}{\partial g_{\mu\nu}} + \partial_\lambda\left\{\frac{\partial\mathcal{L}}{\partial(\partial_\lambda g_{\mu\nu})}\right\}.$$

Then

$$\begin{aligned}
\frac{c}{2}\tilde{T}^{\mu\nu}g_{\mu\nu,\lambda} &= -\frac{\partial\mathcal{L}}{\partial g_{\mu\nu}}g_{\mu\nu,\lambda} + \partial_\kappa\left\{\frac{\partial\mathcal{L}}{\partial(\partial_\kappa g_{\mu\nu})}\right\}g_{\mu\nu,\lambda}\\
&= \partial_\kappa\left\{\frac{\partial\mathcal{L}}{\partial(\partial_\kappa g_{\mu\nu})}g_{\mu\nu,\lambda}\right\} - \left\{\frac{\partial\mathcal{L}}{\partial g_{\mu\nu}}g_{\mu\nu,\lambda} + \frac{\partial\mathcal{L}}{\partial(\partial_\kappa g_{\mu\nu})}g_{\mu\nu,\kappa,\lambda}\right\}\\
&= \partial_\kappa\left\{\frac{\partial\mathcal{L}}{\partial(\partial_\kappa g_{\mu\nu})}g_{\mu\nu,\lambda}\right\} - \frac{\partial\mathcal{L}}{\partial x^\lambda}\\
&= \left\{\frac{\partial\mathcal{L}}{\partial(\partial_\kappa g_{\mu\nu})}g_{\mu\nu,\lambda} - \delta^\kappa_\lambda\mathcal{L}\right\}_{,\kappa}.
\end{aligned} \tag{8.43}$$

Hence (8.42) is

$$(\tilde{T}_\mu{}^\nu + t_\mu{}^\nu)_{,\nu} = 0 \tag{8.44}$$

with

$$t_\mu{}^\nu = c\left\{\delta_\mu{}^\nu\mathcal{L} - \frac{\partial\mathcal{L}}{\partial(\partial_\nu g_{\kappa\lambda})}g_{\kappa\lambda,\mu}\right\}. \tag{8.45}$$

Equation (8.44) is now in a satisfactory form for a conservation equation, and it shows that the 'source' of gravity is not simply $T^{\mu\nu}$ but $T^{\mu\nu} + t^{\mu\nu}$. It is reasonable to suppose that $t^{\mu\nu}$ represents the energy-momentum of the gravitational field itself, but to make this more convincing we must make some adjustments to the Lagrangian (8.21). The important point is that this Lagrangian involves second order as well as first order derivatives of the metric tensor (since the Riemann tensor itself involves these). We want to find a Lagrangian involving only $g_{\mu\nu}$ and its first order derivatives only, and it turns out we can do this by subtracting a total divergence from $\mathcal{L}$.

From (4.23) we have

$$\begin{aligned}\mathcal{L} &= \kappa\sqrt{-g}\,R = \kappa\sqrt{-g}\,g^{\lambda\mu}R_{\lambda\mu} = \kappa\sqrt{-g}\,g^{\lambda\mu}R^{\rho}{}_{\lambda\rho\mu}\\ &= \kappa\sqrt{-g}\,g^{\lambda\mu}\left(\Gamma^{\kappa}{}_{\lambda\mu,\kappa} - \Gamma^{\kappa}{}_{\lambda\kappa,\mu} + \Gamma^{\kappa}{}_{\rho\kappa}\Gamma^{\rho}{}_{\lambda\mu} - \Gamma^{\kappa}{}_{\rho\mu}\Gamma^{\rho}{}_{\lambda\kappa}\right).\end{aligned} \tag{8.46}$$

The first two terms are

$$\begin{aligned}\mathcal{L}_1 &= \kappa\sqrt{-g}\,g^{\lambda\mu}(\Gamma^{\kappa}{}_{\lambda\mu,\kappa} - \Gamma^{\kappa}{}_{\lambda\kappa,\mu})\\ &= \kappa(\sqrt{-g}\,g^{\lambda\mu}\Gamma^{\alpha}{}_{\lambda\mu} - \sqrt{-g}\,g^{\lambda\alpha}\Gamma^{\kappa}{}_{\lambda\kappa})_{,\alpha}\\ &\quad - \kappa(\sqrt{-g}\,g^{\lambda\mu})_{,\alpha}\Gamma^{\alpha}{}_{\lambda\mu} + (\sqrt{-g}\,g^{\lambda\alpha})_{,\alpha}\Gamma^{\kappa}{}_{\lambda\kappa}.\end{aligned} \tag{8.47}$$

In addition, noting that $(\sqrt{-g}g^{\lambda\mu})_{;\alpha} = 0$ and hence (see (3.224))

$$(\sqrt{-g}\,g^{\lambda\mu})_{,\alpha} = -\Gamma^{\lambda}{}_{\kappa\alpha}\sqrt{-g}\,g^{\kappa\mu} - \Gamma^{\mu}{}_{\kappa\alpha}\sqrt{-g}\,g^{\kappa\lambda} + \sqrt{-g}\,g^{\lambda\mu}\Gamma^{\rho}{}_{\rho\alpha},$$

we obtain, after some cancelling and some addition of terms,

$$\begin{aligned}\mathcal{L}_1 &= \kappa(\sqrt{-g}g^{\lambda\mu}\Gamma^{\nu}{}_{\lambda\mu} - \sqrt{-g}g^{\lambda\nu}\Gamma^{\kappa}{}_{\lambda\kappa})_{,\nu}\\ &\quad + 2\kappa\sqrt{-g}g^{\kappa\lambda}(\Gamma^{\mu}{}_{\kappa\nu}\Gamma^{\nu}{}_{\lambda\mu} - \Gamma^{\mu}{}_{\kappa\lambda}\Gamma^{\nu}{}_{\mu\nu}).\end{aligned}$$

This, when substituted into (8.46) gives

$$\mathcal{L} = \mathcal{L}_0 + A^{\nu}{}_{,\nu} \tag{8.48}$$

with

$$\begin{aligned}A^{\nu} &= \kappa\sqrt{-g}(g^{\lambda\mu}\Gamma^{\nu}{}_{\lambda\mu} - g^{\lambda\nu}\Gamma^{\kappa}{}_{\lambda\kappa}),\\ \mathcal{L}_0 &= \kappa\sqrt{-g}g^{\kappa\lambda}(\Gamma^{\mu}{}_{\kappa\nu}\Gamma^{\nu}{}_{\lambda\mu} - \Gamma^{\mu}{}_{\kappa\lambda}\Gamma^{\nu}{}_{\mu\nu}).\end{aligned} \tag{8.49}$$

Making the usual observation that a total divergence in the action has no effect (on the field equations), we may use, instead of $\mathcal{L}$ in the definition (8.45) of $t_{\mu}{}^{\nu}$, the Lagrangian $\mathcal{L}_0$:

$$t_{\mu}{}^{\nu} = c\left\{\delta_{\mu}{}^{\nu}\mathcal{L}_0 - \frac{\partial\mathcal{L}_0}{\partial(\partial_{\nu}g_{\kappa\lambda})}g_{\kappa\lambda,\mu}\right\}. \tag{8.50}$$

The important point about $\mathcal{L}_0$ is that, unlike $R = g^{\mu\nu}R_{\mu\nu}$, it only contains the metric tensor and its *first* derivatives

$$\mathcal{L}_0 = \mathcal{L}_0(g_{\mu\nu}, g_{\mu\nu,\lambda}).$$

The consequence of this is that in a geodesic coordinate system, in which the first derivatives of $g_{\mu\nu}$ vanish, $t^{\mu\nu}$ also vanishes, $t^{\mu\nu} = 0$. In other coordinate systems this is not the case, so $t^{\mu\nu}$ is not a tensor – it vanishes in some coordinate systems but not in others. It is commonly called a 'pseudotensor'. The *covariant* conservation law $T^{\mu\nu}{}_{;\nu} = 0$, in which a 'sham' (to use Schrödinger's word[2]) divergence is used, is replaced by

[2] Schrödinger (1985).

$$(T^{\mu\nu} + t^{\mu\nu})_{,\nu} = 0, \tag{8.51}$$

in which $t^{\mu\nu}$ is a 'sham' tensor.

What is the meaning of $t^{\mu\nu}$? In any system held together by gravity, the gravitational *field* itself contributes to the total mass (equivalently, to the energy), and therefore contributes to its own source (rather like the Yang–Mills field in particle physics). In saying, then, that $T^{\mu\nu} + t^{\mu\nu}$ is the source of the gravitational field, it seems perfectly reasonable, given that $T^{\mu\nu}$ represents the contribution of matter, to identify $t^{\mu\nu}$ as the contribution of the gravitational field to the energy-momentum tensor. It is also to be expected that this contribution is not tensorial, since, by virtue of the equivalence principle, in an inertial frame the gravitational field locally disappears, so of course will carry no energy or momentum; whereas in an arbitrary frame of refernce the energy of the field will not be zero. We cannot, then, identify a *place*, or places, where the gravitational field exists and carries energy, since whether the field carries energy also depends on the frame of reference. Gravitational energy is *not localisable*. We must not, however, forget that, despite the non-covariant nature of some of the quantities we have been concerned with, our whole formalism is relativistically *covariant*. The gravitational field certainly contributes to the mass-energy of a bound system, but this contribution is not localisable. It is, one may reflect, rather neat – Penrose[3] uses the word 'miraculous' – to see how this fits together.

8.3 Cauchy problem

The vacuum field equations

$$R_{\mu\nu} = 0 \tag{8.52}$$

are second order differential equations for the components of the metric tensor, so it is reasonable to suppose that, to find a solution, we must first be given the values of $g_{\mu\nu}$ and its time derivatives $g_{\mu\nu,0}$ on a *spacelike hypersurface S* – for example the hypersurface $x^0 = 0$, as sketched in Fig. 8.2. Note that if $g_{\mu\nu}$ is given over the hypersurface, then the spatial derivatives $g_{\mu\nu,i}$ are already known; in fact, *all* the spatial derivatives of $g_{\mu\nu}$ will be known, and the first order time derivatives are also specified, so the only unknown functions are $g_{\mu\nu,00}$, the second order time derivatives, and these will be found from the field equations (8.52). And then, by differentiation with respect to t, the third and higher order time derivatives may be found, so (assuming that $g_{\mu\nu}$ is an analytic function of t) $g_{\mu\nu}$ can then be found for all time.

We are therefore looking to the field equations to give us expressions for the second order time derivatives $g_{\mu\nu,00}$ – that is for $g_{ik,00}$, $g_{i0,00}$ and $g_{00,00}$. These ten functions are all we need to know. The field equations depend on the Riemann tensor, which, as will now be shown, can be written as

[3] Penrose (2004), p. 468.

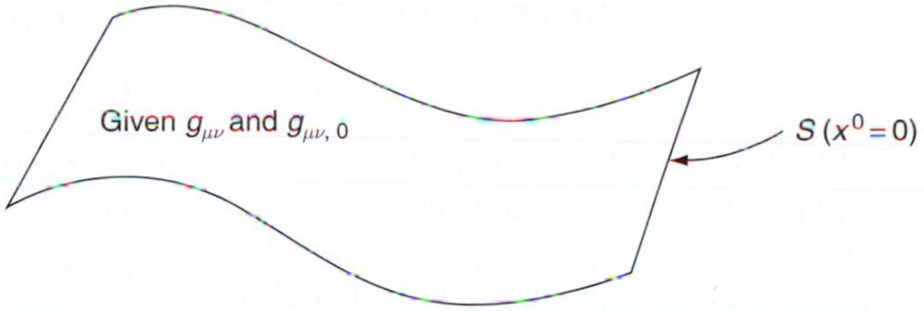

Fig. 8.2 The spacelike hypersurface S, defined by $x^0 = 0$, on which the metric tensor and its first derivatives are given.

$$R_{\sigma\lambda\mu\nu} = {}^1\!/\!_2(g_{\sigma\nu,\lambda\mu} - g_{\sigma\mu,\lambda\nu} + g_{\lambda\mu,\sigma\nu} - g_{\lambda\nu,\sigma\mu}) \\ + g_{\alpha\beta}(\Gamma^\alpha{}_{\sigma\nu}\Gamma^\beta_{\lambda\mu} - \Gamma^\alpha{}_{\sigma\mu}\Gamma^\beta_{\lambda\nu}). \tag{8.53}$$

To see this, note that from (4.31) we have

$$\begin{aligned} R_{\sigma\lambda\mu\nu} &= g_{\sigma\kappa}R^\kappa_{\lambda\mu\nu} \\ &= g_{\sigma\kappa}(\Gamma^\kappa{}_{\lambda\nu,\mu} - \Gamma^\kappa{}_{\lambda\mu,\nu}) + g_{\sigma\kappa}(\Gamma^\kappa{}_{\rho\mu}\Gamma^\rho{}_{\lambda\nu} - \Gamma^\kappa{}_{\rho\nu}\Gamma^\rho{}_{\lambda\mu}) \\ &= {}^1\!/\!_2\, g_{\sigma\kappa}\partial_\mu[g^{\kappa\tau}(g_{\tau\lambda,\nu} + g_{\tau\nu,\lambda} - g_{\lambda\nu,\tau})] \\ &\quad - {}^1\!/\!_2\, g_{\sigma\kappa}\,\partial_\nu[g^{\kappa\tau}(g_{\tau\lambda,\mu} + g_{\tau\mu,\lambda} - g_{\lambda\mu,\tau})] \\ &\quad + g_{\sigma\kappa}(\Gamma^\kappa{}_{\rho\mu}\Gamma^\rho{}_{\lambda\nu} - \Gamma^\kappa{}_{\rho\nu}\Gamma^\rho{}_{\lambda\mu}). \end{aligned}$$

In the first two terms, noting that

$$g_{\sigma\kappa}\,\partial_\mu g^{\kappa\tau} = -g^{\kappa\tau}\,\partial_\mu g_{\sigma\kappa} = -g^{\kappa\tau}(g_{\eta\sigma}\Gamma^\eta{}_{\kappa\mu} + g_{\eta\kappa}\Gamma^\eta{}_{\sigma\kappa})$$

(by explicit calculation), we find, after some algebra and much cancellation of terms, Equation (8.53). The first bracket in this equation contains second derivatives of the metric tensor, and the second bracket only first order derivatives. The second bracket is therefore included in the initial data. The second order time derivatives, because of the antisymmetry properties of the Riemann tensor, come from R_{i0k0}, and these will yield only $g_{ik,00}$, *not* $g_{i0,00}$ or $g_{00,00}$. This could be called a problem of *underdetermination* – the field equations will determine neither $g_{i0,00}$ nor $g_{00,00}$. The components of the Ricci tensor are

$$\begin{aligned} R_{00} &= R^\mu{}_{0\mu 0} = R^i_{0i0} = g^{i\mu}R_{\mu 0 i 0} = g^{ik}R_{0i0k} \\ &= g^{ik}(-{}^1\!/\!_2\, g_{ik,00} + \cdots) \\ &= -{}^1\!/\!_2 g^{ik} g_{ik,00} + (\mathrm{ID})_{00} \end{aligned} \tag{8.54}$$

(where ID stands for initial data),

$$\begin{aligned} R_{0i} &= R^\mu{}_{0\mu i} = g^{\mu\lambda}R_{\lambda 0\mu i} = g^{\mu k}R_{k0\mu i} \\ &= g^{0k}R_{k00i} + (\mathrm{ID})_{0i} \\ &= {}^1\!/\!_2\, g^{0k} g_{ik,00} + (\mathrm{ID})_{0i} \end{aligned} \tag{8.55}$$

and

$$\begin{aligned} R_{ik} &= R^{\mu}{}_{i\mu k} = g^{00}\, R_{0i0k} + (\mathrm{ID})_{ik} \\ &= -{}^1\!/_2\, g^{00} g_{ik,00} + (\mathrm{ID})_{ik}. \end{aligned} \tag{8.56}$$

On the other hand, the field equations (8.52) (= (8.54) + (8.55) + (8.56)) are 10 equations for the 6 unknowns $g_{ik,00}$ – a problem of *overdetermination*. This problem will have to be solved by demanding *compatibility requirements* for the initial data.

Let us summarise the situation so far. To determine the complete evolution of a space-time we want to find the 10 quantities $g_{\mu\nu,00}$. The field equations $R_{\mu\nu} = 0$ only give $g_{ik,00}$ – they do not give $g_{i0,00}$ or $g_{00,00}$: this is Problem 1. On the other hand the field equations are 10 equations in 6 unknowns, so there must be compatibility requirements in the initial data: this is Problem 2.

To solve Problem 1, we shall show that there is a coordinate system in which

$$g_{i0,00} = 0, \quad g_{00,00} = 0. \tag{8.57}$$

These are four equations, which can be made to follow from the four components of a coordinate transformation: we propose the transformation

$$x'^{\mu} = x^{\mu} + \frac{(x^0)^3}{6} A^{\mu}(x). \tag{8.58}$$

The spacelike hypersurface S on which the initial data are specified is $x^0 = 0$, so

$$x'^{\mu} = x^{\mu} \quad \text{on } S. \tag{8.59}$$

From (8.58)

$$\frac{\partial x'^{\mu}}{\partial x^{\nu}} = \delta^{\mu}{}_{\nu} + \delta^{0}{}_{\nu} \frac{(x^0)^2}{2} A^{\mu} + \frac{(x^0)^3}{6} A^{\mu}{}_{,\nu}, \tag{8.60}$$

giving

$$\frac{\partial x'^{\mu}}{\partial x^{\nu}} = \delta^{\mu}{}_{\nu} \quad \text{on } S. \tag{8.61}$$

Differentiating further it is straightforward to see that

$$\frac{\partial^2 x'^{\mu}}{\partial x^{\nu} \partial x^{\lambda}} = 0 \quad \text{on } S, \tag{8.62}$$

and from (8.60), that

$$\left(\frac{\partial x'^{\mu}}{\partial x^0} \right)_{,00} = A^{\mu} \quad \text{on } S. \tag{8.63}$$

Now the metric tensor transforms as

$$g_{\mu\nu} = g'_{\kappa\lambda} \frac{\partial x'^{\kappa}}{\partial x^{\mu}} \frac{\partial x'^{\lambda}}{\partial x^{\nu}}, \tag{8.64}$$

so, from (8.61),

$$g_{\mu\nu} = g'_{\mu\nu} \quad \text{on } S. \tag{8.65}$$

In addition, noting that

$$\begin{aligned} g_{\mu\nu,\lambda} &= \frac{\partial}{\partial x'^{\rho}}\left(g'_{\kappa\sigma}\frac{\partial x'^{\kappa}}{\partial x^{\mu}}\frac{\partial x'^{\sigma}}{\partial x^{\nu}}\right)\frac{\partial x'^{\rho}}{\partial x^{\lambda}} \\ &= g'_{\kappa\sigma\rho}\frac{\partial x'^{\kappa}}{\partial x^{\mu}}\frac{\partial x'^{\sigma}}{\partial x^{\nu}}\frac{\partial x'^{\rho}}{\partial x^{\lambda}} \\ &\quad + g'_{\kappa\sigma}\left(\frac{\partial^2 x'^{\kappa}}{\partial x^{\mu}\partial x^{\lambda}}\frac{\partial x'^{\sigma}}{\partial x^{\nu}} + \frac{\partial x'^{\kappa}}{\partial x^{\mu}}\frac{\partial^2 x'^{\sigma}}{\partial x^{\nu}\partial x^{\lambda}}\right), \end{aligned} \tag{8.66}$$

it follows from (8.61) and (8.62) that

$$g_{\mu\nu,\lambda} = g'_{\mu\nu,\lambda} \quad \text{on } S. \tag{8.67}$$

Equations (8.65) and (8.67) imply that the *initial data are unchanged* in the new coordinate system. We must now show that it is possible to choose A^{μ} in (8.58) such that (8.57) holds. The calculations are actually straightforward and are set as a problem: they involve simply showing that

$$g_{00,00} = g'_{00,00} + 2A^{\mu}g_{\mu 0} \quad \text{on } S, \tag{8.68}$$

$$g_{0i,00} = g'_{0i,00} + g_{\mu i}A^{\mu} \quad \text{on S}, \tag{8.69}$$

and

$$g_{ik,00} = g'_{ik,00} \quad \text{on } S. \tag{8.70}$$

Then, choosing A^{μ} such that

$$\begin{aligned} g_{00,00} &= 2A^{\mu}g_{\mu 0} \quad \text{(1 condition)} \\ g_{0i,00} &= g_{\mu i}A^{\mu} \quad \text{(3 conditions)} \end{aligned}$$

will mean that

$$g'_{00,00} = g'_{0\,i,00} = 0,$$

as claimed in (8.57). We have now solved Problem 1.

Turning to Problem 2, we now regard the equations (see (8.56))

$$R_{ik} = 0 \tag{8.71}$$

as *evolution*, or *dynamical* equations. They are six equations for the six unknown functions $g_{ik,00}$. The remaining four equations

$$R_{i0} = 0, \quad R_{00} = 0 \tag{8.72}$$

can be shown to involve the initial data *only*, and therefore to act as *constraints* on them. For we see that, from the field equations

$$g^{00}R_{00} - g^{ik}R_{ik} = 0, \tag{8.73}$$

so (8.54) and (8.56) give

$$g^{00}(\mathrm{ID})_{00} - g^{ik}(\mathrm{ID})_{ik} = 0. \tag{8.74}$$

This is a constraint (one constraint) on the initial data – expressed in the form (8.73). This equation, however, is equivalent to

$$G^0{}_0 = 0, \tag{8.75}$$

since

$$\begin{aligned} G^0{}_0 &= R^0{}_0 - {}^1\!/\!_2 R = R^0{}_0 - {}^1\!/\!_2(g^{\mu\nu}R_{\mu\nu}) = g^{0\mu}R_{0\mu} - {}^1\!/\!_2(g^{\mu\nu}R_{\mu\nu}) \\ &= {}^1\!/\!_2(g^{00}R_{00} - g^{ik}R_{ik}), \end{aligned}$$

which vanishes, according to (8.73).

In a similar way, the field equations also imply that

$$g^{00}R_{i0} + g^{0k}R_{ik} = 0, \tag{8.76}$$

but from (8.55) and (8.56) this is the same as

$$g^{00}(\mathrm{ID})_{0i} + g^{0k}(\mathrm{ID})_{ik} = 0, \tag{8.77}$$

and therefore acts as three constraints on the initial data. On the other hand

$$G_i{}^0 = R_i{}^0 = g^{0\mu}R_{i\mu} = g^{00}R_{i0} + g^{0k}R_{ik},$$

so (8.76) – and therefore (8.77) – can be stated in the form

$$G_i{}^0 = 0. \tag{8.78}$$

Summarising, the vacuum field equations $R_{\mu\nu}=0$ may be put in the form

$$\begin{aligned} R_{ik} &= 0 \quad \text{evolution equations,} \\ G_i{}^0|s &= 0 \quad \text{constraints on the initial data.} \end{aligned} \tag{8.79}$$

Problem 2 is now solved and the Cauchy, or initial value problem is now cast into the form (8.79) above, where a distinction is made between dynamical and non-dynamical variables. This is as far as we shall go in our account of these matters, but it is pertinent to remark that the point we have reached marks the beginning of a particular approach to the study of the dynamics of General Relativity. This is based on what is known as a 3+1 split, separating time (1) from space (3) by describing the evolution of spacelike hypersurfaces in time. Since the overall formalism is invariant under general coordinate transformations, there is no worry about lack of covariance in any of the calculations. This approach was pioneered by Arnowitt, Deser and Misner (1962), and is also important in the Hamiltonian formulation of General Relativity. It is perhaps also worth remarking that Maxwell's equations of electrodynamics may also be represented by those equations which are truly dynamical and those which act as constraints. This division exists essentially because of the gauge invariance of electrodynamics; and in General Relativity it is general coordinate invariance that plays a role analogous to that of gauge invariance in electrodynamics. Finally, this whole subject comes into its own as one way of approaching quantum gravity. For more details see 'Further reading'.

Further reading

An excellent account of the role of the least action principle in physics is Yourgrau & Mandelstam (1968). The expressions (8.50) and (8.51) for the conservation of energy and the energy-momentum pseudotensor first appear in Einstein (1916a). Detailed accounts of this topic may be found in Anderson (1967), Davis (1970), Papapetrou (1974), as well as in some of the older books: Bergmann (1942), Weyl (1952), Pauli (1958). A good account also appears in Landau & Lifshitz (1971). Very readable and rather enlightening versions of this topic may be found in Mehra (1973), Pais (1982) and Schrödinger (1985).

Good accounts of the Cauchy problem may be found in Bruhat (1962) and Adler *et al.* (1975). Considerably more sophisticated versions may be found in the books of Hawking & Ellis (1973) and of Wald (1984). The seminal work on the dynamics of General Relativity is Arnowitt *et al.* (1962); see also Misner *et al.* (1973). For the Hamiltonian formulation of General Relativity see for example Misner *et al.* (1973), Wald (1984), Poisson (2004); and also Padmanabhan (1989) and Barbashov *et al.* (2001). Introductions to the topic of constraints in quantum field theory may be found in Itzykson & Zuber (1980) and Weinberg (1995); see also Dirac (2001).

Problems

8.1 Prove that under a small variation in the gravitational field

$$g_{\mu\nu} \to g_{\mu\nu} + \delta g_{\mu\nu}$$

the change in the Ricci tensor is

$$\delta R_{\mu\kappa} = {}^1\!/\!_2 g^{\lambda\sigma}[(\delta g_{\sigma\mu})_{;\kappa;\lambda} + (\delta g_{\sigma\kappa})_{;\mu;\lambda} - (\delta g_{\mu\kappa})_{;\sigma;\lambda} - (\delta g_{\lambda\sigma})_{;\mu;\kappa}].$$

8.2 Prove Equation (8.41) from (8.40), using Equation (3.224).

8.3 Prove Equations (8.68) to (8.70).

9 Gravitational radiation

The question whether gravitational radiation exists is of great interest, both theoretically and experimentally. In the weak field approximation Einstein's field equations lead to a wave equation, which, in analogy with the situation with Maxwell's equations of electrodynamics, clearly suggests that gravitational waves exist; and, it would be hoped, again in analogy with the electrodynamic case, that they carry energy. We recall the crucial discovery by Hertz of electromagnetic waves, which convinced him, as well as the general public, of the reality of the field. From the 1960s Weber pioneered experiments to search for gravitational waves, but they have not yet been found. On the other hand there is some very convincing, though indirect, evidence that the gravitational field may radiate energy, which comes from the discovery that the period of the binary pulsar PSR 1913+16 is decreasing. One may feel justified in taking an optimistic view that gravitational radiation exists and might soon be discovered.

On the theoretical side there is a problem which is totally absent in the electromagnetic case. General Relativity is a non-linear theory, and this has the physical consequence that the gravitational field itself carries energy – witness the pseudo-energy-momentum tensor discussed earlier – which then acts as a *source* for more gravitational field. In contrast the electromagnetic field carries no electric charge so is not a source of further field. In the language of quantum theory, there is a graviton–graviton coupling but no photon–photon coupling. So while a wavelike solution to the gravitational field equations might have been found in the weak field approximation – which is linear! – an obviously important question is, does General Relativity, as a complete mathematical structure with no approximation, exhibit radiative solutions to the field equations? This is ultimately a question about the Riemann tensor and, using the insights provided by the Petrov classification, it turns out that it does. These are the topics to be discussed in the present chapter, which we begin with a consideration of the weak field approximation.

9.1 Weak field approximation

The weak field approximation is the linear approximation, which was discussed in Chapter 6, and amounts to the assumption (6.1)

$$g_{\mu\nu} = \eta_{\mu\nu} + h_{\mu\nu}, \quad h_{\mu\nu} \ll 1. \tag{9.1}$$

We recall that in this approximation the energy-momentum tensor $T_{\mu\nu}$ obeys the conservation law (6.6)

$$T^{\mu}{}_{\nu,\mu} = 0. \tag{9.2}$$

The harmonic coordinate condition, defined by (6.18), plays a significant role in this approximation and it is useful to note that it may equivalently be stated in the forms

$$g^{\mu\nu}\Gamma^{\lambda}{}_{\mu\nu} = 0 \tag{9.3}$$

and

$$h^{\mu}{}_{\nu,\mu} = \frac{1}{2} h^{\mu}{}_{\mu,\nu} \tag{9.4}$$

(see Problem 9.1). It is straightforward to see that the field equations (6.19)

$$\Box f_{\mu\nu} = \frac{16\pi G}{c^2} T_{\mu\nu}$$

may be written as

$$\Box h_{\mu\nu} = \frac{16\pi G}{c^2} S_{\mu\nu} \tag{9.5}$$

where

$$S_{\mu\nu} = T_{\mu\nu} - 1/2\, g_{\mu\nu} T^{\lambda}{}_{\lambda}$$

as in (6.5). Equation (9.5) has the retarded potential solution

$$h_{\mu\nu}(x,t) = \frac{4G}{c^2} \int \frac{S_{\mu\nu}(x', t - \frac{|x-x'|}{c})}{|x-x'|} \mathrm{d}^3x'. \tag{9.6}$$

In vacuo the field equation reduces to

$$\Box h_{\mu\nu} = 0. \tag{9.7}$$

So far this is quite general. We want to look for a solution to the field equations which represents a *plane wave* emanating from a particular source. The plane wave is given by

$$h_{\mu\nu} = \varepsilon_{\mu\nu} \exp(\mathrm{i}k_{\lambda}x^{\lambda}) + \varepsilon^*_{\mu\nu} \exp(-\mathrm{i}k_{\lambda}x^{\lambda}). \tag{9.8}$$

The field equation (9.7) is satisfied if

$$k^{\mu}k_{\mu} = 0 \tag{9.9}$$

and the harmonic condition (9.4) requires

$$k_{\mu}\varepsilon^{\mu}{}_{\nu} = \frac{1}{2} k_{\nu}\varepsilon^{\mu}{}_{\mu}. \tag{9.10}$$

Since $h_{\mu\nu}$ is symmetric, so is $\varepsilon_{\mu\nu}$

$$\varepsilon_{\mu\nu} = \varepsilon_{\nu\mu}. \tag{9.11}$$

$\varepsilon_{\mu\nu}$ has 10 independent components and (9.10) amounts to four conditions, so there remain six independent components. These, however, do not all have physical significance, for we may make a coordinate transformation

$$x'^{\mu} = x^{\mu} + \xi^{\mu}(x), \tag{9.12}$$

under which the metric tensor becomes

$$g'_{\mu\nu} = \frac{\partial x^{\rho}}{\partial x'^{\mu}}\frac{\partial x^{\sigma}}{\partial x'^{\nu}} g_{\rho\sigma} = (\delta^{\rho}{}_{\mu} - \xi^{\rho}{}_{,\mu})(\delta^{\sigma}{}_{\nu} - \xi^{\sigma}{}_{,\nu})g_{\rho\sigma} = g_{\mu\nu} - \xi^{\rho}{}_{,\mu}g_{\rho\nu} - \xi^{\sigma}{}_{,\nu}g_{\mu\sigma}$$

to lowest order in ξ; equivalently

$$h'_{\mu\nu} = h_{\mu\nu} - \xi_{\mu,\nu} - \xi_{\nu,\mu}. \tag{9.13}$$

We may write ξ^{μ} in the form (suitable for a plane wave)

$$\xi^{\mu}(x) = \mathrm{i}\, e_{\mu} \mathrm{e}^{ikx} - i\, e^{*}_{\mu} \mathrm{e}^{-ikx} \tag{9.14}$$

and then (9.8) and (9.13) give

$$\varepsilon'_{\mu\nu} = \varepsilon_{\mu\nu} + k_{\mu}e_{\nu} + k_{\nu}e_{\mu}. \tag{9.15}$$

The four parameters e_{μ} mean that the number of independent components of $\varepsilon_{\mu\nu}$ (and therefore $h_{\mu\nu}$) is $6-4=2$. To find these two explicitly, consider a wave travelling in the z direction, so

$$k^3 = k^0 \equiv k(>0), \quad k^1 = k^2 = 0, \tag{9.16}$$

then (9.10) gives, for $\nu = 0, 1, 2, 3$,

$$\begin{aligned} \varepsilon_{30} + \varepsilon_{00} &= -\frac{1}{2}(-\varepsilon_{00} + \varepsilon_{11} + \varepsilon_{22} + \varepsilon_{33}), \\ \varepsilon_{01} + \varepsilon_{31} &= 0, \quad \varepsilon_{02} + \varepsilon_{32} = 0, \\ \varepsilon_{33} + \varepsilon_{03} &= \frac{1}{2}(-\varepsilon_{00} + \varepsilon_{11} + \varepsilon_{22} + \varepsilon_{33}). \end{aligned}$$

These equations enable us to express the four quantities ε_{i0} and ε_{22} in terms of the other six $\varepsilon_{\mu\nu}$:

$$\varepsilon_{01} = -\varepsilon_{31}, \quad \varepsilon_{02} = -\varepsilon_{32}, \quad \varepsilon_{03} = -1/2(\varepsilon_{00} + \varepsilon_{33}), \quad \varepsilon_{22} = -\varepsilon_{11}. \tag{9.17}$$

Under the transformation (9.15) we have

$$\begin{aligned} \varepsilon'_{11} = \varepsilon_{11}, \quad \varepsilon'_{12} = \varepsilon_{12}, \quad \varepsilon'_{13} = \varepsilon_{13} + ke_1, \quad \varepsilon'_{23} = \varepsilon_{23} + ke_2, \\ \varepsilon'_{33} = \varepsilon_{33} + 2ke_3, \quad \varepsilon'_{00} = \varepsilon_{00} - 2ke_0. \end{aligned} \tag{9.18}$$

Then, putting

$$ke_1 = -\varepsilon_{13}, \quad ke_2 = -\varepsilon_{23}, \quad ke_3 = -\frac{1}{2}\varepsilon_{33}, \quad ke_0 = \frac{1}{2}\varepsilon_{00},$$

we find

$$\varepsilon'_{13} = \varepsilon'_{23} = \varepsilon'_{33} = \varepsilon'_{00} = 0,$$

leaving, as the only non-zero components, ε_{11}, $\varepsilon_{12} = \varepsilon_{21}$ and $\varepsilon_{22} = -\varepsilon_{11}$. Hence, dropping the primes, $h_{\mu\nu}$ becomes

$$h_{\mu\nu} = \begin{pmatrix} 0 & 0 & 0 & 0 \\ 0 & h_{11} & h_{12} & 0 \\ 0 & h_{12} & -h_{11} & 0 \\ 0 & 0 & 0 & 0 \end{pmatrix}. \tag{9.19}$$

This expression holds for the propagation of a wave in the z direction. Since $h_{\mu\nu}$ contains only two independent components, it follows that there are two polarisation states of gravitational radiation. This form for $h_{\mu\nu}$ is called the transverse-traceless or TT form. The only non-zero components of $h_{\mu\nu}$ are the space-space ones h_{ik}, and – for propagation in the z direction – $h_{3k}=0$; so $h_{\mu\nu}$ contains only components transverse to the direction of propagation. In addition the trace is zero, so we have

$$h^{\mathrm{TT}}{}_{0i} = h^{\mathrm{TT}\mu}{}_{\mu} = 0. \tag{9.20}$$

It is a straightforward to write an expression for $h_{\mu\nu}$ corresponding to propagation of a wave in the x direction. In place of (9.16) we have

$$k^1 = k^0 \equiv k(>0), \quad k^2 = k^3 = 0;$$

the rest of the algebra proceeds in an analogous way and we finish up with

$$h_{\mu\nu} = \begin{pmatrix} 0 & 0 & 0 & 0 \\ 0 & 0 & 0 & 0 \\ 0 & 0 & h_{22} & h_{23} \\ 0 & 0 & h_{23} & -h_{22} \end{pmatrix}. \tag{9.21}$$

We complete this section by making two observations following from the fact that gravitational radiation has two degrees of polarisation: these concern the spin of the graviton, and the behaviour of matter when it is hit by a gravitational wave.

9.1.1 Spin 2 graviton

To talk about 'gravitons' at all is immediately to introduce the language of quantum theory into what has so far been a completely classical discussion – since, of course, the graviton is defined as being the quantum of the gravitational field, analogous to the photon, the quantum of the electromagnetic field. Although there exists no proper quantum theory of gravity at present it is nevertheless possible, in the weak field approximation, to talk about gravitons as *fields* $h_{\mu\nu}$ existing and propagating in a Minkowski space background. In a fully fledged theory of gravity the metric tensor itself will be subject to some sort of quantisation, but the present considerations are much more limited than this.

Spin was introduced into quantum physics as *intrinsic angular momentum* possessed by electrons, which may be represented by the Pauli spin matrices, which obey the same commutation relations as the generators of the rotation group SO(3), or SU(2).[1] There is a

[1] See any book on quantum mechanics; for example Cohen-Tannoudji *et al.* (1977), Vol. 1, Chapter 4, or Sakurai (1994), Chapter 3.

deep connection between rotations and angular momentum and in quantum theory this is manifested by the fact that under a rotation in parameter space the state in question (for example a spin ½ particle) corresponds to a set of vectors in Hilbert space which act as the basis states for a *representation* of the rotation group. Consider for simplicity a rotation about the z axis through an angle θ, which in coordinate space is given by the matrix

$$R(\theta) = \begin{pmatrix} \cos\theta & \sin\theta & 0 \\ -\sin\theta & \cos\theta & 0 \\ 0 & 0 & 1 \end{pmatrix}, \tag{9.22}$$

so that the transformation

$$R(\theta)\begin{pmatrix} x \\ y \\ z \end{pmatrix} = \begin{pmatrix} x' \\ y' \\ z' \end{pmatrix}$$

gives $x' = x\cos\theta + y\sin\theta$, and so on. The state (e_x, e_y, e_z) transforms according to this (adjoint) representation of the rotation group if, under the same rotation about the z axis

$$\begin{pmatrix} e'_x \\ e'_y \\ e'_z \end{pmatrix} = R(\theta)\begin{pmatrix} e_x \\ e_y \\ e_z \end{pmatrix}, \tag{9.23}$$

or, in index notation, with $(e_x, e_y, e_z) = (e_1, e_2, e_3)$,

$$e_i' = R_{ik}e_k \quad (i, k = 1, 2, 3) \tag{9.24}$$

(and there is no need to distinguish upper and lower indices). The two equations above describe how the state of a spin 1 particle changes under a rotation, according to quantum theory. It is then easy to see that, with the explicit form (9.22) for the rotation matrix

$$e_1' - \mathrm{i}\,e_2' = \mathrm{e}^{\mathrm{i}\theta}(e_1 - \mathrm{i}\,e_2), \quad e_3' = e_3, \quad e_1' + \mathrm{i}\,e_2' = \mathrm{e}^{-\mathrm{i}\theta}(e_1 + \mathrm{i}\,e_2);$$

that is to say, the three states e_-, e_3, e_+ respectively have helicity $h = 1, 0, -1$:

$$R(\theta)\,e(h) = \mathrm{e}^{\mathrm{i}h\theta}e(h). \tag{9.25}$$

This is how the polarisation of the photon is usually described in quantum theory; the photon has three polarisation states, of which two are transverse and one is longitudinal. As is well known, however, the longitudinal state is unphysical. In the language of waves, the helicity of electromagnetic waves is either +1 or −1. In the language of photons, we then state that the photon is a particle with spin 1 but only two polarisation states exist, with $J_z = \pm 1$.

Let us now turn to consider the polarisation states of the gravitational wave (9.20). The states $\varepsilon_{\mu\nu}$ transform not as a vector but as a second rank tensor (since there are two indices) under rotations, so (9.24) becomes replaced by

$$\varepsilon_{ik}' = R_{im}\,R_{kn}\,e_{mn}. \tag{9.26}$$

Then with the only non-zero components being $\varepsilon_{11} = -\varepsilon_{22}$ and $\varepsilon_{12} = \varepsilon_{21}$ we have

$$\begin{aligned}\varepsilon_{11}' &= R_{1m} R_{1n}\, \varepsilon_{mn} = (\cos^2\theta)\, \varepsilon_{11} + (2 \sin\theta\, \cos\, \theta)\, \varepsilon_{12} + (\sin^2\theta)\, \varepsilon_{22} \\ &= (\cos^2\theta - \sin^2\theta)\, \varepsilon_{11} + (2 \sin\theta\, \cos\theta)\, \varepsilon_{12}, \\ \varepsilon_{12}' &= R_{1m}\, R_{2n}\, \varepsilon_{mn} = -(2 \sin\theta \cos\theta)\, \varepsilon_{11} + (\cos^2\theta - \sin^2\theta)\, \varepsilon_{12},\end{aligned}$$

giving

$$\varepsilon_{11}' - \mathrm{i}\, \varepsilon_{12}' = \mathrm{e}^{2\mathrm{i}\,\theta}(\varepsilon_{11} - \mathrm{i}\, \varepsilon_{12}), \tag{9.27}$$

a state *with helicity* $h = 2$. Similarly $\varepsilon_{11} + \mathrm{i}\, \varepsilon_{12}$ has helicity $h = -2$. The two states of polarisation of the gravitational wave have helicity ± 2. In the language of *gravitons*, the graviton is a particle with spin 2, but only the two states $\varepsilon_{11} \pm \mathrm{i}\varepsilon_{12}$ exist, with $J_z = \pm 2$.

This analysis completes the parallels between the photon and the (putative) graviton, and serves as a justification for the statement that the graviton is a particle with spin 2, just as the photon is a particle with spin 1. For the sake of rigour, however, it should be remarked that as far as quantum theory is concerned the analysis above is faulty. The reason is deep and the argument is not particularly easy. It derives from the seminal paper by Wigner[2] in which he analysed the definition of spin from a relativistic standpoint. As stated above, spin, as originally introduced into quantum theory, is described by the rotation group SO(3) (or SU (2)) and this works well for the treatment of electrons in atoms, in the non-relativistic regime. In extending these considerations to the (special) relativistic regime, Wigner realised that the rotation group (describing the invariance of the laws of physics under rotations) had to be enlarged not only to the homogeneous Lorentz group (describing the invariance of the laws of physics also under Lorentz transformations) but actually to the *inhomogeneous* Lorentz group, which includes invariance under space and time *translations*. The generators of these translations are essentially the 4-momentum operators P^μ, and one then defines states according to the value of $P^\mu P_\mu$. According to whether this is >0, $=0$ or <0, the states are *spacelike, null* or *timelike*. The truly surprising feature of Wigner's analysis, however, is that his definition of *spin* depends on the translation operators P^μ, and it turns out that it is only for timelike states that spin is characterised by operators obeying the Lie algebra of O(3), the rotation group. In particular, when $P^\mu P_\mu = 0$, in other words for *massless* particles – like the photon and graviton – spin is not described by the rotation group at all, but by the Euclidean group in the plane, which is a non-compact group. This is presumably why it is only the two states $J_z = \pm J$ that exist for photons and gravitons. The analysis above is therefore faulty – or at least inadequate.

9.1.2 The effect of gravitational waves

We now ask what happens when a gravitational wave strikes matter. Without loss of generality we continue to confine our attention to a wave travelling in the z direction. Consider first the case in which $h_{12} = 0$ (Equation (9.19)). Then, in this region of space-time

$$\mathrm{d}s^2 = -c^2\, \mathrm{d}t^2 + \{1 + h_{11}(ct - z)\}\mathrm{d}x^2 + \{1 - h_{11}(ct - z)\}\mathrm{d}y^2 + \mathrm{d}z^2. \tag{9.28}$$

[2] Wigner (1939).

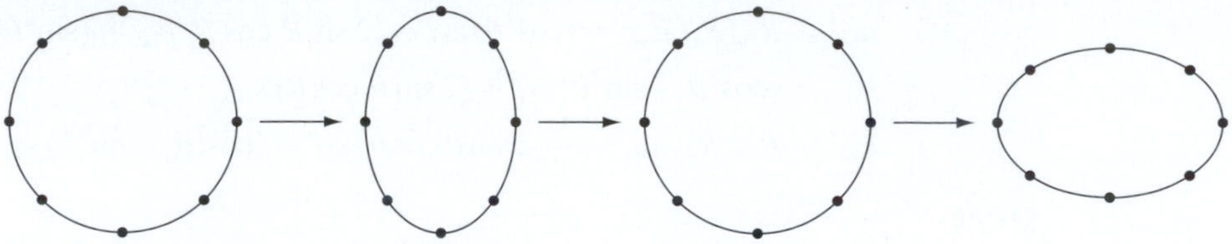

Fig. 9.1 Oscillatory motion of a circle of particles hit by a gravitational wave with polarisation +.

The function $h_{11}(ct-z)$ is an oscillating function, sometimes positive and sometimes negative. Consider some particles arranged round a circle in the xy plane, and hit by a wave with, at some instant, $h_{11}>0$. Then two particles with the same y coordinate have a spatial separation

$$ds^2 = (1+h_{11})\,dx^2$$

and will therefore move apart, whereas two particles with the same x coordinate have a separation

$$ds^2 = (1-h_{11})dy^2$$

and will therefore move closer together. Particles at intermediate points on the circle will move in an intermediate way, so the circle defined by the particles will become 'squashed' as shown in the first part of Fig. 9.1. After a short time h_{11} returns to zero and the circle is restored, after which h_{11} becomes negative and the circle becomes elongated, again later returning to a circular shape. Thus the collection of particles oscillates as shown in the figure. This is clearly a *transverse* motion, and the wave (represented by h_{11} here) is said to have a + polarisation.

Now consider a wave with $h_{11}=0$ – an h_{12} wave. The invariant line element is clearly

$$ds^2 = -c^2\,dt^2 + dx^2 + 2h_{12}\,dx\,dy + dy^2 + dz^2. \tag{9.29}$$

To see what motion this induces in our circular array of particles in the xy plane, it is easiest to rotate through 45° to new axes in the plane:

$$\tilde{x} = x\cos 45^\circ + y\sin 45^\circ = \tfrac{1}{\sqrt{2}}(x+y),$$
$$\tilde{y} = -x\sin 45^\circ + y\cos 45^\circ = \tfrac{1}{\sqrt{2}}(-x+y).$$

Then $dx^2+dy^2 = d\tilde{x}^2 + d\tilde{y}^2$ and $2dx\,dy = d\tilde{x}^2 - d\tilde{y}^2$ and (9.29) becomes

$$ds^2 = -c^2\,dt^2 + (1+h_{12})d\tilde{x}^2 + (1-h_{12})d\tilde{y}^2 + dz^2, \tag{9.30}$$

which is of the same form as (9.28), and therefore corresponds to an oscillatory motion of the same type, but with respect to *rotated* axes, as shown in Fig. 9.2. This is therefore also a transverse wave, and is said to be of polarisation type ×. In this notation (9.19) may be written as

$$h^{\mu\nu} = \begin{pmatrix} 0 & 0 & 0 & 0 \\ 0 & h_+ & h_\times & 0 \\ 0 & h_\times & -h_+ & 0 \\ 0 & 0 & 0 & 0 \end{pmatrix}.$$

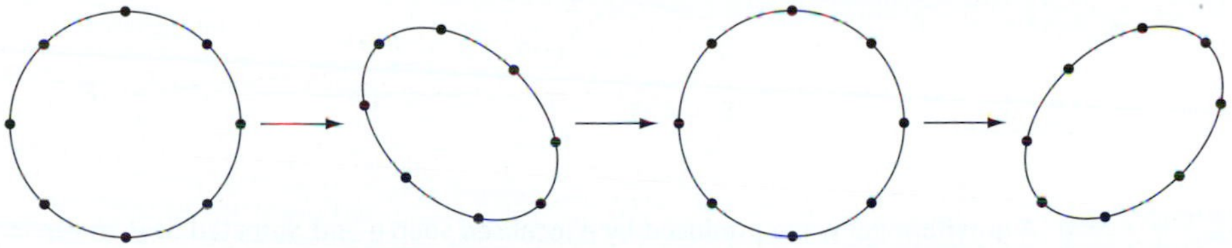

Fig. 9.2 Oscillatory motion induced by a gravitational wave with polarisation $\times$.

9.2 Radiation from a rotating binary source

The quantity $h_{\mu\nu}$ is the amplitude of the gravitational wave produced by the motion of a source, and the source is characterised by the energy-momentum tensor $T_{\mu\nu}$. The expression (9.19) for $h_{\mu\nu}$ has the property $h^{\mu}{}_{\mu}=0$, so from (6.10) and (6.11) $h_{\mu\nu}=f_{\mu\nu}$ and it then follows from (6.22) that

$$h_{\mu\nu}(r,t)=\frac{1}{4\pi}\cdot\frac{16\pi G}{c^2}\int\frac{1}{|r-r'|}T_{\mu\nu}\left(r',t-\frac{|r-r'|}{c}\right)\mathrm{d}^3x',$$

which gives the relation between $h_{\mu\nu}$ and $T_{\mu\nu}$. In the long distance approximation $r\gg r'$ (see Fig. 9.3) so we may put

$$h_{\mu\nu}(r,t)\approx\frac{4G}{c^2}\frac{1}{r}\int T_{\mu\nu}(r',t_{\mathrm{r}})\mathrm{d}^3x', \tag{9.31}$$

where $t_r=t-\dfrac{r}{c}$ is the reduced time. Since we are concerned with weak gravitational fields and slowly moving sources we may assume that $T_{\mu\nu}$ obeys the flat space conservation law

$$T^{\mu\nu}{}_{,\nu}=0. \tag{9.32}$$

Putting $\mu=0$ and differentiating with respect to x^0 then gives

$$T^{00}{}_{,00}=-\frac{\partial}{\partial x^0}\left(\frac{\partial T^{0i}}{\partial x^i}\right)=-\frac{\partial}{\partial x^i}\left(\frac{\partial T^{0i}}{\partial x^0}\right)=-T^{0i}{}_{,0i}. \tag{9.33}$$

Putting $\mu=k$ in (9.32) gives

$$T^{k0}{}_{,0}+T^{ki}{}_{,i}=0,$$

and this together with (9.33) yields

$$T^{00}{}_{,00}=T^{ik}{}_{,ik}. \tag{9.34}$$

Now multiply both sides of this equation by $x^m x^n$ and integrate by parts over all space. The right hand side gives

$$\int\frac{\partial^2T^{ik}}{\partial x^i\,\partial x^k}x^m x^n\,\mathrm{d}^3x=\frac{\partial T^{ik}}{\partial x^i}x^m x^n|_{\Sigma}-\int\frac{\partial T^{ik}}{\partial x^i}(\delta^m{}_k x^n+\delta^n{}_k x^m)\mathrm{d}^3x.$$

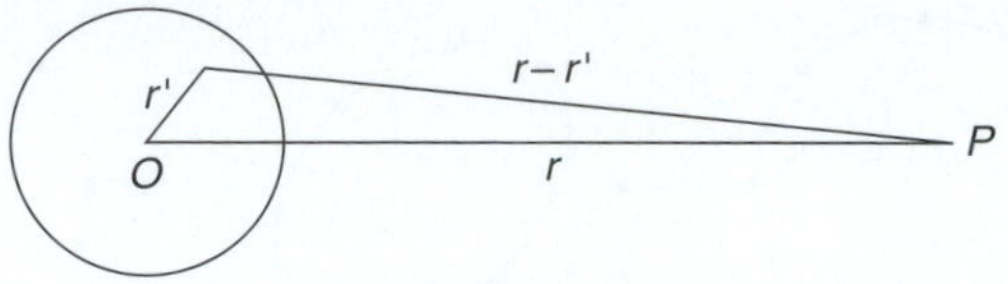

Fig. 9.3 A gravitational wave produced by a localised source and detected at P. In the long distance approximation $r >> r'$.

Fig. 9.4 Two stars of equal mass in orbit around their common centre of mass.

The surface term vanishes for a bounded system and we then integrate by parts once more, again ignoring surface terms, finishing up with

$$2\int T^{mn}\,\mathrm{d}^3x,$$

so (9.34) gives

$$\frac{1}{c^2}\frac{\partial^2}{\partial t^2}\int T^{00}x^m x^n\,\mathrm{d}^3x = 2\int T^{mn}\,\mathrm{d}^3x,$$

which, when substituted in (9.31) gives

$$h^{mn}(r,t) = \frac{2G}{c^4 r}\frac{\partial^2}{\partial t^2}\int T^{00}x^m x^n\,\mathrm{d}^3x.$$

In the non-relativistic (low velocity) limit T^{00} is dominated by the mass density ρ, so we have finally

$$h^{mn}(r,\,t) = \frac{2G}{c^4 r}\ddot{I}^{mn}\,(t_r) \tag{9.35}$$

where $t_r = t - \dfrac{r}{c}$ and

$$I^{mn}\,(t) = \int \rho(t,\,x)\,x^m x^n\,\mathrm{d}^3x. \tag{9.36}$$

This is known as the *quadrupole formula*, since I^{mn} is the second moment (quadrupole moment) of the mass distribution of the source. We should note that although we have used the non-relativistic approximation above, it is a good one for our purposes. Let us now calculate I^{mn} for a binary system.

Suppose that the system consists of two (neutron) stars of approximately the same mass, so in orbiting their common centre of mass they trace out the same circle (Fig. 9.4). The frequency of the gravitational waves emitted is *twice* the orbiting frequency, since after one-half of the period the masses are simply interchanged, which is indistinguishable from the starting configuration. If each star has a mass M and the radius of the orbit is R then by simple Newtonian mechanics the frequency of the orbital motion is

$$\omega = \frac{2\pi}{\tau} = \frac{v}{R} = \left(\frac{GM}{4R^3}\right)^{1/2}. \tag{9.37}$$

At time t the two stars are at the positions

$$(x, y, z) = (R \cos\omega t,\ R \sin\omega t,\ 0) \text{ and } (-R \cos\omega t,\ -R \sin\omega t,\ 0)$$

so

$$I^{mn}(t) = 2Mx^m(t)\,x^n(t);$$

for example

$$I^{xx} = 2MR^2 \cos^2\omega t = MR^2(1 + \cos 2\omega t).$$

The components of the second moment I^{mn} are then

$$I^{mn}(t) = MR^2 \begin{pmatrix} 1+\cos 2\omega t & \sin 2\omega t & 0 \\ \sin 2\omega t & 1-\cos 2\omega t & 0 \\ 0 & 0 & 0 \end{pmatrix}$$

and hence

$$\ddot{I}^{mn}(t) = -4\omega^2 MR^2 \begin{pmatrix} \cos 2\omega t & \sin 2\omega t & 0 \\ \sin 2\omega t & -\cos 2\omega t & 0 \\ 0 & 0 & 0 \end{pmatrix} \tag{9.38}$$

and finally, from (9.36), ignoring the overall minus sign and restoring the formula to a 4×4 form

$$h^{\mu\nu}(t, r) = \frac{8MGR^2\omega^2}{c^4 r} \begin{pmatrix} 0 & 0 & 0 & 0 \\ 0 & \cos 2\omega t_r & \sin 2\omega t_r & 0 \\ 0 & \sin 2\omega t_r & -\cos 2\omega t_r & 0 \\ 0 & 0 & 0 & 0 \end{pmatrix}. \tag{9.39}$$

This is the expression for the gravitational field induced by orbiting stars a distance r away. It is seen, as anticipated above, that the frequency of the oscillating gravitational field is 2ω, twice the frequency of the orbiting motion.

It is also seen that $h^{\mu\nu}$ is of the TT form (9.19) above, so it follows that this form for $h^{\mu\nu}$ already describes radiation emitted in the z direction; it does *not*, however, describe radiation emitted in the x direction, which would have to be of the form (9.21). In particular we need $h_{11} = h_{12} = 0$, so only h_{22} would remain, and the matrix, while transverse, would not then be traceless. This must be rectified, which is not difficult to do: any matrix may be decomposed into a traceless part and a part consisting of the trace. If

$$M = \begin{pmatrix} a & b \\ c & d \end{pmatrix}$$

is a 2×2 matrix, then it can clearly be written as

$$M_{ik} = [M_{ik} - 1/2(\mathrm{tr}\, M)\delta_{ik}] + 1/2(\mathrm{tr}\, M)\delta_{ik},$$

or

$$M = \begin{pmatrix} \frac{a-d}{2} & b \\ c & -\frac{a-d}{2} \end{pmatrix} + \begin{pmatrix} \frac{a+d}{2} & 0 \\ 0 & \frac{a+d}{2} \end{pmatrix}.$$

The first of these matrices is traceless, so in our context will contribute to the gravitational field, whereas the second one does not. We then deduce that the gravitational field corresponding to radiation emitted in the x direction is

$$h^{\mu\nu}(t, r) = \frac{4MGR^2\omega^2}{c^4 r} \begin{pmatrix} 0 & 0 & 0 & 0 \\ 0 & 0 & 0 & 0 \\ 0 & 0 & \cos 2\omega t_r & 0 \\ 0 & 0 & 0 & -\cos 2\omega t_r \end{pmatrix}. \tag{9.40}$$

The expression (9.39) for $h^{\mu\nu}$ contains both diagonal and off-diagonal terms, whereas the version above has only diagonal terms. These terms correspond to the different types of polarisation, denoted h_+ and $h_\times$ above.

9.2.1 Flux

Supposing that the plane of the rotating binary system is the xy plane (Fig. 9.5), let us now calculate the energy flux received from this system, in both the z and in the x directions. This can be done by calculating the energy density in the gravitational field, τ_{00}, either from (8.50), or from the field equations

$$G_{00} = R_{00} - \frac{1}{2} g_{00} R = \frac{8\pi G}{c^2} \tau_{00}. \tag{9.41}$$

We choose the second option, though the two calculations are very similar. The metric tensor is of course

$$g_{\mu\nu} = \eta_{\mu\nu} + h_{\mu\nu},$$

and the calculations will be kept to lowest order in h; so, for example,

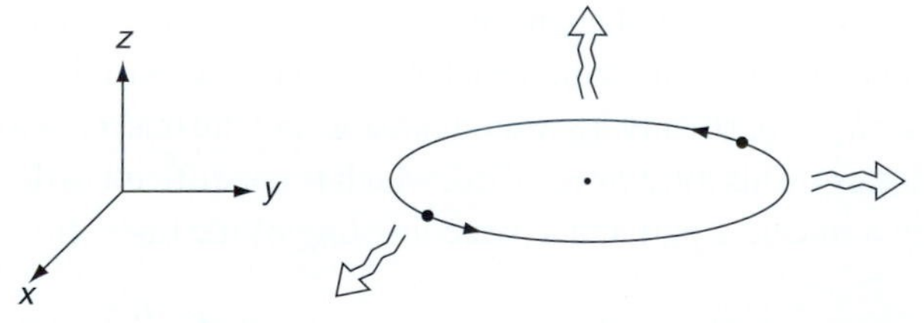

Fig. 9.5 **Orbiting stars in the xy plane. Radiation is emitted in all directions, but not with equal strength.**

$$\Gamma^0{}_{\mu\nu} = \frac{1}{2}\eta^{00}(h_{0\mu,\nu} + h_{0\nu,\mu} - h_{\mu\nu,0}) = \frac{1}{2}h_{\mu\nu,0} + O(h^2).$$

For the sake of definiteness let us proceed with the case of radiation emitted in the z direction. We first calculate the relevant components of the Riemann tensor. We find, for example (after some algebra)

$$R^1{}_{010} = -\frac{1}{2}h_{11,00} - \frac{1}{4}(h_{11,0})^2 - \frac{1}{4}(h_{12,0})^2,$$
$$R^2{}_{020} = \frac{1}{2}h_{11,00} - \frac{1}{4}(h_{11,0})^2 - \frac{1}{4}(h_{12,0})^2,$$
$$R^3{}_{030} = 0,$$

so that

$$R_{00} = R^1{}_{010} + R^2{}_{020} = -\frac{1}{2}\left[(h_{11,0})^2 + (h_{12,0})^2\right].$$

In a similar way we find that $R_{11}=R_{22}=R_{33}=-R_{00}$, and hence, always working to lowest order in h, the curvature scalar R is

$$R = \eta^{\mu\nu}R_{\mu\nu} = -R_{00} + R_{11} + R_{22} + R_{33} = 2\left[(h_{11,0})^2 + (h_{12,0})^2\right]. \tag{9.42}$$

The field equation (9.41) then gives

$$\tau_{00} = \frac{c^2}{16\pi G}\left[(h_{11,0})^2 + (h_{12,0})^2\right], \tag{9.43}$$

which may equivalently be written as

$$\tau_{00} = \frac{c^2}{32\pi G}h_{ij,0}h^{ij}{}_{,0}. \tag{9.44}$$

This is the energy density in the gravitational field. The flux is the amount of energy received per unit area per unit time in a particular direction, which is $c\tau_{00}$. Substituting for h_{11} and h_{22} in (9.39) we find that the flux emitted in the z direction is

$$F_z = \frac{c^3}{16\pi G}\cdot\frac{64M^2G^2R^4\omega^4}{c^8r^2}\cdot 4\omega^4 = \frac{16G}{\pi c^5}M^2R^4\frac{\omega^6}{r^2}. \tag{9.45}$$

In the case of radiation emitted in the x direction the corresponding formula, containing the term $(h_{22,0})^2+(h_{33,0})^2$, will contain a factor $\cos^2 2\omega t$, which clearly oscillates in time, so we should only take the average value of this, which is ½. We then find, for the flux in the x direction

$$F_x = \frac{2G}{\pi c^5}M^2R^4\frac{\omega^6}{r^2}. \tag{9.46}$$

We see that far more energy is emitted in the z direction than the x direction, which is hardly surprising, given the nature of the source. The pattern of energy emission is very anisotropic.

9.2.2 Radiated energy

Since the binary system emits radiation it will clearly lose energy, and this must have observable consequences. We should therefore calculate the total rate of energy emission, but because the emission is not isotropic, the calculation is not entirely straightforward.

We begin by gathering together some relevant formulae. The energy density in the gravitational field at any point is, from (9.44) and (9.35)

$$\tau_{00} = \frac{G}{8\pi c^6 r^2} \dddot{I}_{ij} \dddot{I}_{ij}. \tag{9.47}$$

It depends on the (square of the) third time derivative of the quadrupole moment. We also require, however, that h_{ij}, and therefore I_{ij} (from (9.35)), should be of the TT form. We saw above that for a binary system the form of h_{ij} relevant to emission in the z direction, given by (9.39), is already in the TT form (9.19), but that describing emission in the x direction had to be put in the TT form 'by hand', ending up with Equation (9.40). Since we now have to integrate over *all* directions we must incorporate a mathematical procedure which will guarantee that h_{ij} (equivalently I_{ij}) retains a TT form. We may introduce an initial simplification by starting off with I_{ij} already in traceless form, which is done simply by subtracting a multiple of the unit matrix times the trace, as was done above Equation (9.40). That is, we shall assume that

$$I_{ii} = 0. \tag{9.48}$$

Our aim is to calculate the rate of energy loss through a spherical surface of large radius, centred on the radiating source. The flux of radiation is then perpendicular to the surface. To project out the components of I_{ij} that are transverse to the direction of flow of the energy therefore means projecting out the components in the tangent plane to the surface of the sphere (and neglecting the others). Let $\mathbf{n}$ be the unit normal to the surface at any point, and let $\mathbf{v}$ be some arbitary vector. The *projection operator* P_{ij}

$$P_{ij} = \delta_{ij} - n_i n_j \tag{9.49}$$

projects out the component of $\mathbf{v}$ which is orthogonal to $\mathbf{n}$, i.e. in the tangent plane of the surface. We have

$$P_{ij} v_j = v_j - n_i (\mathbf{v} \cdot \mathbf{n})$$

and hence

$$(P\mathbf{v}) \cdot n = (P_{ij} v_j) n_i = \mathbf{v} \cdot \mathbf{n} - (\mathbf{n} \cdot \mathbf{n})(\mathbf{v} \cdot \mathbf{n}) = 0,$$

as required; $(P_{ij} v_j)$ is orthogonal to $\mathbf{n}$. Note that

$$P_{ij} P_{jm} = P_{im}, \tag{9.50}$$

as expected for projection operators. Now let us upgrade our discussion from vectors v_i to rank 2 tensors M_{ij} (which can also be considered as matrix elements). It is easy to see that

$$N_{kl} = P_{ik}\, P_{jl}\, M_{ij} \tag{9.51}$$

is transverse:

$$N_{kl}\, n_k = N_{kl}\, n_l = 0, \tag{9.52}$$

since

$$N_{kl}\, n_l = (\delta_{ik} - n_i n_k)(\delta_{jl} - n_j n_l) M_{ij} n_l = (\delta_{ik} - n_i n_k)(n_j - n_j\, (\mathbf{n}\cdot\mathbf{n})) M_{ij} = 0,$$

but it is not traceless:

$$N_{kk} = P_{ik}\, P_{jk}\, M_{ij} = P_{ij}\, M_{ij} = P_{ji}\, M_{ij} = \mathrm{tr}(PM),$$

where we have used (9.50) and the fact that $P_{ij} = P_{ji}$. However, the tensor (matrix)

$$R_{kl} = \left(P_{ik} P_{jl} - \frac{1}{2} P_{kl} P_{ij}\right) M_{ij} \tag{9.53}$$

is traceless:

$$R_{kk} = (P_{ik} P_{kj} - 1/2 P_{kk} P_{ij}) M_{ij} = \left(1 - \frac{1}{2} P_{kk}\right) P_{ij} M_{ij},$$

but $P_{kk} = \delta_{kk} - n_k n_k = 3 - 1 = 2$, so

$$R_{kk} = 0. \tag{9.54}$$

R is then the transverse and traceless version of M:

$$R = M^{\mathrm{TT}} = PMP - \frac{1}{2} P(\mathrm{tr}\, PM) \tag{9.55}$$

or

$$M^{\mathrm{TT}}{}_{ij} = \left(P_{ik} P_{jl} - \frac{1}{2} P_{ij} P_{kl}\right) M_{kl}. \tag{9.56}$$

This completes the mathematics we need to find $h^{\mathrm{TT}}{}_{ij}$ at any point. Equation (9.47) will be amended to

$$\tau_{00} = \frac{G}{8\pi c^6 r^2} \dddot{I}^{TT}{}_{ij} \dddot{I}^{TT}{}_{ij}, \tag{9.57}$$

with

$$I^{\mathrm{TT}}{}_{ij} = \left(P_{ik}\, P_{jl} - \frac{1}{2} P_{ij}\, P_{kl}\right) I_{kl}. \tag{9.58}$$

To evaluate (9.57) observe that

$$\begin{aligned} I^{\mathrm{TT}}{}_{ij} I^{\mathrm{TT}}{}_{ij} &= \left(P_{ik} P_{jl} - \frac{1}{2} P_{ij}\, P_{kl}\right)\left(P_{ip}\, P_{jq} - \frac{1}{2} P_{ij}\, P_{pq}\right) I_{kl}\, I_{pq} \\ &= (P_{ik}\, P_{jl}\, I_{kl})(P_{ip}\, P_{jq}\, I_{pq}) - (P_{ik}\, P_{jl}\, I_{kl})(P_{ij}\, P_{pq}\, I_{pq}) \\ &\quad + \frac{1}{4} P_{ij}\, P_{ij}\, (P_{kl}\, I_{kl})(P_{pq}\, I_{pq}). \end{aligned} \tag{9.59}$$

There are three terms on the right hand side of this equation. After some algebra it is seen that the first one is equal to

$$I_{ij}\, I_{ij} - 2\, n_p\, n_p\, I_{ip}\, I_{iq} + n_i\, n_j\, n_p\, n_q\, I_{ij}\, I_{pq},$$

and the second and third terms are, respectively

$$-\, n_k\, n_l\, n_p\, n_q\, I_{kl}\, I_{pq} \quad \text{and} \quad +\frac{1}{2}\, n_k\, n_l\, n_p\, n_q\, I_{kl}\, I_{pq},$$

giving, finally,

$$I^{\mathrm{TT}}{}_{ij}\, I^{\mathrm{TT}}{}_{ij} = I_{ij}\, I_{ij} - 2n_p\, n_q\, I_{ip}\, I_{iq} + \frac{1}{2} n_k\, n_l\, n_p\, n_q\, I_{kl}\, I_{pq}. \tag{9.60}$$

We have, then, that from (9.57) the gravitational energy density is

$$\tau_{00} = \frac{G}{8\pi c^6 r^2} \langle \dddot{I}^{TT}{}_{ij}\, \dddot{I}^{TT}{}_{ij} \rangle \tag{9.61}$$

where the 'averaging' sign is put in to deal with possible oscillating terms; and the energy passing through a spherical surface of radius r in unit time is

$$\frac{\mathrm{d}E}{\mathrm{d}t} = \int c\tau_{00}\, r^2\, \mathrm{d}\Omega = \frac{G}{8\pi c^5} \int \langle \dddot{I}^{TT}{}_{ij}\, \dddot{I}^{TT}{}_{ij} \rangle\, \mathrm{d}\Omega. \tag{9.62}$$

The integral above is, from (9.60)

$$\int \dddot{I}^{TT}{}_{ij}\, \dddot{I}^{TT}{}_{ij}\, \mathrm{d}\Omega = \int \left(\dddot{I}_{ij}\, \dddot{I}_{ij} - 2\, n_p\, n_q\, \dddot{I}_{ip}\, \dddot{I}_{iq} + \frac{1}{2}\, n_k\, n_l\, n_p\, n_q\, \dddot{I}_{kl}\, \dddot{I}_{pq} \right) \mathrm{d}\Omega. \tag{9.63}$$

The factors $\dddot{I}_{ij}$ and so on come outside the integral sign. The remaining angular integrals are

$$\begin{gathered} \int \mathrm{d}\Omega = 4\pi, \quad \int n_i\, n_j\, \mathrm{d}\Omega = \frac{4\pi}{3}\, \delta_{ij}, \\ \int n_i\, n_j\, n_k\, n_l\, \mathrm{d}\Omega = \frac{4\pi}{15}\, (\delta_{ij}\, \delta_{kl} + \delta_{ik}\, \delta_{jl} + \delta_{il}\, \delta_{jk}). \end{gathered} \tag{9.64}$$

These are easily seen: for putting $n_1 = \hat{x} = \sin\theta \cos\phi$, $n_2 = \hat{y} = \sin\theta \sin\phi$, $n_3 = \hat{z} = \cos\theta$ and $\mathrm{d}\Omega = \sin\theta\, \mathrm{d}\theta\, \mathrm{d}\phi$, it is simple to check that

$$\int n_1\, n_2\, \mathrm{d}\Omega = 0, \quad \int n_1^2 \mathrm{d}\Omega = \frac{4\pi}{3}, \quad \int n_1^2\, n_2^2\, \mathrm{d}\Omega = \frac{4\pi}{15},$$

and so on. Hence (9.63) gives (noting (9.48))

$$\int \left\langle \dddot{I}^{TT}{}_{ij}\, \dddot{I}^{TT}{}_{ij} \right\rangle \mathrm{d}\Omega = \langle \dddot{I}_{ij}\, \dddot{I}_{ij} \rangle \left[4\pi - 2 \cdot \frac{4\pi}{3} + \frac{1}{2}\frac{4\pi}{15} \cdot 2 \right] = \langle \dddot{I}_{ij}\, \dddot{I}_{ij} \rangle \frac{24\pi}{15},$$

and, finally, the rate of energy loss is

$$\frac{\mathrm{d}E}{\mathrm{d}t} = \frac{G}{5c^5} \langle \dddot{I}_{ij}\, \dddot{I}_{ij} \rangle. \tag{9.65}$$

Returning to our orbiting binary, from (9.38)

$$\dddot{I}_{ij} = 8MR^2\omega^3 \begin{pmatrix} \sin 2\omega t & -\cos 2\omega t & 0 \\ -\cos 2\omega t & -\sin 2\omega t & 0 \\ 0 & 0 & 0 \end{pmatrix},$$

hence $\dddot{I}_{ij}\, \dddot{I}_{ij} = 128 M^2 R^4 \omega^6$ and

$$\frac{dE}{dt} = \frac{128G}{5c^5} M^2 R^4 \omega^6. \tag{9.66}$$

Substituting further from (9.37) for ω, the frequency of rotation, gives

$$\frac{dE}{dt} = \frac{2G^4}{5c^5} \frac{M^5}{R^5}. \tag{9.67}$$

This is an expression for the *gravitational luminosity* of the source: we may write it as

$$L = \frac{2}{5} \left(\frac{GM}{Rc^2} \right)^5 \frac{c^5}{G}. \tag{9.68}$$

The quantity (GM/Rc^2) is dimensionless and (c^5/G) has the dimensions of luminosity. Note that L depends on $1/R^5$, so most energy is emitted from compact binaries (R is the radius of the binary system).

9.2.3 Spin-up and the binary pulsar PSR 1913+16

The total kinetic energy of the binary system is

$$T = 2 \times \frac{1}{2} M v^2 = \frac{GM^2}{4R},$$

from (9.37). The potential energy is $V = -\dfrac{GM^2}{2R}$ (since the distance between the bodies is $2R$) so the total energy of the binary system, including rest masses, is

$$E = \frac{GM^2}{4R} - \frac{GM^2}{2R} + 2Mc^2 = 2Mc^2 - \frac{GM^2}{4R}. \tag{9.69}$$

As the system emits energy E must decrease, so R must also decrease (assuming M is unaffected) and the orbit will shrink; the system 'spins up' and eventually the stars coalesce. The frequency of the emitted radiation is $f = 2\omega/2\pi$, which from (9.37) is

$$f = \frac{1}{\pi} \left(\frac{GM}{4R^3} \right)^{1/2}. \tag{9.70}$$

As R decreases the pitch of the radiation increases, and the signal is known as a *chirp*, the song of a sparrow.

Let us make a crude estimate of the characteristic time for the radius of the orbit to shrink from R to $R/2$. From above it is clear that an amount of energy $GM^2/4R$ will have been emitted, and *if* we assume that the luminosity L is constant during this time, the characteristic time t_{ch} would be

$$L t_{ch} = \frac{GM^2}{4R},$$

giving, from (9.68),

$$t_{\text{ch}} = \frac{5R}{8c}\left(\frac{GM}{Rc^2}\right)^{-3}. \tag{9.71}$$

This is the *chirp time* for the binary. The factor $5R/8c$ is the time taken for light to travel a distance $5R/8$ and the second factor is dimensionless. It must be admitted that this is a rather crude estimate, but note that because of the $1/R^5$ dependence of the luminosity, the time taken for a further decrease in radius, say from $R/2$ to $R/4$, is very much less, so the chirp time is actually not a bad order-of-magnitude estimate for the time taken for the stars to coalesce.

Now let us be more specific, and consider a system where both stars are neutron stars,

$$M = 1.4M_S, \quad R = 10^6\ \text{km}. \tag{9.72}$$

Then $MG/Rc^2 = 2.06 \times 10^{-6}$ and

$$t_{\text{ch}} = 2.2 \times 10^{17}\ \text{s} \approx 7.1 \times 10^9\ \text{years},$$

comparable to the age of the Universe! This is hardly measurable, but an interesting and measurable quantity is the change in the *period* of an orbit, as a result of its shrinking. The period is, from (9.37),

$$\tau = 4\pi\left(\frac{R^3}{GM}\right)^{1/2} \propto R^{3/2}.$$

As R changes τ will also change, and from above

$$\frac{\mathrm{d}\tau}{\tau} = \frac{3}{2}\frac{\mathrm{d}R}{R}. \tag{9.73}$$

From (9.69) we have $\mathrm{d}E/E = -\,\mathrm{d}R/R$, so

$$\frac{\mathrm{d}\tau}{\tau} = -\frac{3}{2}\frac{\mathrm{d}E}{E}.$$

We want to find $\mathrm{d}\tau/\mathrm{d}t$, the rate of change of the period. From the above equations we find

$$\frac{\mathrm{d}\tau/\mathrm{d}t}{\tau} = -\frac{3}{2}\frac{\mathrm{d}E/\mathrm{d}t}{E} = -\frac{3}{2}\frac{L}{E} = -\frac{12}{5}\frac{G^3M^3}{R^4c^5},$$

hence

$$\frac{\mathrm{d}\tau}{\tau} = -\frac{48\pi}{5}\left(\frac{GM}{Rc^2}\right)^{5/2}. \tag{9.74}$$

With $MG/Rc^2 = 2.06 \times 10^{-6}$ this gives

$$\frac{\mathrm{d}\tau}{\mathrm{d}t} = -1.84 \times 10^{-13};$$

the period of the orbit decreases by 1.84×10^{-13} seconds every second. This is a dramatic prediction of General Relativity, coming about as a consequence of the emission of gravitational radiation from the binary system.

This prediction was verified in 1974 by Russell Hulse and Joseph Taylor who observed the decrease in the period of the pulsar PSR 1913+16 in a binary system like the one considered above: both the pulsar and its companion have masses close to $1.4M_S$.[3] The orbits are tight but eccentric: the semi-major axis is about 9×10^8 m and the eccentricity of the ellipse is 0.62; the formula (9.74) has to be corrected for this. Hulse and Taylor took measurements over many years and concluded that

$$\frac{d\tau}{dt} = (-2.422 \pm 0.006) \times 10^{-12},$$

in agreement with the general relativistic prediction to an accuracy of better than 0.3%. We may thus conclude that we have definite – though indirect – evidence for the existence of gravitational radiation.

9.2.4 Search for gravitational waves

The search for gravitational waves actually has a long history, going back to the pioneering experiments of Weber in the 1960s. These experiments, which in a modified form are still being pursued by many research groups today, consist of a large metal cylinder which, when hit by a gravitational wave, will oscillate, and the oscillations are converted by transducers into electrical signals. Modern versions of this experiment may achieve sensitivities of the order of 10^{-18}. To put this figure into context, let us calculate the amplitude of the signal from PSR 1913+16. Equations (9.37) and (9.39) give

$$h = \frac{2G^2M^2}{c^4 rR}$$

and with $M = 1.4M_S$, $R = 10^9$ m and r, the distance away of the source, equal to 8 kpc $= 2.5 \times 10^{20}$ m, we have $h = 3.4 \times 10^{-23}$, many orders of magnitude down on 10^{-18}! An apparatus of this type would only be able to register signals from much more powerful emitters than this binary pulsar – and of course there are more powerful sources. Even with a sensitivity of 10^{-18}, however, in a bar of length (say) 10 m, we would be looking for a change in length of 10^{-17} m – one hundredth the size of a nucleus! Gravitational wave detection certainly offers an experimental challenge.

More recently a lot of effort is being put into *interferometric* methods of finding gravitational waves. These experimeents are basically of the same design as a Michelson interferometer. Mirrors are placed along perpendicular arms and when a gravitational wave impinges one of the arms contracts while the other expands, followed a short time later by the opposite motion. Thus what is looked for is an oscillation in the path difference and therefore the interference pattern. Some of these experiments are earthbound but the most interesting of them are planned to be set up in space. The LISA (Large Interferometric Space Antenna) project will consist of three spacecraft about 5×10^6 km apart. The expectation is that sensitivities of about 10^{-22} might be achieved in this type of experiment and there is

[3] Hulse & Taylor (1975); see also Weissberg & Taylor (1984).

considerable optimism that, in the words of Bradaschia and Desalvo,[4] the gravitational wave community is poised to prove Einstein right and wrong: right in his prediction that gravitational waves exist, wrong in his prediction that we will never be able to detect them.

9.3 Parallels between electrodynamics and General Relativity: Petrov classification

In considering the question whether gravitational radiation exists, much of our thinking has been guided by the case of electromagnetism. Maxwell's equations give rise to a wave equation whose solutions describe waves carrying energy. Correspondingly, in the linear (weak field) approximation of General Relativity, where the metric tensor differs from its Minkowski value by the small quantity $h_{\mu\nu}$, we saw that this field also obeys a wave equation; and that these waves also carry energy. In the last section we considered convincing, though indirect, evidence that gravitational waves exist, so it might be thought that all that remains is to discover them directly, but this is not true. Even if gravitational waves are found conclusively to exist experimentally, we still need to have a proper theoretical understanding of them – and the weak field approximation does not provide this. In the first place it is *linear*, whereas General Relativity in its complete version is non-linear. In this complete version there is a graviton–graviton coupling, but there is no analogous photon–photon coupling in quantum electrodynamics. The weak field approximation, however, is linear so the obvious question must be asked: what is the status of this approximation? On iteration, does it yield General Relativity, and if it does, how does the non-linearity enter? Is the iteration even convergent? To my knowledge, these questions are not yet satisfactorily answered.

9.3.1 A geometric approach to electrodynamics

Reflections such as these encourage us to look at the question of gravitational radiation in a different way, but still bearing in mind the model of electrodynamics. Consider, for example, two types of electromagnetic field, the Coulomb field and the radiation field, in which the electric and magnetic components have the following behaviour:

$$\begin{aligned} &\text{Coulomb field}: \; E \sim \frac{1}{r^2}, \quad \mathbf{B} = 0, \\ &\text{Plane wave (radiation)}: \; E \sim \frac{1}{r}, \quad |\mathbf{E}| = |\mathbf{B}|, \quad \mathbf{E}\cdot\mathbf{B} = 0. \end{aligned} \tag{9.75}$$

The Coulomb field is quasi-static and carries no energy but the radiation field carries energy – it has a non-vanishing Poynting vector. What we want to show is that these

[4] Bradaschia & Desalvo (2007).

cases correspond to a *classification of the field tensor* $F_{\mu\nu}$ into distinct categories – and then to show that a similar situation exists for the gravitational field and the Riemann tensor.

From the tensor $F_{\mu\nu}$ and its dual $\tilde{F}_{\mu\nu}$ we may form the Lorentz invariant quantities (see Section 2.6)

$$\begin{aligned} P &= \frac{1}{2} F_{\mu\nu} F^{\mu\nu} = -|\mathbf{E}|^2 + |\mathbf{B}|^2, \\ Q &= \frac{1}{2} F_{\mu\nu} \tilde{F}^{\mu\nu} = -\mathbf{E}\cdot\mathbf{B}. \end{aligned} \tag{9.76}$$

Electromagnetic fields with $P=Q=0$ are called *null* fields. The field of a Coulomb (electric) charge has $P<0$, $Q=0$, so is not null. The field of a single magnetic charge is also Coulomb-like, but in this case it is purely magnetic and $P>0$, $Q=0$. In summary

$$\begin{aligned} &\text{Field of a point charge}: \; P \neq 0, \quad Q = 0, \\ &\text{purely electric}, P<0; \quad \text{purely magnetic}, P > 0. \end{aligned} \tag{9.77}$$

On the other hand, a plane wave travelling in the direction $\mathbf{k}$ has

$$\mathbf{E} = \mathbf{E}_0 \exp[\mathrm{i}(\mathbf{k}\cdot\mathbf{x} - \omega t)], \quad \mathbf{B} = \frac{c}{\omega}\mathbf{k}\times\mathbf{E} = \frac{c}{\omega}\mathbf{k}\times\mathbf{E}_0 \exp[\mathrm{i}(\mathbf{k}\cdot\mathbf{x} - \omega t)].$$

In a vacuum $\nabla\cdot\mathbf{E}=0$ so $\mathbf{k}\cdot\mathbf{E}=0$ and

$$|\mathbf{k}\times\mathbf{E}_0|^2 = k^2 {E_0}^2 - (\mathbf{k}\cdot\mathbf{E}_0)^2 = k^2 {E_0}^2$$

so

$$|\mathbf{B}|^2 = \frac{c^2k^2}{\omega^2}{E_0}^2 = {E_0}^2 = |\mathbf{E}|^2 \Rightarrow F_{\mu\nu}F^{\mu\nu} = 0;$$

and in addition

$$\mathbf{E}\cdot\mathbf{B} = \frac{c}{\omega}\mathbf{E}\cdot(\mathbf{k}\times\mathbf{E}) = 0 \Rightarrow F_{\mu\nu}\tilde{F}^{\mu\nu} = 0.$$

We therefore have $P=Q=0$, a null field:

$$\text{Plane wave}: \quad P = 0, \quad Q = 0: \quad \text{null field}. \tag{9.78}$$

We have thus characterised the fields of a point charge and radiation *algebraically*. The next step is to characterise them in the form of an *eigenvalue problem*. This will enable us to make the comparison with gravitational fields.[5]

The eigenvalue problem takes the form

$$F_{\mu\nu}k^\nu = \lambda k_\mu; \tag{9.79}$$

k_μ is the eigenvector and λ the eigenvalue of $F_{\mu\nu}$. For the next part of the argument we give $F_{\mu\nu}$ – an antisymmetric rank 2 tensor – a purely algebraic and geometric interpretation.[6] We describe it as a *bivector* – an object constructed from two vectors. A bivector is *simple* if it can be written as

[5] Much of the following mirrors the work of Frolov (1979).

[6] For a more complete version of the argument that follows see Frolov (1979).

$$F_{\mu\nu} = a_{[\mu}b_{\nu]} = \frac{1}{2}(a_\mu b_\nu - a_\nu b_\mu). \tag{9.80}$$

In general $F_{\mu\nu}$ is not simple, but it can always be decomposed into a pair of simple bivectors:

$$F_{\mu\nu} = a_{[\mu}b_{\nu]} + c_{[\mu}d_{\nu]}. \tag{9.81}$$

It can be shown[7] that a bivector is simple if and only if

$$F_{\mu[\nu}F_{\kappa\lambda]} = 0. \tag{9.82}$$

From the definition

$$F_{\mu\nu}\tilde{F}^{\mu\nu} = \frac{1}{2}\varepsilon^{\mu\nu\kappa\lambda}F_{\mu\nu}F_{\kappa\lambda},$$

it is clear that the right hand side is totally antisymmetric in all its indices so (9.82) implies that

$$F_{\mu\nu}\tilde{F}^{\mu\nu} = 0. \tag{9.83}$$

We now introduce an important transformation, the *dual rotation*:

$$F_{\mu\nu} \to F_{\mu\nu}(\theta) = \cos\theta F_{\mu\nu} + \sin\theta\,\tilde{F}_{\mu\nu}. \tag{9.84}$$

It is actually a transformation which mixes electric and magnetic fields; for example with $(\mu\nu) = (10)$, $E_x \to E_x\cos\theta + B_x\sin\theta$. Since $\tilde{\tilde{F}}_{\mu\nu} = -F_{\mu\nu}$ we have

$$\tilde{F}_{\mu\nu}(\theta) = \cos\theta\,\tilde{F}_{\mu\nu} - \sin\theta\,F_{\mu\nu},$$

so

$$F_{\mu\nu}(\theta)\,\tilde{F}^{\mu\nu}(\theta) = \cos 2\theta\,F_{\mu\nu}\tilde{F}^{\mu\nu} - \sin 2\theta\,F_{\mu\nu}F^{\mu\nu} \tag{9.85a}$$

and

$$F_{\mu\nu}(\theta)\,F^{\mu\nu}(\theta) = \cos 2\theta\,F_{\mu\nu}F^{\mu\nu} + \sin 2\theta\,F_{\mu\nu}\tilde{F}^{\mu\nu}. \tag{9.85b}$$

From (9.85a) it follows that, given an arbitary $F_{\mu\nu}$ there exists a θ such that $F_{\mu\nu}(\theta)\,\tilde{F}^{\mu\nu}(\theta) = 0$, so $F_{\mu\nu}$ is *simple* (by (9.82) and (9.83)). Then by a duality rotation any $F_{\mu\nu}$ may be converted into the form (9.80)

$$F_{\mu\nu}(\theta) = a_{[\mu}b_{\nu]}; \tag{9.86}$$

and since $F_{\mu\nu}(\theta)\,\tilde{F}^{\mu\nu}(\theta) = 0$, $\tilde{F}_{\mu\nu}(\theta)$ is also simple,

$$\tilde{F}_{\mu\nu}(\theta) = c_{[\mu}d_{\nu]}. \tag{9.87}$$

From here on we write $F_{\mu\nu}(\theta)$ simply as $F_{\mu\nu}$ and $\tilde{F}_{\mu\nu}(\theta)$ simply as $\tilde{F}_{\mu\nu}$. Equation (9.86) tells us that the bivector $F_{\mu\nu}$ defines a *plane* $\Pi(\mathbf{a}, \mathbf{b})$ and similarly $\tilde{F}_{\mu\nu}$ defines a plane $\Pi(\mathbf{c}, \mathbf{d})$. And since $\tilde{F}_{\mu\nu} = \frac{1}{2}\,\varepsilon_{\mu\nu\kappa\lambda}a^{[\kappa}b^{\lambda]}$, the plane $\Pi(\mathbf{a}, \mathbf{b})$ is orthogonal to the plane $\Pi(\mathbf{c}, \mathbf{d})$.

[7] Schouten (1954).

With these geometric considerations in mind we now examine more closely the cases $F_{\mu\nu}F^{\mu\nu}=0$ and $F_{\mu\nu}F^{\mu\nu}\neq 0$ (corresponding to radiation and Coulomb-type fields). Consider the following:

Proposition: If $F_{\mu\nu}$ is a simple bivector there exists a pair of vectors p_μ, q_μ such that

$$F_{\mu\nu}=p_{[\mu}q_{\nu]}, \quad p^\mu q_\mu = 0; \tag{9.88}$$

in other words, p_μ and q_μ are *orthogonal.*

Proof: If $F_{\mu\nu}$ is simple then we can write $F_{\mu\nu}=a_{[\mu}b_{\nu]}$, from (9.86), so we have only to find p_μ, q_μ such that $p^\mu q_\mu=0$.

Case (i): if $a_\mu a^\mu \neq 0$, choose

$$p_\mu = a_\mu, \quad q_\mu = b_\mu - \frac{(a_\kappa a^\kappa)}{(a_\lambda a^\lambda)}a_\mu$$

so that

$$p_\mu q^\mu = a_\mu b^\mu - \frac{(a_\kappa a^\kappa)}{(a_\lambda a^\lambda)}a_\mu b^\mu = 0,$$

and clearly

$$p_{[\mu}q_{\nu]} = a_{[\mu}b_{\nu]},$$

(since $a_{[\mu}a_{\nu]}=0$).

Case (ii): if $b_\mu b^\mu \neq 0$, interchange a and b above.

Case (iii): if a_μ and b_μ are both null, $a_\mu a^\mu = b_\mu b^\mu = 0$, then choose

$$p_\mu = \frac{1}{\sqrt{2}}(a_\mu - b_\mu), \quad q_\mu = \frac{1}{\sqrt{2}}(a_\mu + b_\mu),$$

so that

$$p_{[\mu}q_{\nu]} = \frac{1}{2}(a-b)_{[\mu}(a+b)_{\nu]} = \frac{1}{2}\{a_{[\mu}b_{\nu]} - b_\mu a_{\nu]}\} = a_{[\mu}b_{\nu]}$$

and

$$p_\mu q^\mu = \frac{1}{2}(a_\mu - b_\mu)(a^\mu + b^\mu) = \frac{1}{2}(a_\mu a^\mu - b_\mu b^\mu) = 0. \qquad \square$$

If the simple bivector is represented in the canonical form (9.88), then

$$F_{\mu\nu}F^{\mu\nu} = \frac{1}{4}(p_\mu q_\nu - p_\nu q_\mu)(p^\mu q^\nu - p^\nu q^\mu) = \frac{1}{2}(p_\mu p^\mu)(q_\mu q^\mu). \tag{9.89}$$

We may then consider the cases in which $F_{\mu\nu}F^{\mu\nu}<0$ and $F_{\mu\nu}F^{\mu\nu}=0$: in which $F_{\mu\nu}$ is said to be timelike or null. From the mathematical point of view we should also consider the case $F_{\mu\nu}F^{\mu\nu}>0$, but for our present purposes that is of no physical interest.

***Case I*: $F_{\mu\nu} F^{\mu\nu} < 0$** Then from (9.89) one of the vectors (say p_μ) must be timelike and the other spacelike. In the plane $\Pi(p, q)$, however, there must exist a pair of vectors, l_μ and n_μ, with

$$F_{\mu\nu} = l_\mu n_\nu - l_\nu n_\mu, \tag{9.90}$$

such that l_μ and n_μ are *both null*, but they are *not* orthogonal:

$$l_\mu l^\mu = n_\mu n^\mu = 0; \quad l_\mu n^\mu = \alpha \neq 0, \tag{9.91}$$

since then

$$F_{\mu\nu} F^{\mu\nu} = (l_\mu n_\nu - l_\nu n_\mu)(l^\mu n^\nu - l^\nu n^\mu) = -2\alpha^2 < 0.$$

***Case II*: $F_{\mu\nu} F^{\mu\nu} = 0$** Then one of the vectors (say p_μ) must be null and the other spacelike (timelike and null, or two independent null vectors, are not orthogonal). In this case we may write

$$F_{\mu\nu} = l_\mu a_\nu - l_\nu a_\mu \tag{9.92}$$

with l_μ null and a_μ spacelike, and the two vectors orthogonal,

$$l_\mu l^\mu = 0, \quad a_\mu a^\mu > 0, \quad l_\mu a^\mu = 0. \tag{9.93}$$

We now turn to the eigenvalue problem (9.79)

$$F_{\mu\nu} k^\nu = \lambda k_\mu. \tag{9.79}$$

In the case that $F_{\mu\nu}$ is *timelike* we have, from (9.90),

$$\begin{aligned} (l_\mu n_\nu - l_\nu n_\mu) k^\nu &= \lambda k_\mu, \\ l_\mu (\mathbf{n} \cdot \mathbf{k}) - n_\mu (\mathbf{l} \cdot \mathbf{k}) &= \lambda k_\mu, \end{aligned} \tag{9.94}$$

so *either* $k_\mu = l_\mu$ and $(\mathbf{n} \cdot \mathbf{k}) = \lambda = (\mathbf{n} \cdot \mathbf{l})$, i.e. $\lambda = \alpha$, *or* $k_\mu = n_\mu$ and $(\mathbf{l} \cdot \mathbf{k}) = (\mathbf{l} \cdot \mathbf{n}) = -\lambda$, i.e. $\lambda = -\alpha$. In other words, the equation

$$F_{\mu\nu} k^\nu = \lambda k_\mu \tag{9.79}$$

has two solutions

$$k_\mu = l_\mu \text{ or } n_\mu, \text{ with } \lambda = \pm(\mathbf{n} \cdot \mathbf{l}). \tag{9.95}$$

It clearly follows from (9.79) that

$$k_\rho F_{\mu\nu} k^\nu = \lambda k_\rho k_\mu.$$

The right hand side of this equation is symmetric under $\rho \leftrightarrow \mu$ so the antisymmetric part under this interchange vanishes:

$$k_{[\rho} F_{\mu]\nu} k^\nu = 0; \tag{9.96}$$

and we recall that this equations has two solutions in the case where $F_{\mu\nu}$ is timelike.

When $F_{\mu\nu}$ is *null* (9.79) becomes, with (9.92)

$$(l_\mu a_\nu - l_\nu a_\mu)k^\nu = \lambda\, k_\mu$$

or

$$l_\mu(\mathbf{a}\cdot\mathbf{k}) - \alpha_\mu(\mathbf{l}\cdot\mathbf{k}) = \lambda\, k_\mu. \tag{9.97}$$

It might be thought that this equation has two solutions, the first being $k_\mu = l_\mu$, in which case the left hand side of (9.97) is

$$(l_\mu a_\nu - l_\mu a_\mu)l^\nu = l_\mu(\mathbf{a}\cdot\mathbf{l}) - \alpha_\mu(\mathbf{l}\cdot\mathbf{l}) = 0$$

by (9.93), implying that $\lambda = 0$; and the second being $k_\mu = a_\mu$, in which case the left hand side of (9.97) is

$$(l_\mu a_\nu - l_\nu a_\mu)a^\nu = l_\mu(\mathbf{a}\cdot\mathbf{a}) - a_\mu(\mathbf{l}\cdot\mathbf{a}) = (\mathbf{a}\cdot\mathbf{a})\, l_\mu;$$

but this is incompatible with the right hand side, so this second case is not a solution. There is therefore only *one solution* to the eigenvalue problem

$$F_{\mu\nu}k^\nu = \lambda\, k_\mu$$

when $F_{\mu\nu}$ is null, which is

$$k_\mu = A l_\mu\ (A = \text{const}), \quad \text{and} \quad \lambda = 0. \tag{9.98}$$

We *conclude* that in the case $P \neq 0$, $Q = 0$, describing a Coulomb-type field, the eigenvalue equation is (9.96)

$$k_{[\rho}F_{\mu]\nu}k^\nu = 0 \tag{9.99}$$

and this has two solutions. This is called the *non-degenerate* case and the field tensor is denoted $F_{\mu\nu}{}^{[1,1]}$. On the other hand, in the case $P = 0$, $Q = 0$, the null case corresponding to a pure radiation field, the eigenvalue equation is

$$F_{\mu\nu}k^\nu = 0. \tag{9.100}$$

There is only one eigenvalue, $\lambda = 0$; this is the *degenerate* case and the field tensor is denoted $F_{\mu\nu}{}^{[2]}$. The retarded field from an isolated extended source has the asymptotic ($r \to \infty$) behaviour

$$F_{\mu\nu} = \frac{1}{r}F_{\mu\nu}{}^{[2]} + \frac{1}{r^2}F_{\mu\nu}{}^{[1,1]} + \mathrm{O}(r^{-3}). \tag{9.101}$$

9.3.2 Petrov classification

We now move on to the gravitational field, described by the Riemann tensor. The relevant classification is the *Petrov classification*, which applies in the first instance to the *Weyl tensor* $C_{\lambda\mu\nu\rho}$. This is closely related to the Riemann tensor and is defined as follows:

$$
\begin{aligned}
R_{\lambda\mu\nu\rho} = C_{\lambda\mu\nu\rho} &- \frac{1}{2}(g_{\lambda\rho} B_{\mu\nu} + g_{\mu\nu} B_{\lambda\rho} - g_{\lambda\nu} B_{\mu\rho} - g_{\mu\rho} B_{\lambda\nu}) \\
&- \frac{1}{12} R(g_{\lambda\rho} g_{\mu\nu} - g_{\lambda\nu} g_{\mu\rho}),
\end{aligned}
\tag{9.102}
$$

with

$$B_{\mu\nu} = R_{\mu\nu} = -\frac{1}{4} g_{\mu\nu} R. \tag{9.103}$$

Clearly, in a vacuum, where $R_{\mu\nu} = R = 0$,

$$(\textit{in vacuo}) \quad R_{\lambda\mu\nu\rho} = C_{\lambda\mu\nu\rho}. \tag{9.104}$$

Since $g^{\mu\nu} B_{\mu\nu} = R - \frac{1}{4}\delta^{\mu}{}_{\mu} R = 0$, it is clear, multiplying (9.102) by $g^{\lambda\nu}$ that

$$R_{\mu\rho} = g^{\lambda\nu} C_{\lambda\mu\nu\rho} + R_{\mu\rho},$$

hence

$$g^{\lambda\nu} C_{\lambda\mu\nu\rho} = 0. \tag{9.105}$$

The Weyl tensor has the same symmetries as the Riemann tensor: from (9.102)

$$
\begin{aligned}
C_{\lambda\mu\nu\rho} = -C_{\mu\lambda\nu\rho} = -C_{\lambda\mu\rho\nu} = +C_{\nu\rho\lambda\mu}, \\
C_{\lambda\mu\nu\rho} + C_{\lambda\nu\rho\mu} + C_{\lambda\rho\mu\nu} = 0,
\end{aligned}
\tag{9.106}
$$

so $C_{\lambda\mu\nu\rho}$, like $R_{\lambda\mu\nu\rho}$, has 10 independent components; but $g^{\lambda\nu} C_{\lambda\mu\nu\rho}$ is *always* zero, even in the presence of matter. The classification of a tensor depends only on its symmetry properties, so the classification of the Riemann tensor *in vacuo* is the same as the classification of the Weyl tensor anywhere – even in matter.

We shall consider the matter of classification by the method of eigenvectors and eigenvalues, as we did above for the electromagnetic field tensor.[8] If we have, then, a rank 2 tensor T_{ij} we look for a vector V^j with the property

$$T_{ij} V^j = \lambda V_i = \lambda g_{ik} V^k, \tag{9.107}$$

or

$$\left(T_{ij} - \lambda g_{ij}\right) V^j = 0,$$

and the eigenvalues λ are the solutions to the equation

$$|T_{ij} - \lambda g_{ij}| = 0. \tag{9.108}$$

It is most important to note that the classification which we are about to discuss depends crucially on the fact that we choose a *locally inertial frame* at any point P of space-time, at which, therefore, the metric tensor takes on its Minkowski values

$$g_{\mu\nu} = \operatorname{diag}(-1, 1, 1, 1). \tag{9.109}$$

[8] Much of the material below follows the treatments of Landau & Lifshitz (1971) and Papapetrou (1974).

It follows from this that the Petrov classification is a *local* one – in practice, a physical gravitational field may change from one class to another, or a mixture of classes, as we move from one point to another in space.

In the spirit of (9.107) we should define a rank 2 tensor related to the Weyl tensor, and we do this by proceeding, as in Section 4.4, by making the association

$$C_{\lambda\mu\nu\rho} \leftrightarrow C_{AB}, \quad A \sim (\lambda\mu), \quad B \sim (\nu\rho). \tag{9.110}$$

The indices A, B take on the values 1, 2, ..., 6 and in view of (9.106) $C_{AB}=C_{BA}$ is a symmetric tensor in a 6 dimensional space with metric tensor γ_{AB}. This must be symmetric, and is defined by

$$\gamma_{AB} \leftrightarrow g_{\lambda\mu\nu\rho} \equiv g_{\lambda\nu}\, g_{\mu\rho} - g_{\lambda\rho}\, g_{\mu\nu}. \tag{9.111}$$

It has the same symmetries as the Weyl tensor, so in the bivector space is symmetric. The eigenvalue equation is of the form

$$(C_{AB} - \lambda\gamma_{AB})\, W^B = 0. \tag{9.112}$$

It actually corresponds to the equation

$$(C_{\lambda\mu\nu\rho} - \lambda g_{\lambda\mu\nu\rho})\, W^{\nu\rho} = 0,$$

where $W^{\nu\rho}=-W^{\rho\nu}$, but we shall find it easier to work with Equation (9.112). The correspondence between the indices A, B and $\mu\nu$, etc., is

A	$\mu\nu$
1	01
2	02
3	03
4	23
5	31
6	12

so, for example,

$$\gamma_{11} = g_{0101} = g_{00}\, g_{11} - g_{01}\, g_{01} = -1,$$
$$\gamma_{44} = g_{2323} = g_{22}\, g_{33} - (g_{23})^2 = 1,$$
$$\gamma_{12} = g_{0102} = g_{00}\, g_{12} - g_{02}\, g_{01} = 0,$$

and so on, giving

$$\gamma_{AB} = \mathrm{diag}\,(-1, -1, -1, 1, 1, 1). \tag{9.113}$$

An analogous 'bivector' relabelling of the Riemann and Weyl tensors may be carried out, so, for example,

$$R_{\mu\nu\kappa\lambda} \leftrightarrow R_{AB} = R_{BA}. \tag{9.114}$$

Then the relation (see (4.34iv))

$$R_{0123} + R_{0231} + R_{0312} = 0 \tag{9.115}$$

becomes

$$R_{14} + R_{25} + R_{36} = 0, \tag{9.116}$$

and similarly, since the Weyl tensors has the same symmetries,

$$C_{14} + C_{25} + C_{36} = 0. \tag{9.117}$$

To proceed further (noting all the while that the symmetries of the Riemann and Weyl tensors are the same) we now write the 20 independent components of the Riemann tensor as a collection of 3-dimensional tensors[9] M_{ik}, N_{ik} and P_{ik}:

$$M_{ik} = R_{0i0k}, \quad N_{ik} = \frac{1}{2}\varepsilon_{imn}\, R_{0kmn}, \quad P_{ik} = \frac{1}{4}\varepsilon_{imn}\,\varepsilon_{kpq}\, R_{mnpq}, \tag{9.118}$$

noting that in this 3-dimensional locally Minkowski space there is no need to distinguish upper and lower indices. By virtue of the symmetry properties of the Riemann tensor we have

$$M_{ik} = M_{ki}\,(\therefore 6\ \text{components}), \tag{9.119}$$

and

$$P_{ki} = \frac{1}{4}\varepsilon_{kmn}\,\varepsilon_{ipq}\, R_{mnpq} = \frac{1}{4}\varepsilon_{ipq}\,\varepsilon_{kmn}\, R_{pqmn} = P_{ik}\,(\therefore 6\ \text{components}). \tag{9.120}$$

In addition

$$N_{11} = \frac{1}{2}\varepsilon_{1mn}\, R_{01mn} = \frac{1}{2}(R_{0123} - R_{0132}) = R_{0123}; \quad N_{22} = R_{0231},$$
$$N_{33} = R_{0312},$$

hence, from (9.115),

$$N_{ii} = N_{11} + N_{22} + N_{33} = 0. \tag{9.121}$$

Further,

$$N_{12} = R_{0131}, \quad N_{13} = R_{0112}, \quad N_{23} = R_{0331}, \quad N_{32} = R_{0212}, \tag{9.122}$$

and so on. We now make use of the vacuum field equations $R_{\mu\nu} = 0$. We have

$$R_{00} = g^{ik}\, R_{0i0k} = R_{0i0i} = M_{ii},$$

hence

$$M_{ii} = 0. \tag{9.123}$$

[9] That is, tensors whose *indices* take on only the values 1, 2, 3.

We also have

$$P_{12} = \frac{1}{4}\varepsilon_{1mn}\,\varepsilon_{2pq}\,R_{mnpq} = R_{2331} = -R_{3132}.$$

On the other hand,

$$R_{12} = g^{\rho\sigma}R_{\rho 1\sigma 2} = R_{0102} - R_{i1i2} = R_{0102} - R_{3132} = 0,$$

hence, from (9.118) $M_{12} = -P_{12}$, and in general

$$M_{ik} = -P_{ik} \quad (i \neq k). \tag{9.124}$$

In a similar way it follows from $R_{01} = 0$ that $N_{32} = N_{23}$, and in general

$$N_{ik} = N_{ki}. \tag{9.125}$$

Finally, the equations $R_{11} = 0$, $R_{22} = 0$ and $R_{33} = 0$ give, respectively

$$M_{11} = P_{22} + P_{33}, \quad M_{22} = P_{33} + P_{11}, \quad M_{33} = P_{11} + P_{22}, \tag{9.126}$$

which are, on rearrangement,

$$P_{11} = \frac{1}{2}(M_{22} + M_{33} - M_{11}), \quad P_{22} = \frac{1}{2}(M_{11} + M_{33} - M_{22}),$$
$$P_{33} = \frac{1}{2}(M_{11} + M_{22} - M_{33}),$$

from which, using (9.123)

$$P_{ii} = \frac{1}{2}M_{ii} = 0.$$

It then follows from (9.126) that $M_{11} = -P_{11}$ and so on, so (9.124) becomes

$$M_{ik} = -P_{ik} \text{ for all } i,\ k. \tag{9.127}$$

We can now enumerate the number of independent components of these tensors. Because of (9.127) the independent tensors are M_{ik} and N_{ik}. Equations (9.119) and (9.123) imply that M_{ik} has $6-1=5$ independent components, and (9.121) and (9.125) that N_{ik} has 3 off-diagonal and 2 diagonal independent components, making 5 in all. This gives a total of 10 independent components for the Riemann tensor, as expected when $R_{\mu\nu} = 0$.

We can now represent the components of the Riemann tensor as a 6×6 matrix, which in terms of M_{ik} and N_{ik} is

$$C_{AB} = \begin{pmatrix} M & N \\ N & -M \end{pmatrix}, \tag{9.128}$$

with M and N both symmetric traceless 3×3 matrices. The eigenvalues λ of C_{AB} are the roots of

$$\det(C_{AB} - \lambda\gamma_{AB}) = 0, \tag{9.129}$$

i.e.

$$\begin{vmatrix} M_{ik} - \lambda\delta_{ik} & N_{ik} \\ N_{ik} & -M_{ik} - \lambda\delta_{ik} \end{vmatrix} = 0. \tag{9.130}$$

There are six rows and six columns in this matrix. Perform the following operations on it: add to the first column i times the fourth one, to the second column i times the fifth, and to the third i times the sixth. Then add to the fourth *row* i times the first row, to the fifth i times the second and to the sixth i time the third. After these operations (which of course leave the determinant unchanged) the determinant becomes

$$\begin{vmatrix} M_{ik} - \lambda\delta_{ik} + iN_{ik} & N_{ik} \\ 0 & -(M_{ik} - \lambda\delta_{ik} + iN_{ik}) \end{vmatrix} = 0, \tag{9.131}$$

i.e.

$$|M_{ik} - \lambda\delta_{ik} + N_{ik}| = 0, \tag{9.132}$$

involving only the determinant of a 3×3 matrix: the second equation resulting from (9.131) is the complex conjugate of (9.132). Equation (9.132) is cubic in λ:

$$\lambda^3 + a\lambda^2 + b\lambda + c = 0,$$

where a, b and c are functions of M_{ik} and N_{ik}: in particular

$$a = -(M_{ii} + N_{ii}) = 0$$

by virtue of (9.121) and (9.123). However, a is the sum of the roots, $\lambda_1 + \lambda_2 + \lambda_3$, so we have

$$\lambda_1 + \lambda_2 + \lambda_3 = 0, \tag{9.133}$$

and we recall that these roots are in general *complex* numbers. Any *classification* of the Weyl tensor can then begin by observing that (9.133) immediately allows a simple enumeration of three types of solution:

$$\begin{aligned} &\text{(i) type I: 3 roots } \lambda_1\ \lambda_2,\ \lambda_3 \text{ all different} \\ &\text{(ii) type II: 2 roots equal } \lambda_1 = \lambda_2 \neq \lambda_3 \\ &\text{(iii) type III: 3 roots equal, so from (9.133) } \lambda_1 = \lambda_2 = \lambda_3 = 0. \end{aligned} \tag{9.134}$$

When the eigenvalues are known we can find the eigenvectors W^A from (9.112). At this stage, however, we shall abandon a detailed treatment of this topic and simply quote the final results, giving the matrices M and N in 'normal form' for the three types of solution above.

Type I

$$M = \begin{pmatrix} \alpha_1 & 0 & 0 \\ 0 & \alpha_2 & 0 \\ 0 & 0 & \alpha_3 \end{pmatrix}, \quad N = \begin{pmatrix} \beta_1 & 0 & 0 \\ 0 & \beta_2 & 0 \\ 0 & 0 & \beta_3 \end{pmatrix}, \tag{9.135}$$
$$\lambda_i = -(\alpha_i + \mathrm{i}\beta_i) \quad (i = 1, 2, 3).$$

Type II

$$M = \begin{pmatrix} 2\alpha & 0 & 0 \\ 0 & -\alpha + \sigma & 0 \\ 0 & 0 & -\alpha - \sigma \end{pmatrix}, \quad N = \begin{pmatrix} 2\beta & 0 & 0 \\ 0 & -\beta & \sigma \\ 0 & \sigma & -\beta \end{pmatrix}, \tag{9.136}$$
$$\lambda_1 = -2(\alpha + \mathrm{i}\beta), \quad \lambda_2 = \lambda_3 = \alpha + \mathrm{i}\beta.$$

Type III

$$M = \begin{pmatrix} 0 & \sigma & 0 \\ \sigma & 0 & 0 \\ 0 & 0 & 0 \end{pmatrix}, \quad N = \begin{pmatrix} 0 & 0 & 0 \\ 0 & 0 & \sigma \\ 0 & \sigma & 0 \end{pmatrix}. \tag{9.137}$$

Type I is clearly the most general – non-degenerate – solution and may be denoted [1, 1, 1], following the notation of (9.101), Similarly Types II and III may be denoted [2, 1] and [3].

It turns out, however, that the above – algebraic – classification does not do full justice to the problem. An alternative approach is to base the analysis directly on equations analogous to (9.99) and (9.100). It is obvious that even in the degenerate case (9.100), Equation (9.99) also holds, so we can regard the problem as looking for solutions to (9.99) and then finding both the non-degenerate and degenerate solutions $F_{\mu\nu}{}^{[1\ 1]}$ and $F_{\mu\nu}{}^{[2]}$, corresponding respectively to a Coulomb field and a radiation field, with the asymptotic behaviour (9.101). So, in the gravitational case, substituting the Weyl tensor $C_{\kappa\lambda\mu\nu}$ or the Riemann tensor $R_{\kappa\lambda\mu\nu}$ (as long as we are dealing with vacuum solutions) for the electromagnetic tensor $F_{\mu\nu}$, we investigate all possible solutions to the equation

$$k_{[\kappa}R_{\lambda]\mu\,\nu[\rho}k_{\sigma]}\,k^{\mu}k^{\nu} = 0, \quad k^{\mu}k_{\mu} = 0, \tag{9.138}$$

and it is found that there is quite a variety of solutions. In general there are *four* types of Riemann tensor satisfying (9.138). In the most general case the eigenvalues are all distinct. This is the non-degenerate case, labelled [1111] and called Type I. The other types feature degeneracies of varying degrees and may be summarised in the following table:

Type:	I	D	II	III	N
Symbol:	[1111]	[211]	[22]	[31]	[4]

(9.139)

(The notation here is, for example, [31] represents two distinct eigenvalues, one of them 3-fold degenerate, and [211] represents three eigenvalues, one of them 2-fold degenerate.) All the solutions satisfy (9.138). In particular Type I does and we may write

$$k_{[\kappa}\mathrm{I}_{\lambda]\,\mu\,\nu[\rho}k_{\sigma]}\,k^{\mu}\,k^{\nu} = 0.$$

The types with partial or complete degeneracy also satisfy more stringent equations; for example Type II and Type N satisfy

$$\mathrm{II}_{\kappa\,\lambda\,\mu\,[\nu}k_{\rho]}\,k^{\lambda}\,k^{\mu} = 0, \quad N_{\kappa\,\lambda\,\mu\,\nu}k^{\nu} = 0.$$

Finally, the curvature tensor of an isolated distribution of matter has the long distance expansion

$$R_{\kappa\,\lambda\,\mu\,\nu} = \frac{1}{r}\,N_{\kappa\,\lambda\,\mu\,\nu} + \frac{1}{r^2}\,\mathrm{III}_{\kappa\,\lambda\,\mu\,\nu} + \frac{1}{r^3}\,D_{\kappa\,\lambda\,\mu\,\nu} + \cdots. \tag{9.140}$$

Comparing (9.140) with (9.100) it would seem clear that the tensor N corresponds to a *radiative* solution to the field equations; the curvature tensor has a $1/r$ dependence, so the energy of the field may be expected to show a $1/r^2$ dependence, indicating a genuine flux of energy.

It is worth considering, in this context, the Schwarzschild solution. Intuition would lead us to expect that it has similar characteristics to the Coulomb solution – representing the field of a static mass, with, in the Newtonian limit, a $1/r^2$ dependence. In fact, as may be verified from (5.34), the Riemann tensor has the non-zero components

$$R_{0101} = -\frac{2m}{r^3}, \quad R_{0202} = R_{0303} = \frac{m}{r^3}, \quad R_{1212} = R_{1313} = -\frac{m}{r^3}, \quad R_{2323} = \frac{2m}{r^3}.$$

This indicates, from (9.118), that

$$M_{11} = -P_{11} = -\frac{2m}{r^3}, \quad M_{22} = M_{33} = -P_{22} = -P_{33} = \frac{m}{r^3}, \quad N_{ik} = 0$$

(in agreement with (9.124)), so that

$$M_{ik} = \frac{m}{r^3}\begin{pmatrix} -2 & 0 & 0 \\ 0 & 1 & 0 \\ 0 & 0 & 1 \end{pmatrix}, \quad N_{ik} = 0.$$

This therefore corresponds to Type II in (9.136) (with $\beta = 0$) – one of the eigenvalues is degenerate. In the scheme (9.139) it is Type III, with the expected $1/r^2$ fall-off.

We can therefore conclude that an analysis of the Riemann tensor, satisfying Einstein's field equations, indicates that a radiative solution exists, independent of the weak field approximation. It makes the question of the existence of gravitational radiation less problematic from a theoretical point of view. With hopes boosted by the evidence from the binary pulsar, we now await the direct observation of gravitational radiation, which will surely rank as one of the most important discoveries of the twenty-first century.

Further reading

Einstein's papers on gravitational waves are Einstein (1916b) and (1918). Good and relatively modern reviews of gravitational radiation are Douglass & Braginsky (1979) and Thorne (1987). Reviews of laser interferometric searches for gravitational waves are Kawashima (1994) and Barish (2002).

A comparison of the geometric parallels between electromagnetic and gravitational fields may be found in Witten (1962). A full account of the Petrov classification is contained in Petrov (1969). Good lectures on gravitational radiation theory may be found in Sachs (1964) and Pirani (1965). Useful reviews of this topic are Pirani (1962a, b). Somewhat briefer accounts of the geometric classification of fields may be found in Stephani (2004).

Problems

9.1 Show that the harmonic condition (6.17) $f^{\mu\nu}{}_{,\nu} = 0$ may be expressed as

$$g^{\mu\nu}\Gamma^{\lambda}{}_{\mu\nu} = 0.$$

9.2 Taking $n_i = n_3 = (0, 0, 1)$ and a 3×3 matrix M, show that M^{TT} defined by (9.55) is indeed transverse and traceless.

10 Cosmology

In the end the world will be a desert of chairs and sofas … rolling through infinity with no-one to sit on them.

E. M. Forster, *Howards End*

10.1 Brief description of the Universe

Our Sun is one star in a collection of about 10^{11} stars forming our Galaxy. The Galaxy is shaped roughly like a pancake – approximately circular in 'plan' and with thickness much less than its radius – and the Sun is situated towards the outside of this distribution, not far from the central plane. The Galaxy is about 100 000 light years (ly) across. Almost all the stars visible to the naked eye at night belong to our Galaxy, and looking at the Milky Way is looking into its central plane, where the density of stars is greatest. The Andromeda Nebula, also visible to the naked eye, is a separate galaxy about 2 million light years away, and in fact is a member of the Local Group of galaxies. The construction of large telescopes in the first decades of the twentieth century led to the discovery of many galaxies and groups of galaxies and it is now known that there are about 10^{11} galactic clusters in the visible Universe. Considering these clusters as the 'elementary' constituents of the Universe, on scales larger than that of the clusters their distribution in space appears to be *homogeneous and isotropic*. This is the first – and very remarkable – feature of the Universe.

An important cosmological figure is the average density of matter in the Universe. The contribution of the matter contained in galaxies is

$$\rho_b \approx 10^{-31}\,\mathrm{g\,cm^{-3}} = 10^{-28}\,\mathrm{kg\,m^{-3}}. \qquad (10.1)$$

The subscript b refers to the fact that the 'matter' referred to here is *baryonic* matter – made of protons and neutrons. This is the sort of matter of which stars and planets and we ourselves are made. We shall see in due course that there is matter of a different sort – dark matter – in the Universe. The matter referred to in (10.1) is not dark; it constitutes stars, and shines.

The discovery that the Universe is isotropic means that it has *no centre* – there is no privileged point in space. The space (not space-time) of the Universe is a homogeneous space. This is the content of the so-called Cosmological Principle (CP). An extension of this principle is the Perfect Cosmological Principle (PCP) according to which, if it were true, the

Universe would be homogeneous in space *and in time*. It would look the same at all points in space and at all times, which of course implies that it must be of infinite age, in both the past and the future. This principle was proposed by Hermann Bondi and Thomas Gold in 1948 and came to be known as the Steady State Theory.[1] It might be thought that Special Relativity requires that the statement of homogeneity in space be extended to homogeneity in space and time, but this is *not true*, since Special Relativity is a theory about the laws of nature. It demands that the laws of nature be covariant – have the same form – under Lorentz transformations. In doing physics we apply the laws of nature to a particular physical situation – for example studying the motion of a charged particle in an electric field. If the laws of nature are ultimately differential equations – and they are – the situation to which we apply these laws (the motion of the particle) is represented by the *boundary conditions* of these equations (the initial position and speed of the particle). It is indeed a most remarkable circumstance that our knowledge can be divided into these two categories.[2] In the present case the physical system is the Universe as a whole, which therefore plays the role of a 'boundary condition', so the Cosmological Principle is a property of the boundary conditions, not a property of any law of nature, and Special Relativity does not require that it should be extended to a Perfect Cosmological Principle. In fact the Steady State Theory is now discredited; in Big Bang Cosmology the Universe certainly looked very different in the past from how it looks now!

One of the major advances in cosmology was the discovery by Edwin Hubble in the 1920s, working on the big telescopes at the Mount Wilson and Palomar Observatories, that the Universe is *expanding*. This was his interpretation of the fact that the light from far galaxies was *red-shifted* relative to light from nearby galaxies. Hubble interpreted the red-shift as a Doppler shift and concluded that the galaxies are moving away – from us, and from each other. The observed pattern of expansion was that the galaxies are moving away radially, with a speed v proportional to their distance r from us,

$$v = H_0 r. \tag{10.2}$$

The constant H_0 is the present value of *Hubble's constant*, sometimes called the *Hubble parameter*, since it is likely that H_0 varies in time. There is a degree of uncertainty in its present value, which is

$$H_0 = (55\text{–}85)\ \text{km s}^{-1}\ \text{Mpc}^{-1}; \tag{10.3}$$

a galaxy 1 Mpc (= 3.3 ly) away is receding at a speed of around 70 km s^{-1}. It is convenient to write (10.3) as

$$H_0 = h \cdot 100\ \text{km s}^{-1}\ \text{Mpc}^{-1};$$

with h at present in the range 0.55 to 0.85.

There are a number of observations to make about this important discovery. Firstly, it is not a particularly rapid expansion. Since 1 Mpc = 3.09×10^{19} km and 1 year = 3.15×10^{7} s, then, on putting $h = 1$ for definiteness, a galaxy 10^{20} km (about 3 million ly) distant from us

[1] For an account of the Steady State Theory see for example Bondi (1960).

[2] These observations are taken from Wigner's essay 'Symmetry and conservation laws', in Wigner (1967).

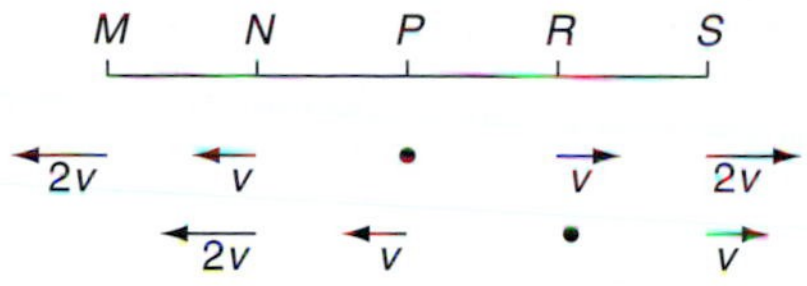

Fig. 10.1 Hubble expansion in one dimension: everyone sees the same pattern.

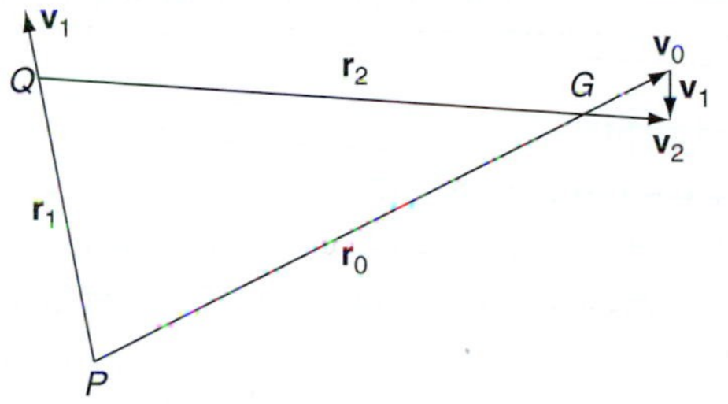

Fig. 10.2 Hubble expansion in three dimensions.

will have receded, after 10^8 years, by about 10^{18} km – an expansion of only 1% in 100 million years. In units of inverse seconds H_0 is

$$H_0 = h\,(3.24 \times 10^{-18})\,\text{s}^{-1}$$

whose inverse is

$$H_0^{-1} = \frac{3.09}{h} \times 10^{17}\,\text{s} \approx \frac{1}{h} \times 10^{10}\,\text{years}. \tag{10.4}$$

If H_0 were a constant in time this figure would represent the age of the Universe – the time that has elapsed since it had 'zero' size. In any case, even if Hubble's parameter is time dependent, its inverse should certainly bear some relation to the age of the Universe.

A second observation is that Hubble's law (10.2) does *not* imply that we are at the centre of expansion (the centre of the Universe), as initially one might be tempted to think. Consider the expansion in one dimension. Suppose we have points M, N, P, R and S a distance d apart from each other (see Fig. 10.1). An observer at P sees himself at rest, sees an observer at R moving to the right with speed $v = Hd$, an observer at S moving to the right with speed $2v$, an observer at N moving to the left with speed v and so on. An observer at R, on the other hand, will see herself at rest, will see S moving to the right with speed v, P moving to the left with speed v, N to the left with speed $2v$, and so on; she sees exactly the same pattern of expansion as P does. Every observer sees the same, so there is no centre of expansion. It is easy to generalise this to three dimensions. Suppose an observer at P sees an arbitrary galaxy G, a distance $\mathbf{r}_0$ away, receding (radially) at speed $\mathbf{v}_0 = H\mathbf{r}_0$, as in Fig. 10.2. What does an observer at Q see? Q moves at velocity $\mathbf{v}_1 = H\mathbf{r}_1$ with respect to P, so by vector addition of velocities will see G, a distance $\mathbf{r}_2$ away, move with velocity

$$\mathbf{v}_2 = \mathbf{v}_0 - \mathbf{v}_1 = H(\mathbf{r}_0 - \mathbf{r}_1) = H\mathbf{r}_2;$$

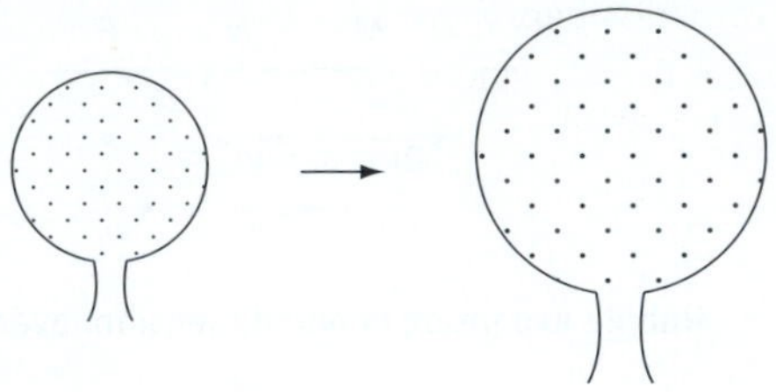

Fig. 10.3 **Dots on a balloon move apart as the balloon inflates.**

Q sees a Hubble expansion, just as P does – everybody sees the same thing. It is clear that Hubble's law (10.2) is the only expansion law compatible with a homogeneous universe: if Hubble had observed that v were proportional to r^2 then we would certainly have been at the expansion centre, and the hypothesis of a homogeneous Universe would have been ruined.

Our third observation involves gravity. Since all the galaxies attract one another (in Newtonian language) this will have the effect of *slowing down* the expansion – we would expect to observe a *deceleration* in the motion of the far galaxies. It has been clear in recent years, however, that the expansion is accelerating. We shall see, in the course of this chapter, that General Relativity, with the Einstein field equations (5.24) also predicts a decelerating universe. To account for an accelerating one involves introducing a 'cosmological constant' Λ into the field equations. We return to this below.

The fourth and final observation is slightly more subtle. It is the statement that the expansion observed by Hubble is not to be interpreted as the motion of galaxies through a fixed, static, space. It is, rather, the observation that the *space itself* is expanding, carrying the galaxies with it. We may visualise the analogy of a balloon being inflated, as in Fig. 10.3. If there are dots painted on the rubber surface of the balloon, then as it is inflated the dots move further apart because they are embedded in the rubber, which is itself stretching. So, on the cosmological scale, in General Relativity, it is the expansion of space itself which is being manifested. After all, in Einstein's vision, space has a reality that it never possessed in physics before. It acquires *curvature* when masses are placed in it; and now we learn that on the scale of the Universe as a whole it is expanding.

10.2 Robertson–Walker metric

Our first task is to translate the Cosmological Principle, that the Universe is homogeneous and isotropic, into an explicitly geometrical condition to be satisfied by the space-time metric – more particularly by its space part. We use a method based on Killing vectors, introduced in Section 6.6. It was shown there that if a space (or space-time) possesses a particular symmetry then the Lie derivative of the Killing vector generating the symmetry vanishes; that is, from (6.133)

$$g_{\mu\nu,\lambda}\,\xi^\lambda + g_{\mu\lambda}\,\xi^\lambda{}_{,\nu} + g_{\lambda\nu}\,\xi^\lambda{}_{,\mu} = 0, \tag{10.5}$$

where ξ^λ is the relevant Killing vector. Let us first implement the requirement of isotropy – that the metric is invariant under spatial rotations. In fact, we have already written down the most general form of *static* metric with spherical symmetry, which is (5.31)

$$ds^2 = -e^{2\nu}c^2\,dt^2 + e^{2\lambda}\,dr^2 + r^2(d\theta^2 + \sin^2\theta\,d\phi^2), \tag{10.6}$$

where $\nu = \nu(r)$, $\lambda = \lambda(r)$; this was a preliminary step to finding the Schwarzschild solution. In the case of cosmology there is no requirement that the metric be static; we shall proceed to implement the requirement of isotropy only, followed by the requirement of homogeneity. We recall from the (6.138) that the Killing vectors for rotations in space are

$$X_4 = -z\frac{\partial}{\partial y} + y\frac{\partial}{\partial z}, \quad X_5 = z\frac{\partial}{\partial x} - x\frac{\partial}{\partial z}, \quad X_6 = -y\frac{\partial}{\partial x} + x\frac{\partial}{\partial y}.$$

Changing notation and writing these as 4-vectors rather than 3-vectors, the Cartesian components of the Killing vectors are

$$\xi^\mu{}_1 = (0, 0, -z, y), \quad \xi^\mu{}_2 = (0, z, 0, -x), \quad \xi^\mu{}_3 = (0, -y, x, 0). \tag{10.7}$$

Then for example, transforming to spherical polars,

$$\begin{aligned}
\boldsymbol{\xi}_1 &= -z\frac{\partial}{\partial y} + y\frac{\partial}{\partial z} \\
&= -r\cos\theta\left(\sin\theta\sin\phi\frac{\partial}{\partial r} + \frac{1}{r}\cos\theta\sin\phi\frac{\partial}{\partial\theta} + \frac{1}{r}\frac{\cos\phi}{\sin\theta}\frac{\partial}{\partial\phi}\right) \\
&\quad + r\sin\theta\cos\phi\left(\cos\theta\frac{\partial}{\partial r} - \sin\theta\frac{\partial}{\partial\theta}\right) \\
&= -\sin\phi\frac{\partial}{\partial\theta} - \cot\theta\cos\phi\frac{\partial}{\partial\phi}; \\
\xi^\mu{}_1 &= (0, 0, -\sin\phi, -\cot\theta\cos\phi),
\end{aligned} \tag{10.8}$$

where in the last line the components are in spherical polar coordinates. Similarly we find

$$\xi^\mu{}_2 = (0, 0, \cos\phi, -\cot\theta\sin\phi), \quad \xi^\mu{}_3 = (0, 0, 0, 1). \tag{10.9}$$

To find the metric tensor we use Killing's equation (10.5). Applied to $\xi^\mu{}_3$ it gives

$$g_{\mu\nu,3} = 0; \quad \frac{\partial}{\partial\phi}g_{\mu\nu} = 0. \tag{10.10}$$

Putting $(\mu\nu) = (00)$, (10.5) applied to ξ_1 gives

$$\begin{aligned}
g_{00,2}(-\sin\phi) + g_{00,3}(-\cot\theta\cos\phi) &= 0, \\
g_{00,2}\sin^2\phi &= -g_{00,3}\cot\theta\cos\phi\sin\phi,
\end{aligned}$$

and applied to ξ_2 gives

$$g_{00,2} = 0; \quad \frac{\partial}{\partial\theta}g_{00} = 0. \tag{10.11}$$

Similarly, for $(\mu\nu) = (01)$, (11), since ξ_1 and ξ_2 are independent of t and r we find

$$g_{10,2} = 0, \quad g_{11,2} = 0. \tag{10.12}$$

In the same vein, with $(\mu\nu)$ = (22), (23), (33), (12), (13), (02), (03), the Killing vector equation for $\xi_1{}^\mu$ gives the seven equations

$$g_{22,2} \sin\phi = -2\, g_{23} \frac{\cos\phi}{\sin^2\theta}, \tag{10.13}$$

$$(g_{23,2} - g_{23} \cot\theta) \sin\phi = \left(-g_{22} + \frac{g_{33}}{\sin^2\theta}\right) \cos\phi, \tag{10.14}$$

$$(-g_{33,2} + g_{33} \cot\theta) \sin\phi = 2g_{32} \cos\phi, \tag{10.15}$$

$$g_{12,2} \sin\phi = -2\, g_{13} \frac{\cos\phi}{\sin^2\theta}, \tag{10.16}$$

$$(-g_{13,2} + g_{13} \cot\theta) \sin\phi = g_{12} \cos\phi, \tag{10.17}$$

$$g_{02,2} \sin\phi = g_{03} \frac{\cos\phi}{\sin^2\theta}, \tag{10.18}$$

$$(g_{03,2} - g_{03} \cot\theta) \sin\phi = g_{02} \cos\phi. \tag{10.19}$$

To these seven equations are to be added the seven similar equations for $\xi_2{}^\mu$, which actually may be found by making the substitutions $\sin\phi \to -\cos\phi$, $\cos\phi \to \sin\phi$; hence

$$g_{22,2} \cos\phi = 2g_{23} \frac{\sin\phi}{\sin^2\theta}, \tag{10.20}$$

$$-(g_{23,2} - g_{23} \cot\theta) \cos\phi = \left(-g_{32} + \frac{g_{33}}{\sin^2\theta}\right) \sin\phi, \tag{10.21}$$

$$(g_{33,2} - g_{33} \cot\theta) \cos\phi = 2g_{32} \sin\phi, \tag{10.22}$$

$$-g_{12,2} \cos\phi = -2\, g_{13} \frac{\sin\phi}{\sin^2\theta}, \tag{10.23}$$

$$(g_{13,2} - g_{13} \cot\theta) \cos\phi = g_{12} \sin\phi, \tag{10.24}$$

$$-g_{02,2} \cos\phi = g_{03} \frac{\sin\phi}{\sin^2\theta}, \tag{10.25}$$

$$-(g_{03,2} - g_{03} \cot\theta) \cos\phi = g_{02} \sin\phi. \tag{10.26}$$

Gathering our results together, (10.10), (10.11) and (10.12) give

$$g_{00} = g_{00}(t,r), \quad g_{01} = g_{01}(t,r), \quad g_{11} = g_{11}(t,r). \tag{10.27}$$

Equations (10.13) and (10.20) give $g_{23}=0$, $g_{22,2}=0$, which together with (10.10) implies

$$g_{22} = g_{22}(t,r), \quad g_{23} = 0. \tag{10.28}$$

This last condition, together with (10.14) gives

$$g_{33} = g_{22}(t,r)\sin^2\theta. \tag{10.29}$$

Finally, (10.18) and (10.25) together give $g_{03}=0$, $g_{02,2}=0$, hence from (10.26) $g_{02}=0$; and (10.17) and (10.24) give $g_{12}=0$, hence (10.23) gives $g_{13}=0$ and we conclude that $g_{\mu\nu}$ has the form

$$g_{\mu\nu} = \begin{pmatrix} g_{00}(r,t) & g_{01}(r,t) & 0 & 0 \\ g_{10}(r,t) & g_{11}(r,t) & 0 & 0 \\ 0 & 0 & g_{22}(r,t) & 0 \\ 0 & 0 & 0 & g_{22}(r,t)\sin^2\theta \end{pmatrix}, \tag{10.30}$$

or

$$\begin{aligned} \mathrm{d}s^2 = &- g_{00}(r,t)c^2\,\mathrm{d}t^2 + 2g_{01}(r,t)\,c\,\mathrm{d}r\,\mathrm{d}t + g_{11}(r,t)\,\mathrm{d}r^2 \\ &+ g_{22}(r,t)\,[\mathrm{d}\theta^2 + \sin^2\theta\,\mathrm{d}\phi^2]. \end{aligned} \tag{10.31}$$

This is the most general metric for an *isotropic* space. On the cosmological scale, however, space is also *homogeneous*. A space is homogeneous if it is isotropic about one point and invaraint under *translations*; this clearly amounts to being isotropic about all points. The Killing vectors for translations are, from (6.138), in Cartesian coordinates

$$\boldsymbol{\eta}_1 = \frac{\partial}{\partial x}, \quad \eta_1{}^\mu = (0,1,0,0),$$

or, in spherical polars

$$\begin{aligned} \boldsymbol{\eta}_1 &= \sin\theta\cos\phi\frac{\partial}{\partial r} + \frac{1}{r}\cos\theta\cos\phi\frac{\partial}{\partial\theta} - \frac{1}{r}\frac{\sin\phi}{\sin\theta}\frac{\partial}{\partial\phi}, \\ \eta_1{}^\mu &= \left(0, \sin\theta\cos\phi,\ \frac{1}{r}\cos\theta\cos\phi,\ -\frac{1}{r}\frac{\sin\phi}{\sin\theta}\right). \end{aligned} \tag{10.32}$$

Similarly

$$\begin{aligned} \eta_2{}^\mu &= \left(0, \sin\theta\sin\phi,\ \frac{1}{r}\cos\theta\sin\phi,\ \frac{1}{r}\frac{\cos\phi}{\sin\theta}\right), \\ \eta_3{}^\mu &= \left(0, \cos\theta,\ -\frac{1}{r}\sin\theta,\ 0\right). \end{aligned} \tag{10.33}$$

Killing's equation (10.5) applied to (10.33) gives, for $(\mu\nu)=(00)$

$$g_{00,1} = 0 \Rightarrow g_{00} = g_{00}(t), \tag{10.34}$$

and with $(\mu\nu)=(22)$

$$g_{22}(r,t) = A(t)\,r^2. \tag{10.35}$$

Finally $(\mu\nu)=(12)$ and (02) give

$$g_{11}=\frac{1}{r^2}g_{22}=A(t),\quad g_{01}=0$$

and these conditions inserted into (10.31) give a line element

$$\mathrm{d}s^2=-g_{00}(t)\,c^2\,\mathrm{d}t^2+A(t)[\mathrm{d}r^2+r^2\,\mathrm{d}\theta^2+r^2\sin^2\theta\,\mathrm{d}\phi^2]. \qquad (10.36)$$

Under a transformation $t\to t'$ with $\mathrm{d}t'^2=g_{00}(t)\,\mathrm{d}t^2$, this may be written, dropping the prime,

$$\mathrm{d}s^2=-c^2\,\mathrm{d}t^2+S^2(t)[\mathrm{d}r^2+r^2\,\mathrm{d}\theta^2+r^2\sin^2\theta\,\mathrm{d}\phi^2]. \qquad (10.37)$$

In this cosmological metric t is a 'universal' time, defined over the whole space-time manifold; it may be identified with proper time at each galaxy (since all galaxies are equivalent). The factor $S^2(t)$ is clearly an 'expansion' factor, affecting the spatial part of the metric, and therefore indicating an expanding universe. We have then proved rather formally that an expanding universe is compatible with a homogeneous and isotropic one – the fact that there is an expansion does *not* mean that there is a 'centre' of expansion. The common analogy, already described, is with the dots painted on the surface of a balloon which is then inflated – see Fig. 10.3 above.

The spatial section in the line element above, however, is *flat*; including the expansion factor $S^2(t)$ it is *conformally* flat. We must generalise this and consider the cases where it is curved, which, in general, it may be. A curved spatial hypersurface, however, may be embedded into a flat 4-dimensional Euclidean space E^4, just as the spherical surface S^2 may be considered embedded into the flat space E^3. The simplest model is a simple generalisation of this case; that the spatial section is essentially a 3-sphere S^3 of radius a with the equation

$$(x^1)^2+(x^2)^2+(x^3)^2+(x^4)^2=a^2, \qquad (10.38)$$

where $x^1, \ldots, x^4$ are the Cartesian coordinates in E^4. This equation is analogous to that of S^2 in E^3: $x^2+y^2+z^2=a^2$ is a spherical surface of radius a. The fourth coordinate x^4 above has, of course, nothing to do with time; E^4 is a ('fictitious') flat space in which is embedded 3-physical-dimensional space. It is convenient to introduce spherical coordinates for S^3,

$$\begin{aligned} x^1&=a\sin\chi\sin\theta\cos\phi,\quad x^2=a\sin\chi\sin\theta\sin\phi,\\ x^3&=a\sin\chi\cos\theta,\quad x^4=a\cos\chi. \end{aligned} \qquad (10.39)$$

The coordinates χ, θ, ϕ label points on S^3, so we are now using a coordinate system

$$(x^0,x^1,x^2,x^3)=(ct,\chi,\theta,\phi). \qquad (10.40)$$

In dealing with the flat space E^3 above we found that the Killing vectors for rotations, (10.8), (10.9), gave us, via Killing's equation (10.5), the isotropic line element (10.30). In the present case we have a slightly different coordinate system so we should check that our new Killing vectors also give an isotropic metric. Analogous to (10.7) the three Killing vectors for rotations in the (23), (31) and (12) planes are

$$\xi_1=x^2\frac{\partial}{\partial x^3}-x^3\frac{\partial}{\partial x^2},\quad \xi_2=x^3\frac{\partial}{\partial x^1}-x^1\frac{\partial}{\partial x^3},\quad \xi_3=x^1\frac{\partial}{\partial x^2}-x^2\frac{\partial}{\partial x^1}. \qquad (10.41)$$

In the coordinate system (10.40) we find

$$\begin{aligned}\xi_1{}^\mu &= \sin\chi\sin\theta\sin\phi\left(0,\cos\chi\cos\theta,-\frac{\sin\theta}{\sin\chi},0\right)\\ &\quad-\sin\chi\cos\theta\left(0,\cos\chi\sin\theta\sin\phi,\frac{\cos\theta\sin\phi}{\sin\chi},\frac{\cos\phi}{\sin\chi\sin\theta}\right)\\ &= (0,0,-\sin\phi,-\cot\theta\cos\phi),\end{aligned}$$

exactly as in (10.8). The other Killing vectors are also given by (10.9). The consequences of isometry under these rotations therefore follow exactly from the equations above and we end up with the metric (10.30), or

$$\begin{aligned}ds^2 &= -g_{00}(\chi,t)\,c^2\,dt^2 + 2g_{01}(\chi,t)\,c\,d\chi\,dt\\ &\quad + g_{11}(\chi,t)\,d\chi^2 + g_{22}(\chi,t)\,[d\theta^2+\sin^2\theta\,d\phi^2],\end{aligned} \tag{10.42}$$

the coordinate χ playing the role of r in the flat case.

We must now extend the symmetry to include isometry under rotations in the (14), (24) and (34) planes. These take the place of translations in the flat case, since there is no translational symmetry in a curved space – see the examples at the end of Section 6.6. The Killing vectors for rotations in the (14), (24) and (34) planes are

$$\boldsymbol{\eta}_1 = x^4\frac{\partial}{\partial x^1}-x^1\frac{\partial}{\partial x^4},\quad \boldsymbol{\eta}_2 = x^4\frac{\partial}{\partial x^2}-x^2\frac{\partial}{\partial x^4},\quad \boldsymbol{\eta}_3 = x^4\frac{\partial}{\partial x^3}-x^3\frac{\partial}{\partial x^4}. \tag{10.43}$$

The components of $\boldsymbol{\eta}_3$ in the basis $\left(\frac{\partial}{\partial t},\frac{\partial}{\partial\chi},\frac{\partial}{\partial\theta},\frac{\partial}{\partial\phi}\right)$ are

$$\begin{aligned}\eta_3{}^\mu &= -\sin\chi\cos\theta(0,-\sin\chi,0,0)+\cos\chi\left(0,\cos\chi\cos\theta,-\frac{\sin\theta}{\sin\chi},0\right)\\ &= (0,\cos\theta,-\cot\chi\sin\theta,0),\end{aligned} \tag{10.44}$$

or

$$\boldsymbol{\eta}_3 = \cos\theta\frac{\partial}{\partial\chi}-\cot\chi\sin\theta\frac{\partial}{\partial\theta}. \tag{10.45}$$

Killing's equation (10.5) for $\eta_3{}^\mu$ gives, for $(\mu\nu)=(00)$,

$$g_{00,1}\cos\theta = 0,$$

hence

$$g_{00} = g_{00}(t). \tag{10.46}$$

For $(\mu\nu)=(22)$ we find

$$g_{22,1} = 2g_{22}\cot\chi,$$

which on integration gives

$$g_{22} = B(t)\sin^2\chi. \tag{10.47}$$

Putting $(\mu\nu)=(12)$ gives

$$g_{22} = g_{11}\sin^2\chi, \tag{10.48}$$

which with (10.47) yields

$$g_{11} = g_{11}(t). \tag{10.49}$$

Finally, $(\mu\nu) = (02)$ gives

$$g_{01} = 0,$$

hence (see (10.30)) $g_{\mu\nu}$ is *diagonal*,

$$g_{\mu\nu} = \mathrm{diag}\{g_{00}(t),\ g_{11}(t), \sin^2\chi\, g_{11}(t),\ \sin^2\chi \sin^2\theta\, g_{11}(t)\},$$

or, redefining t, as before,

$$\mathrm{d}s^2 = -c^2\,\mathrm{d}t^2 + S^2(t)[\mathrm{d}\chi^2 + \sin^2\chi\,\mathrm{d}\theta^2 + \sin^2\chi \sin^2\theta\,\mathrm{d}\phi^2], \tag{10.50}$$

where $S(t)$ is some function of t. Observing that $\sin^2\chi = r^2/a^2$, then

$$\mathrm{d}\chi^2 = \frac{\mathrm{d}r^2}{r^2 - a^2}$$

and, absorbing a factor of $1/a^2$ into $S^2(t)$ we have

$$\mathrm{d}s^2 = -c^2\,\mathrm{d}t^2 + S^2(t)\left(\frac{\mathrm{d}r^2}{1 - \dfrac{r^2}{a^2}} + r^2(\mathrm{d}\theta^2 + \sin^2\theta\,\mathrm{d}\phi^2)\right). \tag{10.51}$$

Instead of the embedding (10.38) of cosmological 3-space into E^4, however, we could have the embedding

$$(x^4)^2 - (x^1)^2 - (x^2)^2 - (x^3)^2 = a^2, \tag{10.52}$$

which describes a *hyperbolic* space. It may be coordinatised by

$$\begin{aligned} x^1 &= a\sinh\chi \sin\theta \cos\phi, \quad x^2 = a\sinh\chi \sin\theta \sin\phi, \\ x^3 &= a\sinh\chi \cos\theta, \quad x^4 = a\cosh\chi, \end{aligned} \tag{10.53}$$

which is just (10.39) but with $\sin\chi \to \sinh\chi$, $\cos\chi \to \cosh\chi$. The Killing vectors may be written down and the analysis goes through very similarly to that above, finishing up with the metric

$$\mathrm{d}s^2 = -c^2\,\mathrm{d}t^2 + R^2(t)\left(\frac{\mathrm{d}r^2}{1 + \dfrac{r^2}{a^2}} + r^2(\mathrm{d}\theta^2 + \sin^2\theta\,\mathrm{d}\phi^2)\right) \tag{10.54}$$

to replace (10.50). This is the cosmological space-time metric for a homogeneous, isostropic universe with a hyperbolic 3-space section. The possible metrics, (10.37), (10.51) and (10.54), may be expressed in the single form

$$\mathrm{d}s^2 = -c^2\,\mathrm{d}t^2 + R^2(t)\left[\frac{1}{1 - Kr^2}\,\mathrm{d}r^2 + r^2(\mathrm{d}\theta^2 + \sin^2\theta\,\mathrm{d}\phi^2)\right], \tag{10.55}$$

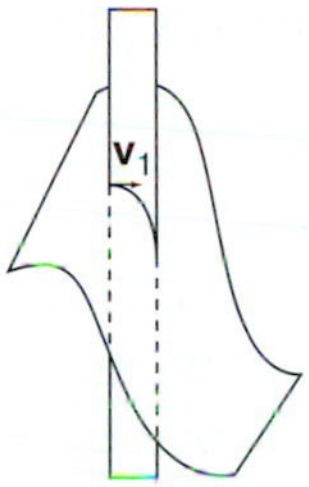

Fig. 10.4 The intersection of a plane with a surface: at a point of intersection the vector $\mathbf{v}_1$ is tangent to a circle of radius r_1.

with

$$K = 0, \quad \frac{1}{a^2} \quad \text{or} \quad -\frac{1}{a^2}, \tag{10.56}$$

respectively. The quantity K is called the *Gaussian curvature*. This quantity is defined for (2-dimensional) surfaces in a 3-dimensional space, in the following way. Consider a 2-dimensional surface, and visualise the intersection of a plane with this surface in some region, as in Fig. 10.4. At a point P on the line of intersection a tangent vector $\mathbf{v}_1$ may be drawn, and in the direction of $\mathbf{v}_1$ the intersection of the surface and the plane defines, to lowest order, a *circle*, with radius say r_1. We then say that the curvature of the surface at P in the direction of the tangent vector $\mathbf{v}_1$ is $k_1 = \frac{1}{r_1}$ (large radius, small curvature and vice versa). In a direction *perpendicular* to $\mathbf{v}_1$ the same is true: in the direction of a tangent vector $\mathbf{v}_2$ the intersection defines a circle with radius r_2, say, and curvature $k_2 = \frac{1}{r_2}$. The Gaussian curvature K at P is

$$K = k_1 k_2 = \frac{1}{r_1 r_2}. \tag{10.57}$$

Gauss showed that K does *not* depend on the choice of the vectors $\mathbf{v}_1$, $\mathbf{v}_2$, or equivalently on the planes of intersection. In this sense K is an *invariant* of the surface at the point considered. It may, however, have a positive or negative sign. Consider, for example, a sphere S^2 of radius a. Because S^2 is a homogeneous space (like S^3 is) the curvature is the same everywhere, so that at any point on the sphere both k_1 and k_2 are equal to $1/a$ and

$$S^2: \quad K = \frac{1}{a^2} > 0; \tag{10.58}$$

the Gaussian curvature is positive. This is because at any point on the sphere the centres of the two circles (defining the curvatures in two perpendicular directions) are on the *same side* of the surface ('inside' it). In contrast, consider a point on the surface of a hyperboloid, as in Fig. 1.8. It is clear that if the plane of intersection with this surface is (zy) the centre of the circle of intersection is *outside* the surface, but the circle resulting from the intersection with the (xz) plane is *inside* it. Since the centres are on opposite sides of the surface the Gaussian curvature is negative,

$$\text{hyperboloid:} \quad K < 0. \tag{10.59}$$

A flat surface (plane), of course, has $K = 0$. In these examples we have considered surfaces (curved or flat) embedded in a *flat* Euclidean space E^3. In the context of cosmology we are considering the Gaussian curvature K of a 3-dimensional space, not a 2-dimensional one, but actually K, or a generalisation of it, may be defined for a space of any dimension. In fact the definition is given in Equation (10.63) below.

It is common to simplify the metric (10.55) by expressing it solely in terms of the *sign* of the Gaussian curvature, rather than the actual magnitude of it. Define

$$k = \frac{K}{|K|} \tag{10.60}$$

so that

$$\begin{aligned} k &= 1 \quad \text{closed space } (S^3), \\ k &= 0 \quad \text{flat space } (E^3), \\ k &= -1 \ \text{open space (hyperbolic)}. \end{aligned} \tag{10.61}$$

Then defining $r' = \sqrt{K}r$, $\dfrac{R^2(t)}{K} = S^2(t)$ and dropping the prime, (10.55) becomes

$$\mathrm{d}s^2 = -c^2\,\mathrm{d}t^2 + S^2(t)\left(\frac{\mathrm{d}r^2}{1 - kr^2} + r^2(\mathrm{d}\theta^2 + \sin^2\theta\,\mathrm{d}\phi^2)\right). \tag{10.62}$$

In this form the metric is known as the *Robertson–Walker metric*, after its original authors.[3] It has been obtained without the use of the field equations, appealing only to the symmetry requirements of isotropy and homogeneity. The factor $S^2(t)$ describes an *expanding* space (or, in principle, a contracting one, though that is not what our Universe is doing at present) and it is noteworthy that this feature arises naturally from the above symmetry requirements. The coordinates above are *comoving*; that is, they move with the matter and are invariant along its world-lines.

It is of great interest to know what the function $S(t)$ is, in the actual Universe, and also to know what k is. As we shall see, these questions may be answered when we consider the *field equations*, which we have not yet done. It is worth noting, however, at this point, that the steps we have taken so far may be expressed in a slightly different way. It was noted in Chapter 6 that a space of dimension n is called a *maximally symmetric space* if it supports $n(n + 1)/2$ Killing vectors. We have seen that imposing the conditions of isotropy and homogeneity implied that the 3-space of the universe allows six Killing vectors, so it follows that this space is a maximally symmetric space, which is also called a space of *constant curvature*. Such a space is characterised by the property[4]

$$R_{\kappa\lambda\mu\nu} = K(g_{\kappa\mu}g_{\lambda\nu} - g_{\kappa\nu}g_{\lambda\mu}), \tag{10.63}$$

where K is the Gaussian curvature, a generalisation of what was discussed above for surfaces in a 3-dimensional space. This quantity appears because at any point P of a manifold we may define the *sectional curvature*[5]

[3] Robertson (1935, 1936), Walker (1936).
[4] See for example Weinberg (1972), chapter 13.
[5] See for example Eisenhart (1926).

$$K_P = \frac{R_{\kappa\lambda\mu\nu} {v_1}^{\kappa} {v_2}^{\lambda} {v_1}^{\mu} {v_2}^{\nu}}{(g_{\kappa\mu}\, g_{\lambda\nu} - g_{\kappa\nu}\, g_{\lambda\mu})\, {v_1}^{\kappa} {v_2}^{\lambda} {v_1}^{\mu} {v_2}^{\nu}}, \tag{10.64}$$

where ${v_1}^{\mu}$ and ${v_2}^{\mu}$ are two orthogonal vectors, as explained above. In the case of a 2-dimensional manifold K_P is the Gaussian curvature at P. A *homogeneous* space, with maximal symmetry, will clearly have $K =$ constant over the space and (10.63) can be shown to imply (10.62). Let us see how this works for a 3-dimensional space. From (10.62), using Latin indices in accordance with our convention for a 3-dimensional space, the Ricci tensor is

$$^{(3)}R_{ik} = g^{lm} R_{ilkm} = K(3g_{ik} - g_{ik}) = 2Kg_{ik}, \tag{10.65}$$

$$[^{(n)}R_{ik} = (n-1)Kg_{ik} \text{ in } n \text{ dimensions}] \tag{10.65a}$$

and the curvature scalar is

$$^{(3)}R = g^{ik}\, {}^{(3)}R_{ik} = 6K = \frac{6}{a^2}, \tag{10.66}$$

$$[^{(n)}R = n(n-1)K \text{ in } n \text{ dimensions}] \tag{10.66a}$$

where in the last step we have used (10.57) for the 3-sphere. This result, (10.65), agrees with a direct calculation for S^3 (Problem 10.1).

We close this section by making some remarks about the geometry of S^3, a contender for the space section of the Universe. Like the 2-sphere S^2 it is a *closed* space, one with no boundary. We write the line element in the form (see (10.51))

$$d\sigma^2 = \frac{dr^2}{1 - r^2/a^2} + r^2(d\theta^2 + \sin^2\theta\, d\phi^2). \tag{10.67}$$

S^3 is a homogeneous space and the origin may be chosen anywhere. The circumference of a circle is found by considering points distinguished only by their θ coordinate, so $dr = d\phi = 0$ and $d\sigma = r\, d\theta$, hence the circumference is $2\pi r$. Likewise the surface area of a 2-sphere in S^3 is

$$A = \int_0^{\pi} r^2 \sin\theta\, d\theta\, d\phi = 4\pi r^2.$$

The 'radius' of a circle, or sphere, is

$$\text{radius} = \int_0^r \left(1 - \frac{r'^2}{a^2}\right)^{-1/2} dr' = a \sin^{-1}(r/a) > r,$$

so

$$\frac{\text{circumference of circle}}{\text{radius of circle}} < 2\pi. \tag{10.68}$$

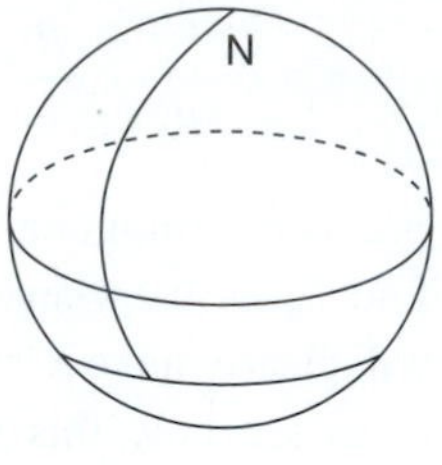

Fig. 10.5 **Circles on the surface of a sphere. Measured from the north pole N the radius of the circle on the equator is the length of the line on the sphere joining them. A circle in the southern hemisphere has a larger radius but a smaller circumference than that at the equator.**

This is also true of a circle S^1 inscribed on a sphere S^2. There are other parallels with S^2. To see these first write the line element (10.67) in polar coordinates (10.52):

$$(S^3):\quad d\sigma^2 = a^2[d\chi^2 + \sin^2\chi(d\theta^2 + \sin^2\theta d\phi^2)]. \tag{10.69}$$

The coordinate χ plays the role of distance; the distance from the origin is $a\chi$. The largest distance in the space is $a\pi$. There is an analogy on the 2-sphere, where

$$(S^2):\quad dl^2 = a^2(d\theta^2 + \sin^2\theta\, d\phi^2). \tag{10.70}$$

Here θ plays a similar role: the distance from any origin is $a\theta$ and $a\pi$ is the largest distance on S^2 – that between the north and south poles.

The surface area of a 2-sphere in S^3 is clearly $4\pi a^2 \sin^2\chi$. As χ increases from 0, this reaches a maximum at $\chi = \pi/2$, at a distance of $a\pi/2$ and thereafter decreases with increasing distance, shrinking to a point as $\chi \to \pi$, a distance $a\pi$ away (the 'opposite pole'). This is completely analogous to the description of circles on a 2-sphere (S^1 on S^2). Starting at, say, the north pole, the circumference of a circle initially increases with increasing distance away ('radius', but measured, of course, on the surface S^2 itself), but it reaches a maximum at the equator, and thereafter, when the circle is in the southern hemisphere, as its distance from the north pole ('radius') increases its circumference *decreases* (see Fig. 10.5), shrinking to a point at the south pole, the maximum distance away. Stepping up the dimensions by 1, then, to consider 2-spheres in S^3, the observation is exactly that made above, that with increasing distance away, the area of the 2-sphere initially increases, but reaches a maximum beyond which it decreases with increasing radius (distance away) – we may say that bigger spheres 'fit inside' smaller spheres.

The volume of the space S^3 is

$$V = \int_0^{2\pi}\int_0^{\pi}\int_0^{\pi} a^3 \sin^2\chi \sin\theta\, d\chi\, d\theta\, d\phi = 2\pi^2 a^3;$$

a *finite* value. This space, like S^3, has no boundary

$$\partial S^3 = 0; \tag{10.71}$$

it is a *closed* space. So, as pointed out by Landau and Lifshitz,[6] Gauss's theorem implies that the total electric charge in the space must be zero, since (see (3.93), (3.95))

$$Q = \int_{S^3} \rho \, \mathrm{d}V = \varepsilon_0 \int_{S^3} \nabla \cdot \mathbf{E} \, \mathrm{d}V = \varepsilon_0 \int_{\partial S^3} \mathbf{E} \cdot \mathbf{n} \, \mathrm{d}\Sigma = 0;$$

the total electric charge in a closed universe must be zero.

10.3 Hubble's law and the cosmological red-shift

The Robertson–Walker metric allows us to make various *kinematic* deductions, independent of the specific form of $S(t)$. The first of these is the Hubble expansion. Suppose that our Galaxy, comoving in the space-time continuum, is located at $r = 0$, and another galaxy is at an arbitary parameter r. Its proper distance L from us at cosmic time t is, from (10.62)

$$L = S(t) \int_0^r \frac{\mathrm{d}r'}{\sqrt{1 - kr'^2}} = S(t) f(r), \tag{10.72}$$

with

$$f(r) = \begin{cases} \sin^{-1}(r) & (k = 1) \\ r & (k = 0) \\ \sinh^{-1}(r) & (k = -1) \end{cases} \tag{10.73}$$

so $L \propto S(t)$: the distance away changes with time (hardly surprisingly!). The *velocity* of recession is

$$v = \frac{\mathrm{d}L}{\mathrm{d}t} = \dot{S}(t) \int_0^r \frac{\mathrm{d}r'}{\sqrt{1 - kr'^2}} = \frac{\dot{S}(t)}{S(t)} L, \tag{10.74}$$

hence

$$v \propto L \tag{10.75}$$

at any instant. This is Hubble's law $v = HL$ ((10.2)) with

$$H(t) = \frac{\dot{S}(t)}{S(t)}. \tag{10.76}$$

Note that Hubble's constant depends on time in general.

Now consider light reaching us, at $r = r_2$, having been emitted from a galaxy at $r = r_1$ – see Fig. 10.6. In particular consider successive crests of light, emitted at times t_1 and $t_1 + \Delta t_1$ and received at times t_2 and $t_2 + \Delta t_2$. Since $\mathrm{d}s^2 = 0$ and the light is travelling radially we have

[6] Landau & Lifshitz (1971), p. 375.

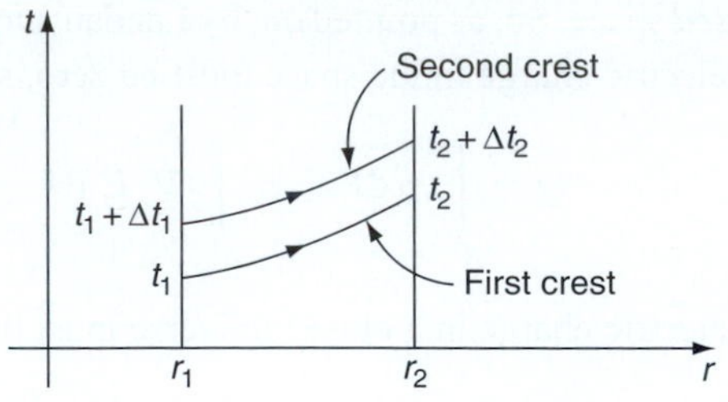

Fig. 10.6 Light emitted from a galaxy at $r = r_1$ reaches us at $r = r_2$.

$$0 = -c^2\,dt^2 + S^2(t)\,\frac{dr^2}{1-kr^2}, \tag{10.77}$$

so for the first crest of light

$$\int_{t_1}^{t_2} \frac{dt}{S(t)} = \frac{1}{c}\int_{r_1}^{r_2} \frac{dr}{\sqrt{1-kr^2}},$$

and for the second crest

$$\int_{t_1+\Delta t_1}^{t_2+\Delta t_2} \frac{dt}{S(t)} = \frac{1}{c}\int_{r_1}^{r_2} \frac{dr}{\sqrt{1-kr^2}};$$

hence

$$\int_{t_1+\Delta t_1}^{t_2+\Delta t_2} \frac{dt}{S(t)} = \int_{t_1}^{t_2} \frac{dt}{S(t)}.$$

Now Δt_1 and Δt_2 are very small so we may assume that $S(t)$ is constant over these intervals. Then, since

$$\int_{t_1+\Delta t_1}^{t_2+\Delta t_2} = \int_{t_1}^{t_2} + \int_{t_2}^{t_2+\Delta t_2} - \int_{t_1}^{t_1+\Delta t_1}$$

we find

$$\frac{\Delta t_1}{S(t_1)} = \frac{\Delta t_2}{S(t_2)} \quad \Rightarrow \quad \frac{\Delta t_1}{\Delta t_2} = \frac{S(t_1)}{S(t_2)}.$$

The emitted and observed wavelengths are

$$\lambda_e = c\,\Delta t_1, \quad \lambda_o = c\,\Delta t_2$$

so the *wavelength shift* is

$$z = \frac{\lambda_o - \lambda_e}{\lambda_e} = \frac{S(t_1)}{S(t_2)} - 1. \qquad (10.78)$$

If the Universe is *expanding*, $S(t_1) > S(t_2)$ so $z > 0$, resulting in a *red-shift*, as observed by Hubble. This is a *cosmological* red-shift and is conceptually distinct from a Doppler shift. In a Doppler shift both the source of the light and the observer are in the same inertial frame, and are moving relatively to each other. In the cosmological setting both the source and the observer are comoving; they are each in their own inertial frame. Their relative motion is due to the expansion of space. And, of course, both of these red-shifts are distinct from the gravitational red-shift.

It is worth making a remark at this point on the Steady State Theory. In this theory, since the Universe presents the same aspect at all times, Hubble's constant must be time independent, so (10.76) gives

$$\frac{dS}{dt} = HS, \quad S(t) = S(0)\,e^{Ht}; \qquad (10.79)$$

the expansion of space is exponential, and $S(t)$ is never zero. Moreover, the curvature of the 3-space (10.67) is $6/a^2 = 6k$, so the curvature of the 3-space part of the Robertson–Walker metric (10.62) is $\frac{6k}{S^2(t)}$. In Steady State Theory this must be *constant* so we must have $k = 0$, and the steady state line element is

$$ds^2 = -c^2\,dt^2 + e^{2Ht}[dr^2 + r^2(d\theta^2 + \sin^2\theta\, d\phi^2)]. \qquad (10.80)$$

Space is *flat* and the metric is determined on *kinematic grounds alone*. This metric is of interest not because the Steady State Theory is of interest but because it has the same form as the de Sitter metric, which we shall meet below.

10.4 Horizons

Let us suppose that the Universe 'began' at time t_B. In most models this will be a finite time in the past, which we may put equal to 0; in the steady state model it is equal to $-\infty$. Forgetting, temporarily, the expansion of the Universe, we may sketch our world-line, as in Fig. 10.7, from t_B to t_0, which is 'now'. In this simple view of things, at t_B there were many galaxies – 'objects' or 'particles' – and it must be true that light from some of these galaxies is only now reaching us for the very first time. These objects are on a 'horizon', called an *object horizon* or *particle horizon*; they are the most distant objects we can now see. From (10.77) the (proper) distance to the particle horizon, r_{PH}, is given by

$$\int_{t_B}^{t_0} \frac{dt}{S(t)} = \frac{1}{c}\int_0^{r_{PH}} \frac{dt}{\sqrt{1-kr^2}} = \frac{1}{c}\sin^{-1} r_{PH} \quad (k = 1) \qquad (10.81)$$

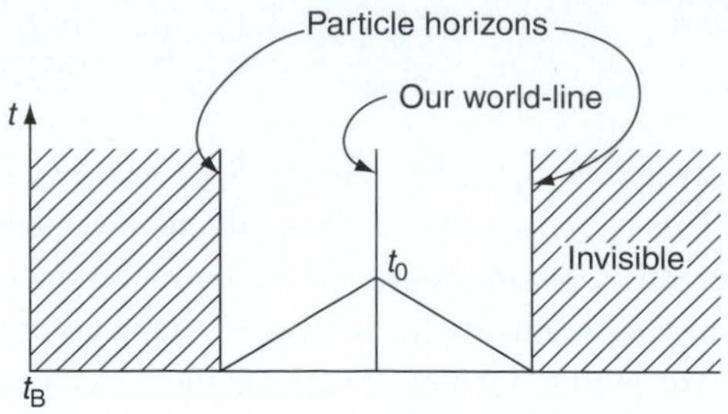

Fig. 10.7 **Particle horizon: t_B is the 'beginning' of the Universe, t_0 is 'now'. In a non-expanding model light from sources further away than ct_0 has not yet reached us.**

or

$$r_{PH} = \sin\left(c\int_{t_B}^{t_0}\frac{dt}{S(t)}\right) \quad (k = 1), \tag{10.82}$$

with analogous expressions for $k=0, -1$. Objects beyond r_{PH} cannot now be seen by us. Clearly r_{PH} increases with t_0; we shall be able to see more in the future than we can see now. In fact the horizon at r_{PH} is like an outgoing spherical wave centred on us. In addition, if there is now someone whom we are able to see for the first time, then 'they' must by symmetry be able to see us for the first time; an outgoing light-front leaving our world-line at t_B will now be reaching them, as also shown in Fig. 10.7.

In certain models we might find $r_{PH}=\infty$, and then the whole Universe can be seen by us, and there is no object horizon. In the $k = 1$ case if

$$c\int_{t_B}^{t_0}\frac{dt}{S(t)} \geq \pi$$

then from (10.80) there is no object horizon, since π is the maximum distance in the S^3 model.

Now let us consider a related situation, but this time concerned with the *end* of the Universe, at t_E (which will be finite in some models, but infinite in others). We ask: what is the coordinate, r_{EH}, of the most distant event occurring *now* (at t_0) that we shall *ever* be able to see? This defines an *event horizon*. Light from the event horizon must reach us before, or as, the Universe ends, at t_E. So

$$\int_{t_0}^{t_E}\frac{dt}{S(t)} = \frac{1}{c}\int_0^{r_{EH}}\frac{dr}{\sqrt{1-kr^2}}$$

or

$$r_{EH} = \sin\left(c\int_{t_0}^{t_E}\frac{dt}{S(t)}\right) \quad (k = 1), \tag{10.83}$$

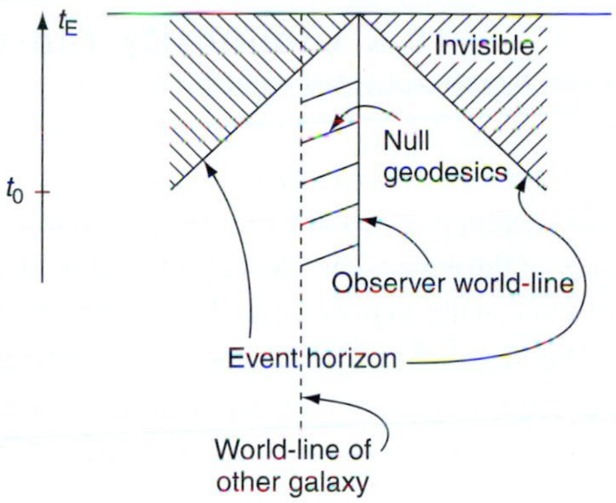

Fig. 10.8 **Event horizon: t_E is the 'end' of the Universe, t_0 is 'now'. Light emitted from galaxies inside the event horizon will reach us before the Universe ends.**

with analogous expressions for the cases $k = 0, -1$. Events occurring now beyond r_{EH} will never be seen by us – see Fig. 10.8. If $r_{EH} = \infty$, every event will at some time be seen by us (by an observer on our world-line). The event horizon is an inward-converging light cone reaching us at $t = t_E$. In Fig. 10.8 is drawn the world-line of another galaxy, with null geodesics, along which light will travel, from that galaxy to our own. It is clear that, as time goes on (as t_0 increases) light from this other galaxy will eventually cease to reach us before the Universe ends at $t = t_E$, so eventually every galaxy passes out of the event horizon of every other galaxy.

The above reasoning was based on the assumption that the Universe is not expanding, but we shall now show that concepts of horizons are basically unaffected by taking the expansion into account. Defining a new time parameter T by

$$\mathrm{d}T = \frac{\mathrm{d}t}{S(t)},$$

the R–W metric (10.62) becomes

$$\mathrm{d}s^2 = S^2(t)\left(-c^2\,\mathrm{d}T^2 + \frac{\mathrm{d}r^2}{1-kr^2} + r^2(\mathrm{d}\theta^2 + \sin^2\theta\,\mathrm{d}\phi^2)\right), \tag{10.84}$$

which is *conformally related* to the non-expanding metric

$$\mathrm{d}\hat{s}^2 = -c^2\,\mathrm{d}T^2 + \frac{\mathrm{d}r^2}{1-kr^2} + r^2(\mathrm{d}\theta^2 + \sin^2\theta\,\mathrm{d}\phi^2);$$

these line elements being obtained from one another by an overall conformal factor, as explained in Section 7.7. As was seen there, this conformal rescaling does not affect null geodesics (since $\mathrm{d}s^2 = 0 \Leftrightarrow \mathrm{d}\hat{s}^2 = 0$); it only has the effect of 'bringing in' points at infinity to the finite domain. Since the phenomena of event and particle horizons depend essentially on null geodesics, qualitative conclusions drawn from a static universe will also hold in an expanding one.

10.5 Luminosity–red-shift relation

The expansion of the Universe is governed by the function $S(t)$ and Hubble's constant $H(t)$ is, as we have seen, proportional to $\dot{S}(t)$. The double derivative $\ddot{S}(t)$ will clearly describe the *deceleration* of the expansion: on simple Newtonian grounds the Universe self-gravitates so we would expect the expansion to be slowing down and $\ddot{S}(t)$ to be non-zero – in fact, negative. The 'deceleration parameter' q, to be defined below, as well as $H(t)$, are quantities that play key roles in cosmology, since they may be derived from theoretical cosmological models but they may also be found by exploring relations between various *observables*. One of these is the one between luminosity and red-shift, which we now discuss.

Let us calculate the light flux received from a distant source. Suppose the light is emitted at (cosmic) time t_e and radial coordinate r_e, and is just now reaching us, at time t_0 and coordinate r_0. This is shown in Fig. 10.9 where we have chosen to put $r_0 = 0$; r decreases as t increases along the light path. Considering, on the other hand, the *source* to be at the centre of a sphere, the light is now crossing the surface of this sphere, whose area is

$$4\pi r_e^2 S^2(t_0).$$

Suppose the source has absolute luminosity L; this is the amount of energy emitted per unit time (i.e. power). The total power *received* per unit area (for example through a telescope) is not simply L divided by the area above, since there are two red-shift factors to be taken into account:

(i) Each *photon* is red-shifted. From (10.78) the ratio of observed and emitted wavelengths and frequencies is

$$\frac{\lambda_0}{\lambda_e} = z + 1 = \frac{S(t_0)}{S(t_e)} = \frac{\nu_e}{\nu_0} = \frac{E_e}{E_0},$$

so the energy of each photon is decreased by $(1 + z)$.

(ii) The *rate of arrival* of photons is decreased by the same factor. Two photons emitted within a time interval δt_e will arive within the time interval

$$\delta t_e \cdot \frac{S(t_0)}{S(t_e)} = \delta t_e(1 + z).$$

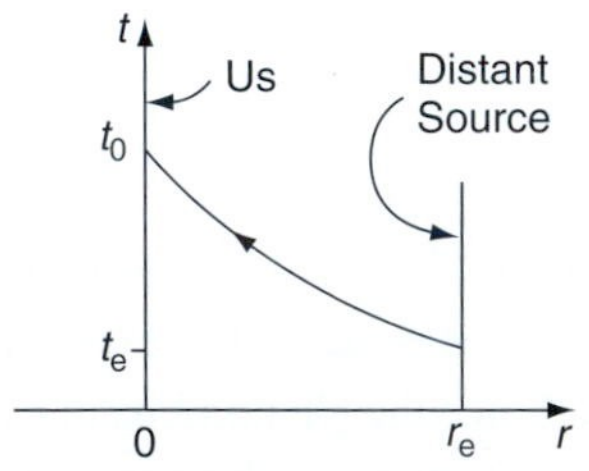

Fig. 10.9 Light emitted from a source at r_e at time t_e reaches us, at a position r_0 (= 0 here) at time t_0.

Hence the power received per unit area, which is the *apparent luminosity l*, is[7]

$$l = \frac{L}{4\pi r_e{}^2 S^2(t_0)} \cdot \frac{1}{(1+z)^2} = \frac{LS^2(t_e)}{4\pi r_e{}^2 S^4(t_0)}. \tag{10.85}$$

A parameter clearly related to apparent luminosity is the *luminosity distance* d_L defined by

$$l = \frac{L}{4\pi d_L{}^2}. \tag{10.86}$$

We now turn to *observables*. Besides Hubble's constant

$$H(t) = \frac{\dot{S}(t)}{S(t)}, \tag{10.87}$$

we also define the *deceleration parameter* $q(t)$,

$$q(t) = -\frac{\ddot{S}(t)}{\dot{S}(t)^2}. \tag{10.88}$$

Expanding about $t = t_0$ and with

$$H_0 = H(t_0), \quad q_0 = q(t_0) \tag{10.89}$$

we have

$$\begin{aligned} S(t) &= S(t_0) + (t - t_0)\dot{S}(t_0) + {}^1/_2\,(t - t_0)^2 \ddot{S}(t_0) + \cdots \\ &= S(t_0)\,[1 + (t - t_0)H_0 - {}^1/_2(t - t_0)^2 q_0 H_0^2 + \cdots]. \end{aligned} \tag{10.90}$$

We now want to express $(t - t_0)$ in terms of the red-shift z, an observable quantity. From (10.78), and using (10.90),

$$\begin{aligned} z = S(t_0)\,S^{-1}(t_e) - 1 &= [1 + (t_e - t_0)H_0 - {}^1/_2(t_e - t_0)^2 q_0\, H_0^2 + \cdots]^{-1} - 1 \\ &= H_0(t_0 - t_e) + \left(1 + \frac{q^0}{2}\right) H_0{}^2 (t_0 - t_e)^2 + \cdots. \end{aligned}$$

Inverting this gives

$$t_0 - t_e = \frac{z}{H_0} - \left(1 + \frac{q_0}{2}\right) H_0 \left(\frac{z}{H_0}\right)^2 = \frac{1}{H_0}\left[z - \left(1 + \frac{q_0}{2}\right) z^2\right]. \tag{10.92}$$

Now let us find an expression for r_e. From (10.62), for light travelling radially we have

$$\int_{t_e}^{t_0} \frac{\mathrm{d}t}{S(t)} = \int_0^{r_e} [1 - kr^2]^{-1/2}\,\mathrm{d}r. \tag{10.93}$$

7 This argument, invoking photons, clearly relies on quantum theory, which is hardly ideal in the context of General Relativity. For an argument for inclusion of the factor $(1+z)^{-2}$, which is based on GR alone, see Robertson (1938).

The left hand side of this is, from (10.90)

$$\text{lhs} = S^{-1}(t_0)\int_{t_e}^{t_0}[1 - H_0(t-t_0) + H_0^2(t-t_0)^2 + \frac{q_0}{2}H_0{}^2(t-t_0)^2 + \cdots]\,\mathrm{d}t$$
$$= S^{-1}(t_0)\,[\,(t_0 - t_e) + {}^1\!/_2\,H_0(t_0 - t_e)^2 + \cdots],$$

while the right hand side is

$$\text{rhs} = r_e + \mathrm{O}(r_e{}^3),$$

so to lowest order in r_e and using (10.93)

$$r_e = \frac{1}{S(t_0)}\left[\,(t_0 - t_e) + \frac{H_0}{2}(t_0 - t_e)^2 + \cdots\right] = \frac{1}{S(t_0)\,H_0}\left[z - \left(1 + \frac{q_0}{2}\right)z^2 + \frac{z^2}{2}\right]$$
$$= \frac{1}{S(t_0)\,H_0}\left[z - {}^1\!/_2\,(1 + q_0)z^2\right]. \tag{10.94}$$

Now substituting successively (10.94), (10.91) and (10.93) into (10.85) gives

$$l = \frac{L}{4\pi}\cdot\frac{S^2(t_e)}{S^2(t_0)}\cdot\frac{H_0{}^2}{z^2}\left[\,1 - {}^1\!/_2(1 + q_0)z\,\right]^{-2}$$
$$= \frac{LH_0{}^2}{4\pi z^2}[\,1 + (t_e - t_0)H_0 + \cdots]^2\,[1 + (1 + q_0)z]$$
$$= \frac{LH_0{}^2}{4\pi z^2}(1 - 2z)\,[\,1 + (1 + q_0)\,z\,]$$
$$= \frac{LH_0{}^2}{4\pi z^2}[1 + (q_0 - 1)z]. \tag{10.95}$$

This is the *luminosity–red-shift relation*. An equivalent relation is the *distance–red-shift relation*, which follows from (10.95) and (10.86):

$$\mathrm{d}_L = \frac{1}{H_0}[z - {}^1\!/_2(q_0 - 1)z^2]. \tag{10.96}$$

From these relations between observable quantities it is, in principle, possible to find H_0 and q_0. Hubble's constant is the easier parameter to find, being simply related to the gradient of the d_L–z graph for small z. Recent estimates give[8]

$$H_0 = (55 - 85)\ \mathrm{km\,s^{-1}\,Mpc^{-1}}, \tag{10.97}$$

or

$$H_0 = h\cdot 100\ \mathrm{km\,s^{-1}\,Mpc^{-1}},\qquad h = (0.55\text{–}0.85). \tag{10.98}$$

The determination of q_0 has, however, seen some rather dramatic developments over the last 10 years. Before about 1998, again working in the area of small red-shifts (z up to about 0.4

[8] Freedman (1997).

or 0.5), q_0 was reckoned to be between 0 and 1; a fairly large uncertainty, but *positive*, corresponding to a *decelerating* universe, understandable through Newtonian intuition.

More recently, however, the data have improved and the reason for this is related to improved estimates of distance. To know the luminosity distance d_L we have to know the absolute luminosity of the source L – see (10.86). In general this gives rise to a problem: if a star is faint it could be because it is *intrinsically* faint, or because it is brighter but further away. How do we know which? We rely on so-called 'standard candles', sources whose absolute magnitudes are known. The classic example is a Cepheid variable. These are stars whose luminosity varies periodically, but with a period dependent on the luminosity. Using Cepheid variables enables us to find d_L, and measuring the red-shift earns the star a place on the d_L–z plot. Beyond a particular distance, however, Cepheid variable stars become too faint. Their role as standard candles, has, however, now been taken over by *type 1a supernovae*. These are binary white dwarf systems, whose heavier partner attracts material (mass) from the lighter partner until its mass exceeds the Chandrasekhar limit (see Section 7.3), resulting in an explosion leading to a supernova – bright, and visible from far away! Crucially, the luminosity of an exploding white dwarf is a fairly good standard candle so the distance of these supernovae can then be estimated and they enable more data to appear on the d_L–z plot, with higher values of d_L and z, so that in particular the *non-linear* term in (10.96) can be found with some accuracy. And, surprisingly, q_0 turns out to be *negative* – the expansion of the Universe is actually *accelerating*, not slowing down. Newtonian intuition fails here, but it turns out that this scenario is not new in cosmology. Einstein introduced his famous cosmological constant Λ to solve a problem that has gone away, but Λ corresponds exactly to an accelerating expansion. We shall return to this topic below.

10.6 Dynamical equations of cosmology

The Robertson–Walker metric (10.62)

$$ds^2 = -c^2\,dt^2 + S^2(t)\left(\frac{dr^2}{1-kr^2} + r^2(d\theta^2 + \sin^2\theta\,d\phi^2)\right) \tag{10.100}$$

contains the 'free' functions $S(t)$ and k, describing the exansion and the curvature of space. The Einstein field equations (5.24)

$$R_{\mu\nu} - {}^1\!/\!{}_2\, g_{\mu\nu}R = \frac{8\pi G}{c^2}\,T_{\mu\nu} \tag{10.101}$$

will yield equations for $S(t)$ and k so enabling us to see what sort of universe General Relativity permits (provided we have an expression for the energy-momentum tensor).

The quickest way to find the left hand side of (10.101) is to use differential forms. The metric (10.100) may be written

$$ds^2 = -(\boldsymbol{\theta}^0)^2 + (\boldsymbol{\theta}^1)^2 + (\boldsymbol{\theta}^2)^2 + (\boldsymbol{\theta}^3)^2 \tag{10.102}$$

which, with basis 1-forms

$$\boldsymbol{\theta}^0 = c\,\mathbf{d}t, \quad \boldsymbol{\theta}^1 = Su^{-1}\,\mathbf{d}r, \quad \boldsymbol{\theta}^2 = Sr\,\mathbf{d}\theta, \quad \boldsymbol{\theta}^3 = Sr\sin\theta\,\mathbf{d}\phi, \tag{10.103}$$

where

$$u(r) = \sqrt{1 - kr^2} \tag{10.104}$$

is an *orthonormal* (anholonomic) basis with metric tensor

$$g_{\mu\nu} = \eta_{\mu\nu} = \mathrm{diag}(-1, 1, 1, 1). \tag{10.105}$$

The exterior derivatives of the basis forms are

$$\mathbf{d}\boldsymbol{\theta}^0 = 0, \tag{10.106a}$$

$$\mathbf{d}\boldsymbol{\theta}^1 = \dot{S}u^{-1}\,\mathbf{d}t \wedge \mathbf{d}r = \frac{\dot{S}}{cS}\boldsymbol{\theta}^0 \wedge \boldsymbol{\theta}^1, \tag{10.106b}$$

$$\mathbf{d}\boldsymbol{\theta}^2 = \dot{S}r\,\mathbf{d}t \wedge \mathbf{d}\theta + S\,\mathbf{d}r \wedge \mathbf{d}\theta = \frac{\dot{S}}{cS}\boldsymbol{\theta}^0 \wedge \boldsymbol{\theta}^2 + \frac{u}{Sr}\boldsymbol{\theta}^1 \wedge \boldsymbol{\theta}^2, \tag{10.106c}$$

$$\begin{aligned}\mathbf{d}\boldsymbol{\theta}^3 &= \dot{S}r\sin\theta\,\mathbf{d}t \wedge \mathbf{d}\phi + S\sin\theta\,\mathbf{d}r \wedge \mathbf{d}\phi + Sr\cos\theta\,\mathbf{d}\theta \wedge \mathbf{d}\phi \\ &= \frac{\dot{S}}{cS}\boldsymbol{\theta}^0 \wedge \boldsymbol{\theta}^3 + \frac{u}{Sr}\boldsymbol{\theta}^1 \wedge \boldsymbol{\theta}^3 + \frac{\cot\theta}{Sr}\boldsymbol{\theta}^2 \wedge \boldsymbol{\theta}^3.\end{aligned} \tag{10.106d}$$

The structure equations (3.182)

$$\mathbf{d}\boldsymbol{\theta}^\mu + \boldsymbol{\omega}^\mu{}_\kappa \wedge \boldsymbol{\theta}^\kappa = 0 \tag{10.107}$$

give, for equation (10.106a),

$$\boldsymbol{\omega}^0{}_1 \wedge \boldsymbol{\theta}^1 + \boldsymbol{\omega}^0{}_2 \wedge \boldsymbol{\theta}^2 + \boldsymbol{\omega}^0{}_3 \wedge \boldsymbol{\theta}^3 = 0, \tag{10.108}$$

where we have used the fact that

$$\boldsymbol{\omega}_{\mu\nu} = -\boldsymbol{\omega}_{\nu\mu}, \tag{10.109}$$

which follows from (10.105) and (3.191). Equations (10.106b) and (10.107) give

$$-\boldsymbol{\omega}^1{}_0 \wedge \boldsymbol{\theta}^0 - \boldsymbol{\omega}^1{}_2 \wedge \boldsymbol{\theta}^2 - \boldsymbol{\omega}^0{}_3 \wedge \boldsymbol{\theta}^3 = \frac{\dot{S}}{cS}\boldsymbol{\theta}^0 \wedge \boldsymbol{\theta}^1,$$

which implies that

$$\boldsymbol{\omega}^1{}_0 = \frac{\dot{S}}{cS}\boldsymbol{\theta}^1 = \frac{\dot{S}}{c}u^{-1}\,\mathbf{d}r. \tag{10.110}$$

Next, (10.106c) gives

$$-\boldsymbol{\omega}^2{}_0 \wedge \boldsymbol{\theta}^0 - \boldsymbol{\omega}^2{}_1 \wedge \boldsymbol{\theta}^1 - \boldsymbol{\omega}^2{}_3 \wedge \boldsymbol{\theta}^3 = \frac{\dot{S}}{cS}\boldsymbol{\theta}^0 \wedge \boldsymbol{\theta}^2 + \frac{u}{Sr}\boldsymbol{\theta}^1 \wedge \boldsymbol{\theta}^2,$$

which implies that

$$\omega^2{}_0 = \frac{\dot{S}}{cS}\theta^2 = \frac{r\dot{S}}{c}\mathbf{d}\theta, \quad \omega^2{}_1 = \frac{u}{Sr}\theta^2 = u\,\mathbf{d}\theta, \tag{10.111}$$

and finally (10.106d) and (10.108) give

$$-\omega^3{}_0 \wedge \theta^0 - \omega^3{}_1 \wedge \theta^1 - \omega^3{}_2 \wedge \theta^2 = \frac{\dot{S}}{cS}\theta^0 \wedge \theta^3 + \frac{u}{Sr}\theta^1 \wedge \theta^3 + \frac{\cot\theta}{Sr}\theta^2 \wedge \theta^3$$

and hence

$$\begin{aligned}
\omega^3{}_0 &= \frac{\dot{S}}{cS}\theta^3 = \frac{\dot{S}}{c}r\sin\theta\,\mathbf{d}\phi,\\
\omega^3{}_1 &= \frac{u}{Sr}\theta^3 = u\sin\theta\,\mathbf{d}\phi,\\
\omega^3{}_2 &= \frac{\cot\theta}{Sr}\theta^3 = \cos\theta\,\mathbf{d}\phi.
\end{aligned} \tag{10.112}$$

Note that equations (10.110)–(10.112) satisfy (10.108)–in fact each term separately vanishes in this equation.

Having found the connection 1-forms we now calculate the curvature 2-forms from (4.45)

$$\Omega^\mu{}_\nu = \mathbf{d}\omega^\mu{}_\nu + \omega^\mu{}_\lambda \wedge \omega^\lambda{}_\nu. \tag{10.113}$$

We have

$$\Omega^1{}_0 = \mathbf{d}\omega^1{}_0 + \omega^1{}_2 \wedge \omega^2{}_0 + \omega^1{}_3 \wedge \omega^3{}_0.$$

The second and third terms vanish and

$$\Omega^1{}_0 = \frac{\ddot{S}}{c}u^{-1}\,\mathbf{d}t \wedge \mathbf{d}r = \frac{\ddot{S}}{c^2S}\theta^0 \wedge \theta^1. \tag{10.114}$$

The Riemann curvature tensor follows from (4.39)

$$\Omega^\mu{}_\nu = {}^1\!/_2 R^\mu{}_{\nu\rho\sigma}\theta^\rho \wedge \theta^\sigma, \tag{10.115}$$

hence

$$R^1{}_{001} = \frac{\ddot{S}}{c^2S} = -R^1{}_{010} \tag{10.116}$$

and all other $R^1{}_{0\rho\sigma}$ vanish. Similarly we have

$$\Omega^2{}_0 = \mathbf{d}\omega^2{}_0 + \omega^2{}_1 \wedge \omega^1{}_0 + \omega^2{}_3 \wedge \omega^3{}_0 = \frac{r\ddot{S}}{c}\mathbf{d}t \wedge \mathbf{d}\theta = \frac{\ddot{S}}{c^2S}\theta^0 \wedge \theta^2 \tag{10.117}$$

and (10.115) gives

$$R^2{}_{002} = \frac{\ddot{S}}{c^2S} = -R^2{}_{020}. \tag{10.118}$$

The same equation holds for $R^3{}_{003}$ so we have

$$R^1{}_{010} = R^2{}_{020} = R^3{}_{030} = -\frac{\ddot{S}}{c^2 S}. \tag{10.119}$$

The remaining components of the Riemann tensor are found using, for example

$$\Omega^2{}_1 = \mathbf{d}\omega^2{}_1 + \omega^2{}_0 \wedge \omega^0{}_1 + \omega^2{}_3 \wedge \omega^3{}_1 = \left(\frac{uu'}{S^2 r} - \frac{\dot{S}^2}{c^2 S^2}\right)\theta^1 \wedge \theta^2, \tag{10.120}$$

from which

$$R^2{}_{121} = \frac{1}{S^2}\left(k + \frac{\dot{S}^2}{c^2}\right). \tag{10.121}$$

Similarly we have

$$R^3{}_{131} = R^3{}_{232} = R^2{}_{121} = \frac{1}{S^2}\left(k + \frac{\dot{S}^2}{c^2}\right). \tag{10.122}$$

From these equations the components of the Ricci tensor are easily found. For example

$$\begin{aligned} R_{00} &= R^\mu{}_{0\mu 0} = R^1{}_{010} + R^2{}_{020} + R^3{}_{030} = -3\frac{\ddot{S}}{c^2 S}, \\ R_{11} &= R^0{}_{101} + R^2{}_{121} + R^3{}_{131} = \frac{\ddot{S}}{c^2 S} + \frac{2}{S^2}\left(k + \frac{\dot{S}^2}{c^2}\right) = R_{22} = R_{33}. \end{aligned} \tag{10.124}$$

The curvature scalar may then be found (recall the metric (10.105))

$$R = \eta^{00} R_{00} + \eta^{11} R_{11} + \eta^{22} R_{22} + \eta^{33} R_{33} = -R_{00} + 3R_{11} = 6\frac{\ddot{S}}{c^2 S} + \frac{6}{S^2}\left(k + \frac{\dot{S}^2}{c^2}\right). \tag{10.125}$$

In addition it is found that

$$R_{12} = R_{13} = R_{23} = R_{01} = R_{02} = R_{03} = 0. \tag{10.126}$$

For the energy-momentum tensor we adopt the model of the *perfect fluid*. From (7.8) this is

$$T_{\mu\nu} = \frac{p}{c^2} g_{\mu\nu} + \left(\rho + \frac{p}{c^2}\right) u_\mu u_\nu. \tag{10.127}$$

In the comoving frame $u^0 = u_0 = 1$, $u^i = u_i = 0$ and $g_{\mu\nu} = \eta_{\mu\nu}$, so

$$T_{00} = \rho, \quad T_{11} = T_{22} = T_{33} = \frac{p}{c^2}, \quad T_{ij} = 0\,(i \neq j), \quad T_{0i} = 0. \tag{10.128}$$

We are now in a position to write down the field equations (10.101). The 'time-time' component

$$R_{00} - {}^1/_2 \eta_{00} R = \frac{8\pi G}{c^2} T_{00}$$

(recall the meric (10.105)!) gives, with (10.123), (10.125) and (10.128)

$$\dot{S}^2 + kc^2 = \frac{8\pi G}{3}\rho S^2 \tag{10.129}$$

and the 'space-space' component

$$R_{11} - {}^1\!/_2 \eta_{11} R = \frac{8\pi G}{c^2} T_{11}$$

gives

$$2S\ddot{S} + \dot{S}^2 + kc^2 = -\frac{8\pi G}{c^2} p\, S^2. \tag{10.130}$$

Equation (10.129) is called the *Friedmann equation*,[9] though more loosely the models obtained by using particular equations of state are also called Friedmann models – see the next section.

There is a compatibilty condition linking these last two equations. Differentiating (10.129) and substituting (for $\ddot{S}$) into (10.130) gives, after a bit of algebra

$$\dot{\rho} + 3\left(\rho + \frac{p}{c^2}\right)\frac{\dot{S}}{S} = 0. \tag{10.131}$$

This is the compatibility condition for (10.129) and (10.130). In what follows it is convenient to treat (10.129) and (10.131) as the fundamental equations. They are differential equations and will be solved below. Firstly, however, it is interesting to consider a Newtonian interpretation of them.

10.6.1 Newtonian interpretation

Consider an expanding Newtonian universe with a centre, and a galaxy of mass m at a distance $r = r_0\, S(t)$ from the centre (see Fig. 10.10). Its kinetic and potential energies are

$$\mathrm{KE} = {}^1\!/_2\, mv^2 = {}^1\!/_2\, mr_0^2\dot{S}^2, \quad \mathrm{PE} = -m\left(\frac{4\pi}{3}r^3\rho\right)\frac{G}{r} = -\frac{4\pi G}{3}m\rho r^2,$$

and conservation of energy gives

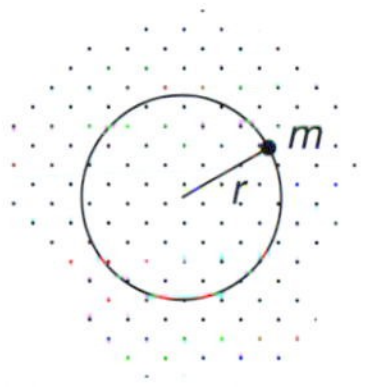

Fig. 10.10 A Newtonian universe. A galaxy of mass m is at a distance r from the centre.

[9] Friedmann (1922).

$$\tfrac{1}{2}\, m r_0^2 \dot{S}^2 - \frac{4\pi G}{3} m \rho {r_0}^2 S^2 = E,$$

$$\dot{S}^2 - \frac{8\pi G}{3}\rho S^2 = \frac{2mE}{{r_0}^2} = \text{const},$$

which is (10.129) with $kc^2 = -\dfrac{2mE}{{r_0}^2}$.

Furthermore, since the expansion of the Universe must be adiabatic, conservation of entropy gives

$$\mathrm{d}U = -p\,\mathrm{d}V$$

where U is the internal energy $= \rho c^2 V$. Hence

$$\frac{\partial}{\partial t}\left(\frac{4\pi}{3} r^3 \rho c^2\right) = -p\frac{\partial}{\partial t}\left(\frac{4\pi}{3} r^3\right)$$

or

$$\frac{\partial \rho}{\partial t} + 3\frac{\dot{S}}{S}\left(\rho + \frac{p}{c^2}\right) = 0,$$

which is (10.131). Equations (10.129) and (10.131) therefore express, in a Newtonian interpretation, conservation of energy and of entropy.

10.6.2 Critical density

Let us now investigate the consequences of Equations (10.129) and (10.131). Subtracting (10.129) from (10.130) gives

$$\ddot{S} = -\frac{4\pi G}{3}\left(\rho + 3\frac{p}{c^2}\right) S, \qquad (10.132)$$

showing that $\ddot{S}$ is *always negative*, so $\dot{S}$ is always decreasing in time – the Universe was expanding faster in the past than it is now. This is of course precisely what our Newtonian intuition would lead us to expect, and has already been mentioned earlier in this chapter. We now see, though, that the slowing expansion is actually a consequence of General Relativity. An expanding universe means that S was smaller in the past than it is now, and there must have been a time when it was 'zero' (zero, that is, in this classical model, in which masses are point masses and quantum effects are ignored). This was the initial singularity that marked the 'birth' of our Universe. Putting $t = 0$ at $S = 0$ and denoting, as before, the present age of the Universe by t_0 we see, by consulting Fig. 10.11, that since the gradient of the tangent at $t = t_0$ is $\dot{S}_0$ then from (10.76) the intercept on the time axis is ${H_0}^{-1}$, which must be *greater* than the age of the Universe. With $h = 0.55$ in (10.98) we have

$$t_0 < 1.8 \times 10^{10} \text{ years}$$

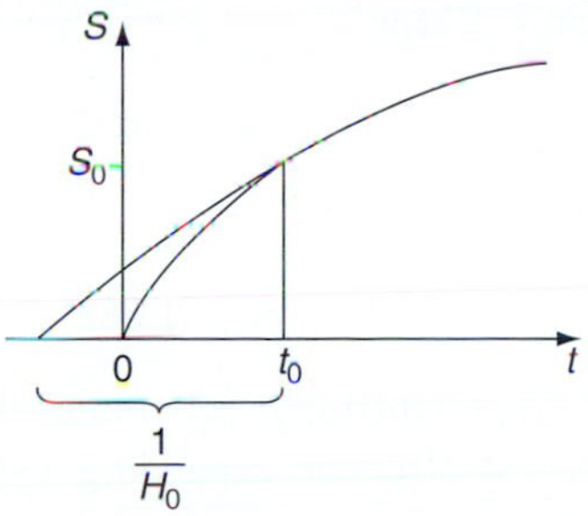

Fig. 10.11 A decelerating universe. Its age is less than the inverse of (the present value of) Hubble's constant.

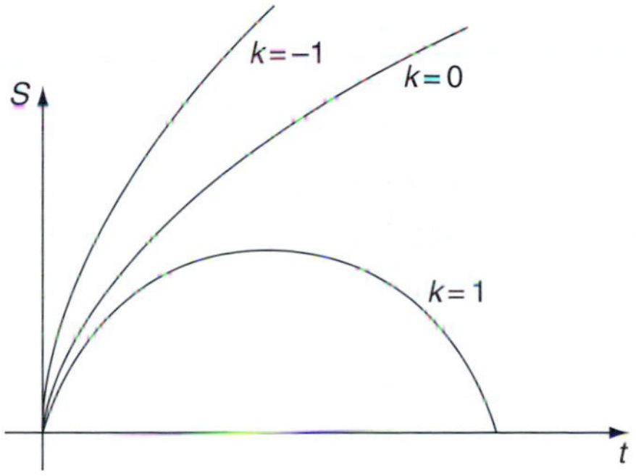

Fig. 10.12 Three types of cosmological model: $k = 1$ – expanding phase comes to an end and is followed by a contracting phase; $k = 0$ – expansion ceases when the Universe is 'infinitely dilute'; $k = -1$ – expansion never ceases – time without end.

as an upper limit. This is consistent with independent estimates of the age of the Earth (about 4.5×10^9 years) and of one of the oldest bright cluster galaxies[10]

$$t_{\mathrm{BCG}} = 13.4^{+1.4}{}_{-1.0} \times 10^9 \text{ years.}$$

From (10.129), if $k=0$ or -1 $\dot{S}^2$ is never zero, so the Universe expands for ever. If $k = 1$, however, the expansion will stop ($\dot{S} = 0$) when ρS^2 reaches the value $3kc^2/8\pi G$. Thereafter, since $\ddot{S} < 0, \dot{S}$ will become negative and the Universe will start to contract. These alternative scenarios are shown in Fig. 10.12. The fate of the Universe, then, depends on k, the sign of the spatial curvature. What is the value of k?

Let us specialise (10.129) to the present moment ($t = t_0$); then

$$kc^2 = \frac{8\pi G}{3}\rho_0 S_0{}^2 - \dot{S}_0{}^2 = \frac{8\pi G}{3} S_0{}^2 \left(\rho_0 - \frac{3H_0{}^2}{8\pi G}\right).$$

The last term in brackets, which is clearly a density, is called the *critical density* ρ_c,

$$\rho_c = \frac{3H_0{}^2}{8\pi G}. \tag{10.133}$$

[10] Ferreras *et al.* (2001).

With $H_0 = 72\,\text{km s}^{-1}\,\text{Mpc}^{-1}$,

$$\rho_c = 0.97 \times 10^{-26}\,\text{kg m}^{-3} \tag{10.134}$$

or more generally

$$\rho_c = 1.88h^2 \times 10^{-26}\,\text{kg m}^{-3}. \tag{10.135}$$

This density is important for it determines the sign of k: if $\rho > \rho_c$, k is positive and the Universe will eventually stop expanding and recontract, but if $\rho < \rho_c$, it will expand for ever – there will not be enough matter in it to halt the collapse. There is thus a connection between the curvature of sapce and the fate of the Universe; but note that these conclusions are limited to the case of a decelerating universe.

What is the *actual* density of matter (and radiation) in the Universe? The average density of *luminous* (or *baryonic*) matter – stars and galaxies – is of the order

$$\rho_b \approx 10^{-28}\,\text{kg m}^{-3}, \tag{10.136}$$

approximately one proton in every 10 cubic metres. This is only a small percentage of the critical density (10.134). There is, however, very good evidence that there is a large amount of mass in the Universe which is *not* luminous; this is called *dark matter.* Dark matter is believed to exist, for example, in galactic haloes – clouds of gas rotating around a galactic nucleus. Suppose the velocity of the gas a distance r from the centre is $v(r)$ and the total mass out to a distance r is $M(r)$. Then from Newtonian mechanics we would expect that

$$\frac{GM(r)}{r^2} = \frac{v^2(r)}{r}.$$

For gas well outside the visible part of a galaxy we would expect that $M(r)$ is roughly constant (independent of r), and we would then have $v(r) \sim r^{-1/2}$. Now the velocity can be found by measuring the Doppler shift of the 21 cm hydrogen line, and what is found is that that $v(r)$ is more or less *constant*, so $M(r)$ is not constant, but grows proportionately to r, $M(r) \sim r$. Hence there must be *invisible* matter in the galaxy, with, as it turns out, a total density much greater than that of the luminous matter. The composition of dark matter is not known, and there is speculation that at least some of it is *non-baryonic* – not made of protons and neutrons, but perhaps other 'exotic' particles. This is an important area of modern cosmology but will not be pursued further here.

A useful parameter for discussing the contributions to the energy density of material in the Universe is

$$\Omega = \frac{\rho}{\rho_c}. \tag{10.137}$$

Recent estimates for the baryonic and matter contributions (and note that, in this definition, 'matter' includes dark matter *and* luminous matter) are[11]

$$\Omega_b = 0.044 \pm 0.004, \tag{10.138}$$

[11] Spergel *et al.* (2003).

$$\Omega_{\mathrm{m}} = 0.27 \pm 0.04. \tag{10.139}$$

As is discussed below (and as the reader is doubtless already aware) there is also thermal radiation in the Universe, at a temperature $T = 2.72$ K. This gives a radiation energy density

$$\rho_{\mathrm{r}} c^2 = \frac{4\sigma}{c} T^4 = 4.16 \times 10^{-14}\,\mathrm{J\,m^{-3}}$$

in energy units, or in mass units

$$\rho_{\mathrm{r}} = 4.62 \times 10^{-31}\,\mathrm{kg\,m^{-3}}, \tag{10.140}$$

giving, for the cosmic thermal radiation,

$$\Omega_{\mathrm{r}} = 4.8 \times 10^{-5}. \tag{10.141}$$

10.7 Friedmann models and the cosmological constant

Let us make some observations about the dynamical equations (10.129) to (10.131) (or (10.132)) above:

$$\dot{S}^2 + kc^2 = \frac{8\pi G}{3} \rho S^2, \tag{10.129}$$

$$2S\ddot{S} + \dot{S}^2 + kc^2 = -\frac{8\pi G}{c^2} p\,S^2, \tag{10.130}$$

$$\dot{\rho} + 3\left(\rho + \frac{p}{c^2}\right)\frac{\dot{S}}{S} = 0. \tag{10.131}$$

These represent, as we have seen, only two independent equations for the three unknowns, ρ, p and S. A complete solution is possible if one or more relations between these quantities – for example an equation of state, relating ρ and p – is known. Models constructed in this way are called *Friedmann models*, after the Russian mathematician and physicist. There are two special cases of interest: dust, modelling ordinary matter, and radiation.

The dust model is a *zero pressure* model. We put $p = 0$ in (10.132), giving

$$\ddot{S} = -\frac{4\pi G}{3} \rho S. \tag{10.142}$$

Differentiating (10.129) gives

$$2\dot{S}\ddot{S} = \frac{8\pi G}{3} (2\rho S\dot{S} + S^2\dot{\rho}), \tag{10.143}$$

which, on substituting (10.142), gives

$$[\text{dust}] \quad \frac{\dot{\rho}}{\rho} = -3\frac{\dot{S}}{S} \Rightarrow \rho S^3 = \text{const} = \rho_0 {S_0}^3. \tag{10.144}$$

This is easy to understand: it is simply the consequence of conservation of mass for non-relativistic matter in an expanding space; $m = \rho V = \text{const}$. This describes a *matter-dominated universe*. Hardly surprisingly, it is a universe in which q, the deceleration parameter, is positive, since from (10.142) we have, for the present value of q

$$q_0 = -\frac{\ddot{S}_0 S_0}{\dot{S}_0^{\,2}} = \frac{4\pi G}{3}\rho_0 \frac{S_0^{\,2}}{\dot{S}_0^{\,2}} = \frac{4\pi G}{3}\frac{\rho_0}{\mathrm{H}_0^{\,2}} > 0.$$

The second case of interest is the *radiation gas model*, in which[12]

$$\rho = \frac{3p}{c^2}. \tag{10.145}$$

Then (10.132) gives

$$\ddot{S} = -\frac{8\pi G}{3}\rho S.$$

Substituting this in (10.143) gives

$$[\text{radiation}] \quad \frac{\dot{\rho}}{\rho} = -4\frac{\dot{S}}{S} \Rightarrow \rho S^4 = \text{const} = \rho_0 S_0^{\,4}. \tag{10.146}$$

This is to be compared with (10.144) for (non-relativistic) matter. To gain an understanding of the extra power of S in the radiation case we could argue that if a box of photons, of volume V, is expanded, the volume of the box increases by a factor S^3, but in addition the energy of each photon $E = h\nu = ch/\lambda$ decreases by a factor $1/S$, since $\lambda \to S\lambda$ under an expansion of space, so the energy *density* decreases by $1/S^4$.

In a universe, therefore, containing matter *and* radiation (though decoupled), their respective densities will behave, under expansion, as

$$\rho_{\mathrm{m}} \propto \frac{1}{S^3}, \quad \rho_{\mathrm{r}} \propto \frac{1}{S^4}, \quad \frac{\rho_{\mathrm{r}}}{\rho_{\mathrm{m}}} \propto \frac{1}{S(t)}. \tag{10.147}$$

This is sketched in Fig. 10.13 (notionally a logarithmic plot). If there is *any* radiation in the Universe at the present time, it must have dominated at early times. The *present* densities of matter and radiation are given by (10.139) and (10.141), so the present Universe is

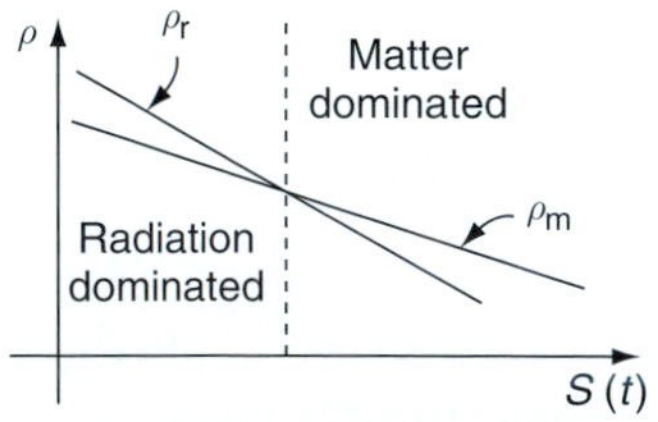

Fig. 10.13 **The density of radiation decreases faster than the density of matter in an expanding universe.**

[12] See for example Feynman *et al.* (1963), Section 39–3.

obviously matter-dominated, but at early enough times it must have been radiation-dominated. This assumption is a key ingredient in studying the physics of the early Universe.

Next, let us consider the question of *static solutions*, thought to be relevant before Hubble's discovery that the Universe is expanding. In a static universe $\dot{S} = \ddot{S} = 0$ and (10.129) and (10.130) then give

$$\rho = -\frac{3p}{c^2}; \tag{10.148}$$

the Universe is occupied by a fluid with negative pressure (if positive energy density). Dismissing this as unphysical we may conclude that the equations have *no static solutions*. For Einstein in 1917 this was a problem, which he addressed by introducing an extra term into the field equations,[13] so that they now read (in modern notation – Einstein used λ rather than Λ)

$$R_{\mu\nu} - {}^1\!/\!{}_2 g_{\mu\nu} R - g_{\mu\nu}\Lambda = \frac{8\pi G}{c^2} T_{\mu\nu}, \tag{10.149}$$

and Λ is called the *cosmological constant*. Equations (10.129) and (10.130) then become

$$\dot{S}^2 + kc^2 = \frac{8\pi G}{3}\rho S^2 + \frac{c^2}{3}\Lambda S^2, \tag{10.150}$$

$$2S\ddot{S} + \dot{S}^2 + kc^2 = -\frac{8\pi G}{c^2} p S^2 + c^2 \Lambda S^2. \tag{10.151}$$

Equation (10.131) is unchanged.

A static universe ($\dot{S} = \ddot{S} = 0$) with $p = 0$ (dust model) now follows if

$$k = \Lambda S^2, \quad \Lambda = \frac{4\pi G}{c^2}\rho. \tag{10.152}$$

This is a closed, static universe, the so-called *Einstein universe*, shown in Fig. 10.14. It is clear that a positive Λ must represent a 'repulsive force' to balance the gravitational attraction of matter. To see this, consider an empty flat universe, $\rho = k = 0$. From (10.150)

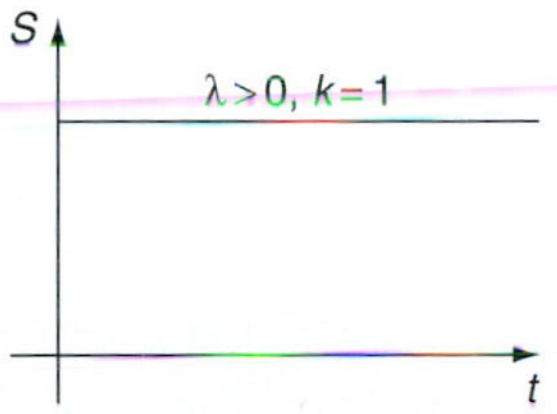

Fig. 10.14 **Einstein universe.**

[13] Einstein (1917).

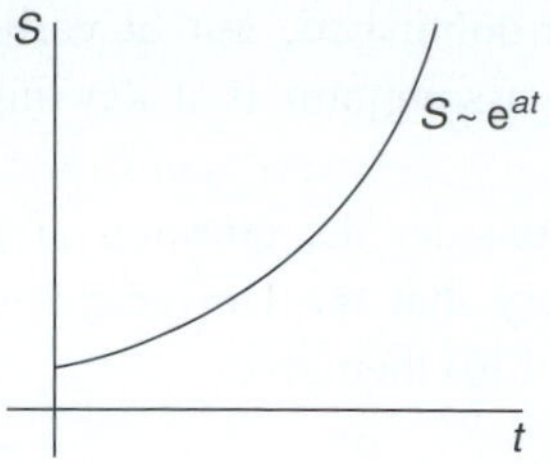

Fig. 10.15 **de Sitter universe.**

$$\dot{S}^2 = \frac{c^2}{3}\Lambda S^2, \quad S(t) = S(0)\exp\left\{\left(\frac{c^2\Lambda}{3}\right)^{1/2} t\right\}. \tag{10.153}$$

This is the *de Sitter universe*, shown in Fig. 10.15. It describes an empty but expanding space. The space expands by 'cosmic repulsion'. This is simlar to the Steady State Theory, except that here $\rho = 0$. In the de Sitter universe the deceleration parameter is negative

$$q = -\frac{\ddot{S}S}{\dot{S}^2} = -1,$$

indicating an accelerating expansion, as observed in our Universe. It is clear then, that this feature of the Universe will be accounted for, in the language of General Relativity, by a positive cosmological constant Λ.

The introduction of the cosmological constant by Einstein, however, was not entirely trouble-free. If we put $\dot{S} = 0$, $k = 0$ and $p = 0$ in (10.151) we find, using (10.152)

$$\ddot{S} = \frac{c^2\Lambda}{2}\mathrm{S} = -4\pi\rho S,$$

hence $\ddot{S} < 0$ and the solution is *unstable*. This, together with Hubble's later discovery that the Universe is expanding after all, led Einstein to abandon the Λ term, calling its introduction the 'greatest blunder' of his scientific life. As seen above, though, the fact that the expansion of the Universe is now seen to be accelerating indicates that there is, after all, a place for Λ, so Einstein's blunder might turn out to be not such a bad idea after all!

Introducing Λ into the picture, however, considerably complicates matters; there are many types of solutions to Equations (10.150) and (10.151), depending on the sign and magnitude of k and Λ. Rather than attempt a general survey we shall consider simply a particular case, which however does seem to stand a reasonable chance of being correct. This is the case in which $k = 0$; the Universe is *flat*. There is in fact evidence for this from the now rather detailed observations of the cosmic background radiation, a topic to be discussed in the next section. The relevant observations are of the *anisotropy* of this radiation. The radiation reaching us now was last scattered some hundreds of thousands of years after the beginning of the universal expansion; and already at that time there were *inhomogeneities* in the cosmic dust, which would in the course of time result in the condensation of dust into galaxies. These inhomogeneities depend on the parameters k and Λ, which govern the dynamics of the evolution – and at the same time can in effect be measured by analysing the inhomogeneities in the scattered radiation. The analysis of these is a rather complicated matter, but Hartle concludes that the evidence, as of 2003, is 'consistent with' a flat universe.[14] Making this assumption and writing (10.150) in the form

$$kc^2 = \frac{8\pi G}{3} S_0{}^2 \left(\rho_0 + \frac{c^2\Lambda}{8\pi G} - \frac{3H_0{}^2}{8\pi G} \right),$$

where the subscript 0 refers to the present time, then putting $k=0$ and $\rho = \rho_r + \rho_m$ we have, with (10.133),

$$\rho_r + \rho_m + \frac{c^2\Lambda}{8\pi G} - \rho_c = 0.$$

We write this as (see (10.137))

$$\Omega_r + \Omega_m + \Omega_\Lambda = 1 \tag{10.154}$$

where

$$\Omega_\Lambda = \frac{c^2\Lambda}{8\pi G} \cdot \frac{1}{\rho_c} = \frac{c^2\Lambda}{3H_0{}^2} \tag{10.155}$$

The term Ω_Λ may be said to represent the contribution of the cosmological constant term to the energy density of the Universe. With Ω_r and Ω_m taking on the values (10.140) and (10.139) it is clear that Ω_Λ provides the dominant contribution (≈ 0.7) to the dynamics of the Universe. The physical nature of this contribution is still unclear, but it goes by the name of *dark energy*.

10.8 Cosmic background radiation

The discovery, in 1965, that the Universe contains thermal radiation, was a landmark in twentieth-century cosmology, equal in importance to Hubble's discovery that the Universe is expanding. This cosmic microwave background (CMB) radiation was discovered (accidentally!) by Arno Penzias and Robert Wilson,[15] who were awarded the 1978 Nobel prize in Physics. Over the years it has been confirmed that this radiation is almost perfectly isotropic and fits a blackbody distribution with a temperature of 2.72 K. The most accurate measurements have been obtained from the COBE satellite (Cosmic Background Explorer) launched in 1989.

Thermal radiation, resulting from a hot beginning of the Universe, had already been predicted by Gamow in 1946 and developed in key papers by Alpher, Bethe and Gamow and by Alpher and Herman in 1948.[16] The model they proposed, now accepted as the 'hot big bang' model, is that in its early moments the Universe contained matter at a high density and

[14] Hartle (2003), p. 410.
[15] Penzias & Wilson (1965).
[16] Alpher, Bethe & Gamow (1948), Alpher & Herman (1948a,b).

high temperature. This presupposes thermal equilibrium and therefore thermal (blackbody) radiation, with equilibrium between matter and radiation being maintained by reactions such as $p + n \leftrightarrow d + \gamma$, $e^+ + e^- \leftrightarrow \gamma + \gamma$ and so on. The idea is that as the Universe has expanded from this early phase the blackbody radiation *has retained its blackbody character* and simply cooled. To demonstrate the feasibility of this model we need to show that it is actually true – that in our model for the expansion of the Universe, blackbody radiation does indeed retain its thermal character as space expands. This is what we shall now demonstrate.

First, note that the relation between density and temperature of blackbody radiation is $\rho = aT^4$ (where $a = 4\sigma/c$ and σ is the Stefan–Boltzmann constant). Then noting from (10.146) that $\rho \propto 1/S^4$ it follows that the variation of T with S is

$$T \propto \frac{1}{S}. \tag{10.156}$$

We shall now show that this is precisely the condition for blackbody radiation to retain its thermal character under space expansion – for the Planck spectrum to be preserved. The proof is based on conservation of photon number. In the Planck distribution the number $\mathrm{d}n(t)$ of photons with frequency between ν and $\nu + \mathrm{d}\nu$ in a volume $V(t_1)$ of space at time t_1 is

$$\mathrm{d}n(t_1) = \frac{8\pi\nu^2 V(t_1)\,\mathrm{d}\nu}{c^3\left[\exp\left(\dfrac{h\nu}{kT(t_1)}\right) - 1\right]}. \tag{10.157}$$

At a later time t_2 the frequency has been red-shifted to (see (10.78))

$$\nu_2 = \nu_1 \cdot \frac{S(t_1)}{S(t_2)} \tag{10.158}$$

and the volume changed to

$$V(t_2) = V(t_1) \cdot \frac{S^3(t_2)}{S^3(t_1)},$$

hence

$$\nu_1{}^2\, V(t_1)\,\mathrm{d}\nu_1 = \nu_2{}^2\, V(t_2)\,\mathrm{d}\nu_2 \tag{10.159}$$

and the – *conserved* – number of photons in the stated frequency range is, from (10.157), using (10.158)

$$\mathrm{d}n(t_2) = \frac{8\pi\nu_1^2\, V(t_1)\,\mathrm{d}\nu}{c^3\left[\exp\left(\dfrac{h\nu_2}{kT(t_2)}\right) - 1\right]}.$$

The condition that this is the same as $\mathrm{d}n(t_1)$ is

$$\frac{\nu_2}{T(t_2)} = \frac{\nu_1}{T(t_1)},$$

i.e. from (10.158)

$$T(t_2) = T(t_1) \cdot \frac{S(t_1)}{S(t_2)},$$

or

$$T(t) \propto \frac{1}{S(t)}, \tag{10.160}$$

which is exactly the same as (10.156). Hence conservation of photon number guarantees that the blackbody nature of the radiation is preserved under space expansion. Finally it is useful to recall that the energy density of the radiation is, from (10.157),

$$\rho_r(t) = \int \frac{h\nu}{V(t)}\, dn(t) = \frac{8\pi h}{c^3} \int_0^\infty \frac{\nu^3\, d\nu}{\exp\left(\frac{h\nu}{kT(t)}\right) - 1} = \frac{8\pi^5 k^4}{15 c^3 h^3} T^4(t) = a\, T^4(t), \tag{10.161}$$

where $a = 4\sigma/c = 7.56 \times 10^{-16}\,\mathrm{J\,m^{-3}\,K^{-4}}$, σ is the Stefan–Boltzmann constant, and we have used the well-known integral[17]

$$\mathrm{I} = \int_0^\infty \frac{z^3\, dz}{e^z - 1} = \frac{\pi^4}{15}. \tag{10.162}$$

10.9 Brief sketch of the early Universe

It seems that almost everyone in the world knows about the Big Bang theory; that the Universe began in a hot and very dense phase, from which it has expanded, and is still expanding, some thousands of millions of years later. It is indeed a most remarkable achievement of twentieth-century physics; that by using the laws of nature, formulated here on Earth in the last three hundred years or so, we are able to give such a detailed account of the first few *minutes* of the Universe. In particular, we are able to 'predict' helium synthesis during these early moments, with very good numerical agreement with the amount of primordial helium found in the Sun and the interstellar medium. Amongst other things, this must imply that the laws of nature have *not changed with time*. This is surely a remarkable fact; the laws we understand at present are something like the absolute truth. In my opinion the Big Bang account of the birth of the Universe is so remarkable that every educated person should be aware of it, and for this reason the present section is devoted to a broad overview of its main features. For more detailed accounts the reader is referred elsewhere.

We have seen (see Fig. 10.13) that if there is any radiation present in the Universe now, in the early moments its contribution to the dynamics of its evolution must have dominated that

[17] See for example Mandl (1988), Appendix A1.

of ordinary matter. And indeed there is thermal radiation in the Universe so we can proceed to investigate Equation (10.129),

$$\dot{S}^2 + kc^2 = \frac{8\pi G}{3}\rho S^2 \tag{10.129}$$

with $\rho \propto 1/S^4$, as in (10.146). It is clear that as we go backwards in time S becomes smaller and smaller, the right hand side of (10.129) becomes larger and larger, so that as $S \to 0$,

$$\frac{8\pi G}{3}\rho S^2 \gg kc^2$$

and (10.129) reduces to

$$\dot{S}^2 = \frac{8\pi G}{3}\rho S^2. \tag{10.163}$$

We see that the *geometry* of space-time is *irrelevant* in the early Universe.

We should note that ρ properly stands for the density of *all relativistic* particles, not only photons ('relativistic' means $E \gg mc^2$). For example, at a temperature of $T = 10^{10}$ K, $kT = 8.7 \times 10^5$ eV ≈ 1 MeV. But the rest mass of an electron is $m_e c^2 = 0.5$ MeV, so electrons are effectively relativistic at this temperature, as are also electron neutrinos ($mc^2 < 3$ eV) and muon neutrinos ($mc^2 < 0.2$ MeV), as well as the antiparticles of all these particles. We note that electrons and neutrinos are fermions, so including their contribution to ρ, the total (mass) density, means that instead of (10.161) we should write

$$\rho = N_{\text{eff}} \cdot \frac{a}{c^2} T^4, \tag{10.164}$$

where N_{eff} is the effective number of relativistic species of particle at a temperature T. It is not the actual number, since fermions and bosons contribute differently.

For electrons, which obey Fermi–Dirac statistics, the energy density per polarisation state is[18]

$$\rho_{\text{f}} = \frac{1}{8\pi^3}\int E(\mathbf{k}) f(E(\mathbf{k}))\, \mathrm{d}\mathbf{k},$$

where $E(\mathbf{k})$ is the energy of an electron with wave vector $\mathbf{k}$ and $f(E(\mathbf{k}))$ is the Fermi function

$$f(E(\mathbf{k})) = \frac{1}{\mathrm{e}^{E/kT} + 1} = [\exp(E/kT) + 1]^{-1}$$

(where the chemical potential has been put equal to 0). Performing the angular integrations

$$\rho_{\text{f}} = \frac{1}{2\pi^2}\int Ek^2[\exp(E/kT) + 1]^{-1}\, \mathrm{d}k.$$

Now $k = (2\pi/h)p$ and in the *relativistic* regime $E = pc$, so

[18] See for example Ashcroft & Mermin (1976), p. 43. Note that the factor $1/4\pi^3$ in Ashcroft & Mermin refers to *two* polarisation states.

$$\rho_{\mathrm{f}} = \frac{4\pi c}{h^3} \int_0^\infty \frac{p^3\,\mathrm{d}p}{\exp(pc/kT) + 1} = \frac{4\pi k^4}{h^3 c^3} \mathrm{I}_1 \cdot T^4 \tag{10.165}$$

with

$$\mathrm{I}_1 = \int_0^\infty \frac{z^3\,\mathrm{d}z}{\mathrm{e}^z + 1}. \tag{10.166}$$

By noting that

$$\frac{2}{\mathrm{e}^{2z} - 1} = \frac{1}{\mathrm{e}^z - 1} - \frac{1}{\mathrm{e}^z + 1}$$

it is easy to show that

$$\mathrm{I}_1 = \frac{7}{8}\mathrm{I} = \frac{7}{120}\pi^4$$

from (10.162), so the energy density per polarisation state of a relativistic fermion is

$$\rho_{\mathrm{f}} = \frac{7}{16}\sigma T^4, \tag{10.167}$$

to be compared with (10.161) for photons.

To return to the early Universe, at a temperature of about 10^{10} K the electron and the positron and the electron and muon neutrinos and their antiparticles will all be relativistic:

$$\mathrm{e}^-, \mathrm{e}^+, \nu_\mathrm{e}, \overline{\nu}_\mathrm{e}, \nu_\mu, \overline{\nu}_\mu.$$

In the usual theory of beta decay, the observed parity violation is ascribed to the fact that ν_e is massless and exists in only one polarisation state (a left-handed state). By an extension of ideas the same is supposed to be true of ν_μ; whereas by contrast e^- and e^+ have 2 (= $2s + 1$) polarisation states so the effective number of relativistic particles is

$$N_{\mathrm{eff}} = 1 + [2 \times 2 + 4 \times 1] \cdot \frac{7}{16} = \frac{9}{2} = 4.5. \tag{10.168}$$

In recent years, however, the theory of *neutrino oscillations* has received much attention since it offers a solution to the solar neutrino problem. The general idea of neutrino oscillations is that while ν_e are emitted from the Sun, for example as products of boron decay $^8\mathrm{B} \rightarrow {}^8\mathrm{Be} + \mathrm{e}^+ + \nu_\mathrm{e}$, by the time these neutrinos reach the Earth they have become a mixture of ν_e, ν_μ and ν_τ; and only the electron neutrinos will be detectable in the laboratory, resulting in a reduced detection rate; this smaller than expected detection rate *is* the solar neutrino problem, which is therefore solved in this theory of oscillations. The theory, however, only works if the electron and muon neutrinos have different masses (the effect depends on the difference in their $(\mathrm{mass})^2$), so they cannot both be massless. Assuming, for the sake of definiteness, that neither of them is massless, they will possess two polarisation states each and instead of (10.167) we will have

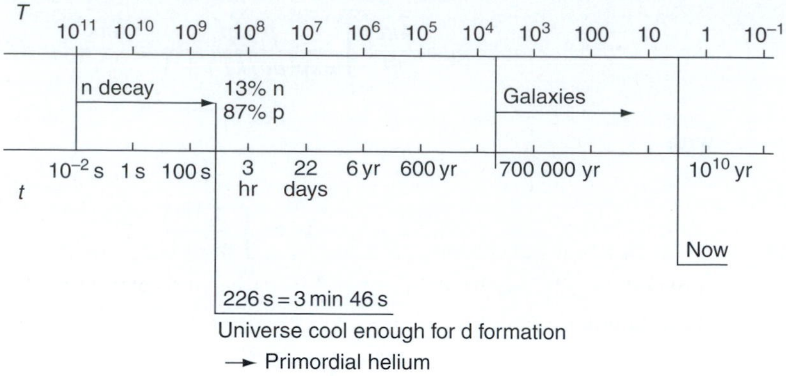

Fig. 10.16 **Thermal history of the Universe.**

$$N_{\text{eff}} = 1 + [2 \times 2 + 4 \times 2] \cdot \frac{7}{16} = \frac{25}{4} = 6.25. \tag{10.169}$$

In any case, substituting (10.163) into (10.164) gives

$$\frac{1}{S^2}\left(\frac{\mathrm{d}S}{\mathrm{d}t}\right)^2 = \frac{8\pi Ga}{3c^2} \cdot N_{\text{eff}} \cdot T^4 \equiv b^2 T^4. \tag{10.170}$$

We want an equation relating T and t. Equation (10.160), however, implies that

$$\frac{1}{S^2}\left(\frac{\mathrm{d}S}{\mathrm{d}t}\right)^2 = \frac{1}{T^2}\left(\frac{\mathrm{d}T}{\mathrm{d}t}\right)^2,$$

so (10.170) gives

$$\left(\frac{\mathrm{d}T}{\mathrm{d}t}\right)^2 = b^2 T^6 \Rightarrow \frac{\mathrm{d}T}{\mathrm{d}t} = -bT^3, \tag{10.171}$$

where the minus sign is chosen since T decreases as t increases. Integrating (10.171) gives

$$T\sqrt{t} = \frac{1}{\sqrt{2b}} = 1.52\, N_{\text{eff}}^{-1/4} \times 10^{10}\,\text{s}^{1/2}\,\text{K}.$$

With $N_{\text{eff}} = 4.5$ this gives

$$T\sqrt{t} \sim 1.04 \times 10^{10}\,\text{s}^{1/2}\,\text{K},$$

or

$$T\sqrt{t} \approx 10^{10}. \tag{10.172}$$

This relates the temperature of the radiation in K to the age of the Universe in seconds, and it crucially allows us to construct a *thermal history of the Universe*, at least in its early stages.

Equation (10.172) is displayed in Fig. 10.16, where both T and t are plotted horizontally on a logarithmic scale. When the universe is 1 second old the temperature is 10^{10} degrees, when $t = 100$ s the temperature has dropped to 10^9 K and so on. It is interesting to see that if one inserts the present age, $t \sim 10^{18}$ s, into (10.172) one finds $T = 10$ K, not too far from the actual value of 2.7 K. But (10.172) was derived for a *radiation-dominated universe*, which the present Universe is *not*, so there is no reason to expect (10.172) to work all that well for the Universe at the present time.

What was the Universe like at early times? If we take a journey backwards in time we see the Universe contract and heat up. When T reaches about 10^4 K, kT is about 1 eV, of the order of atomic ionisation energies, so at this temperature matter, composed of atoms and molecules, will lose its structure and be reduced to atomic nuclei and electrons. When T reaches 10^{10} degrees the thermal energy of the order of MeV is enough to dissociate nuclei, so above this temperature the Universe will consist of protons, neutron, photons, muons, electrons and neutrinos (and antiparticles of these last three). All structure will have disappeared. The Universe is a simple mixture of these particles at a very high temperature. Coming forwards in time, when T drops to about 1000 K the plasma will condense into electrically neutral atoms, so electromagnetic forces will disappear on the large scale and gravity will take over, allowing for galactic condensation, and, in due course, condensation into stars, in which nuclear fusion reactions take place, to produce the chemical elements. The first few of these reactions are

$$\mathrm{p} + \mathrm{n} \rightarrow \mathrm{d} + \gamma, \quad \mathrm{d} + \mathrm{n} \rightarrow {}^3\mathrm{H} + \gamma, \quad {}^3\mathrm{H} + \mathrm{p} \rightarrow {}^4\mathrm{He} + \gamma, \tag{10.173}$$

and these result in helium production. Heavier nuclei are produced in analogous reactions. Given our picture of the early Universe, however, since it consists of protons and neutrons an obvious question is: did the reactions above take place then, producing helium? And if so, how much helium? And were heavier elements produced? Let us consider the first of these questions.[19]

When $t = 0.01$ s, $T = 10^{11}$ K and at that temperature the number of protons is the same as the number of neutrons, $N(\mathrm{p}) = N(\mathrm{n})$. The argument goes as follows. The following reactions convert n into p and vice versa:

$$\mathrm{n} + \nu_e \leftrightarrow \mathrm{p} + \mathrm{e}^-, \quad \mathrm{n} \leftrightarrow \mathrm{p} + \mathrm{e}^- + \overline{\nu}_e,$$

so under conditions of thermal equilibrium the relative population of proton and neutron states will obey a Maxwell–Boltzmann distribution

$$\frac{N(\mathrm{n})}{N(\mathrm{p})} = \exp(-\Delta E/kT), \tag{10.174}$$

where $\Delta E = [m(\mathrm{n}) - m(\mathrm{p})]c^2 = 1.3$ MeV. When $T = 10^{11}$ K, however, $kT = 8.6$ MeV and the above ratio is very close to 1, so at an age of 1/100 second the proton/neutron balance is 50% n, 50% p. Thereafter however, as T drops, an equal population of proton and neutron levels is not guaranteed and the neutrons will start to decay

$$\mathrm{n} \rightarrow \mathrm{p} + \mathrm{e}^- + \overline{\nu}_e. \tag{10.175}$$

[19] I must emphasise once more that this account is greatly simplified, since I am aiming only to give the broadest of outlines of this topic. Much more detail will be found in the texts cited in ‘Further reading’.

The question we are asking is whether the fusion reactions (10.173) will take place in the environment of the early Universe. The answer is that they will not produce stable helium in a hot universe, for the photons will still have enough energy to destroy the deuterium as fast as it is created

$$\mathrm{p} + \mathrm{n} \leftrightarrow \mathrm{d} + \gamma.$$

We have to wait until the photons are not energetic enough for this reaction to go 'backwards'. This energy is easily measured in a laboratory and corresponds to a temperature of about 0.7×10^9 K, which in turn corresponds to a time of 226 seconds. After this time, the 'deuterium bottleneck' has been passed and the further reactions in (10.173) will proceed, producing ^{4}He. During these 226 seconds, however, the neutrons have been decaying and the 50% p, 50% n mixture at $t = 0.01$ seconds has now become, at 226 seconds, 87% p, 13% n. The fusion reactions which then go ahead produce $2 \times 13 = 26\%$ ^{4}He and 74% H; and it is *this* mixture of hydrogen and helium (with a small admixture of heavier elements) which, hundreds of thousands of years later, will constitute the gas forming the raw material of stars and the interstellar medium.

This, in the broadest of terms, is how the 'standard model' of Big Bang Cosmology predicts the primordial synthesis of helium (as well as some heavier elements). The agreement between the theoretical prediction and the observed data is impressive.[20] For example, in HII regions of the interstellar medium[21] the proportion of ^{4}He to all nucleonic matter is predicted to be $Y \approx 0.23$ and is found[22] to be

$$Y = 0.231 \pm 0.006.$$

It should be remarked that one surprising feature of this model is its *simplicity*. It depends on a homogeneous universe, Einstein's field equations applied to a radiation gas, and the well-established fields of thermodynamics and nuclear physics. No new laws of physics have had to be invented. This is surely a remarkable fact – that we can account for the evolution of the Universe from the time when it was only 1/100 second old! From a rather different perspective, what is also remarkable is that the original Big Bang model of Alpher and Herman was for a long time not taken seriously. In fact it was not until the 1960s, when quasars and pulsars were discovered and when in particular John Wheeler (1911–2008) began to urge the theoretical community to take seriously the work of Oppenheimer and his collaborators on gravitational collapse and, in the words of Freeman Dyson, to 'rejuvenate General Relativity; he made it an experimental subject and took it away from the mathematicians',[23] that physicists seriously began to take on board the idea that there actually *was* a time when the Universe was only a few minutes old. In the words of Weinberg:[24]

> This is often the way it is in physics – our mistake is not that we take our theories too seriously, but that we do not take them seriously enough. It is always hard to realise that these numbers and equations we play with at our desks have something to do with the real

[20] For a recent review, see for example Hogan (1997).
[21] Regions of ionised hydrogen: see for example Carroll & Ostlie (1996).
[22] Skillman & Kennicutt (1993).
[23] *New York Times*, 14 April 2008.
[24] Weinberg (1978), p. 128.

world. Even worse, there often seems to be a general agreement that certain phenomena are just not fit subjects for respectable theoretical and experimental effort. Gamow, Alpher and Herman deserve tremendous credit above all to take the early universe seriously, for working out what known physical laws have to say about the first three minutes. Yet even they did not take the final step, to convince the radio astronomers that they ought to look for a microwave radiation background. The most important thing accomplished by the ultimate discovery of the 3 °K radiation background in 1965 was to force us all to take seriously the idea that there *was* an early universe.

10.10 The inflationary universe and the Higgs mechanism

Despite the outstanding success of the Big Bang model there are still some questions left unanswered. Prominent among these is the high degree of *isotropy* of the cosmic background radiation. The temperature (2.72 K) is the same in all directions – including diametrically opposite ones. This creates a difficulty because of the horizon problem, described above. The radiation reaching our apparatus has travelled freely throught the Universe since the recombination era, at about $t = 130\,000$ years. (This is a recombination of protons and electrons into hydrogen, and happens when the radiation is at a temperature T such that kT is rather less than the ionisation energy of hydrogen (13.6 eV), since even at that energy the blackbody spectrum means that in the Wien tail (the high energy end of the spectrum) there will still be enough photons to ionise hydrogen.) The radiation was *thermalised*, however, much earlier, during the first few seconds. Referring to Fig. 10.7, consider a journey backwards in time. As we approach the Big Bang we see that the horizons of a given world-line shrink – other bodies in space become progressively invisible as the light from them has not yet had time to reach us. At the present time we are able to see the light from very many galaxies, but at early times these galaxies – particularly ones at 'opposite ends' of the Universe – must have been definitely *not* within each other's light cone. On the other hand the cosmic background radiation from these directions is at the *same* temperature, which surely argues that these regions must have been in causal contact at very early times, and therefore *within* each other's light cones.

This is known as the *horizon problem*. A solution was proposed by Guth in 1981:[25] all the points in space – even those in diametrically opposite directions – from which we are now receiving light *were* actually *within* each other's light cones at very early times when thermal equilibrium held, and since that time space has expanded by such an enormous factor that these regions now make up the whole of the visible Universe. The situation is sketched in Fig. 10.17. Let us calculate what expansion factor is necessary for this purpose. Guth's argument was developed in the context of particle physics and the reasoning is explained in some detail below. For reasons to be explained there, the expansion is reckoned to have begun at a time related to the 'Grand Unification' energy

$$E_{\mathrm{GUT}} \approx 10^{14}\ \mathrm{GeV}, \tag{10.176}$$

[25] Guth (1981).

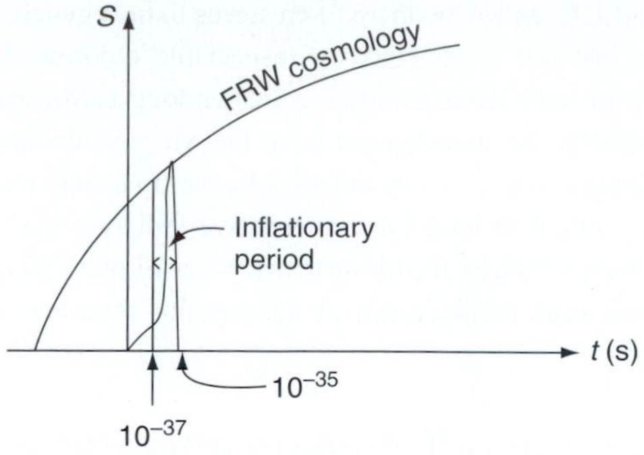

Fig. 10.17 The inflationary phase of expansion, starting at about 10^{-37} seconds and finishing at about 10^{-35} seconds.

which corresponds to a temperature of $T_{\mathrm{GUT}} \approx 10^{27}$ K and therefore to a time, from (10.172)

$$t_{\mathrm{GUT}} \approx 10^{-34}\ \mathrm{s}. \tag{10.177}$$

At this very early time the size of a causally connected region of space is $\sim ct_{\mathrm{GUT}} \approx 10^{-26}$ m. If the expansion of the Universe since that early time had simply followed a Friedmann–Robertson–Walker pattern for a radiation-dominated universe, the ratio of the present value of the expansion factor S_0 to its value at the GUT time, S_{GUT}, is, from (10.156) and (10.172) and noting that the present age of the Universe $\approx 10^{17}$ s,

$$\frac{S_0}{S_{\mathrm{GUT}}} \approx \frac{T_{\mathrm{GUT}}}{T_0} \approx \left(\frac{t_0}{t_{\mathrm{GUT}}}\right)^{1/2} \approx \left(\frac{10^{17}}{10^{-34}}\right)^{1/2} \approx 10^{26}. \tag{10.178}$$

Under this expansion the causally connected region, originally of size $\sim 10^{-26}$ m would now be of size ~ 1 m. Guth's hypothesis, however, is that it should have expanded to the size of the visible Universe, that is to $\sim c \times 10^{17} \approx 10^{26}$ m, which represents a *further* expansion by a factor $\sim 10^{26}$:

$$\text{inflationary factor} \approx 10^{26}. \tag{10.179}$$

Guth's model, known as the *inflationary universe*, is that for a very short time, beginning at t_{GUT} above, the Universe suffered this incredibly large expansion. Exactly how long the expansion lasted (perhaps $100 \times t_{\mathrm{GUT}}$) and, crucially, how the inflationary period *ended* (the so-called 'graceful exit' problem) are more detailed matters which we shall not consider here. But the general idea is that after a (very) short time the expansion of space settled down to the extremely sedate expansion of the Friedmann–Robertson–Walker cosmology that we see now.

The physical motivation for the inflationary model comes from particle physics, in particular the notions of *spontaneous symmetry breakdown* and the *Higgs mechanism*. The key development was the theory of unification of the electromagnetic and weak interactions, into the so-called electroweak interaction. The prototype weak interaction is neutron beta decay

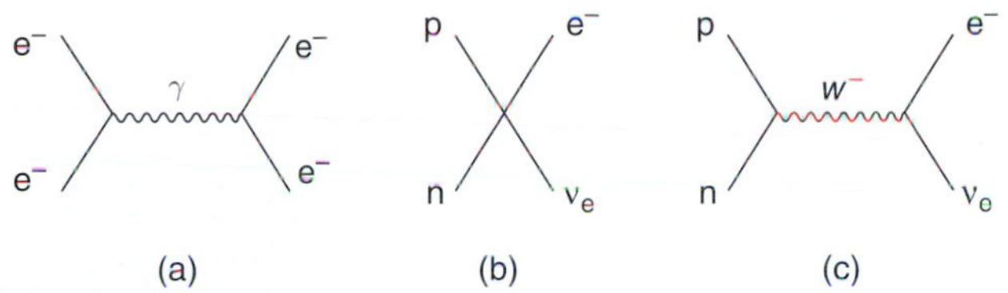

Fig. 10.18 Feynman diagrams for (a) electron–electron interaction via photon exchange, (b) $n + \nu_e \rightarrow p + e^-$ in Fermi's theory, with no field, (c) the same reaction mediated via W exchange.

$$n \rightarrow p + e^- + \overline{\nu}_e.$$

The (quantum field theoretic) amplitude for this decay is essentially the same as that for the related scattering process

$$\nu_e + n \rightarrow e^- + p. \tag{10.180}$$

We may set this alongside the electromagnetic process

$$e^- + e^- \rightarrow e^- + e^- \tag{10.181}$$

of electron–electron scattering. Electrons interact with each other through the electromagnetic field, and, at the quantum level, the basic interaction is represented by the Feynman diagram in Fig. 10.18(a) – two electrons exchange one photon, a quantum of the electromagnetic field.

According to Fermi's theory of weak interactions, on the other hand, the Feynman diagram for the process (10.180) is drawn in Fig. 10.18(b). This is a *contact interaction*. The wave functions for the four particles ν_e, n, e^- and p must all overlap at one point. There is no action-at-a-distance, therefore no need for a field, therefore no field responsible for this weak interaction. Fermi's theory, however, suffered from convergence problems at higher orders in perturbation theory (basically it is 'non-renormalisable'), but these problems are made much less severe if it is supposed that the weak interaction, like the electromagnetic one, is mediated by a *field*. In that case the Feynman diagram of Fig. 10.18(b) is replaced by Fig. 10.18(c); here W^- is the quantum of the weak field. This is the basic motivation behind electroweak theory, and in that theory there are actually four field quanta: γ, W^-, W^+ (its antiparticle) and Z^0. The photon is massless but the others have masses,

$$m_W = 81.8 \text{ GeV}/c^2 = 87\, m_p, \quad m_Z = 92.6 \text{ GeV}/c^2 = 99\, m_p \tag{10.182}$$

(m_p = proton mass). The W and Z particles were predicted at these masses, and found at CERN in 1983.

It is the fact that the masses of W and Z are non-zero that creates the problem, for electroweak theory is in essence a generalisation of Maxwell's electrodynamics, from a gauge group U(1) to the non-abelian group $SU(2) \otimes U(1)$; and in that generalisation the new gauge fields would, like the photon, all have zero mass. The mechanism which allows the W and Z fields to be massive (but γ to remain massless) is known as the *Higgs mechanism*.

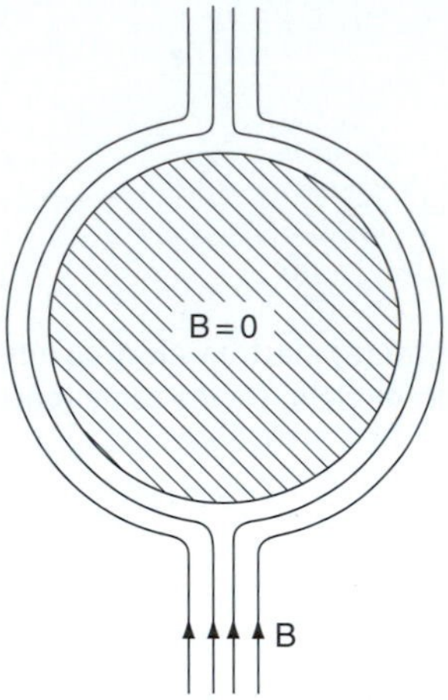

Fig. 10.19 The Meissner effect. Magnetic flux is expelled from a superconductor.

Higgs' work[26] concerned the masses of gauge particles in theories with so-called spontaneously broken symmetry, and a good illustration of this, one in which Higgs was himself interested, is the phenomenon of superconductivity.

The defining characteristic of superconductivity is that at a temperature below a *critical temperature* T_c, some metals lose all electrical resistance, R=0: the resistance is not simply very small, it is *zero*! R is defined as being the proportionality between the electric field **E** and electric current **j**:

$$\mathbf{E} = R\,\mathbf{j}, \tag{10.183}$$

or equivalently

$$\mathbf{j} = \sigma\mathbf{E} \tag{10.184}$$

(where σ is the conductivity). A metal in a superconducting state then exhibits a persistent current even in no field: $\mathbf{j} \neq 0$ when $\mathbf{E} = 0$. The key to understanding superconductivity is to describe the current as a 'supercurrent' $\mathbf{j}_s$, and, in contrast to the equation above, to suppose that this is proportional, not to **E** but to the vector potential **A**:

$$\mathbf{j}_s = -k^2\mathbf{A}, \tag{10.185}$$

with a negative coefficient of proportionality. This is the *London equation*. In a static situation, even if $\mathbf{A} \neq 0$, $\mathbf{E} = -\partial\mathbf{A}/\partial t = 0$, so if R is defined by $\mathbf{E} = R\mathbf{j}_s$, then $R = 0$, as required.

For our purposes the crucial property of superconductors is the *Meissner effect*, which is the phenomenon that magnetic flux is expelled from superconductors, as sketched in Fig. 10.19 – an external magnetic field does not penetrate a superconductor. Higgs' contribution was to show that, suitably transformed into a relativistic theory, this is equivalent to

[26] Higgs (1964a, b, 1966).

saying that the photon has an effective mass. The reasoning goes as follows. First, the London equation explains the Meissner effect; for taking the curl of Ampère's equation

$$\nabla \times \mathbf{B} = \mathbf{j}$$

gives

$$\nabla(\nabla \cdot \mathbf{B}) - \nabla^2 \mathbf{B} = \nabla \times \mathbf{j}\,.$$

Applying this to the supercurrent (10.185) and noting that $\nabla \cdot \mathbf{B} = 0$ (no magnetic monopoles) gives

$$\nabla^2 \mathbf{B} = k^2 \mathbf{B}\,. \tag{10.186}$$

In one dimension the solution to this is

$$B(x) = B(0)\exp(-kx), \tag{10.187}$$

which describes the Meissner effect – the magnetic field is exponentially damped inside the superconductor, only penetrating to a depth of order $1/k$.

Equation (10.186) is equivalent to

$$\nabla^2 \mathbf{A} = k^2 \mathbf{A}\,. \tag{10.188}$$

This equation is, however, non-relativistic. To make it consistent with (special!) relativity ∇^2 should be replaced by the Klein–Gordon operator $\Box$ and $\mathbf{A}$ by the 4-vector $A^\mu = (\phi, \mathbf{A})$ (as in (2.85) above), giving

$$\left(-\frac{1}{c^2}\frac{\partial^2}{\partial t^2} + \frac{\partial^2}{\partial x^2} + \frac{\partial^2}{\partial y^2} + \frac{\partial^2}{\partial z^2} \right) A^\mu = k^2 A^\mu. \tag{10.189}$$

The vector potential is a 'field' quantity, but we are interested in the photon, the quantum of the field, so we make the transition to quantum theory by the usual prescription

$$\frac{\partial}{\partial t} \to -\frac{\mathrm{i}}{\hbar}E, \quad \frac{\partial}{\partial x} \to \frac{\mathrm{i}}{\hbar}p_x, \text{ etc.},$$

giving, for the quantum of the field A^μ

$$E^2 - p^2c^2 = k^2c^2\hbar^2, \tag{10.190}$$

where E is the (total, including rest-mass) energy of the field quantum and p its momentum. Comparison with Einstein's relation $E^2 - p^2c^2 = m^2c^4$ implies that the mass of the quantum in a superconductor is

$$m_\gamma = \frac{k\hbar}{c}; \tag{10.191}$$

the photon behaves as (in effect *is*) a massive particle. This is the import of the Meissner effect.

All this would seem to be a long way from cosmology and the early Universe, but the connection is made by appealing to the Bardeen–Cooper–Schrieffer (BCS) theory of

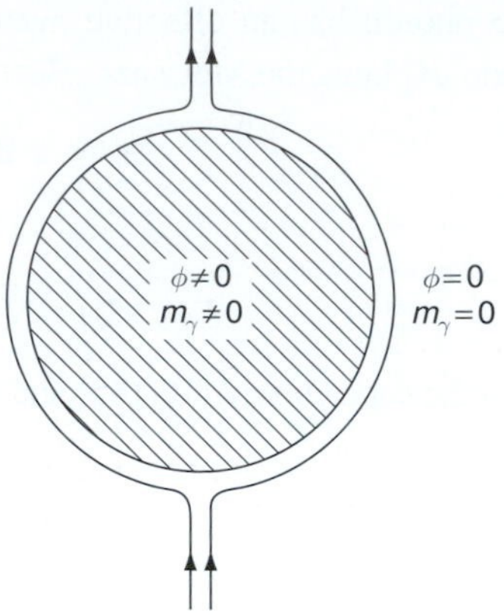

Fig. 10.20 **Photons are massive in a superconductor, where the scalar (BCS) field ϕ is non-zero, but massless outside, where there is no field.**

superconductivity,[27] which is a microscopic theory. At the quantum level the BCS theory accounts for superconductivity by positing a *scalar field* ϕ (i.e. of spin 0) which describes a 'Cooper pair' of electrons (e^-e^-); the pairing is in momentum space rather than coordinate space. Combining this idea with the previous reasoning, we now have the situation that superconductivity is described by a *many-particle wave function (or field)* ϕ, non-zero inside a superconductor, in which also $m_\gamma \neq 0$. Outside the superconductor $\phi = 0$ and $m_\gamma = 0$. This is shown in Fig. 10.20. When a superconductor is *heated* to above the critical temperature, of course it loses its superconductivity, reverting to a 'normal' state. In this transition the field $\phi \to 0$ and the photon becomes massless. This is a thermodynamic phase transition.

A few years after the work of Higgs, and in the context of elementary particle physics, Weinberg[28] and Salam[29] proposed a unified model of weak and electromagnetic interactions. In this model it necessary that the weak field quanta be massive (the reason is essentially that the weak force is of extremely short range, and is therefore carried by massive particles[30]). On the other hand, as has already been remarked, the unified electroweak theory is 'simply' a generalisation of Maxwell's electrodynamics, so the field quanta should all have zero mass. This problem is solved by invoking a mechanism *analogous* to superconductivity. There is a scalar field – now called the Higgs field – which is all-pervasive (unlike the BCS field, which only exists in a superconductor). By virtue of this field the W and Z particles acquire a mass, in the way that the photon acquires a mass in superconductivity, but it can be arranged (somewhat miraculously!) that the photon remains massless. In the Weinberg–Salam theory the Higgs field is a complex scalar field but to illustrate the mechanism at work we may consider a real field ϕ with a potential energy function (sometimes called the Higgs potential)

[27] Bardeen, Cooper & Schrieffer (1957).
[28] Weinberg (1967).
[29] Salam (1968).
[30] The range is proportional to the inverse of the mass of the field quantum; for example electrodynamics has 'infinite' range and the photon is massless.

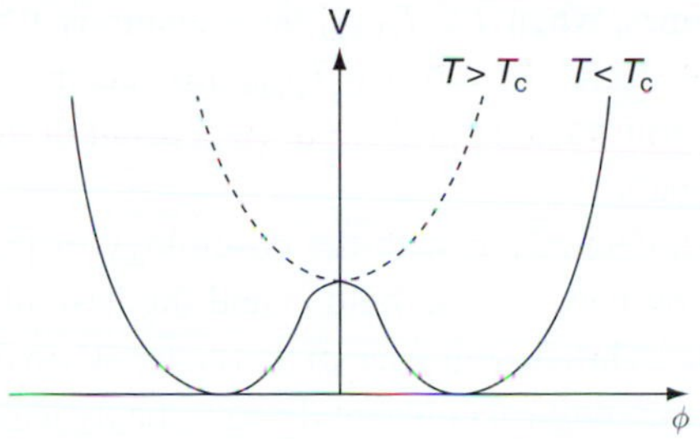

Fig. 10.21 The Higgs potential. When $T < T_c$ it has a maximum at $\phi = 0$ and two minima at $\phi = \pm\, a$. When $T > T_c$ there is only a minimum at $\phi = 0$.

$$V(\phi) = \frac{m^2}{2}\phi^2 + \frac{\lambda}{4}\phi^4. \qquad (10.192)$$

On quantisation m becomes the particle mass and the ϕ^4 term represents a quartic self-interaction. The extremal values of $V(\phi)$, given by $\partial V/\partial\phi = 0$ (differentiation with respect to the *function* ϕ), are clearly

$$\phi = 0,\ \ \pm\sqrt{\frac{-m^2}{\lambda}} \equiv 0, \pm a. \qquad (10.193)$$

In a usual field theory $m^2 > 0$ and the second two possibilities are non-physical but if we are prepared to consider the case $m^2 < 0$ then the function $V(\phi)$ has two minima and a maximum, as shown in Fig. 10.21. The maximum is at $\phi = 0$ and the minima at $\phi = \pm\, a$. A potential function like this begs a question about the nature of the vacuum. Normally the term 'vacuum' has two connotations: it is both the state of lowest energy of a system, and also the state in which no matter and no fields are present. Here, however, these situations become *distinct*: when there is no field, $\phi = 0$, the energy is *not* a minimum, but a maximum; and the state of lowest energy is a state in which the field does not vanish (and is also 2-fold degenerate).

In the case of superconductivity ϕ would represent the BCS condensate, so that $\phi \neq 0$ (and $m_\gamma \neq 0$) when $T < T_c$; but on heating the material to a temperature $T > T_c$, $\phi = 0$ and m_γ also becomes 0. The same assumption is made in particle physics: in the actual world the Higgs field ϕ does not vanish (its vacuum value is non-zero) and the quanta W and Z are massive. But if the temperature were to rise to a value that we may call $T_{\rm el.wk}$ (the subscript standing for 'electroweak') given by (see (10.176))

$$kT_{\rm el.wk} = m_Z c^2 \Rightarrow T_{\rm el.wk} \approx 10^{15}\ {\rm K}, \qquad (10.194)$$

then the vacuum would shift to $\phi = 0$. At temperatures higher than this W, Z and γ would all be massless and the electroweak symmetry becomes an *exact* symmetry; at lower temperatures it is referred to as a *spontaneously broken symmetry*. This, then, would be the situation in the early Universe, at times earlier than (see (10.172))

$$t_{\rm el.wk} = \left(\frac{10^{10}}{T_{\rm el.wk}}\right)^2 \approx 10^{-10}\ {\rm s}. \qquad (10.195)$$

At these times, when $T > T_{el.wk}$, the vacuum is a 'true' minimum of $V(\phi)$, at $\phi = 0$. As the temperature decreases to below $T_{el.wk}$ this becomes a 'false vacuum', and ϕ then 'rolls down' to the new true vacuum at $\phi = \pm a$. As a result of this phase transition the particles W and Z pick up a mass.

The final connection with the cosmology of the early Universe is made by making an identification between the field ϕ and the cosmological constant Λ, as first suggested by Zeldovich.[31] The consequence of this is that when the vacuum expectation value of ϕ is non-zero, the cosmological constant dictates the dynamics, the Universe behaves like a de Sitter universe and *expands exponentially*. So the situation runs along the following lines: at very early times, when the temperature is extremely high (in the present reasoning, higher than 10^{15} K, as above), electroweak symmetry is exact, all field quanta are massless and the vacuum value of the Higgs field is zero – as shown in Fig. 10.21. The expansion parameter goes as $t^{2/3}$ (see Problem 10.3). When the temperature drops to below $T_{el.wk}$ the symmetry is broken, the W and Z quanta become massive and the vacuum value of the scalar field ϕ is non-zero. The Higgs field, in the shape of a cosmological constant, then drives the expansion, which is exponential. This lasts for a short time, after which the expansion reverts to a 'sensible' Friedmann–Robertson–Walker type, and the horizon problem is solved. It should be noted that the *flatness problem* is also solved. It was noted above that an analysis of the inhomogeneities of the CMB radiation spectrum indicated that the Universe is flat, or almost flat. A priori this is difficult to account for, but if the Universe has undergone an inflationary phase then this will result in space becoming almost flat; the more a balloon is inflated the smaller the curvature of its surface becomes.

Finally, to give an account of Guth's model, some adjustments have to be made to the picture above. The first one is that the temperature $T_{el.wk}$ which we have been dealing with is not the temperature assumed in the literature. There the assumption is made that there exists a 'Grand Unified Theory' (GUT) of particle interactions. This goes one better than the unification of electromagnetic and weak interactions achieved by Weinberg and Salam, to embrace QCD, the theory of the strong interaction between quarks, whose quanta are gluons, so this is a unification of strong, electromagnetic and weak interactions. In this theory there are additional field quanta, much heavier than W and Z, so the temperature at which they would become massless is correspondingly higher than $T_{el.wk}$, and the time correspondingly earlier than $t_{el.wk}$. The typical GUT energy is of the order of 10^{14} GeV, corresponding to a temperature of 10^{27} K and a time of about 10^{-34} s, so this would mark the onset of inflation. In honesty it should be pointed out, however, that the status of GUTs is not particularly assured. There are actually several varients of GUT, but the simplest one predicts proton decay

$$p \to e^+ + \pi^0$$

with an estimated half-life of $10^{30\pm1}$ years (though this is model dependent). The experimental figure is $>5 \times 10^{32}$ years.[32] It may be that nature does not recognise a unification of interactions at this level.

[31] Zeldovich (1968).

[32] The reader might wonder at these figures, which far outstrip the age of the Universe. But recall that the deacy of particles obeys a probability law, so a lifetime of 10^{30} years means that in a sample of 10^{30} protons (a decent mountain) one will decay each year.

Further reading

A superb account of cosmology, at a largely non-mathematical level, is Harrison (2000). Treatments of Killing vectors and isometries in their relation to cosmology will be found in Robertson & Noonan (1968), Weinberg (1972), Ryan & Shepley (1975), Ciufolini & Wheeler (1995), McGlinn (2003) and Plebański.& Krasiński (2006). A somewhat older, though still interesting, account may be found in Landau & Lifshitz (1971). A very nice introduction to Gaussian curvature and the differential geometry of curves and surfaces is Faber (1983); see also Struik (1961) and Stoker (1969).

A very good account of horizons is Rindler (2001), Section 17.3. Detailed accounts of the luminosity–red-shift relation may be found in McVittie (1965), Weinberg (1972) and Peebles (1993). For further information about dark matter see Coles & Ellis (1997), Peacock (1999), Harrison (2000), Hartle (2003) and Roos (2003).

Detailed treatments of the early Universe may be found in Weinberg (1972), Börner (1988), Kolb & Turner (1990), Peebles (1993), Coles & Ellis (1997) and Mukhanov (2005). A full account of neutrino oscillations may be found in Raffelt (1996).

Good introductory accounts of superconductivity can be found in Feynman *et al.* (1965), Chapter 21 and Ashcroft & Mermin (1976), Chapter 34. For the Higgs mechanism and its role in electroweak theory see Aitchison & Hey (1982), Cao (1997), Cottingham & Greenwood (1998), Rubakov (2002), Maggiore (2005), Srednicki (2007), Chapters 9 and 10, or Ryder (1996). A particularly attractive and authoritative (and almost non-mathematical) account is Taylor (2001). Detailed accounts of inflationary cosmology appear in Börner (1988), Kolb & Turner (1990), Linde (1990), Mukhanov (2005) and Hobson *et al.* (2006). Recent reviews of the inflationary universe may be found in Guth (2000) and Linde (2000). A good critique of the inflationary model is Penrose (2004), Chapter 28.

Problems

10.1 Prove that the curvature scalar of S^3 is $R = 6/a^2$.

10.2 Prove that in the Steady State Theory there is an event horizon but no particle horizon.

10.3 Show that for a radiation-dominated universe $S \propto t^{1/2}$ and for a matter-dominated one $S \propto t^{2/3}$.

11 Gravitation and field theory

Our picture of the physical world at its most fundamental level, a model that also has a very high degree of experimental support, runs along the following lines. There are only three types of interaction: QCD (Quantum ChromoDynamics), which binds quarks into hadrons, that is, nuclear particles like protons and neutrons, pions and so on; the electroweak interaction, which is the unification of electromagnetism with the weak nuclear force (responsible for beta decay); and gravity. The first two interactions are understood in the context of quantum field theory, more particularly gauge field theory, and the interactions are transmitted by the field quanta, which are gluons (for QCD), and the photon and the W and Z bosons which mediate electroweak interactions. Gravity is described by General Relativity. What is immediately obvious about this statement is that General Relativity is, conceptually, a completely different sort of theory from the other field theories, because of its explicitly geometric nature. The whole enterprise of physics, however, is to reduce the number of fundamental theories and concepts to the absolute minimum, and as a consequence a large number of physicists now work on unification schemes of one sort or another – supergravity, superstring theory, brane worlds, and so on. One guiding principle at work in these endeavours is to unite the three fundamental interactions into one interaction, and another, equally important, aim is to find a *quantum* theory of gravity; it is clear that General Relativity is a classical theory since it never at any point employs notions involving wave–particle duality or Planck's constant. Since the other interactions are all cast in the language of quantum theory, should not General Relativity also be given a 'quantum treatment'?

A full-scale treatment of these matters is well beyond the scope of this book, but in this chapter some topics will be introduced which, while by no means offering anything like a theory of unification, do contain interesting and perhaps relevant considerations to what might lie 'beyond' General Relativity. These topics are gauge theories, the formulation of the Dirac equation in a general space-time and Kaluza–Klein theory, which describes electromagnetism in terms of a fifth dimension to space-time.

It is well known that Einstein spent many years at the end of his life trying to find a unified description of gravity and electromagnetism. It is now apparent that this was a misguided enterprise, for reasons which neither Einstein nor anyone else realised at the time. The attractiveness of the idea (to unite gravity and electromagnetism) was that both are *long range forces*, and as such have a description in classical physics. The strong and weak *nuclear forces* were seen to be something completely different, and particularly in view of their extremely short range, phenomena that could only be treated by using quantum concepts. We now see, however, that this view was rather misleading. For one thing, electrodynamics may be formulated as a quantum field theory – even in Einstein's day

Feynman, Schwinger and Tomonaga completed the work on quantum electrodynamics (QED) for which they later received a Nobel prize. But perhaps the most telling development was in 1967, only twelve years after Einstein's death, when Weinberg proposed his unified theory of *weak* and electromagnetic interactions. Within a few years this theory received dramatic confirmation, with the discovery of the field quanta W and Z, and of so-called neutral current reactions, of the form $\nu + p \rightarrow \nu + p + \pi^0$. Electromagnetism was unified at last, but with the weak (nuclear) interaction, not with gravity. The only reason that electromagnetism has a long range nature is that the photon happens to be massless, whereas W and Z are massive particles. In more recent years we have also seen that protons, neutrons, pions and the other hadrons are not elementary particles, but are bound states of quarks, so the 'true' interaction at the nuclear scale is not that between protons and neutrons, to make atomic nuclei, but that between quarks, to make protons, neutrons and other hadrons. This interaction is called QCD, since it is a generalisation of QED to the case where the symmetry group is SU(3) rather than U(1), as it is for electromagnetism. Hence the fundamental interactions are the electroweak interaction, QCD and gravity. Grand Unified Theories (GUTs) have been proposed to unify QCD and electromagnetism, but there is very little, if any, experimental support for these. Their most interesting prediction is that the proton should be unstable, but no proton decays have yet been found, and the experimental upper limit on the lifetime (about 10^{32} years) exceeds theoretical predictions. The striking thing about both electroweak theory and QCD, however, is that they are both 'gauge theories' – they are both generalisations of Maxwell's theory to cases with an enlarged symmetry group.

So, in a quest for theories beyond General Relativity, where are we to go? The first remark to make is that it is *not* possible to describe the other interactions in geometric terms, so they cannot be made to look like General Relativity in four dimensions. It is quite easy to see this. A geometric formulation of gravity is possible because of the equality of gravitational and inertial mass. Two bodies in a gravitational field experience a *force* proportional to their gravitational mass, and this results in an *acceleration* that is proportional to (the inverse of) their inertial mass. The equality of these masses means that *all* particles accelerate at the same rate in a gravitational field – Galileo's observation – and gravity may be simulated by an accelerating frame of reference. But this state of affairs does not hold for any other of the interations. An electrically charged body will experience a force proportional to its charge, and the acceleration, as before, depends on this force divided by the body's (inertial) mass. The acceleration therefore is proportional to q/m, the charge-to-mass ratio. This is different for different bodies, so the accelerating frame in which the electric force vanishes for one body is not the same as that for another body – there is no universal frame in which electromagnetism 'disappears'. It is therefore not possible to 'geometrise' it. In the 1960s, however, it was found that 'gauging' the Lorentz transformation of Special Relativity did result in a theory extremely like (but not identical with) General Relativity. This was moreover a theory that had been proposed somewhat earlier by Cartan in a purely geometrical context, and on which in fact Cartan and Einstein had exchanged letters. We shall consider this theory below, after an introduction to the gauge idea as exemplified in electromagnetism and Yang–Mills theory.

11.1 Electrodynamics as an abelian gauge theory

Consider a classical scalar complex field $\phi(x^\mu)$ with a Lagrange density[1]

$$\mathcal{L} = (\partial_\mu \phi)(\partial^\mu \phi^*) + m^2 \phi^* \phi, \tag{11.1}$$

and, throughout this section, the space-time is Minkowski,

$$g_{\mu\nu} = \eta_{\mu\nu} = \mathrm{diag}(-1, 1, 1, 1). \tag{11.2}$$

The Euler–Lagrange equations

$$\frac{\partial \mathcal{L}}{\partial \phi} - \partial_\mu \left(\frac{\partial \mathcal{L}}{\partial(\partial_\mu \phi)} \right) = 0 \tag{11.3}$$

yield the equations of motion

$$(\Box - m^2)\phi = 0, \quad (\Box - m^2)\,\phi^* = 0, \tag{11.4}$$

where

$$\Box = -\frac{1}{c^2}\frac{\partial^2}{\partial t^2} + \frac{\partial^2}{\partial x^2} + \frac{\partial^2}{\partial y^2} + \frac{\partial^2}{\partial z^2}.$$

Equation (11.4) is the *Klein–Gordon equation*. It is a second order relativistic wave equation. On the 'quantum' substitutions (in the units $\hbar = c = 1$)

$$\frac{\partial}{\partial t} \to \frac{\mathrm{i}}{\hbar} E, \quad \frac{\partial}{\partial x} \to -\frac{\mathrm{i}}{\hbar} p_x, \quad \text{etc.}$$

it becomes

$$E^2 - p^2 = m^2,$$

which is Einstein's relation, where m is the mass of the quantised version of the field ϕ (and ϕ^*). We now show that the Lagrangian (11.1) possesses a symmetry, as a consequence of which there is a conserved quantity, which we identify with electric charge. The transformation

$$\phi \to \exp(-\mathrm{i}\Lambda)\phi, \quad \phi^* \to \exp(\mathrm{i}\Lambda)\phi^* \tag{11.5}$$

clearly leaves $\mathcal{L}$ invariant: here Λ is a constant parameter and this transformation is called a *gauge transformation of the first kind*. For infinitesimal Λ we have

$$\delta\phi = -\mathrm{i}\Lambda\phi, \quad \delta\phi^* = \mathrm{i}\Lambda\,\phi^*, \tag{11.6}$$

$$\delta(\partial_\mu \phi) = -\mathrm{i}\Lambda(\partial_\mu \phi), \quad \delta(\partial_\mu \phi^*) = \mathrm{i}\Lambda(\partial_\mu \phi^*). \tag{11.7}$$

[1] Readers unfamiliar with classical field theory may consult for example Soper (1976), Itzykson & Zuber (1980), De Wit & Smith (1986) or Ryder (1996).

We then define a current j^μ by

$$\Lambda j^\mu = \frac{\partial \mathcal{L}}{\partial(\partial_\mu \phi)}(\delta\phi) + \frac{\partial \mathcal{L}}{\partial(\partial_\mu \phi^*)}(\delta\phi^*) = (\partial_\mu \phi^*)(-\mathrm{i}\Lambda\phi) + (\partial_\mu \phi)(\mathrm{i}\Lambda\phi^*),$$
$$j^\mu = \mathrm{i}(\phi^* \partial^\mu \phi - \phi \partial^\mu \phi^*). \tag{11.8}$$

This current is conserved:

$$\partial^\mu j_\mu = \mathrm{i}(\phi^* \Box \phi - \phi \Box \phi^*) = 0, \tag{11.9}$$

where we have used (11.4). Consequently

$$\partial_0 \int_V j^0 \,\mathrm{d}^3x = \int_V \nabla \cdot \mathbf{j} \,\mathrm{d}^3x = \int_{\partial V} \mathbf{j} \cdot \mathbf{n} \,\mathrm{d}\Sigma = \mathrm{i} \int_{\partial V} (\phi^* \nabla \phi - \phi \nabla \phi^*) \,\mathrm{d}\Sigma = 0,$$

where Gauss's theorem has been used and it is assumed that the field ϕ vanishes on the boundary ∂V of the volume V. So with

$$Q = \int j^0 \,\mathrm{d}^3x \tag{11.10}$$

we have

$$\frac{\mathrm{d}Q}{\mathrm{d}t} = 0. \tag{11.11}$$

Identifying Q with electric charge we now have an account of a (complex) field ϕ which carries electric charge, a *conserved* quantity.

The transformation (11.5) demands that at every point in space the field ϕ changes by the amount indicated, and all at the same time. This *global* transformation seems to go against the spirit, if not the letter, of Special Relativity. A more reasonable demand is that the parameter Λ should be a function of space-time, so that the transformation of the field ϕ is a *local* one, different at all points in space. Let us then replace (11.5) with the transformation

$$\phi(x) \to \exp\{-\mathrm{i}\,\Lambda(x)\}\phi(x), \quad \phi^*(x) \to \exp\{\mathrm{i}\,\Lambda(x)\}\phi^*(x) \tag{11.12}$$

which is called a *gauge transformation of the second kind*. Under this rule, with infinitesimal $\Lambda(x)$,

$$\delta\phi = -\mathrm{i}\,\Lambda\phi, \quad \delta\phi^* = \mathrm{i}\,\Lambda\phi^*, \tag{11.13}$$

as in (11.6), but (11.7) becomes replaced by

$$\delta(\partial_\mu\phi) = -\mathrm{i}\,\Lambda(\partial_\mu\phi) - \mathrm{i}(\partial_\mu\Lambda)\phi, \quad \delta(\partial_\mu\phi^*) = \mathrm{i}\,\Lambda(\partial_\mu\phi^*) + \mathrm{i}(\partial_\mu\Lambda)\phi^*, \tag{11.14}$$

with extra terms proportional to $\partial_\mu\,\Lambda$. The Lagrangian (11.1), however, is *not* invariant under (11.13) and (11.14):

$$\delta\mathcal{L} = [\delta(\partial_\mu\phi)](\partial^\mu\phi^*) + (\partial_\mu\phi)[\delta(\partial^\mu\phi^*)] + m^2\delta(\phi^*\phi) = (\partial_\mu\Lambda)j^\mu. \tag{11.15}$$

To retain an invariant Lagrangian an extra *field* $A_\mu(x)$ must be introduced into the theory with a corresponding extra term in the Lagrangian

$$\mathcal{L}_1 = -ej^\mu A_\mu. \tag{11.16}$$

The *coupling constant* e has been introduced so that eA_μ has the same dimensions as the operator $\partial/\partial x^\mu$. Then, if under (11.14) we also have

$$A_\mu \to A_\mu + \frac{1}{e}\partial_\mu\Lambda, \tag{11.17}$$

the change in $\mathcal{L}_1$ will be

$$\delta\mathcal{L}_1 = -e(\delta j^\mu)A_\mu - ej^\mu(\delta A_\mu) = -e(\delta j^\mu)A_\mu - e(\partial_\mu\Lambda)j^\mu. \tag{11.18}$$

The second term above cancels (11.15), so

$$\delta\mathcal{L} + \delta\mathcal{L}_1 = -e(\delta j^\mu)A_\mu; \tag{11.19}$$

$\mathcal{L} + \mathcal{L}_1$ is still not invariant! In fact

$$\delta J^\mu = \mathrm{i}\{(\delta\phi^*)\partial_\mu\phi + \phi^*\delta(\partial_\mu\phi) - (\delta\phi)\partial_\mu\phi^* - \phi[\delta(\partial_\mu\phi^*)]\} = 2\partial_\mu\Lambda(\phi^*\phi),$$

so

$$\delta(\mathcal{L} + \mathcal{L}_1) = -2e(\phi^*\phi)(\partial_\mu\Lambda)A^\mu.$$

If we now add a term

$$\mathcal{L}_2 = e^2\phi^*\phi A^\mu A_\mu, \tag{11.20}$$

then

$$\delta\mathcal{L}_2 = 2e^2(\phi^*\phi)(\delta A^\mu)A_\mu = 2e(\phi^*\phi)(\delta_\mu\Lambda)A^\mu$$

and

$$\delta(\mathcal{L} + \mathcal{L}_1 + \mathcal{L}_2) = 0.$$

Having introduced a field A^μ, this field will itself contribute a term to the Lagrangian, independent of $\mathcal{L}_1$ and $\mathcal{L}_2$, which both describe its 'coupling' to the matter field ϕ. Let us define

$$F_{\mu\nu} = \partial_\mu A_\nu - \partial_\nu A_\mu. \tag{11.21}$$

Then under (11.17)

$$\delta(\partial_\mu A_\nu) = \frac{1}{e}\partial_\mu\partial_\nu\Lambda,$$

and hence

$$\delta(F_{\mu\nu}) = 0. \tag{11.22}$$

Then the Lagrangian[2]

$$\mathcal{L}_3 = {}^1\!/_4 F^{\mu\nu} F_{\mu\nu} \tag{11.23}$$

is invariant

$$\delta\mathcal{L}_3 = 0.$$

We have, finally, the total Lagrangian

$$\begin{aligned}\mathcal{L}_{\text{tot}} &= \mathcal{L} + \mathcal{L}_1 + \mathcal{L}_2 + \mathcal{L}_3 \\ &= (\partial_\mu\phi)(\partial^\mu\phi^*) + m^2\phi^*\phi - \mathrm{i}e(\phi^*\partial^\mu\phi - \phi\partial^\mu\phi^*)A_\mu + e^2 A_\mu A^\mu \phi^*\phi \\ &\quad + {}^1\!/_4 F^{\mu\nu}F_{\mu\nu} \\ &= (\partial_\mu\phi + \mathrm{i}eA_\mu\phi)(\partial^\mu\phi^* - \mathrm{i}eA^\mu\phi^*) + m^2\phi^*\phi + {}^1\!/_4 F^{\mu\nu}F_{\mu\nu}.\end{aligned} \tag{11.24}$$

This suggests a definition of *covariant derivatives*

$$D_\mu\phi = (\partial_\mu + \mathrm{i}eA_\mu)\phi, \quad D_\mu\phi^* = (\partial_\mu - \mathrm{i}eA_\mu)\phi^*, \tag{11.25}$$

since under (11.14) and (11.17)

$$\begin{aligned}\delta(D_\mu\phi) &= \delta(\partial_\mu\phi) + \mathrm{i}e(\delta A_\mu)\phi + \mathrm{i}eA_\mu(\delta\phi) = -\mathrm{i}\Lambda(\partial_\mu\phi + \mathrm{i}eA_\mu\phi) \\ &= -\mathrm{i}\Lambda D_\mu\phi;\end{aligned} \tag{11.26}$$

$D_\mu\phi$ transforms in the same way as ϕ (from (11.13)). Then we may write the total Lagrangian as

$$\mathcal{L}_{\text{tot}} = (D_\mu\phi)(D^\mu\phi^*) + m^2\phi^*\phi + {}^1\!/_4 F^{\mu\nu}F_{\mu\nu}. \tag{11.27}$$

We have not quite finished. The conserved current j^μ defined in (11.8) depends on the ordinary, not the covariant derivative. We need a current, call it J^μ, which is defined in terms of the covariant derivative and which, using the total Lagrangian (11.27), is conserved. We might guess that, by comparing with (11.8), it would take the form

$$J^\mu = \mathrm{i}(\phi^* D^\mu\phi - \phi D^\mu\phi^*). \tag{11.28}$$

Indeed, we find, in analogy with the argument leading to (11.8),

$$\Lambda J^\mu = \frac{\partial L_{\text{tot}}}{\partial(D_\mu\phi)}(\delta\phi) + \frac{\partial L_{\text{tot}}}{\partial(D_\mu\phi^*)}(\delta\phi^*) = (D^\mu\phi^*)(-\mathrm{i}\Lambda\phi) + (D^\mu\phi)(\mathrm{i}\Lambda\phi^*),$$

which is precisely (11.28). We now need to show that this current is conserved. Applying the Euler–Lagrange equation (with $\mathcal{L}$ standing for $\mathcal{L}_{\text{tot}}$)

$$\frac{\partial\mathcal{L}}{\partial A^\mu} - \partial_\nu\left(\frac{\partial\mathcal{L}}{\partial(\partial\nu A^\mu)}\right) = 0,$$

[2] This gives, with our metric, $\mathcal{L}_3 = -{}^1\!/_2(E^2 - H^2)$. Most books on field theory use the metric (+, −, −, −) and then $\mathcal{L}_3$ picks up a minus sign.

we have

$$\frac{\partial \mathcal{L}}{\partial A^\mu} = \mathrm{i}\, e\,(\phi\,\partial_\mu \phi^* - \phi^* \partial_\mu \phi) + 2e^2 \phi^* \phi A_\mu, \quad \frac{\partial \mathcal{L}}{\partial(\partial_\nu A^\mu)} = -F_{\nu\mu} = F_{\mu\nu},$$

and hence

$$\begin{aligned}\partial_\nu F^{\mu\nu} &= -\mathrm{i}\, e(\phi\,\partial_\mu \phi^* - \phi^*\,\partial_\mu \phi) + 2e^2 \phi^* \phi A_\mu = -\mathrm{i}\, e\,(\phi D_\mu \phi^* - \phi^* D_\mu \phi) \\ &= -e\, J^\mu.\end{aligned} \tag{11.29}$$

Because of the antisymmetry in $(\mu\nu)$ it then follows that

$$\partial_\mu J^\mu = 0; \tag{11.30}$$

the 'covariant' current J^μ is conserved.

Let us summarise what we have done.

(i) The Lagrangian (11.1) for a complex scalar field has a *symmetry* – a rotation in the complex plane. By virtue of this there is a conserved current j^μ and hence a conserved quantity which may be identified with electric charge Q.

(ii) On making the symmetry a *local* symmetry – $\Lambda(x^\mu)$ – the Lagrangian $\mathcal{L}$ is no longer invariant, but invariance may be restored by introducing a field A^μ and associated 4-dimensional curl

$$F_{\mu\nu} = \partial_\mu A_\nu - \partial_\nu A_\mu.$$

This is the *electromagnetic field*, whose *source* is electric charge.

This version of electromagnetism is of course not geometrical, but nevertheless there are some parallels with General Relativity. The reader might be inclined to think that calling $D_\mu \phi$ a 'covariant' derivative is unwarranted; that this is simply an attempt to *make* electromagnetism look like General Relativity. But this objection is not really justified. It will be recalled (see Equations (4.56) and (4.57)) that the commutator of two covariant derivatives in General Relativity is proportional to the curvature tensor: from (4.57)

$$[\nabla_\mu, \nabla_\nu]\, \mathbf{e}_\kappa = R^\rho{}_{\kappa\mu\nu}\, \mathbf{e}_\rho. \tag{11.31}$$

What about the analogous commutator in electrodynamics? It follows from (11.25) that

$$[D_\mu, D_\nu]\, \phi = [\partial_\mu + \mathrm{i}\, e\, A_\mu,\ \partial_\nu + \mathrm{i}\, e\, A_\nu]\, \phi = \mathrm{i}\, e\,(\partial_\mu A_\nu - \partial_\nu A_\mu)\phi$$

or

$$-\frac{\mathrm{i}}{e}[D_\mu,\ D_\nu]\, \phi = F_{\mu\nu}\, \phi. \tag{11.32}$$

This would imply an analogy between curvature and field strength. In fact, let us write (11.21) without indices:

$$F_{..} = \partial_{[.} A_{.]} \tag{11.33}$$

(where the brackets [. .] stand for antisymmetrisation). This may be compared with Equation (4.31), also written without indices

$$R^{\cdot}\ldots = \partial_{[\cdot}\Gamma_{\cdot]} + [\Gamma, \Gamma]. \tag{11.34}$$

We may then claim that the potential A in electrodynamics plays a role analogous to that of the connection coefficient Γ in differential geometry; and the field strength F is analogous to the curvature R. It is true that the second term on the right hand side of (11.34) does not have a counterpart in (11.33), but even this situation is changed when we generalise electrodynamics to the case of a non-abelian gauge symmetry, which we do in the next section.

Let us write some of the above equations in a form that suggests a generalisation of this treatment of electrodynamics. Equation (11.5) may be written

$$\phi \to U\phi, \quad \phi^* \to U^\dagger \phi^*, \tag{11.35}$$

with

$$U = \mathrm{e}^{\mathrm{i}\Lambda}, \quad U^\dagger U = 1, \tag{11.36}$$

and $U^\dagger$ is the Hermitian conjugate of U. Here U is a unitary (1-dimensional!) matrix – simply a phase factor. Under a *local* transformation, $\Lambda = \Lambda(x^\mu)$, we have

$$\partial_\mu U = \mathrm{i}(\partial_\mu \Lambda)U. \tag{11.37}$$

The transformations (11.17) and (11.22) may be written

$$A_\mu \to A_\mu - \frac{\mathrm{i}}{e} U^\dagger \partial_\mu U, \quad F_{\mu\nu} \to F_{\mu\nu}. \tag{11.38}$$

If two 'matrices' U_1 and U_2 obey (11.36), then so does their product U_1U_2. Furthermore each matrix U has an inverse U^{-1} which obeys the same condition so these matrices form a *group*, the group U(1) of unitary matrices in one dimension. This is the symmetry group of electrodynamics. It is also, however, isomorphic to the group O(2) of orthogonal matrices in two dimensions. It is easy to see this, for putting

$$\phi = \frac{1}{\sqrt{2}}(\phi_1 + \mathrm{i}\phi_2), \quad \phi^* = \frac{1}{\sqrt{2}}(\phi_1 - \mathrm{i}\phi_2),$$

with ϕ_1, ϕ_2 real, the transformation (11.35) is

$$\begin{pmatrix}\phi_1\\ \phi_2\end{pmatrix} \to \begin{pmatrix}\phi_1'\\ \phi_2'\end{pmatrix} = \begin{pmatrix}\cos\Lambda & \sin\Lambda\\ -\sin\Lambda & \cos\Lambda\end{pmatrix}\begin{pmatrix}\phi_1\\ \phi_2\end{pmatrix}, \tag{11.39}$$

a *rotation* in the (ϕ_1–ϕ_2) plane through an angle Λ. Rotations in 2-dimensional space are described by orthogonal matrices; they leave the distance from the origin unchanged, so

$$(\phi_1')^2 + (\phi_2')^2 = \phi_1^2 + \phi_2^2.$$

Then, putting $\phi_i' = R_{ij}\phi_j$,

$$\phi_i'\,\phi_i' = R_{ij}\,R_{ik}\,\phi_j\,\phi_k$$

and hence

$$R_{ij} R_{ik} = \delta_{jk},$$

which is the condition for orthogonal matrices. They also form a group; if R and S are orthogonal, so is RS, and so are R^{-1} and S^{-1}. Moreover $RS = SR$ so the group is *abelian* – all its elements commute; clearly $R(\alpha)\, R(\beta) = R(\alpha + \beta) = R(\beta)\, R(\alpha)$, where $R(\alpha)$ is a rotation through an angle α.

All of the above outline theory is summarised in the statement that electrodynamics is a U(1) *gauge theory*; the term 'gauge' refers to the *local* nature of the transformation (11.12) – that Λ depends on x^μ. A way to generalise electrodynamics – and this is the path to electroweak theory – is to enlarge the symmetry group U(1) to SU(2) (or equivalently SO(2) to SO(3)). (The S here means the matrices have unit determinant.) This is the subject of the next section.

Before leaving our account of electrodynamics, however, it is worth making the following observation. It is crucial to both Special and General Relativity that the speed of light is an *absolute* speed. As a statement about *photons* this is the observation that they must have no mass; particles with non-zero mass may be brought to rest, but photons (light) are never at rest. This requirement of zero mass actually follows from gauge invariance. If photons were to have a mass there would be a term in the Lagrangian of the form

$$\mathcal{L}_m = m^2 A_\mu A^\mu. \tag{11.40}$$

The Euler–Lagrange equation would then give an equation of motion

$$m^2 A^\mu + \partial_\nu F^{\mu\nu} = 0,$$

which in the Lorenz gauge $\partial_\nu A^\nu = 0$ would reduce to

$$\Box A^\mu = m^2 A^\mu$$

rather than the usual $\Box A^\mu = 0$. The Lagrangian (11.40) is however not invariant under the gauge transformation (11.17), so *gauge invariance guarantees zero mass for the photon*, which is crucial for relativity theory.

11.2 Non-abelian gauge theories

It might well be interesting to enlarge the symmetry group U(1) of electrodynamics, to 'see what happens', but why should anyone want to do this? What is the physical motivation? As often in the history of physics the original motivation for taking this step proved to be initially unfruitful – though eventually it turned out to be extremely fruitful. The original idea, due to Yang and Mills,[3] was conceived in the context of nuclear physics, and in particular was concerned with what is now called isospin, but was then known as isotopic, or sometimes isobaric, spin. And the origin of this idea was the observation that particles with

[3] Yang & Mills (1954). This is reprinted with a commentary in Yang (1983).

nuclear interactions came in ‘families’ with very similar properties. Thus the proton and neutron have almost the same mass, and the same spin and baryon number. They could then be thought of as two *states* of one particle, the nucleon:

$$\mathrm{N} = \begin{pmatrix} \mathrm{p} \\ \mathrm{n} \end{pmatrix}. \tag{11.41}$$

The nucleon is said to have isospin $I = ½$, the proton having isospin ‘up’, $I_3 = +½$, and the neutron having isospin ‘down’, $I_3 = -½$. This is in direct analogy with spin; the mathematics is the same, though the physics completely different. Spin is mathematically connected with rotations in 3-dimensional parameter space, and isospin therefore with an ‘abstract’ 3-dimensional space. I_3 is the component of the isospin along the ‘third’ axis of this space. In a similar way, as well as there being two nucleons there are also three pions, π^+, π^0 and π^-. They have very similar masses (π^+ and π^- have the same mass, being particle and antiparticle), the same spin (zero) and baryon number (zero). They therefore are the three components of a ‘pion’ with isospin $I = 1$:

$$\pi = \begin{pmatrix} \pi^+ \\ \pi^0 \\ \pi^- \end{pmatrix}, \tag{11.42}$$

with π^+, π^0 and π^- respectively having $I_3 = 1, 0, -1$, again in mathematical analogy with the states of a spin 1 particle. Under an arbitrary rotation in isospin space, p, for example, will change into a mixture of p and n, and π^+ will similarly change into a mixture of all three pions. Now if the two nucleons had *exactly* the same mass, and the three pions also had exactly the same mass, this would give an extra *symmetry* to the nuclear interactions. It was this that was the focus of Yang and Mills’ attention.

What is the relation between this symmetry and electrodynamics? The point is that, as seen from the two examples above, there is a relation between I_3 and Q, electric charge. It is obvious that for the nucleon and pions multiplets we have

$$\mathrm{N}\colon\ Q = I_3 + {}^1\!/_2; \qquad \pi\colon\ Q = I_3.$$

Isospin is a ‘vector’ quantity in some space whose third dimension is connected with electric charge. If electric charge is the source of the electromagnetic field, with all the apparatus of Maxwell’s equations and the U(1) gauge symmetry we have been considering, might not isospin be understood by simply generalising the gauge group from SO(2) to SO(3) – from the group of rotations in two dimensions to rotations in three dimensions? Unlike SO(2), however, SO(3) is *non-abelian* – its different elements do not commute. This may be observed in a simple experiment, as seen in Fig. 11.1. Rotate an object first around the x axis and then around the y axis, in each case through an angle $\pi/2$; and then perform these rotations in the reverse order. The final configurations are different, so

$$R_x(\pi/2)\, R_y(\pi/2) \neq R_y(\pi/2)\, R_x(\pi/2)\,.$$

This is true for any angles of rotation. Under a rotation about the z axis through an angle α a vector transforms as

$$V_x \to \cos\alpha\, V_x + \sin\alpha\, V_y, \quad V_y \to -\sin\alpha\, V_x + \cos\alpha\, V_y, \quad V_z \to V_z, \tag{11.43}$$

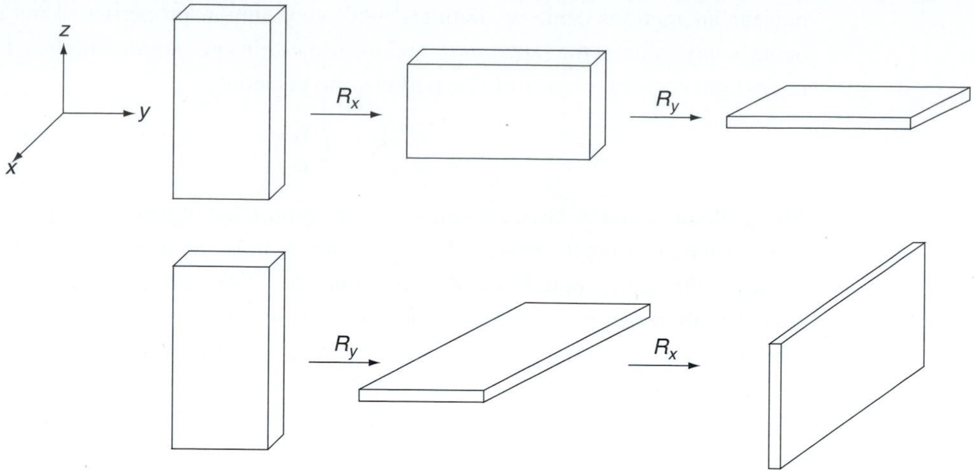

Fig. 11.1 **Rotations about the *x* axis and about the *y* axis do not commute.**

so the rotation matrix is

$$R_z(\alpha) = \begin{pmatrix} \cos\alpha & \sin\alpha & 0 \\ -\sin\alpha & \cos\alpha & 0 \\ 0 & 0 & 1 \end{pmatrix}. \tag{11.44}$$

Similarly the rotation matrices for rotations about the x and y axes are

$$R_x(\beta) = \begin{pmatrix} 1 & 0 & 0 \\ 0 & \cos\beta & \sin\beta \\ 0 & -\sin\beta & \cos\beta \end{pmatrix}, \quad R_y(\gamma) = \begin{pmatrix} \cos\gamma & 0 & -\sin\gamma \\ 0 & 1 & 0 \\ \sin\gamma & 0 & \cos\gamma \end{pmatrix}, \tag{11.45}$$

and it is clear that they do not commute,

$$R_x(\beta)\,R_z(\alpha) \neq R_z(\alpha)\,R_x(\beta). \tag{11.46}$$

A rotation through an infinitesimal angle α is, to lowest order

$$R_z(\alpha) = \begin{pmatrix} 1 & 0 & 0 \\ 0 & 1 & 0 \\ 0 & 0 & 1 \end{pmatrix} + \alpha \begin{pmatrix} 0 & 1 & 0 \\ -1 & 0 & 0 \\ 0 & 0 & 1 \end{pmatrix}.$$

Writing this as

$$R_z(\alpha) = 1 + \mathrm{i}J_z\,\alpha,$$

the *generator* of rotations about the z axis is

$$J_z = \begin{pmatrix} 0 & -\mathrm{i} & 0 \\ \mathrm{i} & 0 & 0 \\ 0 & 0 & 0 \end{pmatrix}. \tag{11.47}$$

In a similar way,

$$J_x = \begin{pmatrix} 0 & 0 & 0 \\ 0 & 0 & -\mathrm{i} \\ 0 & \mathrm{i} & 0 \end{pmatrix}, \quad J_y = \begin{pmatrix} 0 & 0 & \mathrm{i} \\ 0 & 0 & 0 \\ -\mathrm{i} & 0 & 0 \end{pmatrix}. \tag{11.48}$$

These matrices do not commute,

$$[J_x, J_y] = \mathrm{i} J_z \text{ and cyclic permutations.} \tag{11.49}$$

A rotation through a finite angle (for example about the z axis) is given by

$$R_z(\alpha) = \exp(\mathrm{i} J_z \alpha), \tag{11.50}$$

(see Problem 11.1) or, in general, a rotation about an axis $\mathbf{n}$ through an angle α is

$$R_{\mathbf{n}}(\alpha) = \exp(\mathrm{i}\, \mathbf{n} \cdot \mathbf{J}\, \alpha). \tag{11.51}$$

A vector is an object with three components which transforms under rotations in the same way as the coordinates (x, y, z), and the three states of the pion, (11.42), transform as a vector in isospin space. For reasons of our own, however, we are more interested in the nucleon, which has two components. This is a *spinor*, and strictly speaking is not a basis for a representation of the rotation group SO(3), but of SU(2), the group of unitary matrices in two dimensions, with unit determinant. A general 2×2 unitary matrix is given by

$$U = \begin{pmatrix} a & b \\ -b^* & a^* \end{pmatrix}, \quad |a|^2 + |b|^2 = 1, \tag{11.52}$$

so that

$$UU^\dagger = U^\dagger U = 1, \quad \det U = 1, \tag{11.53}$$

as may easily be checked. If U_1 and U_2 are unitary, so is $U_1 U_2$; unitary matrices, like orthogonal ones, form a group (but Hermitian ones do not). The *structure* of SU(2) is (apart from global considerations[4]) the same as that of SO(3). In an analogous way to (11.51) we may write

$$U = \exp[(\mathrm{i}/2)\, \mathbf{n} \cdot \boldsymbol{\sigma}\, \alpha], \tag{11.54}$$

where $\boldsymbol{\sigma}$ are the Pauli matrices

$$\sigma_x = \begin{pmatrix} 0 & 1 \\ 1 & 0 \end{pmatrix}, \quad \sigma_y = \begin{pmatrix} 0 & -\mathrm{i} \\ \mathrm{i} & 0 \end{pmatrix}, \quad \sigma_z = \begin{pmatrix} 1 & 0 \\ 0 & -1 \end{pmatrix}. \tag{11.55}$$

It is easy to check that

$$\left[\frac{\sigma_x}{2}, \frac{\sigma_y}{2} \right] = \mathrm{i} \frac{\sigma_z}{2} \text{ and cyclic perms,} \tag{11.56}$$

so the *generators* of SU(2) obey the same commutation relations as the generators of SO(3), Equation (11.49). It is straightforward to show that (Problem 11.2)

[4] For a very nice account of these global considerations see for example Speiser (1964).

$$U = \exp\left(\frac{i}{2}\,\mathbf{n}\cdot\boldsymbol{\sigma}\,\alpha\right) = \cos\frac{\alpha}{2} + i\,\mathbf{n}\cdot\boldsymbol{\sigma}\,\sin\frac{\alpha}{2}, \qquad (11.57)$$

so for example if $\mathbf{n} = n_z = (0, 0, 1)$,

$$U_z(\alpha) = \cos\frac{\alpha}{2} + \begin{pmatrix} i & 0 \\ 0 & -i \end{pmatrix} \sin\frac{\alpha}{2} = \begin{pmatrix} e^{i\alpha/2} & 0 \\ 0 & e^{-i\alpha/2} \end{pmatrix}. \qquad (11.58)$$

Under this transformation a *spinor* will transform as $\psi \to \psi' = U\psi$, or[5]

$$\psi = \begin{pmatrix} \psi_1 \\ \psi_2 \end{pmatrix} \to \begin{pmatrix} \psi_1{}' \\ \psi_2{}' \end{pmatrix} = \begin{pmatrix} e^{i\alpha/2}\psi_1 \\ e^{-i\alpha/2}\psi_2 \end{pmatrix}. \qquad (11.59)$$

The *outer product* of two spinors is

$$\psi\psi^\dagger = \begin{pmatrix} \psi_1 \\ \psi_2 \end{pmatrix} (\psi_1^* \ \psi_2^*) = \begin{pmatrix} \psi_1\psi_1^* & \psi_1\psi_2^* \\ \psi_2\psi_1^* & \psi_2\psi_2^* \end{pmatrix}. \qquad (11.60)$$

Under (11.59) it is easy to see that

$$\begin{aligned} (\psi_1\psi_2^* + \psi_2\psi_1^*) &\to \cos\alpha(\psi_1\psi_2^* + \psi_2\psi_1^*) + \sin\alpha\{i(\psi_1\psi_2^* - \psi_2\psi_1^*)\}, \\ i(\psi_1\psi_2^* - \psi_2\psi_1^*) &\to \cos\alpha\{i(\psi_1\psi_2^* - \psi_2\psi_1^*)\} - \sin\alpha(\psi_1\psi_2^* + \psi_2\psi_1^*), \end{aligned} \qquad (11.61)$$

which may be compared with the transformation of a vector, (11.43). Hence

$$(\psi_1\psi_2^* + \psi_2\psi_1^*) \sim V_x, \quad i(\psi_1\psi_2^* - \psi_2\psi_1^*) \sim V_y,$$

and spinors may be said to transform like the 'square roots' of vectors.

So far in this section we have given an outline of the group SU(2) which, unlike U(1), the symmetry group of electromagnetism, is non-abelian, and so possesses more structure – is more interesting! We have also deliberately chosen to consider fields with spin ½ (spinors), because in the next section we want to consider the consequences of gauging the *Lorentz group* (which is of course non-abelian), when it acts on a spin ½ particle. It is precisely this which turns out to have very similar consequences to General Relativity. Our immediate task now is to consider a *spinor field* ϕ with Lagrangian (cf. (11.1))

$$\mathcal{L} = \partial_\mu\phi^\dagger\,\partial^\mu\phi + m^2\phi^\dagger\phi, \qquad (11.62)$$

with

$$\phi = \begin{pmatrix} \phi_1 \\ \phi_2 \end{pmatrix}, \quad \phi^\dagger = (\phi_1^* \ \phi_2^*). \qquad (11.63)$$

This is invariant under

$$\phi \to U\phi, \quad \phi^\dagger \to \phi^\dagger U^\dagger, \qquad (11.64)$$

[5] A noteworthy feature of this formula is the appearance of the half-angle. It means that under a spatial rotation through 2π a spinor changes sign, $\psi \to -\psi$. This has actually been observed experimentally for the neutron (a spin ½ particle) by Werner *et al.* (1975). See also Rauch & Werner (2000).

where U is a matrix of the form (11.57). This is the generalisation of (11.5). For *constant* parameters α it is a gauge transformation of the first kind. Our first job is to find the conserved current, analogous to (11.8), in this non-abelian theory. The second (and more crucial) job is to consider the case in which α is position dependent, $\alpha = \alpha\,(x^\mu)$, and then to find the generalisation of the covariant derivatives and invariant Lagrangian, (11.25), (11.27), in this non-abelian case.

To take the first task first, writing U, given by (11.54), in the form

$$U = \exp\left(\mathrm{i}\frac{\sigma^a}{2}\alpha^a\right), \qquad a = 1,\, 2,\, 3 \tag{11.65}$$

(where $\sigma^1 = \sigma_x$, $\sigma^2 = \sigma_y$, $\sigma^3 = \sigma_z$ and there is an implied summation over a), the transformation (11.64) for infinitesimal α^a is

$$\phi \to \left(1 + \frac{\mathrm{i}}{2}\sigma^a\alpha^a\right)\phi, \tag{11.66}$$

so

$$\delta\phi = \mathrm{i}\frac{\sigma^a}{2}\alpha^a\phi, \quad \delta(\partial_\mu\phi) = \mathrm{i}\frac{\sigma^a}{2}\alpha^a(\partial_\mu\phi). \tag{11.67}$$

In addition (noting that $\sigma^{a\dagger} = \sigma^a$)

$$\phi^\dagger \to \phi^\dagger\left(1 - \frac{\mathrm{i}}{2}\,\sigma^a\,\alpha^a\right),$$

so

$$\delta\phi^\dagger = -\mathrm{i}\phi^\dagger\frac{\sigma^a}{2}\alpha^a, \quad \delta(\partial_\mu\phi^\dagger) = -\mathrm{i}(\partial_\mu\phi^\dagger)\frac{\sigma^a}{2}\alpha^a. \tag{11.68}$$

The current is constructed from (cf. (11.8))

$$\begin{aligned}\frac{\partial\mathcal{L}}{\partial(\partial_\mu\phi)}(\delta\phi) + \delta\phi^\dagger\frac{\partial\mathcal{L}}{\partial(\partial_\mu\phi^\dagger)} &= \frac{\mathrm{i}}{2}\partial_\mu\phi^\dagger\,(\sigma^a\alpha^a)\,\phi - \frac{\mathrm{i}}{2}\phi^\dagger(\sigma^a\alpha^a)\,\partial_\mu\phi \\ &= \mathrm{i}\alpha^a\left[\partial_\mu\phi^\dagger\frac{\sigma^a}{2}\phi - \phi^\dagger\frac{\sigma^a}{2}\partial_\mu\phi\right],\end{aligned}$$

so putting

$$\alpha^a j_\mu^a = \frac{\partial\mathcal{L}}{\partial(\partial_\mu\phi)}(\delta\phi) + \delta\phi^\dagger\frac{\partial\mathcal{L}}{\partial(\partial_\mu\phi^\dagger)}$$

gives

$$j_\mu{}^a = \mathrm{i}[\partial_\mu\phi^\dagger\frac{\sigma^a}{2}\phi - \phi^\dagger\frac{\sigma^a}{2}\partial_\mu\phi]. \tag{11.69}$$

This current, as well as having four space-time components, also has three 'internal' (isospin, or spin) components labelled by a.

It is easily seen that the equations of motion of ϕ and $\phi^\dagger$ are

$$(\Box - m^2)\phi = 0, \quad (\Box - m^2)\phi^\dagger = 0, \tag{11.70}$$

and that, as a consequence, the current $j_\mu{}^a$ is conserved,

$$\partial^\mu j_\mu{}^a = 0. \tag{11.71}$$

Now to the second question: what happens when the parameters α^a are 'gauged', $\alpha^a \to \alpha^a(x^\mu)$? In analogy with the electromagnetic case we shall expect to have to introduce a 'gauge potential' $A_\mu{}^a$ (which will now carry an extra index a), that the derivative $\partial_\mu\phi$ will have to be replaced by a covariant derivative $D_\mu\phi$, and that there will be a field tensor analogous to $F_{\mu\nu}$. We are considering an isospin rotation

$$U = \exp\left\{\mathrm{i}\frac{\sigma^a}{2}\alpha^a(x)\right\} \tag{11.72}$$

and for infinitesimal $\alpha^a(x)$ the change in ϕ is given by (11.67)

$$\delta\phi(x) = \mathrm{i}g\frac{\sigma^a}{2}\alpha^a(x)\phi(x). \tag{11.73}$$

The covariant derivative $\phi_{;\mu}(x)$ must by definition transform in the same way

$$\delta\phi_{;\mu} = \mathrm{i}g\frac{\sigma^a}{2}\alpha^a(x)\phi_{;\mu}. \tag{11.74}$$

We shall show that this follows if $\phi_{;\mu}$ is defined by

$$\phi_{;\mu} = \phi_{,\mu} - \mathrm{i}gA_\mu{}^a\frac{\sigma^a}{2}\phi, \quad \text{or } D_\mu\phi = \partial_\mu\phi - \mathrm{i}gA_\mu{}^a\frac{\sigma^a}{2}\phi, \tag{11.75}$$

while A_μ^a transforms as

$$A_\mu{}^a \to A_\mu{}^a + \partial_\mu\alpha^a - g\varepsilon_{abc}\alpha^b A_\mu{}^c. \tag{11.76}$$

To show this note that the right hand side of (11.74) is

$$\mathrm{i}g\frac{\sigma^a}{2}\alpha^a\left(\phi_{,\mu} - \mathrm{i}gA_\mu{}^b\frac{\sigma^b}{2}\phi\right) = \mathrm{i}g\frac{\sigma^a}{2}\alpha^a\phi_{,\mu} + \frac{g^2}{4}\alpha^a A_\mu{}^a\phi + \frac{\mathrm{i}g^2}{2}\varepsilon_{abc}\alpha^c A_\mu{}^b\frac{\sigma^c}{2}\phi, \tag{11.77}$$

where we have used the relation

$$\sigma^a\sigma^b = \delta_{ab} + \mathrm{i}\varepsilon_{abc}\sigma^c, \tag{11.78}$$

which holds for the Pauli matrices (as the reader can easily check); note that there is no distinction between upper and lower positions for the indices a, b, c. On the other hand, from (11.73), (11.75) and (11.76), with a small amount of algebra,

$$\begin{aligned}\delta\phi_{;\mu} &= \delta\phi_{,\mu} - \mathrm{i}g\frac{\sigma^a}{2}[(\delta A_\mu{}^a)\phi + A_\mu{}^a(\delta\phi)] \\ &= \mathrm{i}g\frac{\sigma^a}{2}\alpha^a\phi_{,\mu} + \frac{g^2}{4}\alpha^a A_\mu{}^a\phi + \frac{\mathrm{i}g^2}{2}\varepsilon_{abc}\frac{\sigma^a}{2}\alpha^b A_\mu{}^c\phi\end{aligned}$$

which is the same as (11.77). Hence (11.74) is proved.

It is useful to write down the finite form of the transformation (11.76), which is

$$A_\mu \to U A_\mu U^{-1} - \frac{\mathrm{i}}{g}(\partial_\mu U)U^{-1}, \tag{11.79}$$

where

$$A_\mu = A_\mu{}^a \frac{\sigma^a}{2} \tag{11.80}$$

and U is given by (11.72). Indeed, to lowest order in α^a,

$$\begin{aligned} U A_\mu U^{-1} &= \left(1 + \mathrm{i}g\frac{\sigma^a}{2}\alpha^a\right) A_\mu{}^b \frac{\sigma^b}{2}\left(1 - \mathrm{i}g\frac{\sigma^c}{2}\alpha^c\right) \\ &= A_\mu + \frac{\mathrm{i}g}{4} A_\mu{}^b (\sigma^a \sigma^b \alpha^a - \sigma^b \sigma^c \alpha^c) \\ &= A_\mu + \frac{\mathrm{i}g}{4} A_\mu{}^b [\sigma^a, \sigma^b]\, \alpha^a \\ &= A_\mu - g\varepsilon_{abc}\,\alpha^a A_\mu{}^b \frac{\sigma^c}{2}, \end{aligned}$$

and

$$(\partial_\mu U)U^{-1} = \mathrm{i}g\frac{\sigma^a}{2}\alpha^a{}_{,\mu},$$

so (11.79) is

$$A_\mu{}^a \frac{\sigma^a}{2} \to A_\mu{}^a \frac{\sigma^a}{2} - g\,\varepsilon_{abc}\,\alpha^a A_\mu{}^b \frac{\sigma^c}{2} + \alpha^a{}_{,\mu} \frac{\sigma^a}{2},$$

i.e.

$$A_\mu{}^a \to A_\mu{}^a + \alpha^a{}_{,\mu} - g\,\varepsilon_{abc}\,\alpha^b A_\mu{}^c, \tag{11.81}$$

which is (11.76).

In the notation of (11.80), (11.75) is

$$D_\mu \phi = \partial_\mu \phi - \mathrm{i}g A_\mu \phi,$$

or, as an operator equation for the covariant derivative,

$$D_\mu = \partial_\mu - \mathrm{i}g A_\mu, \tag{11.82}$$

which is the generalisation of (11.25) to the non-abelian case, A_μ above being a 2×2 matrix in this SU(2) theory. Since we now have a potential A_μ in the non-abelian case, we must enquire what the generalisation of the field tensor $F_{\mu\nu}$, Equation (11.21), is to this case. The best way to proceed is to *define* the non-abelian field strength as

$$F_{\mu\nu} = \frac{\mathrm{i}}{g}[D_\mu, D_\nu]. \tag{11.83}$$

From (11.82) this becomes

$$F_{\mu\nu} = \partial_\mu A_\nu - \partial_\nu A_\mu - \mathrm{i}g[A_\mu, A_\nu], \tag{11.84}$$

which after some algebra gives

$$F_{\mu\nu} = F_{\mu\nu}{}^{a}\frac{\sigma^a}{2} \tag{11.85}$$

with

$$F_{\mu\nu}{}^{a} = \partial_\mu A_\nu{}^{a} - \partial_\nu A_\mu{}^{a} + g\,\varepsilon_{abc} A_\mu{}^{b} A_\nu{}^{c}. \tag{11.86}$$

In the non-abelian case we should note that the field strength $F_{\mu\nu}$ has acquired a term *quadratic* in A_μ, which is absent in pure electromagnetism. Under a gauge transformation U (see (11.72)) it can be seen, after some algebra, that

$$F_{\mu\nu} \to U F_{\mu\nu} U^{-1} \tag{11.87}$$

(Problem 11.3). Thus $F_{\mu\nu}$ is not gauge invariant, as the equivalent quantity is in electromagnetism (see (11.38)), and neither is $F_{\mu\nu}F^{\mu\nu}$, which would correspond to the final term in the Lagrangian for the electromagnetic field, Equation (11.27). But since, in this case,

$$F_{\mu\nu}F^{\mu\nu} \to U F_{\mu\nu}F^{\mu\nu} U^{-1},$$

the *trace* of this quantity will indeed be invariant (since traces are unchanged under cyclic permutation of the matrices). With (11.85) we have

$$\mathrm{tr}\,(F_{\mu\nu}F^{\mu\nu}) = 1/4\, F_{\mu\nu}{}^{a} F^{\mu\nu b}\,\mathrm{tr}(\sigma^a\sigma^b) = 1/2\, F_{\mu\nu}{}^{a} F^{\mu\nu a}$$

(see (11.78)), so

$$1/2\,\mathrm{tr}\,(F_{\mu\nu}F^{\mu\nu}) = 1/4 F_{\mu\nu}{}^{a} F^{\mu\nu a}. \tag{11.88}$$

We have, finally, that the Lagrangian for an isospinor field ϕ with a *local* ('gauged') SU(2) symmetry is

$$\mathcal{L} = (D^\mu\phi^\dagger)(D_\mu\phi) + m^2\phi^\dagger\phi - 1/2\,\mathrm{tr}\,(F_{\mu\nu}F^{\mu\nu}). \tag{11.89}$$

There is a conclusion of considerable physical significance to be drawn from this Lagrangian. Since, as we have seen, $F_{\mu\nu}$ contains terms (linear and) quadratic in A_μ, the final term in $\mathcal{L}$ above will contain terms cubic and quartic in A_μ, resulting in the primitive vertices shown in Fig. 11.2. The significance of these is that the gauge field is

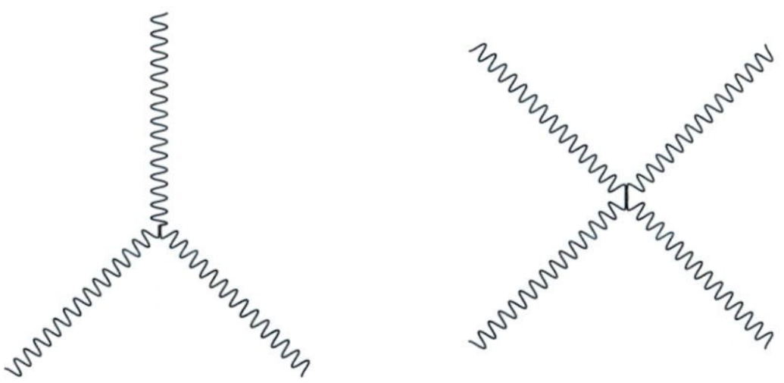

Fig. 11.2 **Primitive vertices for non-abelian vector field gauge theories.**

Table 11.1 Correspondances between non-abelian gauge theories and General Relativity

Non-abelian gauge theory	General Relativity
Gauge potential A_μ	Connection coefficient $\Gamma^\kappa{}_{\lambda\mu}$
Covariant derivative $D_\mu\phi=(\partial_\mu-\mathrm{i}gA_\mu)\phi$	Covariant derivative $V_{\mu;\rho}=V_{\mu,\rho}-\Gamma^\kappa{}_{\mu\rho}\,V_\kappa$
Field strength $F_{..}=\partial_{[.}\,A_{.]}+[A,A]$	Curvature $R^{\cdot}{}_{...}=\partial_{[.}\,\Gamma_{.]}+[\Gamma,\Gamma]$

'self-coupling'; in propagating through space-time it can emit or absorb another quantum of the field – it acts as its own source. Actually, this can be seen already from the fact that the field $A_\mu{}^a$ carries the isospin index a: it has three components, those of an 'isovector'. The whole spirit of gauge theories is that fields (particles) carrying the conserved quantity (isospin) are sources of the isovector field. Thus the field $A_\mu{}^a$ is a source for itself. This is in contrast with the case of electromagnetism. There, charged particles act as sources for the electromagnetic field, but the electromagnetic field itself carries no charge. Photons are not charged. This feature of non-abelian gauge theories has a parallel in General Relativity. Gravitational waves carry energy (albeit non-localised), and anything carrying energy (or equivalently, mass) acts as the source of a gravitational field. Gravitational waves therefore generate an 'extra' gravitational field. This is an aspect of the non-linearity of General Relativity, and is shared by non-abelian gauge theories. In the language of quantum theory, just as there is a 3-boson vertex in SU(2) gauge theory, there should be, in some future quantum theory of gravity, a 3-graviton (and perhaps a 4-graviton) vertex, analogous to those of Fig. 11.2.

To conclude this account of non-abelian gauge theories let us summarise some of the relevant formulae of this section and point out parallel formulae in General Relativity. The correspondances are quite striking and are shown in Table 11.1. The analogies are between the gauge potential A_μ and connection coefficients $\Gamma^\kappa{}_{\lambda\mu}$ on the one hand, with corresponding definitions of 'covariant derivative' in the two theories, and, on the other hand, between the field strength $F_{\mu\nu}$ and curvature $R^\kappa{}_{\lambda\mu\nu}$, defined in terms of the potential and connection coefficient. The formulae for electromagnetism are the same except that the second, commutator, term $[A, A]$ in the definition of F is absent in this case, since it is an *abelian* gauge theory. (The general relativistic formulae are all given in a coordinate basis.) It is hard to believe that these similarities are mere coincidences.

11.3 Gauging Lorentz symmetry: torsion

In the previous sections we have discussed a complex scalar field ϕ whose Lagrangian $\mathcal{L}$ is invariant under the phase transformation

$$\phi \to e^{\mathrm{i}\alpha}\phi.$$

If α is made *space-time dependent* then $\mathcal{L}$ is no longer invariant but invariance may be restored if a 'gauge potential' A_μ is introduced. This is the 4-vector potential of Maxwell's electrodynamics. As an extension of these ideas, if ϕ is taken to be a *two-component complex field* (a 'spinor'), the phase transformation above becomes

$$\phi \rightarrow U\phi$$

where U is a unitary 2×2 matrix with unit determinant – an element of the group SU(2). When the elements of U are made space-time dependent a similar situation to the previous one develops; $\mathcal{L}$ is no longer invariant but invariance may be restored by introducing a gauge potential, this time a 2×2 matrix $A_\mu = \frac{\sigma^a}{2} A_\mu{}^a$. The resulting theory is a generalisation of electromagnetism which turns out to be recognised in nature as electroweak theory, rather than as isospin symmetry in nuclear physics, as originally envisaged by Yang and Mills. Here ϕ corresponds, for example, to the spinor

$$\phi = \begin{pmatrix} \nu_e \\ e^- \end{pmatrix} \tag{11.90}$$

and the three components $A_\mu{}^a$ ($a = 1, 2, 3$) correspond to the $W^\pm$ bosons carrying the weak interaction, and a linear superposition of the photon and the Z boson (recall that this is a quantum theory, so linear superpositions of states are allowed).

In this section we develop the gauge idea further, but in a slightly different direction. Instead of taking ϕ to be a multi-component field whose members differ in their *electric charge*, we take ϕ to be a spinor, but this time in actual 'spin space', whose components differ in their spin projection in a particular direction in space. The characteristics of these fields (mass, spin) derive from space-time itself, not from any 'additional' attributes like electric charge. And the way these properties are understood – see any book on quantum field theory – is to assume *Lorentz invariance*. Our method of procedure, then, is to assume precisely this, but add the crucial ingredient that the *parameters* of the Lorentz transformations ('boost' velocities, rotation angles) are not constant but are functions of space-time. This 'gauging' of Lorentz invariance has the consequence that derivatives must be replaced by 'covariant' derivatives, involving a 'gauge potential'. This covariant derivative, if used in the Dirac equation, a *wave* equation and therefore a differential equation, allows a description of (say) electrons in a Riemannian space-time. This is the general programme for this section, and we begin, for the benefit of readers not already familiar with it, by deriving the Dirac equation.

The Schrödinger equation

$$-\frac{\hbar^2}{2m}\nabla^2\psi + V(x)\,\psi = \mathrm{i}\hbar\frac{\partial\psi}{\partial t} = E\,\psi, \tag{11.91}$$

which was so successful in solving many problems in atomic physics in the first part of the twentieth century, is nevertheless non-relativistic, since it involves second order derivatives in spatial coordinates but a first order derivative with respect to time. A relativistically covariant equation should be consistently first order or consistently second order in both space and time coordinates. Let us take the view that ψ is a field, which when quantised will 'become' a particle of mass m (as in the above equation). Then if this particle has total energy (rest-mass plus kinetic) E and momentum p, Einstein's relation

$$E^2 - c^2p^2 = m^2c^4 \tag{11.92}$$

will hold. Using the 'quantum' substitutions

$$E \to -\mathrm{i}\hbar\frac{\partial}{\partial t}, \quad p_x \to \mathrm{i}\hbar\frac{\partial}{\partial x}, \quad \text{etc.,} \tag{11.93}$$

we may convert (11.92) to a wave equation

$$\left(-\frac{1}{c^2}\frac{\partial^2}{\partial t^2} + \nabla^2\right)\psi = \left(\frac{mc}{\hbar}\right)^2\psi, \tag{11.94}$$

or, with

$$\Box = -\frac{1}{c^2}\frac{\partial^2}{\partial t^2} + \nabla^2 \tag{11.95}$$

and in the units $\hbar = c = 1$,

$$\Box\psi = m^2\psi, \tag{11.96}$$

the Klein–Gordon equation, as seen in (11.4) above. This is clearly a second order relativistic wave equation, but Dirac wanted a first order equation. The easiest way of finding this is simply to suppose that it exists and is of the form

$$\mathrm{i}\gamma^\mu\partial_\mu\psi = -m_\mathrm{D}\psi, \tag{11.97}$$

where the γ^μ are coefficients of the first order derivative operators ∂_μ, m_D is the mass of the Dirac particle and we are using the units $\hbar = c = 1$ for simplicity. Applying the 'operator' $\mathrm{i}\gamma^\nu\partial_\nu$ to both sides of this equation gives the *second order* equation

$$-\left[\left(\gamma^0\right)^2\frac{\partial^2}{\partial t^2} + \left(\gamma^1\right)^2\frac{\partial^2}{\partial x^2} + \cdots\right]\psi + \left[\left(\gamma^0\gamma^1 + \gamma^1\gamma^0\right)\partial_0\partial_1 + \cdots\right]\psi = m_\mathrm{D}^2\psi.$$

This must be the Klein–Gordon equation so we must have

$$(\gamma^0)^2 = 1, \quad (\gamma^1)^2 = (\gamma^2)^2 = (\gamma^3)^2 = -1,$$

$$\{\gamma^0,\gamma^1\} = \{\gamma^0,\gamma^2\} = \cdots = \{\gamma^1,\gamma^2\} = \cdots = 0, \tag{11.98}$$

where

$$\{\gamma^\mu,\gamma^\nu\} \equiv \gamma^\mu\gamma^\nu + \gamma^\nu\gamma^\mu. \tag{11.99}$$

($\{A,B\} = AB + BA$ is often called the *anticommutator* of A and B.) The second set of equations (11.98) means that the coefficients γ^μ cannot be pure numbers, but they could be matrices. In fact they can be 4×4 matrices (but not 2×2 ones), and a specific solution to (11.98) is

$$
\gamma^0 = \begin{pmatrix} 1 & 0 & 0 & 0 \\ 0 & 1 & 0 & 0 \\ 0 & 0 & -1 & 0 \\ 0 & 0 & 0 & -1 \end{pmatrix} \equiv \begin{pmatrix} 1 & 0 \\ 0 & -1 \end{pmatrix},
$$

$$
\gamma^1 = \begin{pmatrix} 0 & 0 & 0 & 1 \\ 0 & 0 & 1 & 0 \\ 0 & -1 & 0 & 0 \\ -1 & 0 & 0 & 0 \end{pmatrix} \equiv \begin{pmatrix} 0 & \sigma^1 \\ -\sigma^1 & 0 \end{pmatrix},
$$

$$
\gamma^2 = \begin{pmatrix} 0 & 0 & 0 & -\mathrm{i} \\ 0 & 0 & \mathrm{i} & 0 \\ 0 & \mathrm{i} & 0 & 0 \\ -\mathrm{i} & 0 & 0 & 0 \end{pmatrix} \equiv \begin{pmatrix} 0 & \sigma^2 \\ -\sigma^2 & 0 \end{pmatrix}, \tag{11.100}
$$

$$
\gamma^3 = \begin{pmatrix} 0 & 0 & 1 & 0 \\ 0 & 0 & 0 & -1 \\ -1 & 0 & 0 & 0 \\ 0 & 1 & 0 & 0 \end{pmatrix} \equiv \begin{pmatrix} 0 & \sigma^3 \\ -\sigma^3 & 0 \end{pmatrix}.
$$

(In these equations *all* the matrices are 4×4: the form of them given on the right is simply easier to deal with.) With the definition (11.99), Equations (11.98) may be written

$$
\{\gamma^\mu, \gamma^\nu\} = -2\eta^{\mu\nu} \tag{11.101}
$$

(see (11.2)). Readers will doubtless enjoy checking the above equations and also convincing themselves that there is no 2×2 solution to (11.101). Since the gamma matrices are 4×4 the wave function ψ must have four components. Assuming the Dirac equation describes spin ½ particles (which it does, though this derivation does not make this clear) one might expect ψ to have two components (spin up and spin down, $j_z = \pm\, \hbar/2$), so what is the significance of the other two solutions? The answer is famous: they describe *antiparticles* – that is, particles of the same mass and spin (½), but opposite electric charge (or other label, for example lepton number L). The Dirac equation in fact describes particles *and* antiparticles together (electrons and positrons, neutrinos and antineutrinos, etc.); it is not a single particle wave equation.

We now want to consider how the Dirac spinor ψ transforms under Lorentz transformations. It is useful first to express the commutation relations between the generators J_i and K_i of rotations and Lorentz boost transformations, in a single equation. These generators were introduced in Chapter 2 and the relevant commutation relations are shown in Equation (2.35). Defining $J_{\mu\nu}$ as follows:

$$
J_{\mu\nu}: \quad J_{ij} = \varepsilon_{ijk} J_k, \quad J_{0i} = K_i, \tag{11.102}
$$

these commutation relations take the form

$$
\mathrm{i}\,[J_{\kappa\lambda}, J_{\mu\nu}] = -\eta_{\kappa\mu} J_{\lambda\nu} + \eta_{\kappa\nu} J_{\lambda\mu} - \eta_{\lambda\nu} J_{\kappa\mu} + \eta_{\lambda\mu} J_{\kappa\nu}. \tag{11.103}
$$

Homogeneous infinitesimal Lorentz transformations are of the form

$$x^\mu \to x'^\mu = x^\mu + \delta x^\mu, \tag{11.104}$$

$$\delta x^\mu = \omega^\mu{}_\nu x^\nu, \quad \omega^{\mu\nu} = -\omega^{\nu\mu}. \tag{11.105}$$

For example an infinitesimal rotation about the z axis gives

$$x' = x + y\theta, \quad y' = -x\theta + y \Rightarrow \delta x = y\theta, \quad \delta y = -x\theta \Rightarrow \omega_{12} = \theta = -\omega_{21}.$$

A Lorentz boost along the x axis, on the other hand,

$$x' = \gamma(x + vt), \quad t' = \gamma(t + vx/c^2),$$

has the infinitesimal form

$$\delta x^1 = \frac{v}{c} x^0, \quad \delta x^0 = \frac{v}{c} x^1 \Rightarrow \omega^{01} = \frac{v}{c} = -\omega^{10},$$

and in both cases $\omega^{\mu\nu}$ is antisymmetric, as in (11.105).

An infinitesimal Lorentz transformation on a *scalar field* $\phi(x)$ is

$$\phi'(x) = \left(1 - \frac{\mathrm{i}}{2}\omega^{\mu\nu} J_{\mu\nu}\right)\phi(x), \tag{11.106}$$

where, with $\hbar = 1$,

$$J_{\mu\nu} = -\mathrm{i}\,(x_\mu \partial_\nu - x_\nu \partial_\mu) \tag{11.107}$$

is the (relativistic) orbital angular momentum operator. For *Dirac fields*, however, with four components, there is also a matrix contribution to $J_{\mu\nu}$, acting on and rearranging these components. We denote this $\Sigma_{\mu\nu}$:

$$J_{\mu\nu} = -\mathrm{i}\,(x_\mu \partial_\nu - x_\nu \partial_\mu) + \Sigma_{\mu\nu}. \tag{11.108}$$

$\Sigma_{\mu\nu}$ must obey the commutation relations (11.103) and it is straightforward to check that

$$\Sigma_{\mu\nu} = \frac{\mathrm{i}}{4}[\gamma_\mu, \gamma_\nu] \tag{11.109}$$

does this: it represents the 'intrinsic spin' operator for the Dirac field. Our aim now is to write down an expression for $\delta\psi(x)$, where $\psi(x)$ is the Dirac spinor, under Lorentz transformations; and then let the parameters $\omega^{\mu\nu}$ become functions of x^μ, hence finishing up, as above, with a covariant derivative to replace ∂_μ.

But at this point it is necessary to take a step back and ask a rather basic question: *how are we to treat spinor fields in General Relativity?* In particular, how are we to find covariant derivatives for them? We are perfectly familiar with the construction of these for co- and contra-variant vectors (vectors and 1-forms) and for mixed tensors of arbitary rank, but what about spinors? This is a non-trivial problem whose origin lies in the fact that the group of general coordinate transformations, which lies at the absolute foundation of General Relativity, has vector and tensor representations, but not a spinor representation. In a sense there was a similar situation with regard to rotations and Lorentz transformations:

rotations are described by the group SO(3), of which there are vector and tensor, but not spinor representations. The group SU(2), however, possesses spinor representations and is homomorphic to SO(3). What is more, nature 'recognises' these spinor representations – see Footnote 5 above. Unfortunately there is no such easy solution for General Relativity. The group of general coordinate transformations is an infinite parameter group and there is no way of finding a similar group with spinor representations. So we return to the question: how do we treat spinor fields in General Relativity? The answer was provided by Weyl.[6] One constructs, at every point in space-time, a vierbein (or tetrad) field $h_\mu{}^\alpha$. This is a set of four orthonormal vectors h_μ, the label α telling which vector, as discussed in Section 6.5 above. These vectors define a *frame*, and because of the Equivalence Principle this frame can be made *inertial* at every point. We have the relations

$$g_{\mu\nu}(x) = h_\mu{}^a(x)\, h_\nu{}^b(x)\; \eta_{ab} = h_\mu{}^a(x)\, h_{\nu a}(x), \tag{11.110}$$

$$eta_{ab} = h^\mu{}_a(x)\, h^\nu{}_b(x)\, g_{\mu\nu}(x) = h^\mu{}_a(x)\, h_{\mu b}(x), \tag{11.111}$$

the raising and lowering of the indices $a, b, \ldots$ being performed with η_{ab} and η^{ab}, and of the indices $\mu, \nu, \ldots$ with $g_{\mu\nu}$ and $g^{\mu\nu}$, the inverses being defined in the usual way. These two types of index have a different significance. The Greek indices (μ, ν etc.) are *world indices*, just as x^μ is a coordinate in the (curved) 'world' space-time. The Latin indices (a, b etc.) are *tangent space* indices. The tangent space is flat – Minkowski – with metric η_{ab}. Physical quantities have separate transformation properties in world space and tangent space. The vierbein $h^\mu{}_a$ is a contravariant vector in world space and a covariant vector in tangent space.

Weyl's proposal was that a Dirac spinor ψ transforms like a *scalar* with respect to 'world' transformations

$$\delta\psi = -\xi^\mu\, \partial_\mu \psi, \tag{11.112}$$

with $\xi^\mu = \omega^\mu{}_\nu\, x^\nu$, but a spinor *with respect to Lorentz transformations in tangent space*

$$\delta\psi = -\frac{\mathrm{i}}{2}\omega^{ab}\, \Sigma_{ab}\, \psi. \tag{11.113}$$

In the general case, then

$$\delta\psi = -\xi^\mu\, \partial_\mu \psi - \frac{\mathrm{i}}{2}\omega^{ab}\, \Sigma_{ab}\, \psi. \tag{11.114}$$

Following the general philosophy of gauge theories we now take the parameters ξ^μ and ω^{ab} to be space-time dependent, though here we shall concentrate only on ω^{ab}, which then becomes $\omega^{ab}(x^\mu)$. In general terms this programme was inspired by the work of Yang and Mills and was embarked on by Utiyama, Sciama and Kibble.[7] We shall outline the results here; the reader is referred to these papers for more details. A Dirac spinor changes by an amount (11.114) under Lorentz transformations, but when $\omega^{ab} \to \omega^{ab}(x^\mu)$ the derivative of ψ will transform as

[6] Weyl (1929, 1950). See also Weinberg (1972), Section 12.5.
[7] Utiyama (1956), Kibble (1961), Sciama (1962).

$$\delta(\psi_{,\mu}) = -\frac{i}{2}\left\{\omega^{ab}\Sigma_{ab}(\psi_{,\mu}) + \omega^{ab}{}_{,\mu}\,\Sigma_{ab}\,\psi\right\}, \tag{11.115}$$

acquiring an anwanted term in $\omega^{ab}{}_{,\mu}$. We then replace this derivative by a 'covariant derivative', denoted $\psi_{|\mu}$ and assumed to be of the form

$$\psi_{|\mu} = \psi_{,\mu} + {}^1\!/\!_2 A^{ab}{}_{\mu}\,\Sigma_{ab}\,\psi \tag{11.116}$$

such that under Lorentz transformation

$$\delta(\psi_{|\mu}) = {}^1\!/\!_2\,\omega^{ab}\,\Sigma_{ab}(\psi_{|\mu}). \tag{11.117}$$

These equations can be solved for $A^{ab}{}_{\mu}$, allowing the Dirac equation to be written down. What is claimed is the the programme of *gauging* Lorentz symmetry results in a theory with the same – or almost the same – structure as General Relativity; that it is a 'back-door' way of arriving at space-time curvature. In support of this view is the fact that the *commutator* of two covariant derivatives (11.116) turns out to be

$$\psi_{|\mu\nu} - \psi_{|\nu\mu} = -{}^1\!/\!_2\,R^{ab}{}_{\mu\nu}\,\Sigma_{ab}\,\psi \tag{11.118}$$

where

$$R^a{}_{b\mu\nu} = A^a{}_{b\nu,\mu} - A^a{}_{b\mu,\nu} + A^a{}_{c\mu}A^c{}_{b\nu} - A^a{}_{c\nu}A^c{}_{b\mu}. \tag{11.119}$$

This expression is of precisely the same form as the Riemann tensor. Assuming that it actually *is* the Riemann tensor, or rather, that

$$R^{\kappa}{}_{\lambda\mu\nu} = h_a{}^{\kappa}h^b{}_{\lambda}R^a{}_{b\mu\nu} \tag{11.120}$$

is the Riemann tensor, it turns out that we have reproduced the key feature of Riemannian geometry – the curvature tensor. It also, however, turns out that this approach has an additional feature which General Relativity lacks, which is that the connection coefficients $\Gamma^{\lambda}{}_{\mu\nu}$ are not necessarily symmetric in their lower indices, $\Gamma^{\lambda}{}_{\mu\nu} \neq \Gamma^{\lambda}{}_{\nu\mu}$. In that case we would have a space-time with *torsion* as well as with curvature, the torsion tensor being defined by

$$S_{\mu\nu}{}^{\lambda} = {}^1\!/\!_2\,(\Gamma^{\lambda}{}_{\mu\nu} - \Gamma^{\lambda}{}_{\nu\mu}). \tag{11.121}$$

Moreover, just as the curvature of space-time is induced by mass, or more generally the energy-momentum tensor,

$${}^1\!/\!_2\,g\,T^{\mu\nu} = \frac{\partial\mathcal{L}}{\partial g_{\mu\nu}},$$

the torsion of space-time is, in some theories, induced by *spin*

$$g\,\tau_{\mu}{}^{\kappa\lambda} = \frac{\partial\mathcal{L}}{\partial K_{\lambda\kappa}{}^{\mu}},$$

where $K_{\lambda\kappa}{}^{\mu} = S_{\lambda}{}^{\mu}{}_{\kappa} + S_{\kappa}{}^{\mu}{}_{\lambda} - S_{\lambda\kappa}{}^{\mu}$ is the so-called *contorsion tensor* and $\tau_{\mu}{}^{\kappa\lambda}$ is the spin angular momentum density. Theories of this type result in a new spin–spin force of purely gravitational origin. Trautman has even speculated that such forces may avert gravitational singularities.[8]

It is worth noting, as a final remark on gravitational theories with torsion (often called Einstein–Cartan theories) that they often play a role in *quantum theory.* This should be clear from the treatment above: the Dirac field is a four component field which serves as a basis for a representation of the Lorentz group. That is, it belongs to a *vector space*, and vectors (belonging to vector spaces) can be *added* – they obey the linear *superposition principle*, which lies at the foundation of quantum mechanics. Consider even something as elementary as the two-slit experiment, as explicated by Feynman:[9] or, more simply, just note that the spin operator is proportional to Planck's constant – non-relativistically it is $(\hbar/2)\sigma^i$.

11.4 Dirac equation in Schwarzschild space-time

The Dirac equation in Minkowski space-time is (11.97), which on restoring factors of $\hbar$ and c is

$$\mathrm{i}\hbar\gamma^\mu\partial_\mu\psi = -m_\mathrm{D}\,c\,\psi. \tag{11.122}$$

We must now replace the derivative $\partial_\mu\psi$ by a covariant one, as in (11.116). Let us first write $\partial_\mu\psi$ in terms of a 'spinor 1-form'

$$\mathbf{d}\psi = \partial_\mu\psi\,\boldsymbol{\theta}^\mu, \tag{11.123}$$

where the $\boldsymbol{\theta}^\mu$ are basis 1-forms. The consequences of Weyl's prescription (11.114) is that instead of (11.123) we now have

$$\mathbf{d}\psi \to \mathbf{D}\psi = \mathbf{d}\psi - \frac{\mathrm{i}}{2}\boldsymbol{\omega}^{\kappa\lambda}\Sigma_{\kappa\lambda}\,\psi, \tag{11.124}$$

where $\boldsymbol{\omega}^{\kappa\lambda}$ is a connection 1-form, sometimes called the spinor connection (and is *not* to be confused with the *parameters* $\omega^{\kappa\lambda}$ of a Lorentz transformation!). In terms of the basis forms it may be expanded, as in Equation (3.171)

$$\boldsymbol{\omega}^\kappa{}_\lambda = \Gamma^\kappa{}_{\lambda\mu}\boldsymbol{\theta}^\mu. \tag{11.125}$$

With (11.109), then we have

$$\mathbf{D}\psi = \{\partial_\mu\psi + {}^1\!/\!_8\Gamma_{\kappa\lambda\mu}[\gamma^\kappa,\gamma^\lambda]\,\psi\}\boldsymbol{\theta}^\mu, \tag{11.126}$$

where

$$\Gamma_{\kappa\lambda\mu} = g_{\kappa\rho}\,\Gamma^\rho{}_{\lambda\mu}, \tag{11.127}$$

so the Dirac equation becomes

$$\mathrm{i}\hbar\gamma^\mu\{\partial_\mu + {}^1\!/\!_8\Gamma_{\kappa\lambda\mu}[\gamma^\kappa,\gamma^\lambda]\}\,\psi = m_\mathrm{D}\,c\,\psi. \tag{11.128}$$

[8] Trautman (1973a). See also Kopczyński & Trautman (1992), Chapter 19.

[9] Feynman *et al.* (1965), Chapter 1.

There is one further modification to make, which is relevant to the way we shall perform the calculation in Schwarzschild space. This is to replace the vector ∂_μ, defined in a *holonomic* (coordinate) basis, by e_μ, defined in an anholonomic one. In fact we shall work in an orthonormal basis, so (11.128) becomes

$$i\hbar\,\gamma^\mu(e_\mu + \Gamma_\mu)\,\psi = m_D\,c\,\psi, \tag{11.129}$$

where

$$\Gamma_\mu = {}^1\!/_8\Gamma_{\kappa\lambda\mu}[\gamma^\kappa, \gamma^\lambda] \tag{11.130}$$

and $\Gamma_{\kappa\lambda\mu}$ is given by (3.259).

Let us now find the Dirac equation in Schwarzschild space-time, with line element (5.37)

$$ds^2 = -\left(1 - \frac{2m}{r}\right)c^2\,dt^2 + \left(1 - \frac{2m}{r}\right)^{-1} dr^2 + r^2(d\theta^2 + \sin^2\theta\,d\phi^2). \tag{5.37}$$

Write this as

$$ds^2 = -(\boldsymbol{\theta}^0)^2 + (\boldsymbol{\theta}^1)^2 + (\boldsymbol{\theta}^2)^2 + (\boldsymbol{\theta}^3)^2,$$

with

$$\boldsymbol{\theta}^0 = c\left(1 - \frac{2m}{r}\right)^{1/2}\mathbf{d}t, \quad \boldsymbol{\theta}^1 = \left(1 - \frac{2m}{r}\right)^{-1/2}\mathbf{d}r, \quad \boldsymbol{\theta}^2 = r\mathbf{d}\theta, \quad \boldsymbol{\theta}^3 = r\sin\theta\,\mathbf{d}\phi \tag{11.131}$$

and

$$g_{\mu\nu} = \eta_{\mu\nu}, \tag{11.132}$$

an orthonormal basis. The dual basis is clearly

$$\mathbf{e}_0 = \frac{1}{c}\left(1 - \frac{2m}{r}\right)^{-1/2}\frac{\partial}{\partial t}, \quad \mathbf{e}_1 = \left(1 - \frac{2m}{r}\right)^{1/2}\frac{\partial}{\partial r}, \quad \mathbf{e}_2 = \frac{1}{r}\frac{\partial}{\partial\theta}, \quad \mathbf{e}_3 = \frac{1}{r\sin\theta}\frac{\partial}{\partial\phi} \tag{11.133}$$

(and note that $m = \dfrac{MG}{c^2}$, not the mass of the Dirac particle, which we have denoted m_D). Then the commutators of these vectors may be found; for example

$$[\mathbf{e}_0, \mathbf{e}_1] = \frac{m}{r^2}\left(1 - \frac{2m}{r}\right)^{-1/2}\mathbf{e}_0 \Rightarrow C_{01}{}^0 = \frac{m}{r^2}\left(1 - \frac{2m}{r}\right)^{-1/2}. \tag{11.134}$$

The non-zero commutators turn out to be

$$\begin{aligned} C_{010} &= -\frac{m}{r^2}\left(1 - \frac{2m}{r}\right)^{-1/2} = -C_{100}, \\ C_{122} = -C_{212} &= -\frac{1}{r}\left(1 - \frac{2m}{r}\right)^{1/2} = C_{133} = -C_{313}, \\ C_{233} = -C_{323} &= -\frac{\cot\theta}{r}. \end{aligned} \tag{11.135}$$

The quantities $\Gamma_{\mu\nu\lambda}$ are then found from (3.259), appropriate to this orthonormal basis,

$$\Gamma_{\mu\nu\lambda} = -{}^1\!/\!_2\,[C_{\mu\nu\lambda} + C_{\nu\lambda\mu} - C_{\lambda\mu\nu}], \tag{11.136}$$

giving

$$\begin{aligned}
\Gamma_{010} = -\Gamma_{100} &= \frac{m}{r^2}\left(1-\frac{2m}{r}\right)^{-1/2}, \qquad \Gamma_{001} = 0,\\
\Gamma_{122} = -\Gamma_{212} = \Gamma_{133} = -\Gamma_{313} &= \frac{m}{r^2}\left(1-\frac{2m}{r}\right)^{1/2}, \qquad \Gamma_{221} = \Gamma_{331} = 0,\\
\Gamma_{233} = -\Gamma_{323} &= \frac{\cot\theta}{r}, \qquad \Gamma_{332} = 0.
\end{aligned} \tag{11.137}$$

The quantites Γ_μ may then be found from (11.130):

$$\begin{aligned}
\Gamma_0 &= \frac{m}{2r^2}\left(1-\frac{2m}{r}\right)^{-1/2}\alpha^1,\\
\Gamma_1 &= 0, \quad \Gamma_2 = -\frac{\mathrm{i}m}{2r^2}\left(1-\frac{2m}{r}\right)^{1/2}\Sigma^3,\\
\Gamma_3 &= \frac{\mathrm{i}m}{2r^2}\left(1-\frac{2m}{r}\right)^{1/2}\Sigma^2 - \frac{\mathrm{i}\cot\theta}{2r}\Sigma^1,
\end{aligned} \tag{11.138}$$

where

$$\alpha^1 = \begin{pmatrix} 0 & \sigma^1 \\ \sigma^1 & 0 \end{pmatrix}, \quad \Sigma^i = \begin{pmatrix} \sigma^i & 0 \\ 0 & \sigma^i \end{pmatrix}$$

are 4×4 matrices. Substituting (11.138) and (11.133) into (11.129) and noting that

$$\gamma^0\alpha^1 = -\gamma^1, \quad \gamma^2\Sigma^3 = -\gamma^3\Sigma^2 = \mathrm{i}\gamma^1, \quad \gamma^3\Sigma^1 = \mathrm{i}\gamma^2$$

gives the equation

$$\begin{aligned}
\mathrm{i}\hbar\Bigg\{&\left(1-\frac{2m}{r}\right)^{-1/2}\left[\gamma^0\frac{1}{c}\frac{\partial\psi}{\partial t} - \frac{m}{2r^2}\gamma^1\psi\right] + \left(1-\frac{2m}{r}\right)^{1/2}\gamma^1\frac{\partial\psi}{\partial r}\\
&+\gamma^2\frac{1}{r}\frac{\partial\psi}{\partial\theta} + \frac{m}{r^2}\left(1-\frac{2m}{r}\right)^{1/2}\gamma^1\psi + \gamma^3\frac{1}{r\sin\theta}\frac{\partial\psi}{\partial\phi} + \frac{\cot\theta}{2r}\gamma^2\psi\\
&= m_{\mathrm{D}}\,c\,\psi,
\end{aligned} \tag{11.139}$$

for a Dirac particle of mass m_{D} in a Schwarzschild field.

11.5 Five dimensions: gravity plus electromagnetism

In the early 1920s a number of people constructed models of 'unified field theories' – that is, unifications of gravity and electromagnetism. Most of these have not stood the test of time, largely because we now realise that unification is a greater task than was then thought; it should now include QCD and electroweak theory. In 1921 Kaluza,[10] however, constructed a unified model (of gravity and electromagnetism) by extending the space-time manifold to five dimensions, earning, in the process, praise from Einstein: 'At first glance I like your idea enormously', he wrote. Klein later speculated that the fifth dimension might have something to do with quantisation,[11] and this joint theory became known as Kaluza–Klein theory. Because the number of dimensions may in principle be extended beyond five, to accomodate the other interactions beyond electromagnetism, enlarged versions of Kaluza–Klein theory play a part in contemporary string theory. In this section we shall outline Kaluza's original proposal, restricted to five dimensions.

The rationale is to extend the line element

$$\mathrm{d}s^2 = g_{\mu\nu}\,\mathrm{d}x^\mu\,\mathrm{d}x^\nu \quad (\mu, \nu = 0, 1, 2, 3)$$

to a 5-dimensional manifold

$$\mathrm{d}s^2 = \gamma_{mn}\,\mathrm{d}x^m\,\mathrm{d}x^n \quad (m, n = 0, 1, 2, 3, 5) \tag{11.140}$$

where the new metric tensor is

$$\gamma_{mn} = \begin{pmatrix} \gamma_{00} & \gamma_{01} & \gamma_{02} & \gamma_{03} & \gamma_{05} \\ \gamma_{10} & \gamma_{11} & \gamma_{12} & \gamma_{13} & \gamma_{15} \\ \gamma_{20} & \gamma_{21} & \gamma_{22} & \gamma_{23} & \gamma_{25} \\ \gamma_{30} & \gamma_{31} & \gamma_{32} & \gamma_{33} & \gamma_{35} \\ \gamma_{50} & \gamma_{51} & \gamma_{52} & \gamma_{53} & \gamma_{55} \end{pmatrix} = \begin{pmatrix} \gamma_{\mu\nu} & \gamma_{\mu 5} \\ \gamma_{5\nu} & \gamma_{55} \end{pmatrix}. \tag{11.141}$$

Under general 5-dimensional coordinate transformations

$$\gamma'_{mn} = \frac{\partial x^p}{\partial x'^m}\frac{\partial x^q}{\partial x'^n}\,\gamma_{pq}. \tag{11.142}$$

Consider translations in the fifth dimension

$$x'^\mu = x^\mu, \quad x'^5 = x^5 - f(x^\mu), \tag{11.143}$$

then

$$\frac{\partial x^\nu}{\partial x'^\mu} = \delta^\nu{}_\mu, \quad \frac{\partial x^\mu}{\partial x'^5} = 0, \quad \frac{\partial x^5}{\partial x'^\mu} = \partial_\mu f, \quad \frac{\partial x^5}{\partial x'^5} = 1,$$

and hence

$$\gamma'_{\mu 5} = \gamma_{\mu 5} + (\partial_\mu f)\gamma_{55}. \tag{11.144}$$

Similarly we find

$$\gamma'_{55} = \gamma_{55}.$$

[10] Kaluza (1921).
[11] Klein (1926).

Putting

$$\gamma_{\mu 5} = \gamma_{5\mu} = A_\mu, \quad \gamma_{55} = 1 \tag{11.145}$$

then (11.144) becomes

$$A'_\mu = A_\mu + \partial_\mu f, \tag{11.146}$$

a *gauge transformation* in electrodynamics, under which $F_{\mu\nu}$, and therefore the electric and magnetic fields, are invariant. We also find, under (11.143),

$$\gamma'_{\mu\nu} = \gamma_{\mu\nu} + A_\nu(\partial_\mu f) + A_\mu(\partial_\nu f) + (\partial_\mu f)(\partial_\nu f).$$

But also $g'_{\mu\nu} = g_{\mu\nu}$ so $(g_{\mu\nu} + A_\mu A_\nu)$ transforms in this way, and we may therefore identify

$$\gamma_{\mu\nu} = g_{\mu\nu} + A_\mu A_\nu \tag{11.147}$$

and write the 5×5 metric tensor γ_{mn} in the form

$$\gamma_{mn} = \begin{pmatrix} g_{\mu\nu} + A_\mu A_\nu & A_\mu \\ A_\nu & 1 \end{pmatrix}. \tag{11.148}$$

We have succeeded then in relating translations in the fifth dimension to gauge transformations in electrodynamics. We now have to consider the theories of General Relativity and electrodynamics themselves and discover how this 5-dimensional formulation will relate them. We shall work at the level of the integrands of the *actions* of the theories. As we saw in (8.21) the integrand of the Einstein action is the curvature scalar R, which we shall here denote $\bar{R}$, the overbar denoting that the quantity is a strictly 4-dimensional one. The symbol R will now refer to the 5-dimensional scalar, and what we shall show is that

$$R = \bar{R} + {}^1\!/\!_4 F_{\mu\nu} F^{\mu\nu}, \tag{11.149}$$

the second term being the Maxwell action – see (11.23) above. A direct calculation with γ_{mn} above is rather messy; it is better to work with differential forms in an orthonormal basis.[12]

The 5-dimensional line element is

$$\begin{aligned} ds^2 &= \gamma_{mn}\, dx^m\, dx^n = (g_{\mu\nu} + A_\mu A_\nu)\, dx^\mu\, dx^\nu + 2A_\mu\, dx^\mu\, dx^5 + (dx^5)^2 \\ &= g_{\mu\nu}\, dx^\mu\, dx^\nu + (dx^5 + A_\mu\, dx^\mu)^2. \end{aligned}$$

Writing $g_{\mu\nu}$ in terms of the vierbein $h_\mu{}^a$,

$$g_{\mu\nu} = h_\mu{}^a\, h_\nu{}^b \eta_{ab}, \tag{11.150}$$

and putting

$$\boldsymbol{\theta}^a = h_\mu{}^a\, \mathbf{d}x^\mu, \quad \boldsymbol{\theta}^5 = \mathbf{d}x^5 + A_\mu\, \mathbf{d}x^\mu \equiv \mathbf{d}x^5 + A_a\, \boldsymbol{\theta}^a \tag{11.151}$$

gives

$$ds^2 = \boldsymbol{\theta}^a\, \boldsymbol{\theta}^b\, \eta_{ab} + (\boldsymbol{\theta}^5)^2 = \gamma_{AB}\, \boldsymbol{\theta}^A\, \boldsymbol{\theta}^B, \tag{11.152}$$

12 We follow here the method of Thirring (1972).

with

$$\gamma_{AB} = \begin{pmatrix} \eta_{ab} & 0 \\ 0 & 1 \end{pmatrix}. \tag{11.153}$$

The indices

$$A, B = 0, \ldots, 5 \text{ and } a, b = 0, \ldots, 3 \tag{11.154}$$

are cotangent space indices. We must now find the connection forms $\boldsymbol{\omega}_{AB}$ as a preliminary to finding the curvature tensor. We use the *Cartan structure equations* (3.182)

$$\mathbf{d}\boldsymbol{\theta}^A + \boldsymbol{\omega}^A{}_B \wedge \boldsymbol{\theta}^B = 0, \quad \boldsymbol{\omega}_{AB} = -\boldsymbol{\omega}_{BA}. \tag{11.155}$$

With $A = 5$ they give

$$\mathbf{d}\boldsymbol{\theta}^5 = -\boldsymbol{\omega}^5{}_a \wedge \boldsymbol{\theta}^a, \tag{11.156}$$

but on the other hand, from (11.151)

$$\mathbf{d}\boldsymbol{\theta}^5 = A_{a,b}\,\boldsymbol{\theta}^b \wedge \boldsymbol{\theta}^a = {}^1\!/\!{}_2\,F_{ab}\,\boldsymbol{\theta}^a \wedge \boldsymbol{\theta}^b, \tag{11.157}$$

from which we find the connection form

$$\boldsymbol{\omega}^5{}_a = {}^1\!/\!{}_2\,F_{ab}\,\boldsymbol{\theta}^b. \tag{11.158}$$

The basis form $\boldsymbol{\theta}^a$ in (11.152) is intrinsically 4-dimensional, so we may write

$$\boldsymbol{\theta}^a = h_\mu{}^a\,\mathbf{d}x^\mu = \bar{\boldsymbol{\theta}}^a, \tag{11.159}$$

an overbar, as before, denoting an intrinsically 4-dimensional quantity. On taking the exterior derivative of this equation we have, from (11.155)

$$\begin{aligned}\mathbf{d}\boldsymbol{\theta}^a &= -\boldsymbol{\omega}^a{}_A \wedge \boldsymbol{\theta}^A = -\boldsymbol{\omega}^a{}_b \wedge \boldsymbol{\theta}^b - \boldsymbol{\omega}^a{}_5 \wedge \boldsymbol{\theta}^5 \\ &= -\boldsymbol{\omega}^a{}_b \wedge \boldsymbol{\theta}^b + {}^1\!/\!{}_2\,(F^a{}_b\,\boldsymbol{\theta}^b) \wedge \boldsymbol{\theta}^5 \\ &= -(\boldsymbol{\omega}^a{}_b + {}^1\!/\!{}_2\,F^a{}_b\boldsymbol{\theta}^5) \wedge \boldsymbol{\theta}^b.\end{aligned} \tag{11.160}$$

On the other hand we may put

$$\mathbf{d}\bar{\boldsymbol{\theta}}^a = -\bar{\boldsymbol{\omega}}^a{}_b \wedge \bar{\boldsymbol{\theta}}^b = -\bar{\boldsymbol{\omega}}^a{}_b \wedge \boldsymbol{\theta}^b,$$

thus *defining* the intrinsically 4-dimensional connection form $\bar{\boldsymbol{\omega}}^a{}_b$. A comparison of these equations yields

$$\boldsymbol{\omega}^a{}_b = -\bar{\boldsymbol{\omega}}^a{}_b - {}^1\!/\!{}_2\,F^a{}_b\,\boldsymbol{\theta}^5. \tag{11.161}$$

Now use the second Cartan structure equation

$$\mathbf{d}\boldsymbol{\omega}_{AB} + \boldsymbol{\omega}_{AC} \wedge \boldsymbol{\omega}^C{}_B = {}^1\!/\!{}_2\,R_{ABCD}\,\boldsymbol{\theta}^C \wedge \boldsymbol{\theta}^D, \tag{11.162}$$

where R_{ABCD} is the 5-dimensional Riemann curvature tensor. From (11.161) we have, using (11.157)

$$\mathbf{d}\boldsymbol{\omega}_{ab} = \mathbf{d}\bar{\boldsymbol{\omega}}_{ab} - {}^1\!/\!_2\,\mathbf{F}_{ab,c}\,\boldsymbol{\theta}^c \wedge \boldsymbol{\theta}^5 - {}^1\!/\!_2\,\mathbf{F}_{ab}({}^1\!/\!_2\,\mathbf{F}_{cd})\,\boldsymbol{\theta}^c \wedge \boldsymbol{\theta}^d,$$

so the left hand side of (11.162) is, with $A = a$ and $B = b$, and using (11.158),

$$\begin{aligned}
&\mathbf{d}\boldsymbol{\omega}_{ab} + \boldsymbol{\omega}_{ac} \wedge \boldsymbol{\omega}^c{}_b + \boldsymbol{\omega}_{a5} \wedge \boldsymbol{\omega}^5{}_b \\
&= \mathbf{d}\,\bar{\boldsymbol{\omega}}^a{}_b - {}^1\!/\!_2\,F_{abc}\,\boldsymbol{\theta}^c \wedge \boldsymbol{\theta}^5 - {}^1\!/\!_2\,F_{ab}\,F_{cd}\,\boldsymbol{\theta}^c \wedge \boldsymbol{\theta}^d \\
&\quad + (\bar{\boldsymbol{\omega}}_{ac} - {}^1\!/\!_2\,F_{ac}\,\boldsymbol{\theta}^5) \wedge (\bar{\boldsymbol{\omega}}^c{}_b - {}^1\!/\!_2\,F^c{}_b\,\boldsymbol{\theta}^5) + (-{}^1\!/\!_2\,F_{ac}\,\boldsymbol{\theta}^c) \wedge ({}^1\!/\!_2\,F_{bd}\,\boldsymbol{\theta}^d).
\end{aligned} \tag{11.163}$$

The right hand side of (11.162) is, again with $A = a$ and $B = b$,

$$ {}^1\!/\!_2\,R_{abcd}\boldsymbol{\theta}^c \wedge \boldsymbol{\theta}^d + R_{abc5}\,\boldsymbol{\theta}^c \wedge \boldsymbol{\theta}^5. \tag{11.164}$$

Comparing these equations we may find R_{abcd} by looking at the coefficients of $\boldsymbol{\theta}^c \wedge \boldsymbol{\theta}^d$. Writing the contribution to R_{abcd} which comes purely from 4-dimensional quantities as $\bar{R}_{abcd}$, we can separate out the contribution which comes from the fifth dimension:

$${}^1\!/\!_2\,R_{abcd} = {}^1\!/\!_2\bar{R}_{abcd} - {}^1\!/\!_4\,F_{ab}\,F_{cd} - {}^1\!/\!_8(F_{ac}\,F_{bd} - F_{ad}\,F_{bc})$$

and hence

$${}^1\!/\!_2\,R^{ab}{}_{ab} = {}^1\!/\!_2\,\bar{R}^{ab}{}_{ab} - {}^1\!/\!_4\,F^{ab}\,F_{ab} + {}^1\!/\!_8\,F^{ab}\,F_{ab} = {}^1\!/\!_2\,\bar{R}^{ab}{}_{ab} - {}^1\!/\!_8\,F^{ab}\,F_{ab}. \tag{11.165}$$

In a similar way, putting $A = a$, $B = 5$ in (11.162) gives, for the left hand side

$$\begin{aligned}
&\mathbf{d}\boldsymbol{\omega}_{a5} + \boldsymbol{\omega}_{ab} \wedge \boldsymbol{\omega}^b{}_5 \\
&= -{}^1\!/\!_2\,F_{ab,c}\,\boldsymbol{\theta}^c \wedge \boldsymbol{\theta}^d - {}^1\!/\!_2\,F_{ab}(-\bar{\boldsymbol{\omega}}^b{}_c \wedge \boldsymbol{\theta}^c) + (\bar{\boldsymbol{\omega}}_{ab} - {}^1\!/\!_2\,F_{ab}\,\boldsymbol{\theta}^5) \wedge (-{}^1\!/\!_2\,F^b{}_c\,\boldsymbol{\theta}^c).
\end{aligned}$$

The right hand side is

$${}^1\!/\!_2\,R_{a5cd}\,\boldsymbol{\theta}^c \wedge \boldsymbol{\theta}^d + R_{a5c5}\,\boldsymbol{\theta}^c \wedge \boldsymbol{\theta}^5,$$

and following the same logic as before, looking at the coefficients of $\boldsymbol{\theta}^c \wedge \boldsymbol{\theta}^5$ in these equations gives

$$R_{a5c5} = \bar{R}_{a5c5} - {}^1\!/\!_4\,F_{ab}\,F^b{}_c,$$

hence

$$R^{a5}{}_{a5} = \bar{R}^{a5}{}_{a5} - {}^1\!/\!_4\,F^a{}_b\,F^b{}_a = \bar{R}^{a5}{}_{a5} + {}^1\!/\!_4\,F^{ab}\,F_{ab}. \tag{11.166}$$

The Lagrangian in the 5-dimensional theory is

$$R = R^{AB}{}_{AB} = R^{ab}{}_{ab} + 2\,R^{a5}{}_{a5}. \tag{11.167}$$

This will receive contributions from the 4-dimensional sector and from the fifth dimension. Distinguishing these, as above, by an overbar gives, from (11.165) and (11.166)

$$R = \bar{R} - {}^1\!/\!_4\,F^{ab}\,F_{ab} + {}^1\!/\!_2\,F^{ab}\,F_{ab} = \bar{R} + {}^1\!/\!_4\,F^{ab}\,F_{ab}. \tag{11.168}$$

This, then, is the result of this theory: the action of the 5-dimensional theory is simply the *sum* of the actions for General Relativity and Maxwell's electrodynamics. This is, of course, a remarkable result, but a disappointment is that it does not amount to a *unified* theory – gravity and electrodynamics separate out, like oil and water. There is no 'coupling' between them. If there were cosmic 'switches' allowing the different interactions in nature to be switched on and off at will, gravity could still be switched off without affecting electromagnetism, and vice versa. Pauli is reputed to have remarked, as an ironic variation on the words of the priest at a wedding service, 'Let not man unite what God has put asunder'.

In this chapter we have considered a few of the topics that have appeared on the agenda since Einstein's day, as a consequence of his great theory. There are, of course other questions raised by General Relativity, perhaps the most famous of which is quantum gravity. How should a quantum theory of gravity be constructed? There is as yet no generally agreed answer to this question, but many clever people have devoted many years to thinking about it. At the end of an introduction to Einstein's theory, however, it is best not immediately to start thinking about the next challenge. Like a climber who has arrived at the top of his mountain, we should simply sit down and admire the view. Is it not absolutely remarkable that Einstein was able to create a new theory of gravity in which the geometry of space itself became a part of physics? Whatever would Euclid have thought?

Further reading

A good account of gauge invariance in electrodynamics is to be found in Aitchison & Hey (1982). Readers who wish to familiarise themselves with some particle physics will find good accounts in Burcham & Jobes (1995), Perkins (2000) and, at a slightly higher level, Cottingham & Greenwood (1998). An excellent, though largely non-mathematical account of gauge fields will be found in Taylor (2001) and an account of the mathematical basis of gauge field theory, including their fibre bundle formulation, is Healey (2007).

For more information on the relation between SO(3) and SU(2) and between spinors and vectors see for example Sakurai (1994), Chapter 3, or Ryder (1996) Chapter 2. Good accounts of non-abelian gauge theory appear in Rubakov (2002) and Maggiore (2005); see also Aitchison & Hey (1982), Cheng & Li (1984), Ryder (1996) and Srednicki (2007). An extended treatment of the formal mathematical approach to gauge theories and General Relativity may be found in Frankel (1997); see also Göckeler & Schücker (1987) and Nakahara (1990). A review of Poincaré gauge theory appears in Blagojević (2002). Good accounts of the Dirac equation may be found in, for example, Bjorken & Drell (1964), Itzykson & Zuber (1980), Brown (1992), Gross (1993) or Huang (1998). Good introductions to theories of space-time with torsion are Hehl & von der Heyde (1973) and Hammond (1994, 1995). More complete accounts may be found in Hehl (1973, 1974); see also Trautman (1973b), Hehl *et al.* (1976) and Shapiro (2002). The Dirac equation in a gravitational field is treated by Brill & Cohen (1966), Chapman & Leiter (1976) and Sexl & Urbantke (1983).

There are accounts of Kaluza–Klein theory in Bergmann (1942), Pauli (1958) and Pais (1982). A review of the generalisations of this theory to higher dimensions is Bailin & Love (1987).

Problems

11.1 Show that, with $R_z(\alpha)$ and J_z given by (11.44) and (11.47),

$$R_z(\alpha) = \exp(\mathrm{i}J_z\alpha).$$

11.2 Show that

$$\exp[\mathrm{i}\,\mathbf{n}\cdot\boldsymbol{\sigma}\,(\alpha/2)] = \cos(\alpha/2) + \mathrm{i}\,\mathbf{n}\cdot\boldsymbol{\sigma}\,\sin(\alpha/2).$$

11.3 Show that under the gauge transformation U, given by (11.72), the field tensor $F_{\mu\nu}$, given by (11.84), transforms as in (11.87):

$$F_{\mu\nu} \to U\,F_{\mu\nu}\,U^{-1}.$$

11.4 Find the Dirac equation in Minkowski space in spherical polar coordinates.

References

Adler, R., Bazin, M. & Schiffer, M. (1975), *Introduction to General Relativity* (2nd edn), New York: McGraw-Hill

Aitchison, I. J. R. & Hey, A. J. G. (1982), *Gauge Theories in Particle Physics*, Bristol: Adam Hilger

Alpher, R. A., Bethe, H. & Gamow, G. (1948), The origin of chemical elements, *Physical Review* **73**, 803–804

Alpher, R. A. & Herman, R. C. (1948a), Evolution of the Universe, *Nature* **162**, 774–775

Alpher, R. A. & Herman, R. C. (1948b), On the relative abundance of the elements, *Physical Review* **74**, 1737–1742

Anderson, J. L. (1967), *Principles of Relativity Physics*, New York: Academic Press

Anderson, R., Bilger, H. R. & Stedman, G. E. (1994), 'Sagnac' effect: a century of Earth-rotated interferometers, *American Journal of Physics* **62**, 975–985

Arfken, G. (1970), *Mathematical Methods for Physicists* (2nd edn), New York: Academic Press

Arnowitt, R., Deser, S. & Misner, C. W. (1962), The dynamics of General Relativity, in L. Witten (ed.), *Gravitation: An Introduction to Current Research*, New York: Wiley; http://arXiv.org:gr-qc/0405109

Ashcroft, N. W. & Mermin, N. D. (1976), *Solid State Physics*, Philadelphia: Saunders

Bacry, H. (1977), *Lectures on Group Theory and Particle Theory*, New York: Gordon and Breach

Bailin, D. & Love, A. (1987), Kaluza-Klein theories, *Reports on Progress in Physics* **50**, 1087–1170

Barbashov, B. M., Pervushin, V. N. & Pawlowski, M. (2001), Time-reparametrization-invariant dynamics of relativistic systems, *Элементарных Частиц и Атомного Ядра* (Dubna) **32**, 546

Barbour, J. B. & Pfister, H. (1995), *Mach's Principle: From Newton's Bucket to Quantum Gravity*, Boston: Birkhäuser

Bardeen, J., Cooper, L. N. & Schrieffer, J. R. (1957), Theory of superconductivity, *Physical Review* **108**, 1175–1204

Bardeen, J. M., Carter, B. & Hawking, S. W. (1973), The four laws of black hole mechanics, *Communications in Mathematical Physics* **31**, 161–170

Barish, B. C. (2002), Gravitational waves: the new generation of laser interferometric detectors, in *Proceedings of the Ninth Marcel Grossmann Meeting on General Relativity* (V. G. Gurzadyan, R. T. Jantzen & R. Ruffini, eds), Singapore: World Scientific

Bażański, S. (1962), The problem of motion, in *Recent Developments in General Relativity*, Oxford: Pergamon Press, and Warszawa: PWN Polish Scientific Publishers

Bekenstein, J. D. (1980), Black-hole thermodynamics, *Physics Today*, January 1980, 24–31
Bergmann, P. G. (1942), *Introduction to the Theory of Relativity*, New York: Prentice-Hall
Berry, M. V. (1976), *Principles of Cosmology and Gravitation*, Cambridge: Cambridge University Press
Bethe, H. (1939), On energy generation in stars, *Physical Review* **55**, 434–456
Bjorken, J. D. & Drell, S. D. (1964), *Relativistic Quantum Mechanics*, New York: McGraw-Hill
Blagojević, M. (2002), *Gravitation and Gauge Symmetries*, Bristol, Philadelphia: Institute of Physics Publishing
Bondi, H. (1960), *Cosmology* (2nd edn), Cambridge: Cambridge University Press
Bondi, H. & Samuel, J. (1997), The Lense–Thirring effect and Mach's principle, *Physics Letters A* **228**, 121–126
Bonnor, W. B. & Steadman, B. R. (1999), The gravitomagnetic clock effect, *Classical and Quantum Gravity* **16**, 1853–1861
Börner, G. (1988), *The Early Universe: Facts and Fiction*, Berlin: Springer-Verlag
Bousso, R. (2002), The holographic principle, *Reviews of Modern Physics* **74**, 825–874; http://arXiv.org:hep-th/0203101
Boyer, R. H. & Lindquist, R. W. (1967), Maximal analytic extension of the Kerr metric, *Journal of Mathematical Physics* **8**, 265–281
Bradaschia, C. & Desalvo, R. (2007), A global network listens for ripples in space-time, *CERN Courier*, December, p. 17
Brault, J. (1963), Gravitational red shift of solar lines, *Bulletin of the American Physical Society* **8**, 28
Brill, D. B. & Cohen, J. M. (1966), Cartan frames and the general relativistic Dirac equation, *Journal of Mathematical Physics* **7**, 238–243
Brown, L. S. (1992), *Quantum Field Theory*, Cambridge: Cambridge University Press
Bruhat, Y. (1962), The Cauchy problem, in L. Witten (ed.), *Gravitation: An Introduction to Current Research*, New York: Wiley
Burcham, W. E. & Jobes, M. (1995), *Nuclear and Particle Physics*, Harlow: Longman Scientific and Technical
Cao, Tian Yu (1997), *Conceptual Developments of 20th Century Field Theories*, Cambridge: Cambridge University Press
Carroll, B. W. & Ostlie, D. A. (1996), *An Introduction to Modern Astrophysics*, Reading, Massachusetts: Addison-Wesley
Cartan, E. (2001), *Riemannian Geometry in an Orthogonal Frame*, Singapore: World Scientific
Cartan, H. (1967, 1971), *Formes Différentielles: Applications Élémentaires au Calcul et à la Théorème des Courbes et des Surfaces*, Paris: Hermann (1967); *Differential Forms*, London: Kershaw Publishing Company (1971)
Carter, B. (1979), The general theory of the mechanical, electromagnetic and thermodynamic properties of black holes, in S. W. Hawking & W. Israel (eds.), *General Relativity: An Einstein Centenary Survey*, Cambridge: Cambridge University Press
Chandrasekhar, S. (1931a), The density of white dwarf stars, *Philosophical Magazine* **11**, 592–596

Chandrasekhar, S. (1931b), The maximum mass of ideal white dwarfs, *Astrophysical Journal* **74**, 81–82

Chandrasekhar, S. (1969), Some historical notes, *American Journal of Physics* **37**, 577–584

Chandrasekhar, S. (1972), The increasing role of general relativity in astronomy, *Observatory* **92**, 160–174

Chapman, T. C. & Leiter, D. J. (1976), On the generally covariant Dirac equation, *American Journal of Physics* **44**, 858–862

Cheng, T-P. & Li, L-F. (1984), *Gauge Theory of Elementary Particle Physics*, Oxford: Clarendon Press

Choquet-Bruhat, Y. (1968), *Géométrie Différentielle et Systèmes Extérieurs*, Paris: Dunod

Choquet-Bruhat, Y., DeWitt-Morette, C. & Dillard-Bleick, M. (1982), *Analysis, Manifolds and Physics* (rev. edn), Amsterdam: North-Holland

Chow, W. W. *et al.* (1985), The ring laser gyro, *Reviews of Modern Physics* **57**, 61–104

Christodoulou, D. (1970), Reversible and irreversible transformations in black hole physics, *Physical Review Letters* **25**, 1596–1597

Ciufolini, I. (1995), Dragging of inertial frames, gravitomagnetism and Mach's Principle, in J. B. Barbour & H. Pfister (eds.), *Mach's Principle: From Newton's Bucket to Quantum Gravity*, Boston: Birkhäuser

Ciufolini, I. & Pavlis, E. C. (2004), A confirmation of the general relativistic prediction of the Lense–Thirring effect, *Nature* **431**, 958–960

Ciufolini, I. & Wheeler, J. A. (1995), *Gravitation and Inertia*, Princeton: Princeton University Press

Clarke, C. J. S. (1993), *The Analysis of Space-Time Singularities*, Cambridge: Cambridge University Press

Cohen-Tannoudji, C., Diu, B. & Laloë, F. (1977), *Mécanique Quantique*, Paris: Hermann; *Quantum Mechanics*, New York: Wiley

Colella, R., Overhauser, A. W. & Werner, S. A. (1975), Observation of gravitationally induced quantum interference, *Physical Review Letters* **34**, 1472–1474

Coles, P. & Ellis, G. F. R. (1997), *Is the Universe Open or Closed?: The Density of Matter in the Universe*, Cambridge: Cambridge University Press

Cottingham, W. N. & Greenwood, D. A. (1998), *An Introduction to the Standard Model of Particle Physics*, Cambridge: Cambridge University Press

Crampin, M. & Pirani, F. A. E. (1986), *Applicable Differential Geometry*, Cambridge: Cambridge University Press

Davis, W. R. (1970), *Classical Fields, Particles, and the Theory of Relativity*, New York: Gordon & Breach

De Felice, F. & Clarke, C. J. S. (1990), *Relativity on Curved Manifolds*, Cambridge: Cambridge University Press

De Wit, B. & Smith, J. (1986), *Field Theory in Particle Physics*, Amsterdam: North-Holland

Dicke, R. H. (1964), Experimental relativity, in C. DeWitt & B. S. DeWitt (eds.), *Relativity, Groups and Topology*, London: Blackie; New York: Gordon & Breach

d'Inverno, R. (1992), *Introducing Einstein's Relativity*, Oxford: Clarendon Press

Dirac, P. A. M. (2001), *Lectures on Quantum Mechanics*, New York: Dover Publications

Doughty, N. A. (1990), *Lagrangian Interaction*, Reading, Mass.: Addison-Wesley

Douglass, D. H. & Braginsky, V. B. (1979), Gravitational-radiation experiments, in S. W. Hawking & W. Israel (eds), *General Relativity: An Einstein Centenary Survey*, Cambridge: Cambridge University Press

Drever, R. W. P. (1960), A search for anisotropy of inertial mass using a free precession technique, *Philosophical Magazine* **6**, 683–687

Eddington, A. S. (1924), *Nature* **113**, 192

Eguchi, T., Gilkey, P. B. & Hanson, A. J. (1980), Gravitation, gauge theories and differential geometry, *Physics Reports* **66**, 213–393

Einstein, A. (1905a), Zur Elektrodynamik bewegter Köper, *Annalen der Physik* 17, 891–921; translated as On the Electrodynamics of moving bodies, in H. A. Lorentz, A. Einstein, H. Minkowski & H. Weyl (1952), *The Principle of Relativity*, 35–65, New York: Dover Publications; and in *The Collected Papers of Albert Einstein* (English transl.) **2**, 140–171, Princeton: Princeton University Press (1989)

Einstein, A. (1905b), Ist die Trägheit eines Körpers von seinem Energie inhalt abhängig?, *Annalen der Physik* **18**, 639–641; translated as Does the inertia of a body depend upon its energy content?, in H. A. Lorentz, A. Einstein, H. Minkowski & H. Weyl (1952), *The Principle of Relativity*, 67–71, New York: Dover Publications; and in *The Collected Papers of Albert Einstein* (English transl.) **2**, 172–174, Princeton: Princeton University Press (1989)

Einstein, A. (1911), Über die Einfluss der Schwerkraft auf die Ausbreitung des Lichtes, *Annalen der Physik* **35**, 898–908; translated as On the influence of gravitation on the propagation of light, in H. A. Lorentz, A. Einstein, H. Minkowski & H. Weyl (1952), *The Principle of Relativity*, 97–108, New York: Dover Publications; and in *The Collected Papers of Albert Einstein* (English transl.) **3**, 379–387, Princeton: Princeton University Press (1993)

Einstein, A. (1916a), Hamiltonsches Prinzip und allgemeine Relativitätstheorie, *Sitzungsberichte der Königlich Preußische Akademie der Wissenschaften*; translated as Hamilton's Principle and the General Theory of Relativity, in H. A. Lorentz, A. Einstein, H. Minkowski & H. Weyl (1952), *The Principle of Relativity*, 165–173, New York: Dover Publications; and in *The Collected Papers of Albert Einstein* (English transl.) **6**, 240–246, Princeton: Princeton University Press (1997)

Einstein, A. (1916b), Näherungsweise Integration der Feldgleichungen der Gravitation, *Sitzungsberichte der Königlich Preußische Akademie der Wissenschaften, Sitzung der physikalisch-mathematischen Klasse* 688; translated as Approximative integration of the field equations of gravitation, in *The Collected Papers of Albert Einstein* (English transl.) **6**, 201–210, Princeton: Princeton University Press (1997)

Einstein, A. (1917), Kosmologische Betrachtungen zur allgemeinen Relativitätstheorie, *Sitzungsberichte der Königlich Preußische Akademie der Wissenschaften, Sitzung der physikalisch-mathematischen Klasse*; translated as Cosmological considerations in the General Theory of Relativity, in H. A. Lorentz, A. Einstein, H. Minkowski & H. Weyl (1952), *The Principle of Relativity*, pp. 175–188 New York: Dover Publications; and in *The Collected Papers of Albert Einstein* (English transl.) **6**, 421–432, Princeton: Princeton University Press (1997)

Einstein, A. (1918), Über Gravitationswellen, *Sitzungsberichte der Königlich Preußische Akademie der Wissenschaften, Sitzung der physikalisch-mathematischen Klasse* 154

Einstein, A. & Rosen, N. (1933), The Particle Problem in the General Theory of Relativity, *Physical Review* **48**, 73–77

Einstein, A. & Infeld, L. (1949), On the Motion of Particles in General Relativity Theory, *Canadian Journal of Mathematics* **1**, 209–241

Einstein, A. & Straus, E. G. (1946), A Generalisation of the Relativistic Theory of Gravitation II, *Annals of Mathematics* **47**, 731–741

Eisenhart, L. P. (1926), *Riemannian Geometry*, Princeton: Princeton University Press

Ellis, G. F. R. & Williams, R. (1988), *Flat and Curved Space-times*, Oxford: Clarendon Press

Everitt, C. W. F. *et al.* (2001), Gravity Probe B: countdown to launch, in C. Lämmerzahl, C. W. F. Everitt & F. W. Hehl (eds.), *Gyros, Clocks, Interferometers...: Testing Relativistic Gravity in Space*, Berlin: Springer-Verlag

Faber, R. L. (1983), *Differential Geometry and Relativity Theory: an Introduction*, New York: Dekker

Fairbank, J. D., Deaver, B. S. Jr., Everitt, C. W. F. & Michelson, P. F. (1988), *Near Zero: New Frontiers of Physics*, New York: Freeman

Ferreras, I., Melchiarri, A. & Silk, J. (2001), Setting new constraints on the age of the Universe, *Monthly Notices of the Royal Astronomical Society* **327**, L47–L51

Feynman, R. P., Leighton, R. B. & Sands, M. (1963), *The Feynman Lectures on Physics*, vol. 1, Reading, Massachusetts: Addison-Wesley

Feynman, R. P., Leighton, R. B. & Sands, M. (1964), *The Feynman Lectures on Physics*, vol. 2, Reading, Massachusetts: Addison-Wesley

Feynman, R. P., Leighton, R. B. & Sands, M. (1965), *The Feynman Lectures on Physics*, vol. 3, Reading, Massachusetts: Addison-Wesley

Finkelstein, D. (1958), Past-future asymmetry of the gravitational field of a point particle, *Physical Review* **110**, 965–967

Flanders, H. (1989), *Differential Forms with Applications to the Physical Sciences*, New York: Dover Publications

Frankel, T. (1979), *Gravitational Curvature*, San Francisco: Freeman

Frankel, T. (1997), *The Geometry of Physics: An Introduction*, Cambridge: Cambridge University Press

Freedman, W. L. (1997), Determination of the Hubble Constant, in N. Turok (ed.), *Critical Dialogues in Cosmology*, Singapore: World Scientific

Friedmann, A. (1922), Über die Krümmung des Raumes, *Zeitschrift für Physik* **10**, 377

Frolov, V. P. (1979), The Newman–Penrose method in the General Theory of Relativity, in N. G. Basov (ed.), *Problems in the General Theory of Relativity and Theory of Group Representations*, New York: Consultants Bureau

Fulton, T., Rohrlich, F. & Witten, L. (1962), Conformal invariance in physics, *Reviews of Modern Physics* **34**, 442–457

Gamow, G. (1970), *My World Line*, New York: Viking Press

Geroch, R. (1960), What is a singularity in general relativity?, *Annals of Physics* **48**, 526–540

Göckeler, M. & Schücker, T. (1987), *Differential Geometry, Gauge Theories, and Gravity*, Cambridge: Cambridge University Press

Goldstein, H. (1950), *Classical Mechanics*, Reading, Massachusetts: Addison-Wesley

Gronwald, F., Gruber, E., Lichtenegger, H. I. M. & Puntigam, R. A., in *Proceedings of the Alpbach Summer School 1997: Fundamental Physics in Space*, organised by the Austrian and European Space Agency, ed. A. Wilson, http://arxiv:gr-qc / 9712054

Gross, F. (1993), *Relativistic Quantum Mechanics and Field Theory*, New York: Wiley

Gürsey, F. (1965), Group combining internal symmetries and spin, in C. DeWitt & M. Jacob (eds.), *High Energy Physics*, New York: Gordon and Breach

Guth, A. H. (1981), Inflationary Universe: A possible solution to the horizon and flatness problems, *Physical Review D* **23**, 347–356

Guth, A. H. (2000), Inflation and eternal inflation, *Physics Reports* **333–334**, 555–574

Hafele, J. C. & Keating, R. E. (1972), Around-the-world atomic clocks: predicted relativistic time gains, *Science* **177**, 166–168

Hammond, R. (1994), Spin, torsion, forces, *General Relativity and Gravitation* **26**, 247 – 263

Hammond, R. (1995), New fields in General Relativity, *Contemporary Physics* **36**, 103–114

Harrison, E. (2000), *Cosmology* (2nd edn.), Cambridge: Cambridge University Press

Hartle, J. B. (1978), Bounds on the mass and moment of inertia of non-rotating neutron stars, *Physics Reports* **46**, 201–247

Hartle, J. B. (2003), *Gravity*, San Francisco: Addison-Wesley

Hawking, S. W. (1971), Gravitational radiation from colliding black holes, *Physical Review Letters* **26**, 1344

Hawking, S. W. (1975), Particle creation by black holes, *Communications in Mathematical Physics*, **43**, 199–220; also in C. J. Isham, R. Penrose & D. W. Sciama (eds.), *Quantum Gravity: An Oxford Symposium*, Oxford: Clarendon Press

Hawking, S. W. & Ellis, G. F. R. (1973), *The Large Scale Structure of Space-time*, Cambridge: Cambridge University Press

Hawking, S. W. & Penrose, R. (1969), The singularities of gravitational collapse and cosmology, *Proceedings of the Royal Society of London A* **314**, 529–548

Healey, R. (2007), *Gauging What's Real*, Oxford: Oxford University Press

Hehl, F. W. (1973), Spin and torsion in General Relativity I: Foundations, *General Relativity and Gravitation* **4**, 333–349

Hehl, F. W. (1974), Spin and torsion in General Relativity II: Geometry and field equations, *General Relativity and Gravitation* **5**, 491–516

Hehl, F. W. & von der Heyde, P. (1973), Spin and the structure of space-time, *Annales de l'Institut Henri Poincaré* **19**, 179–196

Hehl, F. W., von der Heyde, P. & Kerlick, G. D. (1976), General relativity with spin and torsion: foundations and prospects, *Reviews of Modern Physics* **48**, 393–416

Helgason, S. (1978), *Differential Geometry and Symmetric Spaces* (2nd edn), New York: Academic Press

Higgs, P. W. (1964a), Broken symmetries, massless particles and gauge fields, *Physics Letters* **12**, 132–133

Higgs, P. W. (1964b), Broken symmetries and the masses of gauge bosons, *Physical Review Letters* **13**, 508–509

Higgs, P. W. (1966), Spontaneous symmetry breakdown without massless bosons, *Physical Review* **145**, 1156–1163

Hobson, M. P., Efstathiou, G. & Lasenby, A. N. (2006), *General Relativity: An Introduction for Physicists*, Cambridge: Cambridge University Press

Hoffmann, B. (1983), *Relativity and Its Roots*, New York: Scientific American Books – Freeman

Hogan, C. J. (1997), Big bang nucleosynthesis and the observed abundances of light elements, in N. Turok (ed.), *Critical Dialogues in Cosmology*, Singapore: World Scientific

Huang, K. (1998), *Quantum Field Theory: From Operators to Path Integrals*, New York: Wiley

Hughes, V. W., Robinson, H. G. & Beltran-Lopez, V. (1960), Upper limit for the anisotropy of inertial mass from nuclear resonance experiments, *Physical Review Letters* **4**, 342–344

Hulse, R. A. & Taylor, J. M. (1975), Discovery of a pulsar in a binary system, *Astrophysical Journal Letters* **195**, L51–L53

Israel, W. (1986), Third law of black-hole dynamics: a formulation and proof, *Physical Review Letters* **57**, 397–399

Israel, W. (1987), Dark stars: the evolution of an idea, in S. W. Hawking & W. Israel (eds.), *Three Hundred Years of Gravitation*, Cambridge: Cambridge University Press

Itzykson, C. & Zuber, J-B. (1980), *Quantum Field Theory*, New York: McGraw-Hill

Jackson, J. D. (1975), *Classical Electrodynamics*, New York: Wiley

Jammer, M. (2000), *Concepts of Mass in Contemporary Physics and Philosophy*, Princeton: Princeton University Press

Kaluza, T. (1921), Zum Unitätsproblem der Physik, *Sitzungsberichte der Königlich Preußische Akademie der Wissenschaften, Sitzung der physikalisch-mathematischen Klasse* 966–972

Kawashima, N. (1994), The laser interferometric gravitational wave antenna: present status and future plan, *Classical and Quantum Gravity* **11**, A83–A95

Kerr, R. P. (1963), Gravitational field of a spinning mass as an example of algebraically special metrics, *Physical Review Letters* **11**, 237–238

Kibble, T. W. B. (1961), Lorentz invariance and the gravitational field, *Journal of Mathematical Physics* **2**, 212–221

Kibble, T. W. B. & Berkshire, F. H. (1996), *Classical Mechanics* (4th edn), Harlow: Addison Wesley Longman

Kilmister, C. W. (1973), *General Theory of Relativity*, Oxford: Pergamon Press

Klein, O. (1926), *Zeitschrift für Physik* **37**, 895; The atomicity of electricity as a quantum theory law, *Nature* **118**, 516; Generalisations of Einstein's theory of gravitation considered from the point of view of quantum field theory, *Helvetica Physica Acta Supplementum* **4**, 58–71

Kolb, E. W. & Turner, M. S. (1990), *The Early Universe*, Redwood City, California: Addison-Wesley

Kopczyński, W. & Trautman, A. (1992), *Spacetime and Gravitation*, Chichester, New York: Wiley; Warszawa: PWN Polish Scientific Publishers

Kruskal, M. D. (1960), Maximal extension of Schwarzschild metric, *Physical Review* **119**, 1743–1745

Lämmerzahl, C. & Neugebauer, G. (2001), The Lense–Thirring effect: from the basic notions to the observed effects, in C. Lämmerzahl, C. W. F. Everitt & F. W. Hehl (eds.),

Gyros, Clocks, Interferometers...: Testing Relativistic Gravity in Space, Berlin: Springer-Verlag

Lamoreaux, S. K. *et al.* (1986), New limits on spatial anisotropy from optically pumped ^{201}Hg and ^{199}Hg, *Physical Review Letters* **57**, 3125–3128

Landau, L. D. & Lifshitz, E. M. (1971), *The Classical Theory of Fields*, Oxford: Pergamon Press

Lebach, D. E. *et al.* (1995), Measurements of the solar gravitational deflection of radio waves using very-long-baseline interferometry, *Physical Review Letters* **75**, 1439–1442

Levi-Civita, T. (1927), *The Absolute Differential Calculus*, London: Blackie

Lichnérowicz, A. (1958), *Géométrie des Groupes de Transformations*, Paris: Dunod

Linde, A. (1990), *Particle Physics and Inflationary Cosmology*, Chur, Switzerland: Harwood Academic Publishers

Linde, A. (2000), Inflationary cosmology, *Physics Reports* **333–334**, 575–591

Lorentz, H. A., Einstein, A., Minkowski, H. & Weyl, H. (1952), *The Principle of Relativity*, New York: Dover Publications

Ludvigsen, M. (1999), *General Relativity: A Geometric Approach*, Cambridge: Cambridge University Press

Mach, E. (1919), *The Science of Mechanics* (4th edn), La Salle, Illinois: Open Court

Maggiore, M. (2005), *A Modern Introduction to Quantum Field Theory*, Oxford: Oxford University Press

Mandl, F. (1988), *Statistical Physics* (2nd edn), Chichester: Wiley

Martin, D. (1991), *Manifold Theory: an Introduction for Mathematical Physicists*, New York: Ellis Horwood

Mashhoon, B. (1993), On the gravitational analogue of Larmor's theorem, *Physics Letters* A **173**, 347–354

Mashhoon, B. & Santos, N. O. (2000), Rotating cylindrical systems and gravitomagnetism, *Annalen der Physik* **9**, 49–63

Mashhoon, B., Hehl, F. W. & Theiss, D. S. (1984), On the gravitational effects of rotating masses: the Thirring–Lense papers, *General Relativity and Gravitation* **16**, 711–750

Mashhoon, B., Gronwald, F. & Lichtenegger, H. I. M. (2001), Gravitomagnetism and the Clock Effect, in C. Lämmerzahl, C. W. F. Everitt & F. W. Hehl (eds.), *Gyros, Clocks, Interferometers...: Testing Relativistic Gravity in Space*, Berlin: Springer-Verlag

McGlinn, W. D. (2003), *Introduction to Relativity*, Baltimore, Maryland: Johns Hopkins University Press

McVittie, G. C. (1965), *General Relativity and Cosmology*, Urbana, Illinois: University of Illinois Press

Mehra, J. (1973), Einstein, Hilbert and the Theory of Gravitation, in J. Mehra (ed.), *The Physicist's Conception of Nature*, Dordrecht: Reidel

Misner, C. W. (1964), Differential geometry, in C. DeWitt and B. S. DeWitt (eds.), *Relativity, Groups and Topology*, London: Blackie; New York: Gordon & Breach

Misner, C. W., Thorne, K. S. & Wheeler, J. A. (1973), *Gravitation*, San Francisco: Freeman

Møller, C. (1972), *The Theory of Relativity*, Oxford: Clarendon Press

Moore, W. (1989), *Schrödinger: Life and Thought*, Cambridge: Cambridge University Press

Morin, D. (2007), *Introduction to Classical Mechanics*, Cambridge: Cambridge University Press

Mukhanov, V. (2005), *Physical Foundations of Cosmology*, Cambridge: Cambridge University Press

Nakahara, M. (1990), *Geometry, Topology and Physics*, Bristol and New York: Adam Hilger

Newman, E. T. & Janis, A. I. (1965), Note on the Kerr spinning-particle metric, *Journal of Mathematical Physics* **6**, 915–917

Newman, E. T. & Penrose, R. (1962), An approach to gravitational radiation by a method of spin coefficients, *Journal of Mathematical Physics* **3**, 566–578

Nieto, M. M., Hughes, R. J. & Goldman, T. (1989), Actually, Eötvös did publish his results in 1910, it's just that no-one knows about it... *American Journal of Physics* **57**, 397–404

Nordstrøm, G. (1918), On the energy of the gravitational field in Einstein's theory, *Proc. Kon. Ned. Akad. Wet.* **20**, 1238–1245

Novikov, I. D. & Frolov, V. P. (1989), *Physics of Black Holes*, Dordrecht: Kluwer Academic Publishers

Okun, L. B., Selivanov, K. G. & Telegdi, V. (2000), On the interpretation of the redshift in a static gravitational field, *American Journal of Physics* **68**, 115–119

Oppenheimer, J. R. & Snyder, H. (1939), On continued gravitational contraction, *Physical Review* **56**, 455–459

Oppenheimer, J. R. & Volkoff, G. M. (1939), On massive neutron cores, *Physical Review* **55**, 374–381

Padmanabhan, T. (1989), Some fundamental aspects of semiclassical and quantum gravity, *International Journal of Modern Physics A*, **4**, 4735–4818

Pais, A. (1982), *Subtle is the Lord*, New York: Oxford University Press

Panofsky, W. K. H. & Phillips, M. (1962), *Classical Electricity and Magnetism* (2nd edn), Reading, Massachusetts: Addison-Wesley

Papapetrou, A. (1974), *Lectures on General Relativity*, Dordrecht: Reidel Publishing Company

Pauli, W. (1958), *Theory of Relativity*, Oxford: Pergamon Press

Pauli, W. (1965), Continuous groups in quantum mechanics, *Ergebnisse der Exakten Naturwissenschaften* **37**, 85–104

Peacock, J. A. (1999), *Cosmological Physics*, Cambridge: Cambridge University Press

Peebles, P. J. E. (1993), *Principles of Physical Cosmology*, Princeton: Princeton University Press

Penrose, R. (1964), Conformal treatment of infinity, in C. DeWitt & B. S. DeWitt (eds.), *Relativity, Groups and Topology* London: Blackie; New York: Gordon & Breach

Penrose, R. (1965), Gravitational collapse and space-time singularities, *Physical Review Letters* **14**, 57–59

Penrose, R. (1969), Gravitational collapse: the role of general relativity, *Rivista del Nuovo Cimento* **1**, 252–276

Penrose, R. (1998), The question of cosmic censorship, in R. Wald (ed.), *Black Holes and Relativistic Stars*, Chicago: Chicago University Press

Penrose, R. (2004), *The Road to Reality*, London: Jonathan Cape

Penzias, A. A. & Wilson, R. W. (1965), A measurement of excess antenna temperature at 4080 Mc/s, *Astrophysical Journal* **142**, 419–421

Perkins, D. H. (2000), *Introduction to High Energy Physics* (4th edn), Cambridge: Cambridge University Press

Petrov, A. Z. (1969), *Einstein Spaces*, Oxford: Pergamon Press

Pirani, F. A. E. (1962a), Survey of gravitational radiation theory, in *Recent Developments in General Relativity*, Oxford: Pergamon Press; Warszawa: PWN Polish Scientific Publishers

Pirani, F. A. E. (1962b), Gravitational radiation, in L. Witten (ed.), *Gravitation: An Introduction to Current Research*, New York: Wiley

Pirani, F. A. E. (1965), Introduction to gravitational radiation theory, in Trautman, A., Pirani, F. A. E. & Bondi, H. *Lectures on General Relativity*, (Brandeis Summer Institute 1964), Englewood Cliffs, New Jersey: Prentice-Hall

Plebański, J. & Krasiński, A. (2006), *An Introduction to General Relativity and Cosmology*, Cambridge: Cambridge University Press

Poisson, E. (2004), *A Relativist's Toolkit: The Mathematics of Black-Hole Mechanics*, Cambridge: Cambridge University Press

Popper, D. M. (1954), Red shift in the spectrum of 40 Eridani B, *Astrophysical Journal* **120**, 316–321

Post, E. J. (1967), Sagnac effect, *Reviews of Modern Physics* **39**, 475

Pound, R. V. & Rebka, G. A. (1960), Apparent weight of photons, *Physical Review Letters* **4**, 337–341

Pound, R. V. & Snider, J. L. (1964), Effect of gravity on nuclear resonances, *Physical Review Letters* **13**, 539–540

Raffelt, G. G. (1996), *Stars as Laboratories for Fundamental Physics: The Astrophysics of Neutrinos, Axions and Other Weakly Interacting Particles*, Chicago: Chicago University Press

Raine, D. J. (1981), Mach's principle and space-time structure, *Reports on Progress in Physics* **44**, 1151–1195

Rauch, H. & Werner, S. A. (2000), *Neutron Interferometry: Lessons in Experimental Quantum Mechanics*, Oxford: Clarendon Press

Reissner, H. (1916), Über die Eigengravitation des elecktrischen Feldes nach der Einsteinschen Theorie, *Annalen der Physik* **50**, 106–120

Rindler, W. (1977), *Essential Relativity* (2nd edn), Berlin: Springer Verlag

Rindler, W. (1994), The Lense–Thirring effect exposed as anti-Machian, *Physics Letters A* **187**, 236–238

Rindler, W. (1997), The case against space dragging, *Physics Letters A* **233**, 25–29

Rindler, W. (2001), *Relativity*, Oxford: Oxford University Press

Robertson, H. P. (1935), Kinematics and world structure, *Astrophysical Journal* **82**, 248–301

Robertson, H. P. (1936), Kinematics and world structure, *Astrophysical Journal* **83**, 187–201

Robertson, H. P. (1938), The apparent luminosity of a receding nebula, *Zeitschrift für Astrophysik* **15**, 69–81

Robertson, H. P. & Noonan, T. W. (1968), *Relativity and Cosmology*, Philadelphia: Saunders

Roll, P. G., Krotkov, R. & Dicke, R. H. (1964), The equivalence of inertial and passive gravitational mass, *Annals of Physics* **26**, 442–517

Roos, M. (2003), *Introduction to Cosmology* (3rd edn), Chichester: Wiley

Rubakov, V. (2002), *Classical Theory of Gauge Fields*, Princeton: Princeton University Press

Ruggiero, M. L. & Tartaglia, A. (2002), Gravitomagnetic effects, http://arXiv:gr-qc/0207065

Ryan, M. P. Jr., & Shepley, L. C. (1975), *Homogeneous Relativistic Cosmologies*, Princeton: Princeton University Press

Ryder, L. H. (1996), *Quantum Field Theory* (2nd edn), Cambridge: Cambridge University Press

Ryder, L. H. (1999), Relativistic spin operator for Dirac particles, *General Relativity and Gravitation*, **31** 775–780

Sachs, R. K. (1964), Gravitational radiation, in C. DeWitt & B. S. DeWitt (eds), *Relativity, Groups and Topology* London: Blackie; New York: Gordon & Breach

Sagnac, G. (1913a), L'éther lumineux démontré par l'effet du vent relatif d'éther dans un interférométre en rotation uniforme, *Comptes Rendues de l'Académie des Sciences* **157**, 708–710

Sagnac, G. (1913b), Sur la preuve de la réalité d'éther lumineux par l'expérience de l'interférographe tournant, *Comptes Rendues de l'Académie des Sciences* **157**, 1410–1413

Sakurai, J. J. (1994), *Modern Quantum Mechanics* (rev. edn), Reading, Massachusetts: Addison-Wesley

Salam, A. (1968), Weak and electromagnetic interactions, *Elementary Particle Physics: Nobel Symposium No 8*, (N. Svartholm, ed.), Stockholm: Almqvist Wiksell

Schiff, L. I. (1939), A question in general relativity, *Proceedings of the National Academy of Sciences (USA)* **25**, 391–395

Schiff, L. I. (1960), Possible new experimental test of general relativity theory, *Physical Review Letters* **4**, 215–217

Schmidt, B. G. (1971), A new definition of singular points in general relativity, *General Relativity and Gravitation* **1**, 269–280

Schouten, J. A. (1954), *Ricci-Calculus*, Berlin: Springer

Schreiber, M. (1977), *Differential Forms: A Heuristic Introduction*, Berlin: Springer-Verlag

Schröder, U. E. (2002), *Gravitation: Einführung in die Allgemeine Relativitätstheorie*, Frankfurt am Main: Verlag Harri Deutsch

Schrödinger, E. (1985), *Space-Time Structure*, Cambridge: Cambridge University Press

Schutz, B. (1980), *Geometrical Methods of Mathematical Physics*, Cambridge: Cambridge University Press

Schwarzschild, K. (1916a), Über das Gravitationsfeld eines Massenpunktes nach der Einsteinschen Theorie, *Sitzungsberichte der Königlich Preußische Akademie der Wissenschaften, Physik-Math Klasse*, 189–196

Schwarzschild, K. (1916b), Über das Gravitationsfeld einer Kugel aus inkompressibler Flüssigkeit nach der Einsteinschen Theorie, *Sitzungsberichte der Königlich Preußische Akademie der Wissenschaften, Physik-Math Klasse*, 424–434

Sciama, D. W. (1953), On the origin of inertia, *Monthly Notices of the Royal Astronomical Society* **113**, 34–42

Sciama, D. W. (1962), On the analogy between charge and spin in general relativity, in *Recent Developments in General Relativity*, Oxford: Pergamon Press; Warszawa: PWN Polish Scientific Publishers

Sciama, D. W. (1969), *The Physical Foundations of General Relativity*, Garden City, New York: Doubleday

Sexl, R. U. & Urbantke, H. K. (1976), *Relativität, Gruppen, Teilchen*, Vienna: Springer-Verlag

Sexl, R. U. & Urbantke, H. K. (1983), *Gravitation und Kosmologie*, Mannheim: Bibliographisches Institut

Shankar, R. (1980), *Principles of Quantum Mechanics*, New York: Plenum Press

Shapiro, I. L. (2002), Physical aspects of the space-time torsion, *Physics Reports* **357**, 113–213

Skillman, E. & Kennicutt, R. C. (1993), Spatially resolved optical and near-infrared spectroscopy of 1Zw18, *Astrophysical Journal* **411**, 655–666

Smarr, L. (1973), Mass formula for Kerr black holes, *Physical Review Letters* **30**, 71–73

Smolin, L. (2001), *Three Roads to Quantum Gravity*, London: Basic Books

Soper, D. E. (1976), *Classical Field Theory*, New York: McGraw-Hill

Speiser, D. (1964), Theory of compact Lie groups and some applications to elementary particle physics, in F. Gürsey (ed.), *Group Theoretical Concepts and Methods in Elementary Particle Physics*, New York, London: Gordon & Breach

Spergel, D. N. *et al.* (2003), First year Wilkinson Microware Anisotropy Probe (WMAP) observations, *Astrophysical Journal Supplement* **148**, 175

Spivak, M. (1970), *A Comprehensive Introduction to Differential Geometry*, vol. I, Brandeis University

Srednicki, M. (2007), *Quantum Field Theory*, Cambridge: Cambridge University Press

Stedman, G. E. (1997), Ring-laser tests of fundamental physics and geophysics, *Reports on Progress in Physics* **60**, 615–688

Stephani, H. (1982), *General Relativity*, Cambridge: Cambridge University Press

Stephani, H. (2004), *Relativity: An Introduction to Special and General Relativity* (3rd edn), Cambridge: Cambridge University Press

Stephani, H., Kramer, D., MacCallum, M., Hoenselaers, C. & Herlt, E. (2003), *Exact Solutions of Einstein's Field Equations* (2nd edn), Cambridge: Cambridge University Press

Stoker, J. J. (1969), *Differential Geometry*, New York: Wiley

Straumann, N. (1991), *General Relativity and Relativistic Astrophysics*, Berlin: Springer-Verlag

Struik, D. J. (1961), *Lectures on Classical Differential Geometry*, Reading, Massachusetts: Addison-Wesley

Susskind, L. & Lindesay, J. (2005), *An Introduction to Black Holes, Information and the String Theory Revolution: The Holographic Universe*, New Jersey: World Scientific

Synge, J. L. (1964), *Relativity: The General Theory*, Amsterdam: North-Holland

Szekeres, G. (1960), On the singularities of a Riemannian manifold, *Publ. Mat. Debrecen* **7**, 285–301

Tartaglia, A. (2002), General treatment of the gravitomagnetic clock effect, in *Proceedings of the Ninth Marcel Grossmann Meeting on General Relativity* eds. V. G. Gurzadyan, R. T. Jantzen, R. Ruffini, Part B, p. 969, New Jersey: World Scientific

Taylor, J. C. (2001), *Hidden Unity in Nature's Laws*, Cambridge: Cambridge University Press

Thirring, H. & Lense, J. (1918), Über den Einfluss der Eigenrotation der Zentralkörper auf die Bewegung der Planeten und Monde nach der Einsteinschen Gravitationstheorie, *Physikalisches Zeitschrift* **19** 156–163

Thirring, W. (1972), Five dimensional theories and CP violation, *Acta Physica Austriaca Supplementum* **9**, 256–271

Thomas, L. W. (1926), The motion of the spinning electron, *Nature* **117**, 514

Thomas, L. W. (1927), The kinematics of an electron with an axis, *Philosophical Magazine* (7th series) **3**, 1–22

't Hooft, G. (1997), *In Search of the Ultimate Building Blocks*, Cambridge: Cambridge University Press

Thorne, K. S. (1994), *Black Holes and Time Warps*, London: Picador

Thorne, K. S. (1987), Gravitational radiation, in *Three Hundred Years of Gravitation*, Hawking, S. W. & Israel, W. (eds.), Cambridge: Cambridge University Press

Tolman, R. C. (1939), Static solutions of Einstein's field equations for spheres of fluid, *Physical Review* **55**, 364–373

Tonnelat, M-A. (1964), *Les Vérifications Expérimentales de la Relativité Générale*, Paris: Masson

Torretti, R. (1996), *Relativity and Geometry*, New York: Dover Publications

Trautman, A. (1973a), Spin and torsion may avert gravitational singularities, *Nature Physical Science* **242**, 7

Trautman, A. (1973b), On the structure of the Einstein–Cartan equations, *Symposia Mathematica* **12**, 139–162, Istituto Nazionale di Alta Matematica, Bologna

Tung, Wu-Ki (1985), *Group Theory in Physics*, Philadelphia, Singapore: World Scientific

Utiyama, R. (1956), Invariant interpretation of interaction, *Physical Review* **101**, 1597–1607

Vessot, R. F. C. & Levine, M. W. (1979), A test of the equivalence principle using a space-borne clock, *General Relativity and Gravitation* **10**, 181–204

Vessot, R. F. C. *et al.* (1980), Test of relativistic gravitation with a space-borne hydrogen maser, *Physical Review Letters* **45**, 2081–2084

Wald, R. M. (1984), *General Relativity*, Chicago: Chicago University Press

Wald, R. M. (1994), *Quantum Field Theory in Curved Spacetime and Black Hole Thermodynamics*, Chicago: Chicago University Press

Wald, R. M. (2001), The thermodynamics of black holes, *Living Reviews in Relativity*, www.livingreviews.org/Articles/Volume4/2001–6wald

Walker, A. G. (1936), On Milne's theory of world-structure, *Proceedings of the London Mathematical Society* **42**, 90–127

Weinberg, S. (1967), A model of leptons, *Physical Review Letters* **19**, 1264–1266

Weinberg, S. (1972), *Gravitation and Cosmology*, New York: Wiley

Weinberg, S. (1978), *The First Three Minutes: A Modern View of the Origin of the Universe*, Glasgow: Fontana/Collins

Weinberg, S. (1995), *The Quantum Theory of Fields*, vol. 1, Cambridge: Cambridge University Press

Weissberg, J. M. & Taylor, J. M. (1984), Observations of post-Newtonian timing effects in the binary pulsar PSR 1913+16, *Physical Review Letters* **52**, 1348–50

Werner, S. A., Colella, R., Overhauser, A. W. & Eagen, C.F. (1975), Observation of the phase shift of a neutron due to precession in a magnetic field, *Physical Review Letters* **35**, 1053–1055

Wess, J. (1960), The conformal invariance in quantum field theory, *Il Nuovo Cimento* **18**, 1086–1107

Westenholtz, C. von (1978), *Differential Forms in Mathematical Physics*, Amsterdam: North-Holland

Weyl, H. (1929), Elektron und Gravitation, *Zeitschrift für Physik* **56**, 330–352

Weyl, H. (1950), A remark on the coupling of gravitation and electron, *Physical Review* **77**, 699–701

Weyl, H. (1952), *Space-Time-Matter*, New York: Dover Publications

Wheeler, J. A. (1964), Geometrodynamics and the issue of the final state, in C. DeWitt, & B. S. DeWitt (eds.), *Relativity, Groups and Topology*, London: Blackie; New York: Gordon & Breach

Wightman, A. S. (1960), L'invariance dans la mécanique quantique relativiste, in C. DeWitt & R. Omnès (eds.), *Relations de Dispersion et Particules Elémentaires*, Paris: Hermann; New York: Wiley

Wigner, E. P. (1939), On unitary representations of the inhomogeneous Lorentz group, *Annals of Mathematics* **40**, 149–204

Wigner, E. P. (1964), Unitary representations of the inhomogeneous Lorentz group including reflections, in F. Gürsey (ed.), *Group Theoretical Concepts and Methods in Elementary Particle Physics*, New York: Gordon and Breach

Wigner, E. P. (1967), *Symmetries and Reflections: Scientific Essays*, Cambridge, Massachusetts: MIT Press

Will, C. M. (1993), *Theory and Experiment in Gravitational Physics*, (rev. edn) Cambridge: Cambridge University Press

Will, C. M. (2001), The Confrontation between General Relativity and Experiment, gr-qc/0103036

Witten, L. (1962), A geometric theory of the electromagnetic and gravitational fields, in L. Witten (ed.), *Gravitation: An Introduction to Current Research*, New York: Wiley

Yang, C. N. (1983), *Selected Papers 1945–1980 With Commentary*, San Francisco: Freeman

Yang, C.N. & Mills, R.L. (1954), Conservation of isotopic spin and isotopic gauge invariance, *Physical Review* **96**, 191–195

Yourgrau, Y. & Mandelstam, S. (1968), *Variational Principles in Dynamics and Quantum Theory*, (3rd edn), London: Pitman

Zel'dovich, Ya. B. (1968), The cosmological constant and the theory of elementary particles, *Soviet Physics Uspekhi* **11**, 381–393

Zel'dovich, Ya.B. & Novikov, I.D. (1996), *Stars and Relativity*, New York: Dover Publications

Index